ENTOMOLOGY

nipa
Browse Subject
eBooks / eChapters / Articles
eBooks
eBooks / eChapters / Articles
eBooks
Browse, Search, Read & Buy...
Subject Catalogues
eChapters
Publishers
Print Books
Forthcoming
eBooks
New Titles
scan for catalogue
NIPA
GENX
ONLINE RESOURCES
Publishing Books and Journals
Ebooks and e articles, Current Affairs
Reasoning, Logic and Aptitude
Competitive Examination Preparation
Language Learning and Development Programme
Document Quality Checker and Improvement Tool
Effective Public Speaking, Presentation and Interpersonal Skills
Online Programmes for Professional Development
Personality Development and Human Values
UPI
PayPal
PAY USING

ENTOMOLOGY
Novel Approaches

Editors

Dr. P.C. Jain (Retired)
Dean, S.K.N. College of Agriculture,
Rajasthan Agricultural University Campus, Jobner

Dr. M.C. Bhargava
Associate Professor, Department of Entomology
S.K.N. College of Agriculture, Rajasthan Agricultural University Campus, Jobner
&
Principal Investigator of ICAR project entitled Bio-ecology and Management of Stored Grain Pests on Different Commodities in Rajasthan

2007

New India Publishing Agency

Pitam Pura, New Delhi- 110 088

ISBN 81-89422-32-4

Typeset at:

Laxmi Art Creations
Shiva Mkt., Pitam Pura, New Delhi
Cell : 9811482328

Printed at :

Jai Bharat Printing Press
Rohtas Nagar, Shahdara, Delhi

Published by:
Sumit Pal Jain *for*

New India Publishing Agency

101, Vikas Surya Plaza, CU Block, L.S.C. Mkt.,
Pitam Pura, New Delhi- 110 088, (INDIA)
Phone: 011-27341717, Fax: 011-27341616
E-mail: newindiapublishingagency@gmail.com
Web: www.bookfactoryindia.com

Maharana Pratap University of Agriculture
and Technology, Udaipur - 313 001 (Raj)
Phone : 0294-2471101 71101 (O)
0294-2427691 7691 (R)
Fax : 0294-2470682- 240294

Prof. S. L. Mehta
Vice-Chancellor

Foreword

In the development of agricultural technologies in modern times, the concept of pest management has replaced the earlier one of pest control. The modern concept is inclusive of genetic engineering, semiochemicals, economic, biological and managerial issues like economic threshold levels, pest resurgence and integrated plant protection strategies. It is therefore, of paramount importance to develop and adopt the strategies of pest management to overcome the problems of pesticidal hazards and to provide the farming community with an economical and simple approach to pest management. This will only be possible if these information are available in appropriate form. Opportunities for recurrent education have become absolutely essential for every one wishing to keep scientifically alive.

A multidirectional approach was initiated by the editors for introducing the publication of recent advances in integrated pest management and related fields. The present efforts, hence is an intra-disciplinary approach to utilize the knowledge and expertise modern day entomologists to aware with recent advances made under integrated pest management system, genetic engineering, biotechnological approaches, semiochemicals etc. and present status of pesticidal contamination in environment.

I am sure that this book entitled "*Entomology: Novel Approaches*" edited by **Dr. P.C. Jain**, Dean (Retd.) and **Dr. M.C. Bhargava**, Associate Professor and Principal Investigator, Department of Entomology, S.K.N. College of Agriculture, Rajasthan Agricultural University, Campus Jobner, which cover some of modern concepts of pest management, would be of great value to students, teachers, researchers and extension workers.

(S. L. Mehta)

Central Agricultural University
Imphal-795 004, Manipur (India)
Tel. : (0385) 2415933 (O)
: 2410450 (R)
Gram : Agrivarsity
Fax : 2410450 (R)

Dr. S.N. Puri, NAAS
Vice-Chancellor

Foreword

It was a great pleasure to go through this book entitled ***"Entomology: Novel Approaches"*** edited by **Prof. P.C. Jain**, Former Dean, S.K.N. College of Agriculture, Jobner and **Dr. M.C. Bhargava**, Associate Professor and Principal Investigator, Department of Entomology, S.K.N. College of Agriculture, Rajasthan Agricultural University Campus: Jobner.

The field of entomology is a vast and the aspects, which have been covered in the present book, are update in a lucid, easily understandable and appropriate form. The recent developments made in the field of genetic engineering, semiochemicals, post harvest grain technology, insect pest management in different crops and pesticide contamination have been included in this book. Besides, it has many other chapters making the publication broad based.

I am confident, the book will serve as a valuable reference source and would be of great help for the students, teachers, researchers and extension workers. It will also help to evolve ecologically compartible and economically sound pest management programme for a sustainable agriculture.

Dr. S.N. Puri

Preface

Plant protection is an integral part of crop husbandry without which the desired production can not be achieved. Production of food grain has tremendously increased in the last 50 years and it has touched the level of about 210 million tones presently, which is a matter of pride for Indian agriculture. We have not only enough food production but are also in a position to export it to other countries.

One can not deny the fact that heavy application of pesticides was made to save the crops from ravages of pests, diseases and weeds. But the indiscriminate use of pesticides has polluted the environment and affected our health adversely beside deleterious effects on non-target organisms. More and more and more insects are being reported resistant to different insecticides year after year posing a serious threat to agriculture. However, there has been work also on alternate control measures like biopesticides, varietal resistance, transgenic plants etc. for a better pest management strategy.

The present book **"Entomology: Novel Approaches"** gives a vast knowledge about the progress made in India on different entomological aspects. Scientists from different disciplines of entomology will only be able to collaborate effectively if they have sufficient knowledge and understanding of newer technologies of pest management. This will only be possible if the necessary information is available in appropriate form. It has been our intension that this book should provide a source of information on the research techniques and practices of pest management, advanced genetic and biotechnological researches, new pest management technologies on different crops, pesticidal contamination status in environment. The aim of this book is to provide an overview of insect pest management, highlighting the recent developments, for teachers, students, researchers and extension workers engaged in pest management strategies. Views expressed and information provided in the present book are contributor's opinion. We are thankful to all scientists and experts who have responded positively and have contributed the chapters.

We are extremely thankful to Hon'ble Vice-Chancellors, Prof. Parmatma Singh (RAU, Bikaner), Prof. S.L. Mehta (MPUA&T, Udaipur) and Dr. S.N. Puri (CAU, Imphal) for their constant help, guidance and encouragement during the editing of this book. We would like to thank Dr. R.A. Singhania, Dean, Dr. B.R. Chhipa, Professor (Soil Science), Dr. A. K. Soni, Assistant Professor (Horticulture) and Sh. R. K. Choudhary, S.R.F. (Entomology) for going through the proof and making necessary improvement. Our special thanks are due to all our colleagues in the Department of Entomology, S.K.N. College of Agriculture, Jobner for their active involvement and whole hearted support. We also wish to thank Smt. Prem Lata Jain and Smt. Alka Bhargava for their calm inspiration, buoyant temperament.

Date: 12-08-2006

P.C. Jain
M.C. Bhargava

Contents

Entomology : Novel Approaches, 2007
Eds.: *P.C. Jain and M.C. Bhargava,*(Page 1-18)
New India Publishing Agency, New Delhi
E-mail: newindiapublishingagency@gmail.com
web: www.bookfactoryindia.com

CHAPTER 1

HISTORY OF INTEGRATED PEST MANAGEMENT

R.S. Jaglan and Ram Singh
Department of Entomology, CCS Haryana Agricultural University, Hisar –125 004

INTRODUCTION

The history of Integrated Pest Control goes back much farther than the mere coining of the term in the early fifties and the subsequent elaboration and clarification of the concept. The origins are deeply rooted in the evolution of pest control practices as developed by Entomologists and Plant Pathologists in the Nineteenth Century.

Human history is a series of attempts to gain increasing control over the environment. At first this control was minimal to the degree that poor shelter and unstable food supplies imposed severe population constraints. The gradual gain in man's capacity to control his environment parallels the gradual rise of civilization. But as man aggregated into villages near rivers and planted crops near by, he encountered increasingly severe attacks by pests on the crops. For thousands of years, man could do nothing about these pests but appeal to the power of magic and a variety of gods. For the most part, early humans had to live with and tolerate the ravages of plant diseases and insects, but gradually they learned how to improve their conditions through "trial and error" experiences. These improvements included the beginnings of pest control or pest management, the most preferred term today.

Prior to the emergence of the crop protection sciences and even before the broad outlines of the biology of pests and the casual nature of plant diseases were understood, humans evolved many culture and physical control practices for protection of their crops. Many of these practices were subsequently proven to be scientifically valid even though originally they were derived mostly through empirical methods. The appropriate use of these cultural methods reduced the damage potential to crops of essentially all pests and all provided satisfactory economic control of some. But there were many pests of high damage potential that could not be controlled adequately by early agriculturists with any combination of known cultural control methods.

Since the dawn of agriculture at about 10,000 B.C., severe food shortages and famines have been a constant threat to human populations around the world. Food shortages with the associated malnutrition increase susceptibility to disease, which has been largely responsible for the very slow growth of human population during most of history. The potential for such catastrophic loss still exists. The greatest crop losses were caused by inclement weather and diseases. There were, and still are, periodic insect outbreaks, such as locusts, army-worms, stem borers, and planthoppers, that cause problems from time to time. The early literature of China and Japan contains records of insect pest outbreaks and methods for their control. Weeds have always been an important problem in agricultural production, but probably never were a cause of famines, because weeds can be controlled by manual labour. Vertebrate pests were an ever-present source of food loss.

As biological knowledge grew during the 18th and 19th centuries and as pest problems became more severe because of an intensification of agriculture and the introduction of pests into new areas, man became increasingly preoccupied with the search for more effective pest control measures. The discovery during 1850's that fungi can produce diseases of plants opened the way for the scientific study of agents to control diseases and the principal search was for chemical compounds. Various chemicals and concoctions were recommended for the control of insects and diseases as early as the 18th century. Much progress was made in the techniques of chemical control of both plant diseases and insects during the last quarter of 19th century. As early as the turn of 20th century, there were optimistic expectation that both diseases and insects would ultimately be controlled by chemical pesticides alone.

Major improvements in the knowledge about pests occurred over the past 200 years. Much of the technology used in current pest management is the re-sult of advances over the last 60 years or so, and was not available prior to 1940. Current developments in molecular biology and precision agriculture will pro-vide technological advances that promise to further revolutionize pest control.

EVOLUTION OF PEST CONTROL

Ancient Times

During this period, people in different parts of the world domesticated the major crop species, such as wheat, barley, oats, rice, maize, soybean, beans, potato, sweet potato, yam, manioc, date palm, coconut, sorghum, grapes and cotton. Harvested produce, mainly cereal grains, was stored for various lengths of time. Both spoilage and destruction by rats, mice and insect pests were constant problems.

Lacking in understanding of the biology of crop production and associated pests, humans were subject to the vagaries of nature for the whole of this period. There are numerous historical accounts of pest problems during this time, and pest outbreaks were often considered as retribution visited on humans by angry gods.

Pathogens also damaged and killed crops, and, like insects, most could not be controlled. Diseases dictated what crops people could grow, and because plant disease is a function of environment, plant diseases could even influence where people lived.

The consumption of stored grain products by vertebrates, especially rats and mice, was especially important. Rodents living in proximity to humans created other problems, because they acted as hosts for vectors of human diseases. The bubonic plague was a devastating disease transmitted by a flea from house rats.

Middle Ages

Not much is known about agricultural developments during this period, because little was recorded. What is known, however, is that plagues of rats, cockroaches, locusts, and disease epidemics leading to famines were frequent occurrances.

Although people recognized the importance of pests, the lack of under-standing, even of basic pest biology, limited their ability to develop control tactics. Without the tools to see microscopic organisms, and lacking an understanding of genetics, the progress was minimal. Some of the early attempts to control nonweed pests were based on superstition.

Seventeenth Century

The seventeenth century saw the beginnings of the scientific method and the invention of scientific equipment, such as the microscope, which permitted major advances to occur. Control of most pests remained at the folklore level. An important advance was the concept that insects arose from eggs rather than spontaneously from decaying matter. The seventeenth century also saw an increased need for agriculture due to the expansion of urbanization.

Eighteenth Century

Carl Linnaeus (1708-1778) introduced the Latin binomial system of nomenclature for all living organisms, and thus greatly facilitated progress in pest control.

Two additional discoveries occurred that further advanced the foundations for the scientific control of arthropod pests. The connection was made between heat summation, or "degree days," and insect development, which are a key to understanding rates of individual growth, development, and reproduction. These are key factors in regulating population dynamics. The existence of plant natural defenses against insects was also recognized, which led to increased development and use of botanical insecticides such as pyrethrin, rotenone, and nicotine. These botanical insecticides are still in use. The realization that plants could be resistant to insects coincided with world exploration, resulting in collection of plant material that had insecticidal properties.

Nineteenth Century

During the nineteenth century, societal assumption of control over the environment accelerated, as did the experimental approach to solving agricultural problems, with the establishment of agricultural experimental stations in Eu-rope and North America.

In 1845, M. J. Berkeley argued that an Oomycete, considered to be a fungus at the time, was the cause of potato late blight disease. This was the first convincing argument that a microorganism caused disease. The realization that fungi caused plant disease was a major advance in understanding diseases and their management. Discovery of bacterial pathogens was also made. In the 1860s, Louis Pasteur provided experimental evidence that convincingly demonstrated living organisms did not arise from nonliving matter, but rather, living organisms arise from other living organisms. Pasteur used heat to kill the microbes in his experiments, thus disproving the theory of spontaneous generation. Killing microbes with heat resulted in sterilization procedures that provided effective microbe control.

In 1876, Robert Koch discovered that he could take blood from a sick animal and transmit disease to a healthy animal, and that this process could be continued from one animal to another. In this way he proved that bacteria could cause disease. He formulated a set of criteria, now referred to as Koch's postulates, which are still followed to demonstrate that a particular organism causes a particular disease.

The latter part of the nineteenth century saw the initial examples of chemical therapy to control pests, and set in motion the whole concept of the "silver bullet" approach to pest control. The major discovery that a mixture of copper sulfate and lime controlled grape downy mildew was an important breakthrough for disease control. The first pesticide application equipment was developed in France to apply this fungicide.

Table 1 Chronological events in the history of evolution of IPM

Important Dates	Events
4700 B.C.	The Chinese domesticate the silkworm, an indication of understanding basic knowledge of how insects grow and reproduce.
2500 B.C.	Insecticidal properties of sulfur were discovered, and the element was used to control insects and mites by the Sumerians.
1000 B.C.	Egyptians and Chinese learnt of "botanical insecticides" to protect stored grains. These plant-derived chemicals showed insecticidal properties, such as nicotine and pyrethrin.
Middle Ages	
400-Circa 800	In China, silkworm droppings were used for control of rice pests.The connection was made between crop phenology (developmental stage) and insect attack, which led to the concept of crop planting date as a means to avoid an insect problem. Another important development in China during this time was the use of ants for biological control in Citrus. The Chinese even constructed "bridges" of bamboo between trees, so the ants could readily move from tree to tree. Ant nests were sold near Guangzhou for biological control purposes.
900	The Chinese used arsenic to control garden pests.
1100 circa 1200	In China, soaps were used to control some insects. Albertus Magnus observed that mistletoes are parasites, and found that the "sick" host recovered when the parasite was removed. His discovery was not recognized and, as noted by Horsfall and Cowling, humans were doomed to another 650 years of suffering due to plant parasites.
Seventeenth Century	
1665	Robert Hooke observed the teliospores of a rust fungus for the first time. Hooke mistook what he saw to be the result of the disease, not the cause. This error persisted for nearly 200 years.
1666	Encyclopedias of natural history were published in Japan with abundant information on insect biology and descriptions of species.
1690	Infusions of tabacco were discovered to be insecticidal. Arsenic was used as an insecticide. Although toxicity problems were encountered, people continued to use arsenicals well into the twentieth century as fear of hunger was more powerful than fear of poisons.

Mid 1600s	About this time, farmers in France realized the connection between the barberry bush and incidence of rust disease in wheat, leading to legislation requiring removal of barberry bushes for disease management. This was probably the first example of what is now called areawide IPM.
Eighteenth Century	
1729	Micheli noted that dust from a diseased melon causes the disease to develop on previously disease-free melons. He thought the dust was "seeds" of the fungus.
1743	Needham first observed plantparastic nematodes in galled wheat kernels.
1755	Tillet noted that smut disease is worse if he inoculates wheat with smut "dust", and that treating seeds with copper sulfate before planting reduces the disease. The fungus causing the disease was later named after him. This was the first unequivocal demonstration of a contagious plant disease, although Tillet thought that a poisonous principal caused the disease, not the microorganisms.
1762	*Mynah* (a bird) from India was imported for the control of locusts in Mauritius.
1782	"Underhill" variety of wheat reported resistant to Hessian fly in USA.
Nineteenth Century	
1831	First record of a plant resistant to an insect pest. The apple variety 'Winter Majetin' was resistant to the wooly apple aphid in England.
1845-1846	Devastating epidemics of late blight on potatoes in Ireland resulted in famine in Ireland and mass emigration to the United States. Berkeley argued convincingly that late blight was caused by a microbial pathogen, but this concept was still not accepted.
1848	Derris (Rotenone) reported to be used in insect control in Asia.
1850-1860s	Berkeley observed root-knot nematode galls on the roots of greenhouse-grown cucumbers, bulb nematodes on teasel, and cyst nematodes on sugar beet roots.
1855	A. Fitch reported the role of lady bird beetles, green lacewings and other predaceous insects in the control of insect pests of crops.
1878	Downy mildew of grapes reached France from America. This biological invasion resulted in a devastating disease epidemic on grapes and almost destroyed the French wine industry.
1878	First demonstration that bacteria cause diseases in plants; fire blight disease of pear, discovered by Burrill in the United States, was caused by a bacterium; anthrax was found in cattle 2 years earlier by Pasteur and Koch.
1882	Millardet, an observant grape researcher in France, noticed that grapes treated with a mixture of lime and copper sulfate to deter pilferers have less downy mildew than adjacent rows that have not been treated.
1885	Millardet perfected a mixture of copper sulfate and hydrated lime to control the mildew. The mixture was called Bordeaux mixture because it was first used in the Bordeaux region of France. This discovery gave great impetus to research on plant diseases, because of the possibility that they could be controlled.

1887	Balfour published his book *The Agricultural Pests of India and of Eastern and Southern Asia*. It was the first publication dealing with agricultural insect pest control in this region.
1888	The Vedalia beetle was introduced into California from Australia for biological control of the cottony cushion scale that had been accidentally introduced into Citrus growing areas in 1868. By 1880, the scale was threatening to eliminate the citrus industry. The Vedalia beetle importation and release was the first example of effective classical, biological control. This biocontrol system is still working today, although during a period of heavy pesticide use in the 1960s, the beetle was nearly eliminated from California groves.
1889	Biological control of cottony cushion scale on citrus in the USA by use of Vedalia beetle imported from Australia.
1890	Control of grape phylloxera in Europe by grafting of European grapevine scions to resistant North American rootstocks.
1890s	Several pathogens were shown to be vectored by insects. It was conceptually an important idea to know that insects could transmit the disease-causing agents. The cotton boll weevil invaded southern Texas from Mexico. The weevil spread to the rest of the cotton belt in the following 30 years almost destroying the cotton industry in the South.
1892	Lead arsenate used for the control of Gypsy moth in the USA.
1898	Beijerink showed that the disease agent with which Mayer and Ivanowski had worked could be killed by heat and thus must be a "contagious living fluid." He coined the term *virus* for it.
1895 to early twentieth century	Arguments continued to rage about whether bacteria caused or were the result of plant disease. Record of this controversy is one of the best documented scientific disagreements in the literature.
1898	The coccinellid, *Cryptolaemus montrouzieri* Mulsant from Australia was released against coffee green scale, *Coccus viridis* (Green) in India. It established but failed to control the scale.
Twentieth Century	
Early 1900s	Inorganic chemicals, such as copper sulfate, sodium chlorate, and iron sulfate, were used to kill plants. The problem was the treatments killed most plants, and the application rates were high. These chemicals were never widely used except in situations like railroad ballast. Selective chemicals for weed control in crops such as wheat and barley also found limited use. Insecticide dusts (many based on lead arsenate) were widely used in crops such as cotton for boll weevil control, and apples for codling moth control. Problems included low efficacy, poor safety to workers and crops, and soil contamination. In some situations so much arsenic accumulated in the soil of treated orchards that trees replanted in the orchards died.
1905	The first discovery that resistance to diseases is inherited (e.g., rust on wheat) lays the foundation for breeding pest-resistant crops.

1911	First demonstration that there is host-range variability among isolates of a pathogen. Such variability is now the basis for defining races, pathovars, biotypes, or strains.
1911	Cotton Pest Act was enforced in Madras State. Under the Act cotton stalks had to be removed by 1st August every year to minimize the incidence of pink bollworm.
1912	Concept of control by legislative action was established in the United States with development of quarantines to prevent introduction of foreign pests.
1913	Riehm used organo-mercury compounds as seed treatments to prevent diseases. Such treatments were widely used until the 1960s when the toxicity of mercury was recognized, and all, uses were phased out.
1914	First report was given of resistance in insects to an insecticide: the San Jose scale to lime sulfur.
1919	The term *biological control* was introduced, based on the idea that natural predators and parasites could control pest organisms.
1923	Multiple component suppression techniques involving the use of resistant varieties, santiation practices and need-based application of insecticides developed for the control of bollweevil in the USA.
1931	The cottony cushion scale attacking wattle of commerce, *Acacia decurrens* was controlled in India by release of predatory beetle, *Rodolia cardinalis* Mulsant from California.
1934	Tisdale discovered the dithiocarbamate fungicide Thiram. This discovery led to rapid development of many synthetic contact fungicides. It was not until over 30 years later in 1966 that a systemic fungicide was discovered.
1930 to 1940s	First discovery occurred of organic chemicals that kill plants. The early ones (e.g., dinitro-ortho-cresol) were not selective and thus found limited use. Discovery of 2,4-D during World War II changed everything by allowing farmers to kill weeds in the crop without killing the crop. Selectivity of the herbicides is fundamental to modern crop production. The concept of selectivity was later extended to insecticides with the introduction of the principles of integrated control.
1939	Swiss chemist Paul Mueller discovered insecticidal properties of DDT (although the compound had been synthesized during the nineteenth century). Mueller was awarded the Nobel Prize in 1948. The discovery of DDT, and of benzene hexachloride, marked the onset of the insecticide era. The absolute dominance of insecticides for insect control lasted until the 1960s when disillusionment set in because of problems documented by entomologists who pioneered the integrated control concept, as compiled by Rachel Carson in her book, *Silent Spring*. Killing insects had become a preoccupation rather than producing crops or maintaining environments.
1940	Work on the development of jassid-resistant varieties was undertaken in India resulting in the release of varieties like LSS, 4F, 289F, etc., during 1940s.
1941	Insecticidal activity of Hexachlorocyclohexane (HCH) discovered in France.
1946	Parathion, the first organo-phosphatic insecticide developed.

1948	Use of DDT and HCH on agricultural crops in India."Doom" based on *Bacillus popilliae* Dutky and *B. lentimorbus* registered in USA for the control of Japanese beetle larvae on turf.
1950s	The use of antibiotics for bacterial disease control became widespread. Resistance to new classes of pesticides started to appear in insect populations.
1950s to present	Chemical industry synthesized many new pesticides.
1951	R.H. Painter published his classic book *Insect Resistance in Crop Plants.* Introduction of first carbamate insecticide, Isolan.
1952	First plant to produce HCH established at Rishra (India).
1953	Watson and Crick described the structure of DNA, which sets the stage for modern molecular biology with implications for plant breeding and genetic engineering.
1955	E. Knipling, of the United States Department of Agriculture, first described a method for suppressing insect reproduction in the field by sterile insect release.
1959	Karlson and Butenandt identified the first insect sex pheromone, for silkworm moth. Sex pheromones are key elements of the reproductive behavior of many pest insects and have been successfully used in the tactic called behavioral control. By the mid-1980s, over 1,000 sex pheromones had been identified.
1959	Concept of integrated control involving integration of chemical and biological control introduced.Concept of economic injury level and economic threshold developed by V.M. Stern and coworkers.
1962	Rachel Carson wrote *Silent Spring,* a book that brought the problems caused by pesticides to the attention of the public and forever changed the way pesticides are viewed by the general public and dealt with by scientists.
Late 1960s	Resistance to a herbicide was first documented in a weed population, and the number of weeds showing resistance increased rapidly in the next two decades, to the extent that resistance has now started to change the herbicide paradigm. First documented field resistance to fungicides also occurred fast. The problem became more widespread as new fungicides with site specific activity were introduced over the next 30 years.
1964	Publication of the book *Biological Control of Insect Pests and Weeds* by Paul DeBach, which established biological control as a separate discipline in Entomology.
1973	Development of first photo-stable pyrethroid, permethrin.
1975	Elcar (*Helicoverpa* NPV) registered for the control of bollworm and tobacco budworm on cotton.First insect growth regulator (Methoprene) registered for commercial use in USA.Publication of the book *Introduction to Insect Pest Management* by R.L. Metcalf and W.H. Luckmann which was the first comprehensive treatise on IPM and established the concept on a firm footing.
1980s to present	Rapid development occurred of molecular biology and application of transgenic crops (also known as genetically modified organisms, or (GMOs) to pest management.

1989	Development of first transgenic plant, reported by M. Vaeck and coworkers of Belgian Biotechnology Company, Plant Genetic Systems by transferring *B. thuringiensis* ä-endotoxin gene to tobacco for the control of *Manduca sexta* (Johannsen).
1992	Concept of Environmental Economic Injury Levels proposed by L.P. Pedigo and L.G. Higley.Dr. Edward F. Knipling and Dr. Raymond C. Bushland were awarded the World Food Prize for developing sterile insect technique.United Nations Conference on Environment and Development (Rio de Janeiro, Argentina) assigned a pivotal role to IPM in the agricultural programmes and policies envisaged as part of its Agenda 21.
1994	First transgenic squash varieties were released with resistance to viruses. First glyphosate-resistant transgenic varieties of corn and soybean crops were released.
1994	A task force consisting of FAO, the World Bank, UNDP and UNEP co-sponsored the establishment of the Global IPM Facility with the Secretariat located at FAO, Rome.
1995	Release of insect-resistant cotton, corn, and potatoes using genes derived from *Bacillus thuringiensis* that encode endotoxin production.
1995	Dr Hans R. Herren was awarded the World Food Prize for developing and implementing the world's largest biological control project for cassava mealybug which had almost destroyed the entire cassava crop of Africa.
1997	Dr. Ray F. Smith and Dr. Perry L. Adkisson were awarded the World Food Prize for their pioneering work in development and implementation of integrated pest management (IPM) concept.
Mid 1990s to present	Societal concerns about engineered transgenic crops (GMOs) increased in several regions of the world, especially Europe. Public pressure dramatically slowed adoption of the technology in IPM systems.

Source: Norris *et al.*, 2003; Dhaliwal and Arora, 2001

A major change in thinking occurred during this period following the realization that pesticides had serious limitations, and in some cases pro-duced numerous undesirable side effects. The search for silver bullet chemical solutions to pest problems began to be replaced with a more balanced ecological view of pest control. This change continues to this day.

Resistance of all types of pests to pesticides was observed during this period, rendering many pesticides ineffective for controlling the intended target pest. The problems with pesticide resistance were initially observed, and were more severe in insect pests, but later resistance proved to be a problem for all types of pests, with the possible exception of nematodes. Pesticide resistance became so severe that all areas of pest management had to devote considerable effort to develop a new approach to the use of pesticides, called resistance man-agement. The phenomenon of pest resurgence and release of minor pests due to killing of beneficial insects was documented.

At about the same time that pesticide resistance became a problem, ecologists also observed that pesticides might have effects on non target species, such as eggshell thinning in several bird species and earthworm kill. The problem of bio-magnification of organochlorine insecticides in the food web was recognized. So far there is no evidence of herbicide or nematicide bio-magnification in food webs, but their movement through the soil profile into groundwater became an important pollution problem. The multitude of problems associated with careless pesticide use forced a change in the pest control paradigm from silver bullet to an understanding of the biology and ecology of crop-pest interactions. The biological and ecological information is used to de-sign integrated management strategies that we now call integrated pest management (IPM).

After nearly 25 years, the acronym IPM has become household term but oftenly misused by professionals and lay persons alike without giving much thought to implications of the expression or its impact on modern agriculture (Kogan, 1998). The word 'pest' in IPM includes insects, weeds, pathogens and non-arthropod animals also. The focus on insect pests in present context does not be little the contributions of plant pathologists and weed scientists to IPM.

ORIGIN OF IPM

Basic tactics of IPM were proposed and used to defend crop plants against the ravages of pests long before the expression was coined (Jones, 1973). Throughout the late nineteenth and early twentieth century, in the absence of powerful pesticides, crop protection specialists relied on knowledge of pest biology and cultural practices to produce multitactical control strategies some of them proved basis of modern IPM system (Gains, 1957). During first half of 20th Century, economic entomology was responsible for research and teaching about economically important insect species (pests) and means to control them. 'Pest control' was understood as the set of actions taken to avoid or delay the impact of pests on crops. However, in early 1960's with the advent of organo synthetic insecticide, plant protection scientists started paying maximum attention to testing of chemicals, to the detriment of studying pest biology and non-insecticidal methods of control. The period of late 1940s through-mid-1960s has been called the dark ages of pest control in the world. By the late 1950s warnings began to hear about the risks involved in the use of insecticides in pest control in USA, Canada and Europe.

Integrated Control

The idea of integrated control appeared in the paper of Hoskin *et al.* (1939) but the recognized originator role is attributed to Michelbacher and Bacon (1952) who emphasized the importance of considering the entire entomological picture in developing a treatment for any particular pest.

Series of papers published by Smith and Allen established integrated control as a new trend in economic entomology. Towards the end of 1960s integrated control was at its peak both in literature and in practice to control pests, although by this time 'pest management' started gaining popularity.

Rise of Integrated Control Concept

Failure of new organosynthetic insecticides, resistance, resurgence of primary pests, upsurge of secondary pests and overall environmental contamination were the primary factors leading to integrated control concept. Publication of Silent Spring (1962) further accentuated the acceptance of Integrated Control Concept.

In its original formulation, the scope of integrated control was broad. Smith and Allen (1954) emphasized that integrated control will utilize all the resources of ecology and give us the most permanent, satisfactory and economic insect control. However, in subsequent publications, integrated control was more narrowly defined as applied control which combines and integrates biological and chemical control (Stern *et al.*, 1959). This definition stood through much of the late 1950s and early 1960s but began to change again in the early 1960s as the concept of pest management gained acceptance among crop protection specialists.

Pest Management

Pest Management gained considerable exposure at the 12th International Congress of Entomology, London. The Australian ecologists who coined the expression contented that "Control" as in pest control, also includes the effect of elements that act independently of human interference. Population are naturally controlled by biotic and abiotic factors, even if at levels intolerable to humans. Management on the other hand implies human interference. In 1966 Geier favoured the term pest management. Subsequently, two publications, in 1969 (NAS, 1969) and in 1970 (Rabb and Guthrie, 1970) contributed significantly to the acceptance by American Entomologists of the Australian "pest management" in lieu of the American "Integrated control". By the mid-1970s both integrated control and pest management coexisted essentially as synonyms.

Integrated Control v/s Pest Management

In 1965 "symposium convened in Rome, Italy by United Nations FAO became a landmark in the advancement of the integrated control concept in its broadest sense. Participants at the symposium recommended establishment of the FAO Panel of experts in IPC. The definition of IPC proposed by Smith and Reynolds (1966) was adopted with modifications at the first session of the Panel of Experts in 1967. The concept of pest management based on the ecological background evolved from Australian's work and ideas became popular in USA and worldwide (Rabb and Guthrie, 1970) but was resisted by the FAO Panel of experts being dominated by Californians (R.E Smith, H. Reynolds, V. Stern and R. Van den Bosch). At its 3rd meeting (FAO, 1970), the panel agreed that integrated control is not synonymous with pest management which is more equivalent to pest control in usage. But at the 4th meeting in 1972 they again changed their view to consider IPC and pest management as equivalent on the basis that though they are different but pest management is difficult to translate in most of the languages as compared to integrated control or integrated plant protection.

Integrated Pest Management

The debates in 1960s resulted in substantial agreement that

(I) 'Integration' meant the harmonious use of multiple methods to controls single pest as well as the impact of multiple pests.

(II) 'Pests' were any organism detrimental to humans, including invertebrate and vertebrate animals, pathogens and weeds.

(III) IPM was a multidisciplinary endeavour.

(IV) 'Management' referred to a set of decision rules based on ecological principles and economic and social considerations. The backbone for the management of pests in agricultural system was the concept of economic injury level 'EIL'.

In February 1972 a message from President R. Nixon to the US Congress for environmental protection was transmitted which contained a paragraph on IPM. Hence a first use of Integrated Pest Management in print. Thereafter in 1972, a report on Integrated Pest Management was prepared by the Council on Environmental Quality.

Definitions

The search for a perfect definition of IPM has endured since integrated control was first defined. A survey recorded 64 definitions of integrated control, pest management, or integrated pest management. The most often cited definition is still of Stern *et al.* (1959) which was adopted by FAO panel of experts (FAO, 1970) broadly as "Integrated Pest Control is a pest management system that in the context of the associated environments and the population dynamics of the pest species, utilizes all suitable techniques and methods in as compatible a manner as possible and maintains the pest population at levels below those causing economic injury. "

Based on analysis of definitions spanning the past 35 years the following definition can become an ideal current thought (Kogan, 1998).

IPM is a decision support system for the selection and use of pest control tactics, singly or harmoniously coordinated into a management strategy, based on cost/benefit analysis that take into account the interest of and impacts on producers, society and the environment.

PROGRESS IN IPM

The Huffaker Project

US-IBP (International Biological Programme) for IPM (1972) was later on designated as Huffaker Project. The principles, strategies and tactics of pest population regulation and control in major crop ecosystems were analysed. Six crops-alfalfa, citrus, cotton, pines, pome and stone fruits and soybean were selected following criteria that included current level of insecticide use (very high in cotton and citrus, very low in soybean), potential for successful biological control (alfalfa, citrus, pome and stone fruit) and representation of a nonagricultural system (pines). An anticipated outcome of the program was 40-50 per cent reduction in the use of the more environmentally polluting insecticides within a 5 year period and 70-80 per cent in 10 years. The project spanned 1972-1978. Advances were made in many aspects implementing improved IPM strategies for all systems, but insufficient involvement of extension specialists may have reduced the potential of the programme to gain greater UN adoption.

CIPM

After the success of Huffaker Project, a group of scientists under the leadership of PL Adkisson secured funding for the second large scale IPM project in the United States entitled "Consortium for Integrated Pest Management" (Frisbie, 1985). Major crops were alfalfa, apple, cotton and soybean. It was claimed that average adoption of IPM for the crops was 66 per cent in 5.76 million ha. The main indicators of adoption were the use of economic thresholds and economic injury levels for spray decisions, use of selective pesticides or application of lower dosages of broad spectrum insecticides. A significant achievement of the programme was the genuine attempt to integrate weed science and plant pathology and the emphasis on economic assessment of IPM adoption.

USDA Regional IPM Program

With the CIPM project conclusion in 1985, the USDA-CSRS (Cooperative States Research Stations) redirected some IPM funds to management through the newly formed National IPM Coordinating Committee (NIPMCC), which gave deans and experiment station directors at land grant universities control over the direction of IPM funding. Under the banner of NIPMCC and CSRS, the first national IPM symposium/workshop was organised in 1989. Two following national symposia, in 1994 and 1996, were instrumental in promoting a national agenda for IPM and ushering in the Clinton Administration's National IPM Initiative.

The National IPM Initiative

In September, 1993, the Clinton Administration submitted that implementing IPM practices on 75 per cent of the Nation's crop acres by the year 2000 was a national goal. This has culminated into a National IPM Forum to formulate Areawide IPM Programme.

Areawide IPM Systems

The concept of "areawide pest management" was introduced by Knipling and Rohwer in a proposal to the North American Plant Protection Organisation". They argued that pest management projects (a) must be conducted on geographic area: (b) should be co-ordinated by organisation rather than by individual farmers, (c) should focus on reducing and maintaining key pest population at acceptable low densities, although eradication may be involved if practical and advantageous, (d) may involve a mandatory component "to ensure full participation in the programme. Besides, Cooperative Extension Service (CES) also played a key role in diffusion of IPM in the United States.

IPM in FAO

The role of FAO in the diffusion of IPM (or IPC-integrated pest control) has been well documented. The concepts and ideas developed by the panel of experts between the first meetings in 1968, through fifth, in 1974 formulated a new programme - the Cooperative Global Programme. This programme sets practices for crops or cropping systems and regions, for adoption of IPM strategies, implementation, research, and training and education. Key for the success of FAO Sponsored Intercountry IPM-Rice Programme was demonstration of the relationship between outbreaks of the BPH (brown plant hopper) with over use of broad-spectrum insecticides.

The latest development in the support for IPM at FAO is the establishment of the Global IPM facility with cosponsorship of FAO, UNDP, UNEP and world bank. This facility will serve as a co-ordinating, 'consulting, advising and promoting entity for the advancement of IPM worldwide.

IPM in Europe

IOBC played a major role in the development and implementation of integrated plant protection in Europe (Table 2). The starting point of review on the evolution of plant protection methods that has been established in 1977 but still retains its basic validity.

Some 25 years ago scientists have described 4 steps in the development of plant protection with the conclusion that integrated plant protection is the most advanced step that can be reached. They separated step 4 from step 5 (Integrated Production) by a solid line indicating a sort of final destination. However, a major improvement has been made in recent years as there is common agreement that plant protection has to be removed from this isolation and put into the context of all farm operations. Therefore, solid line in the Table has been broken to emphasise that Integrated Plant Protection is and has to become an integral part of Integrated Production.

Definition of Integrated Plant Protection (FAO definition modified by IOBC, 1977)

All economically, ecologically and toxicologically defensible methods will be applied to keep damaging organisms below economic damage levels whilst conscious exploitation of natural control factors is emphasized.

This definition leaves open space for a broad spectrum of interpretations. Many illustrations in text books show Integrated Plant Protection as a large range of plant protection measures arranged around the crop. This suggests that we can make any given combination (=integration) of control methods according to our personal taste and declare it an integrated protection programme. An extreme case found in the literature declares that "the principle (of integrated plant protection) consists of a combination of biological and chemical control methods". Obviously, this view does no longer reflect a modern concept of plant protection in the context of a sustainable agriculture.

Table 2. The Evolution of Plant Protection Methods (IOBC 1977, modified)

Blind chemical control	General schematic and routine applications of the most pesticides; Advice industry.
Chemical control based on advice	Application of usually broad spectrum pesticides after consultation with an official advisory service.
Specific control	Introduction of the concept of the "economic threshold levels"
Transitory phase	Application of pesticides with no negative side-effects : Protection of beneficial organisms
Integrated plant protection *	Like specific control, in addition

Contd. ...

Table 2 Contd. ...

Dynamic phase	Integration of biological and biotechnical methods and method of good agricultural practice : Chemical control strongly regulated.
Integrated agricultural production	Like integrated plant protection in addition. Observance, integration and exploitation of all positive factors in the agro ecosystem according to ecological principles.

* In the original Table step 4 was separated clearly from step 5 by a solid line. The solid line has replaced by a broken line to indicate that in the modern concept integrated plant protection is removed from its isolation and put into the context of all farm operations.

The modern concept that evolved in the 1980s puts emphasis on the agro-ecosystem as one of the key elements of Integrated and Organic Farming.

A clear hierarchy of priorities replaces the free combination of control methods

It cannot be the main task of plant protection to repair damages caused by inadequate farming practices. Based on these considerations IOBC has adopted a clear concept of priorities for plant protection in the context of a sustainable agriculture (IOBC, 1993).

The holistic systems approach gives highest priority to preventive measures that can be summarised as indirect plant protection. This first element includes, (1) the optimal use of natural resources already in the planning stage of a new crop, (2) the elimination of all farm operations with negative impact on the agro-ecosystem (i.e. causing or enhancing plant protection problems), and (3) the protection and augmentation of natural antagonists.

Monitoring and forecasting systems as important second element provide the necessary instruments for the decision if and when the third element, namely direct plant protection (=control measures) has to be applied. Hence, the use of pesticides is not per se an integral part of integrated plant protection but the last option when prevention alone does not produce acceptable results.

Variations in the Basic Theme

Although success of IPM worldwide over the past 10 years can not be disputed, however, concern exists that (a) the rate of adoption of programme has been very slow (b) many programme still rely primarily on timely applications of pesticides as the principal management tactic; and (c) the majority of those programme focus on pests within categories (arthropods, pathogens or weedy plants) with little consideration of multiple pest interactions. Still alternative proposals are offered to strengthen IPM.

Biological Intensive IPM

Because of its entomological roots, the use of EILS and scouting for the assessment of pest population has been the main criteria of IPM adoption. Based on these criteria, programmes that rely entirely on pesticidal controls may qualify as IPM, if treatments are made following scouting to determine whether pest population have reached the EIL. As these programmes have been endorsed by the agrochemical industry, the perception exists that IPM has become "agrochemically intensive". Furthermore, despite nearly three decades of IPM educational and implementation programmes, use of pesticides has not declined. In view of this trend. Frisbie and Smith (1991) proposed a switch to "biologically intensive" IPM or "biointensive" IPM. Biointensive IPM would rely on host plant resistance, biological control and cultural control and the industry would develop biorational pesticides more easily integrated into biointensive IPM systems.

Ecologically Based IPM

Ecologically based pest management (EBPM) is proposed to be (a) safe for growers, farm workers and consumers, (b) cost effective and easy to adopt and integrate with other production practices, (c) long-term sustainable and without adverse environmental, economic or safety consequences and (d) with ecosystems , as the ecological focus. However, in principle, IPM programme has also the same objectives except that in practice IPM is still dominated by the use of pesticides.

Sustainable Agriculture

The term sustainable agriculture first appeared in the literature in 1978 (McIsaac, 1994) but was formally introduced in 1985 when the US Congrees "enacted the Food Security Act that initiated a program in 'Low Input Sustainable Agriculture" to help farmers use resources more efficiently, protect the environment, and preserve rural communities" (McIsaac, 1994). With a theoretical foundation in agroecology (Gilessman, 1990), proponents of the sustainability concept for crop production have found great affinity with principles and approaches of IPM. Indeed, IPM provided both a conceptual approach and an implementation paradigm for sustainable agriculture (Gilessman, 1990; McIsaac & Edwards, 1994; Olson, 1992). From an IPM perspective, the concept of sustainable agriculture provides a platform for launching IPM to higher levels of integration (Kogan, 1988).

IPM is the component of sustainable agriculture with the most robust ecolog-ical foundation. IPM not only contributes to the sustainability of agriculture, it also serves as a model for the practical application of ecological theory and provides a paradigm for the development of other agricultural system compo-nents. The IPM paradigm provides that pests and their management exist at the interface of three multidimensional universes: ecological and agricultural, in hierarchical order, with ascending levels of complexity and expanding spatial scales. Based on these considerations, IPM can be conceived as interactive systems with multiple levels of integration.

Success of IPM programs often has been measured by the overall reduction in the volume of pesticides used to control prevalent pests. Although reduction in pesticide usage is a desirable consequence of IPM, it cannot be the only measure of success. There are special circumstances in which to maintain a viable agricultural production. The issue is pesticide use within the principles of IPM, i.e. selective use after maximizing effectiveness of natural controls.

CONCLUSION

The foregoing discussion has traced the history of integrated pest management back into the 19th century and even earlier. The concept of integrated pest management is not a disjunctive development in crop protection, but is an evolutionary in the history of pest control. Many components of IPM concepts were developed in the late 19th and early 20th century, but IPM as it is now conceived is unique because is based on ecological principles and integrates multidisciplinary methodologies in developing agro ecosystems management strategies that are practical, effective, economical and protective of both public health and the environment.

References

Carson, R. 1962. *Silent Spring. Boston,* MA: Houghton Mifflin: 368 pp.

Dhaliwal, G.S. and Arora, R. 2001. *Integrated Pest Management: Concepts and Approaches.* Kalyani Publishers, New Delhi, 427 pp.

Food and Agriculture Organization, 1970. *Rep. FAO Panel of Experts on Integrated Pest Control, 3rd, Sept. 10-16.* Rome, Italy : FAO-UN, Meet. Rep. 1970/M/7. 38pp.

Frisbie, R.E. 1985. Consortium for Integrated Pest Management (CIPM) – organization and administration. In: *Integrated Pest Management on Major Agricultural Systems* (Frisbie, R.E. and Adkisson, P.L. eds.) Texas A&M Univ., Texas Agric. Exp. Stn. MP 1616, pp. 1-9.

Frisbie, R.E. and Smith, J.W. Jr. 1991. Biologically intensive integrated pest management: the future. In: *Progress and Perspectives for the 21st Century,* (Menn, J.J. and Steinhauer, A.L., eds.) Entomol. Soc. Am., Cent. Symp. Lanham, MD pp. 151-64.

Gaines, J.C. 1957. Cotton insects and their control in the United States. *An. Rev. Entomol.* 2: 319-38.

Geier, P.W. 1966. Management of insect pest. *Annu. Rev. Entomol.* 11 : 471-90.

Gilessman, S.R., ed. 1990. *Agroecology; Research in the Ecological Basis for Sustainable Agriculture.* New York: Springer 380 pp.

Hoskins, W.M., Borden A.D. and Michelbacher A.E. 1939. Recommendations for a more discriminating use of insecticides. *Proc. 6th Pac. Sci. Congr.* 5: 119-23.

Jones, D.P. 1973. Agricultural entomology. In : *History of Entomology.* (Smith, R.F., Mittler, T.E., Smith, C.N. eds.), Palo Alto, CA, pp. 307-32.

Kogan, M. 1988. Integrated pest management theory and practice. *Entomol. Exp. Appl.* 49: 59-70.

Kogan, M. 1998. Integrated pest management historical perspectives and contemporary developments. *An. Rev. Ent.* 43 : 243-270.

McIsaac, G. 1994. Sustainability and sustainable agriculture: conceptual evolution, competing paradigms, and possible consensus. In: *Sustainable Agriculture in the Midwest : Lessons from the Past, Prospects for the Future* (McIsaac, G. and Edawards, W.R., eds.) Urbana IL: Univ. Press. pp 9-34.

McIsaac, G. and Edawards, W.R., eds., 1994. *Sustainable Agriculture in the Midwest : Lessons from the Past, Prospects for the Future.* Urbana, IL Univ. Press 291 pp.

Michelbacher, A.E. and Bacon, O.G. 1952. Walnut insect and spider mite control in Northern California. *J.Econ. Entomol.* 45: 1020-27.

National Academy of Sciences, 1969. *Insect-Pest Management and Control.* Washington, DC: Natl. Acad. Sci. Publ. 1695. 508 pp.

Norris, R.F., Caswell-Chen, E.P. and Kogan, M. 2003. *Concepts in Integrated Pest Management.* Prentice Hall, New Jersy, 586 pp.

Olson, R.K. ed. 1992. *Integrating Sustainable Agriculture , Ecology, and Environmental Policy*, New York Food Products. 161 pp.

Rabb, R.L. and Guthrie, F.E., eds. 1970. *Concepts of Pest Management*. Raleigh, NC: N.C. State Univ. Press. 242 pp.

Smith R.F. and Reynolds H.T. 1966. Principles, definitions and scope of integrated pest control. In : *Proc. FAO Symp. Integrated Pest Control* (Food and Agriculture Organization) Rome, pp. 11-17.

Smith, R.F. and Allen, W.W. 1954. Insect control and the balance of nature. *Sci. Am.* 190(6): 38-92.

Stern, V.M., Smith, R.F., van den Bosch, R. and Hagen, K.S. 1959. The integrated control concept. *Hilgardia* 29: 81-101.

Entomology : Novel Approaches, 2007
Eds.: *P.C. Jain and M.C. Bhargava,*(Page 19-42)
New India Publishing Agency, New Delhi
E-mail: newindiapublishingagency@gmail.com
web: www.bookfactoryindia.com

CHAPTER 2

SAMPLING, SURVEILLANCE AND FORECASTING OF PESTS

R.K. Saini and P.R. Yadav
Department of Entomology, CCS Haryana Agricultural University, Hisar – 125 004

INTRODUCTION

Management of several of the important insect-pests of crops is becoming increasingly difficult owing to a number of factors by which pests have been able to offset the effect of control interventions. Several of the present day pest management problems may be ascribed to the excessive and indiscriminate use of chemical pesticides during the last about three decades. Routine application of insecticides for controlling insect-pests, without taking into consideration whether economic damage to the crops is being inflicted by the pest or not, has resulted into several problems like pest resistance, pest resurgence, pest replacement, killing of natural enemies of pests, presence of toxic residues in food, and soil and environmental pollution, etc., apart from increase in the cost of plant protection. Entomologists have been trying to find solution to these problems through integrated pest management of which monitoring of insect activity through regular surveys is an important component. In fact, knowledge of insect activity is indispensable since pest management system can not operate without accurate estimates of pests and natural enemies population or without reliable assessment of plant damage and its effect on yield.

Different activities of insects such as their populations on the crop, damage inflicted to the plants, insect stage present, local movement, migration and dispersal etc. are documented through surveillance. The decision, whether the control interventions are needed or not, depends upon the accurate estimation of numbers. Further, based on the study of interrelations of pests populations with various environmental factors, we should be able to predict the future population trends or outbreaks of pests so that appropriate control measures are initiated when required. For example, when chemical control is to be adopted we must be able to forecast the need to spray or to predict the most effective spray date in the light of likelihood of pest population reaching economic threshold level.

DEFINITION

Survey is a planned activity to collect some data. When survey of the same place or locality is carried out at regular interval to record some observation or to ascertain the changes in the subject of study, it is called 'surveillance'. The word 'surveillance' and 'monitoring' have been used in the literature synonymously as the regular monitoring of pests becomes surveillance. The basic principles of pest management are that no action is to be taken against a pest unless (i) the pest is actually present, and (ii) its population has multiplied in sufficient numbers to cause economic loss.

OBJECTIVES OF PEST SURVEILLANCE

In relation to pest management, the major objective of pest surveillance is to monitor the pests population and/or the damage caused by them to the crop regularly in order to decide when to undertake control interventions. The data so obtained for several years in the background of varied environmental conditions may help in working out pest-environment relationships or interactions to aid in pest forecasting with reasonable precision. Further, endemic areas of various pests may also be marked. Since pest surveillance is quite time consuming and expensive, it would be cost effective to gather information on other parameters also, which affect the success of a pest management programme. For example, surveillance programme may be planned in such a way that could include information on natural enemies of pests also. Similarly, surveillance could also be undertaken for monitoring of build up of insecticide resistance in insects. To achieve the above objectives, one should have a thorough knowledge of the diagnostic symptoms of damages caused by the pests on the crops and also of pests' life history and behaviour.

A survey may be qualitative which is aimed to identify the different insect species present over an area or their density whether it is abundant, common or rare/in traces, or it may be quantitative which may give estimate of one or more species of insects in time and space. It may help in predicting future population trends or the corresponding potential damage to the crops or both.

SAMPLING INSECT POPULATIONS

It is not possible or even desirable to count all the insects in a habitat. Therefore, to estimate the population density of a pest or the damage caused by it to the crop, one has to resort to sampling. The basic principles of pest sampling and advances made in sampling have been comprehensively reviewed by Morris (1960), Cochran (1977), Southwood (1978), Kuno (1991) and Binns and Nyrop (1992) who have emphasized that randomization and choice of sample unit are the fundamentals of sampling. The total number of samples to be taken depends on the degree of precision required. This may be expressed either in terms of achieving a standard error of pre-determined size or in probability terms. It would be desirable to understand some of the common terms related to sampling:

Sampling unit

A sampling unit is a proportion of the habitat from which insect counts are to be made. The sampling unit may be a plant, branch of a plant, a certain number of leaves or fruiting bodies, a hill, a clump, a micro-plot of one square meter or number of plants in one meter of row length, etc. The size of the sampling unit can also be determined by a sampler. However, the sampling unit must be distinct and should not over lap.

Sample

A group of sampling units is called the sample from which an estimate is made.

Sampling universe

It is the habitat in which the population occurs. It is important that no sampling units are to be taken in non-inhabited areas.

Sampling programme

It is the procedure which employs the sampling technique to obtain a sample and make a density estimate. A sampling programme would envisage insect stage to be sampled, number of sampling units required, spatial pattern to obtain sampling units and timing of samples.

In general, sampling programme is of two types, extensive and intensive. Extensive programme is conducted over broad areas to determine the species distribution or the status of injurious insect stages. In this, usually only a single insect stage is sampled and only one or a few samples are taken per season. Here, only moderate levels of precision are required. On the other hand, intensive programmes are conducted as part of research in population ecology and dynamics. In this case, sampling is done frequently, often more than once per week in a small area (a few fields). Usually all or most stages in the life cycle of an insect are sampled and a high degree of precision is sought.

Population intensity

It is the number of insects per habitat unit and it relates closely to crop injury. Estimation of population intensity helps in establishing economic injury level and may be expressed as number of aphids per leaf, number of borers per stem, number of larvae per plant, etc.

COMMON SAMPLING TECHNIQUES AND THEIR APPLICATION

Sampling technique involves how and from where the samples are to be taken? It is the way how insects are counted, whether with naked eye or taking help of some equipment. There is no single sampling technique which can work for all categories of pests. Several workers have given a good account of various common sampling techniques in insect pest management (Cochran, 1977; Thompson, 1983; Atwal and Singh, 1990; Matthews, 1984; Karel, 1996; Pedigo, 1996). Each technique has certain advantages and disadvantages in terms of precision, efficiency, cost, ease etc. The selection of technique also depends on the pest's biology and ecology. Some of the most commonly used techniques and their application are discussed below:

In situ counts

It involves direct counts or direct observation and often requires no special equipment but rely on a good eye. For example, making counts of leafhopper nymphs on cotton leaves or recording of egg or larval population of *Helicoverpa armigera* etc. This technique is useful for counting large and conspicuous insects usually on a part of the plant (e.g. leaves) or the whole plant (if the plant is small). However, for smaller insects like aphids and thrips or for mites, help of a hand lens may be taken in making the counts (Fig. 1). Further, when these insects are in large number, 1 cm^2 windows may be cut out in a thick paper sheet and placed on the leaf surface for counting the insects visible through windows (Fig. 2). A quadrat (1 m^2) or a square metal frame can also be used to take observations from the plants falling in the frame. In some cases, plants in an area are cut off at the ground level, bagged and the insect counts are made later. In the case of large plants such as shrubs and trees, only certain number of leaves, stems, flowers, buds or pods may observed, or number of insect per shoot or branch may be counted. Efficiency of this technique varies with the insect activity and the observer.

Fig. 1: *In situ* counting of smaller arthropods with the aid of hand lens

Fig. 2: Counting of minute insects through 1 cm^2 windows

Knockdown

Knockdown is similar to *in situ* counting, except that the insects are dislodged from the plants causing them to fall into either a tray, a funnel or on a sheet or a piece of cloth and are subsequently counted. The method of dislodgement may be jarring, chemicals or heating:

Jarring: It is probably the most common method of knocking down insects from the plants and particularly useful in collecting the white grub beetles (*Holotrichia* spp.) from the trees soon after these emerge from soil during night after first heavy rains. Here, a piece of cloth is placed on the ground and a branch is given jerks several times with the help of an iron hook attached to a stick to dislodge the beetles. Beetles falling on the cloth are counted. Similarly, gram pod borer (*Helicoverpa armigera*) larvae may be sampled by shaking chickpea plants vigorously over a piece of cloth or a chart sheet and the larvae falling on the sheet may be counted. This is also known as drop sheet method. However, jarring may not be effective for species concealed within vegetation e.g. galls, leaf-miners, etc.

Beat-bucket method (Knutson and Wilson, 1999) also uses similar technique in which top 25 cm portion of cotton plant is bent in a plastic bucket and is beaten against the sides of the bucket to dislodge pests as well as arthropod predators.

Chemicals: In certain cases, the whole plant or several plants may be enclosed in a polyethylene envelope and treated with a quick-knockdown insecticide such as pyrethrum or dichlorvos. After some time, the plants may be shaken to remove dead or immobilized insects on a cloth or other container for counting.

Heating: Heating envisages forcing the insects to come out of their dwellings by increasing the temperature of their habitat. Here, the plant sample harbouring insects may be placed in a special device, the Berlese funnel, that heats the sample. Finding the plant habitat inhospitable, insects move out of the plant material, fall through a funnel, and are retrieved from a receptacle for counting. This approach saves time and effort but sometimes insects may die during the extraction process. Thus, insects may not come out and not be counted.

The other techniques of dislodging insects from their substrates include washing insects from plants and plant parts with soap, alcohol or other solutions. To remove mites from the leaves, a special brushing machine may be used.

Netting

Netting of insects is one of the most widely used techniques which is relatively simple and inexpensive. It is particularly useful in the case of small insects and those with poor clinging habit. However, the major disadvantage of this technique is that there is great sampling variability. This technique is closely related to jarring when used to sample insects in a plant canopy. Netting may be of several types:

Sweep netting: Here, a muslin net is swept through the crop canopy (generally 20 cm deep) and the plant is jarred so that the insects present on the plant may fall into the net (Fig. 3). After 20-25 swings or sweeps, plant debris is removed and the insects are counted. Efficiency of sweep net

Fig. 3: A sweep net in action

varies with species, crop height, wind speed, time of the day and the person doing sweeping. The sweep catch is expressed as the mean number of insects per sweep. The limitation with sweep netting may be that if there are changes in vertical distribution of the pest within the vegetation, the pest may not be detected at all. Further, various crops may require different sweeping technique which need to be standardized.

Vacuum netting: In this technique, engine power is used to create strong vacuum which sucks the insects from plant canopy. Such nets have been used for sampling insects like leafhoppers.

Aerial netting: Aerial nets are used for sampling air-borne insects. A true aerial net is a mesh-bag net with a handle which is swept through the air to capture an insect. Other aerial nets include rotary net and tow net. The rotary net consists of one or two nets at the ends of beams that are rotated by a motor-driven shaft. The insects collected are counted after a given period of time. Tow nets are attached to the moving aircraft, automobiles or ships to detect insects.

Trapping

Insects are attracted to different visual, chemical or olfactory cues to communicate with each other or find suitable hosts. This weakness of insects has been exploited through the use of traps. Traps are particularly useful for detecting the initial appearance of a pest and can give an indication for a correct time of initiation of control measures. Seasonal activity of various pests can also be monitored through the use of traps. Different types of traps have been used to sample insects. Traps are installed at specific locations and the insects trapped in them are counted at regular intervals. Sometimes, some lures/attractants are placed in the traps so that specific insects get attracted and trapped. However, some limitations with such traps are that it is difficult to define the area of influence and that sometimes insect responses may change with time. Use of different types of traps has been made to monitor the activity of various pests:

Light trap: It is the most widely used visual trap employed for sampling agricultural pests, particularly moths, hoppers and beetles, etc. A light trap essentially consists of a light source above a funnel and a container below to collect the catch (Fig. 4a and Fig 4b). It is covered with a protective roof. The light source is generally an oil lamp or electric bulb or a fluorescent tube. Insecticides with quick knockdown or fumigant action, such as dichlorvos, are sometimes added in the container for killing of the trapped insects. Many moths (particularly noctuids) are attracted to short wavelengths of the light spectrum or the blacklight. So, blacklight lamps emitting ultraviolet light are widely used in trap design. Some modifications in trap design may be made depending upon pest species to be trapped. The New Jersey trap combines a blacklight and dry ice (CO_2) for attraction of mosquitoes and fan for sucking the insects into the trap when they fly near. The Manitoba trap uses a black or red sphere suspended under a transparent cone for collection of horse flies. Other commonly used light traps include, Rothamsted trap, Robinson trap, Pennsylvanian trap, etc. In an improved version of light trap, insects which approach the light source cannot escape, turning the trap almost a no-return device (Chatterjee *et al.*, 2001). Light traps have been used for many years to monitor long term population changes in nocturnal flying insects, particularly moths. They also provide information on their time of arrival in a particular locality e.g. *Spodoptera exempta* in East Africa. In Britain, the Rothamsted Insect Survey operates a network of 195 light traps which sample about 600 species of moths. Light traps have been used in the past for mass trapping of insects like red hairy caterpillar (*Amsacta moorei*) (Patel *et al.*, 1981; Saini and Verma, 1991). The major problem with light traps is that, sometimes, large number of non-target species, which may include useful ones also, get trapped which not only make sorting cumbersome but also disturbs ecological balance.

Fig. 4a: An omni-directional light trap

Bait trap: Bait traps rely on an insect olfaction, or sense of smell, for attraction. A common attractant is food; e.g. a mixture of yeast and molasses in a cone trap to sample corn maggot (*Delia platura*) adults. The utility of such traps is often short lived due to infection of the bait by microorganisms. Therefore, the bait needs to be changed frequently to keep the trap active. For monitoring sorghum shoot fly (*Atherigona soccata*), fish-meal trap is quite effective (Singh and Verma, 1988) (Fig. 5).

Fig. 4b

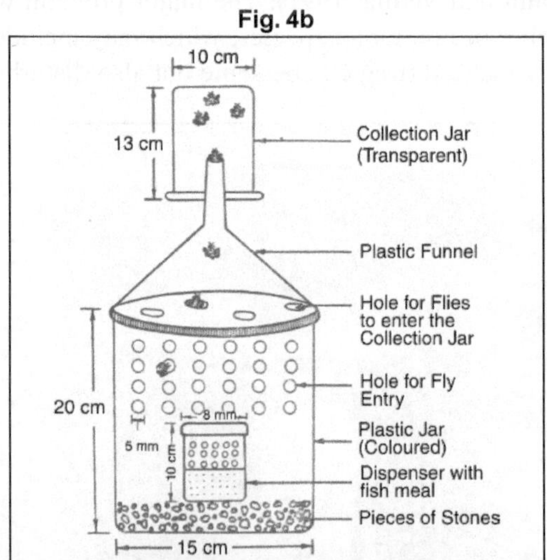

Fig. 5: Fish-meal trap for sorghum shoot fly
(Courtesy Dr. S.P. Singh, Senior Scientist, Department of Entomology, CCSHAU, Hisar,)

Pheromone traps: Pheromone traps have been widely used in detecting or sampling fruit flies (*Dacus* spp.), spotted bollworms (*Earias* spp.), cotton bollworm (*Helicoverpa armigera*), codling moth (*Cydia pomonella*) and pink bollworm (*Pectinophora gossypiella*) (Fig. 6). Since pheromone traps are specifically attractive to the target species, there is no sorting or identification problem. Further, no power is required as in the case of light traps. So, these can be installed in any field. On the other hand, the major disadvantage with such traps is that almost all traps detect only the adult males. For monitoring of *Helicoverpa armigera*, traps suspended at 1.8 m from the ground and spaced at 50 and 100 m at initial stage and peak pest activity, respectively were found to be most effective (Krishna *et al.*, 1999). In pigeonpea, 6 traps/ha were found to be adequate for monitoring of *H. armigera* (Gowda *et al.*, 2002).

Fig. 6: A pheromone trap ('Biotrap') for cotton bollworms

Malaise trap: The malaise trap is basically a tent made of cotton or nylon mesh with one side open that intercepts flying insects. The roof of the tent slopes upwards to a peak where a container with some preservative is located. Insects tend to move upward and into the container while flying and get collected in the container.

Suction trap: A suction trap consists of a wire-gauze funnel leading to a collecting jar and a motor-driven fan is situated below the funnel to create the suction. Such traps are useful for sampling winged aphids and leafhoppers.

Window trap: This is a kind of interception trap consisting of a large sheet of glass that sits in a collecting trough supported by wooden legs. Insects flying into the glass are knocked down into the trough containing soapy water.

Water trap: It consists of a shallow open pan or tray mounted on a wooden post which is filled with water having some detergent or soap or an oil film to aid in wetting or drowning the insects. An omni-directional baffle is usually set in the pan for the interception of insects. The baffle and interior of pan are sometimes painted (e.g. yellow for aphid) to add physical attraction to the trap. Water traps are useful in sampling rice pests e.g. brown planthopper. However, these require protection from weather and catch is to be removed and traps refilled regularly.

Sticky trap: These traps are installed on a wooden stack or a bamboo stick at various heights above a plant canopy. The insects get stuck in the adhesive applied to the trap's surfaces. Sticky traps are useful only against the actively flying insects and provide useful information on height of flight as well as on direction. Insects like aphids, hoppers, flies, hymenopterans and beetles, etc. get trapped in such traps. The pheromone traps are a kind of baited sticky traps. The catch requires regular removal at each observation, particularly if it is high, so that the surface maintains its sticking properties.

Pitfall trap: It captures ground moving insects. The pitfall trap consists of a smooth sided container or a jar or bottle sunk in the ground with a funnel at the soil surface that empties into the container. The trap collects both mature and immature insects which fall accidentally into it. Pitfall traps are more useful in dry areas since in other areas rain may flood them and render them non-functional. However, it cannot detect immobile insects. It is useful for monitoring ground dwelling predators such as carabids, staphylinid beetles, spiders and predatory mites.

Removal trapping: It involves reduction in pest population by removing individuals through repeated catches so that catch per unit time decreases. The population size is estimated from the rate at which the catch decreases.

Extraction from soil

Sampling of insects from soil is often complex and expensive since the insects usually can not be seen or easily extracted from it. About 90 per cent of insect species spend at least one stage in the soil or at the soil surface. Therefore, soil sampling can give useful information on the insects population. Sampling for soil pests may be done by digging in a quadrat or drilling the soil with some equipment and insect counts are made from a fixed volume of soil. Common techniques used for extraction of insects from soil samples are: using Berlese funnels, sieving, washing and flotation. Soil sample may be taken with cores, golf hole bores, bulb cutters, shovels, trowels, metal frames, or other equipment. Berlese funnels are more frequently used to sample small, abundant arthropods in the soil e.g. mites, springtails and beetles. For termites, the insects are attracted by placing some paper on the soil or fixing some wooden blocks or pieces in the soil.

Sieving: Sieving techniques do not depend on insect movement for extraction. Both dry and wet sieving techniques are used to sample insects. In dry sieving, one or more sieves of varying coarseness, beginning with a coarse mesh to screen out large soil particles and other debris and ending with a fine mesh are employed. Standardized soil sieves may be purchased or may be fabricated depending on requirement with respect to the insects body size. Wet sieving follows the same principle as dry sieving, except that water is used to facilitate the movement of soil particle through the sieves.

Flotation: Flotation technique is another widely used extraction procedure that may be used alone or in conjunction with sieving and other techniques. The efficiency of flotation could be increased by adding some salt such as magnesium sulphate to increase the medium's specific gravity.

Mark, release and recapture technique

This technique is useful for estimating populations of mobile insects. In this technique, the insects are first captured from the field, marked with some suitable paint or dye, released in the field and recaptured along with the unmarked individuals.

The population can be estimated by using Lincoln index :

$$P = \frac{an}{r}$$

where, P is the population size; a, total number of marked and released individuals; n, total number of individuals recaptured; and r, the number of marked individuals recaptured (Karel, 1996).

Indirect techniques

In indirect techniques, insect population are estimated by measuring the effect of insects on crop plants or through some indirect sampling methods called population indices e.g. number of leaves mined by leaf miners, percentage defoliation by hairy caterpillars and armyworms, percentage of plant attacked, percentage of fruiting forms or bolls damaged, leafhopper injury grade in cotton, number of wilted or dead plants by termites, plants with "dead hearts" caused by borers, etc. The sampling technique involved with most of these measures is direct observation of results of insect injury.

Sometimes, pest populations are assessed with the help of insect products which include larval and pupal skins, frass, honey dew (e.g. secreted by aphid, whitefly) and nests of colonial insects. The measurement of frass drop in a collecting tray has been used to assess population size of several forest pests.

Pedigo and Buntin (1994) described ratio classification method, based on pest-natural enemy ratio to monitor the effectiveness of biological control and presented a FORTRAN computer programme to analyze the sampling plan.

Remote sensing

Remote sensing techniques such as radar can automatically monitor the height, horizontal speed, direction, orientation, body mass and the shape of arthropods intercepting the radar beam. It can provide information of aerial migration of pests and natural enemies. It can be particularly useful for monitoring locust swarms. Radar entomology was first used in 1968 and since then comprehensive and intensive studies have been conducted in the UK, USA, Australia and China and it was predicted that fully automatic, season long and real time monitoring will be feasible with the vertical-looking radar (Zhai, 1999). Remote sensing technique relies on changes in the absorbance or reflectance of plants in response to pest attack. An instrument sensitive to specific wave lengths of radiation is used to detect such changes. Remote sensing in conjunction with '3S' technique can help in achieving three-dimensional real time visualization of insect pest populations (Wang *et al.*, 2003).

Imagery provided by remote sensing satellites could be utilized in identifying pest affected areas and intensity of pest damage. This could be particularly useful for pests which produce visible symptoms of crop damage over large area e.g. hopper burn symptoms in paddy, blacking of cotton leaves caused by sooty mould growing on honey dew secreted by aphid and whitefly, etc. Similarly satellite data have also been used to identify areas of vegetation capable of supporting desert locusts. Further, such data can also find application in studying the effect of environmental changes on build up, long distance migration and flight behaviour of air-borne pests.

The Australian Centre for Remote Sensing (ACRES) has introduced a new service to provide satellite data for real time application. The STAR (Speedy Transmission After Reception) service provides access to digital satellite data on various aspects which includes monitoring of pest infestations (Thankappan, 2001).

KINDS OF ESTIMATES

Generally, estimates can be divided into two broad categories *viz.*, absolute and relative.

Absolute estimates

Absolute estimates measure actual numbers in the insect population. Such estimates are expressed in numbers per ground surface area; for examples, number per acre, number per hectare, number per square meter. These are often the most difficult estimates to make and frequently the most costly. These are of utmost importance in research in insect population dynamics, but because of their high cost, these are not used widely in pest management practices.

Relative estimates

Relative estimates differ greatly from absolute estimates because they do not give definite numbers per unit area. Relative estimates are based on the kind of sampling technique used such as number of insects per sweep or insects per some fixed number of sweeps. These estimates do not directly answer the question of how many insects are present per acre, per plant or per meter of row length. Rather, they are used to compare population sizes in time and space (e.g. at different locations). Many techniques have been used to make relative estimates e.g. several kinds of netting, all types of trapping, some *in situ* methods, the knockdown approach, etc. Relative methods have the advantage of being inexpensive and are often the most appropriate for insect pest management.

Population indices are usually employed similarly to relative estimates for population comparisons e.g. cast skins, webbings, root feeding, or leaf feeding. Percentage defoliation of a plant canopy is one of the most popular population indices in insect pest management.

SURVEY EQUIPMENT

Some of the equipment/material which could be required by scouts or the surveillance unit include: hand lens, tape measure, thermometers, elastic bands, hair brushes, plastic bags, plastic vials, paper bags, field cloth, muslin, rain suit, rubber boots, utility knife, sweep net, traps of different types, data recording sheets, quadrat, cloth bag, bucket, killing bottle, setting board, insect box, naphthalene bolls, thread bolls, etc.

PARAMETERS OF SAMPLING PROGRAMME

A sampling programme requires the following considerations:

i) **Insect stage to be sampled :** The insect stage most often sampled is the one causing damage and that with long developmental time. However, when reasonable time is available before control measures are initiated, a stage prior to the damaging stage may be selected so that carly predictions can be made. For example, sampling the non-feeding adult stage can help predict the subsequent damaging larval population. There should be a minimum of 10 replications.

ii) **Number of sampling units :** The number of sampling units depends on the degree of precision needed and the cost. When greater precision is desired, the number of sampling units may be increased. The relative variance (RV) should remain around 10 per cent for research purpose and upto 25 per cent for IPM programme. The following formula can be used for calculating the number of sampling units (Pedigo, 1996):

$$N = \{(t \times s)/(D \times \text{`}x)\}^2$$

where,

N = number of sampling units required

'x = mean density

D = required precision (RV) expressed as a decimal (for example, 0.25)

s = standard deviation

t = student's t value = a value obtained from statistical tables (usually the value of t for 0.05 probability and sample number used to calculate the mean).

Morris (1955) laid the following criteria for sample unit and sample size:

a) The size of sample be such that all units of the universe have an equal chance of selection.
b) The sampling unit must be easily delineated in the field.
c) The sampling unit should of such size as to provide a reasonable balance between the variance and cost.

iii) **Time to sample:** Sampling time depends on pest as well as crop phenology. It is most important for the sampling to coincide with the time of peak numbers of the stage. Such timing may be predicted using degree-day or other phenological models or by trial sampling. For trial sampling, the samples are taken regularly (once or twice per week) until a peak is obtained from the cumulative data. Time of the day at which samples are taken is also important e.g. flight activity of some insects may be at dusk or dawn, so samples should be taken during that activity period. Further, diurnal rhythms of insects may cause them to move from one part of the habitat to another.

iv) **Pattern of sampling :** Depending on the uniformity of the habitat and population dispersion, the sampling pattern may vary. The most common sampling patterns are the X, W, U, diagonal, zig-zag diagonal, micro-plots, a fixed row length, etc. Common sampling patterns are illustrated by Fig. 7. Sampling units should be spread out over an area from which samples are to be taken at random. The edges of a growing area (first 5 meters or so) are usually avoided in making estimates unless species distribution demands so. For example, in the case of armyworms, grasshoppers, stalk borers, initial aphid incidence, etc. samples may also be required to be taken from boarder areas.

1. Random. 2. Along one diagonal 3. 'X' pattern or along two diagonals 4. Zig-zag diagonal 5. In 'W' pattern 6. In micro-plots of 1 m^2 7. In one meter of row length 8. In 'U' pattern

v) **Type of sampling:**. Sometimes, it may be necessary to examine weeds and other alternate hosts of the pests while in other cases, the pest may overwinter in the surrounding areas.

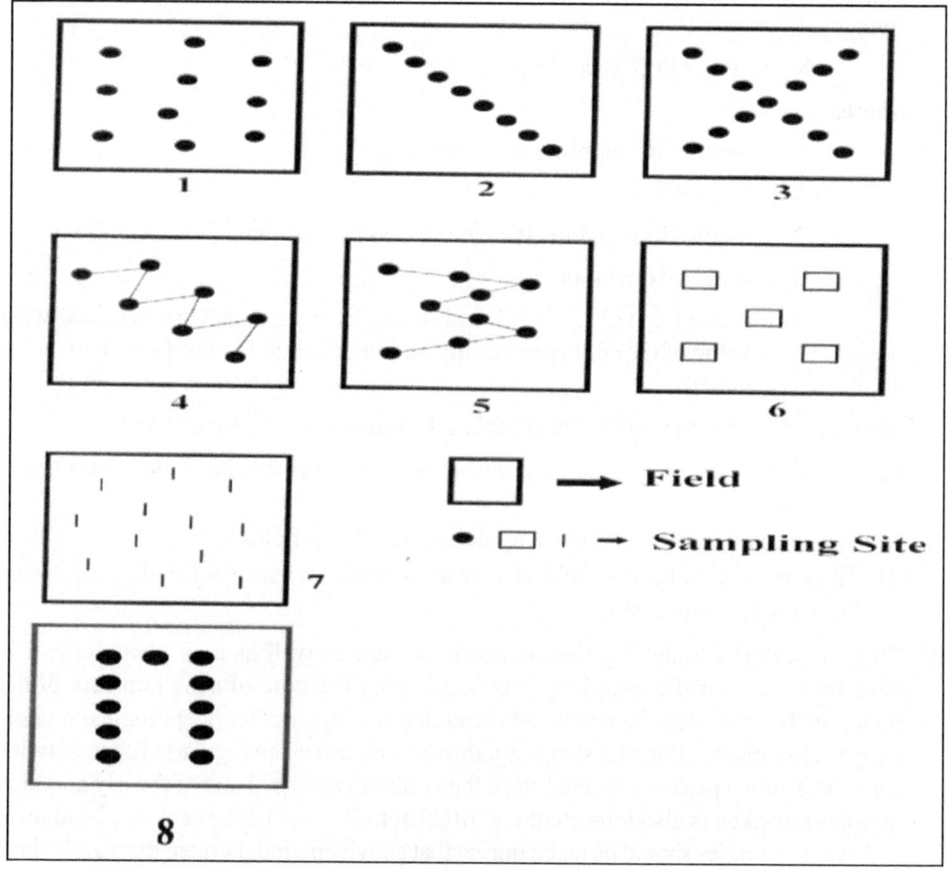

Fig. 7: Common sampling patterns for insect sampling:

Therefore, depending upon pest dispersion in the crop and the surrounding habitat, an appropriate sampling needs to be selected:

a) **Random sampling:** It involves selecting a number of samples from a population such that every sample has an equal chance of selection. For this, random number tables can be used.

b) **Stratified random sampling :** In this case the population is divided into different strata from which random samples are then taken. The strata are sub-divisions of the samples based on knowledge of the distribution of the population.

c) **Systematic sampling :** It involves taking samples at fixed interval. The first sample is taken at the reference point and subsequent samples at successive intervals. For this, usually a prescribed route through a growing area is selected taking samples regularly along this route.

d) **Trap sampling:** It involves the use of various types of traps in collecting insect populations.

e) **Sequential sampling:** It is based on insect dispersion patterns and economic decision levels. With sequential sampling, the total number of samples to be taken in an area is unknown when sampling is begun. The sampling is continued until a pre-established upper or lower infestation level is found. Sequential sampling tells us how many sampling units to take before we can be confident about making a pest management decision. This type of sampling usually allows the decision to be made with fewer samples, thus, saving time and money.

SAMPLING TECHNIQUES FOR SOME IMPORTANT CROP PESTS

Some commonly used sampling techniques employed for some of the important crop pests are summarized in Table 1. However, these may be suitably modified keeping in view the objectives and degree of precision required.

PEST SURVEILLANCE IN INDIA

In India, the pest surveillance work is mostly carried out by the government agencies, namely, the respective state agricultural university, state agriculture department and some institutions under central government. At the central level, this programme is mainly undertaken by the Directorate of Plant Protection, Quarantine and Storage, Ministry of Agriculture, Department of Agriculture and Cooperation, Government of India with headquarter at Faridabad (Haryana) through its Central Integrated Pest Management Centres spread in different states and union territories of the country. It publishes a monthly report on pest and disease survey. In addition, Directorate of Indian Council of Agricultural Research also carries out pest surveillance through rapid roving surveys. In pursuance of the recommendation of the central government, several state departments of agriculture and the state agricultural universities have established the pest and disease surveillance units in their respective states. Central government agencies are coordinating the surveillance work with the states through National Level Monitoring Teams while the states have also formed their own State Level Monitoring Teams. After each survey, the survey reports highlighting the pest situation in different areas of the state/country are issued. Different states have developed surveillance methodology suited to their requirements. For example, Ghosh *et al.* (1992) have developed a practical guide to methods of pest monitoring for irrigated rice in eastern India. The Directorate of Plant Protection, Quarantine and Storage (Faridabad) has brought proforma for rice pests/diseases roving survey and has provided methodology for surveying of various pests and diseases (Anonymous, 1994).

Table 1 Sampling techniques for major insect pests of some crops

Crop	Pest	Economic threshold	Method of sampling
1	2	3	4
Cotton	(i) Leafhopper (jassid) (*Amrasca biguttula biguttula*) leaves from margins.	2 nymphs/leaf or yellowing and curling of 20%	Count leafhopper nymphs from underside of three fully developed leaves in the upper canopy of each of 20 random plants or count leaves showing yellowing and curling from margins and healthy leaves of 20 random plants in a field.
	(ii) Whitefly (*Bemisia tabaci*)	6-8 adults/leaf	Count whitefly adults as above.
	(iii) Spotted bollworms (*Earias* spp.)	10% drooping shoots or 10% infested fruiting bodies	Count drooping and healthy shoots of 25 random plants or examine all green fruiting bodies of the above plants for spotted bollworm induced holes or damage.
	(iv) Pink bollworm (*Pectinophora gossypiella*)	10% infested fruiting bodies	Count rosetted flowers and examine all fruiting bodies for damage by pink bollworm on 25 random plants.
	(v) American bollworm (*Helicoverpa armigera*)	10% infested fruiting bodies or one larva/2 plants.	Examine all fruiting bodies of 20-25 random plants for pest damageand also count number of larvae present.
	(vi) Aphid (*Aphis gossypii*)	10-15% infested plants	Examine presence of aphid or its symptoms on 20-25 random plants.
	(vii) Thrips (*Thrips tabaci*)	10 thrips/leaf or 25% infested plants	Same as for aphid
Paddy	(i) Green leafhopper (*Nephotettix nigropictus* & *N. virescens*)/ white backed plant hopper (*Sogatella furcifera*)/ brown leafhopper (*Nilaparvata lugens*)	5-15 insects/hill	Select 5 micro-plots of 1m^2 each in a field and shake vigorously plants in 5 hills/plot or shake vigorously 25 random plants and count leafhoppers fallen on water

Contd. ...

Table 1 Contd. ...

	(ii) Stem borer (*Scirpophaga incertulas*)	5-10% plants with dead-hearts or 2% white ears or one egg mass or moth/m^2.	Count infested and healthy tillers in 25 random plants.
	(iii) Leaf-folder (*Cnaphalocrocis medinalis*)	2 damaged leaves/ plant or one larva/hill	Count infested and healthy leaves among 25 random plants or count number of larvae in 25 plants.
	(iv) Root weevil (*Hydronomidius molitor*)	2 grubs/hill	Same as above
	(v) Rice gundhi bug (*Leptocorisa acuta*)	1-2 insects/hill	Count the insect on 25 random plants.
Sugar-cane	(i) Early shoot borer (*Chilo infuscatellus*)	Dead hearts in 15-20% tillers	From 5 random rows in a field, examine 100 tillers/row for dead-hearts.
	(ii) Top borer (*Scirpophaga excerptalis*)	Dead hearts in 10% tillers	Same as for early shoot borer
	(iii) Gurdaspur borer (*Acigona steniella*)	Drying of 10% canes	Count dried-up and healthy canes as above.
	(iv) Pyrilla (*Pyrilla perpusilla*)	3-5 insects/leaf	Count pyrilla nymphs and adults on 10 random plants taking 6 (2 upper, 2 middle and 2 lower) leaves/plant.
	(v) Black bug (*Cavelerius sweetii*)	25 insects/plant	Count nymphs and adults on 20-25 random plants.
Sorghum	(i) Shoot fly (*Atherigona soccata*)		
	a) After one week of crop germination	One egg/plant or presence of eggs on 5% plants	Examine 30 plants for the presence of eggs.
	b) After two weeks of germination	Dead-hearts in 15% plants	Examine 30 plants for the presence of eggs or count dead hearts.
	(ii) Stem borer (*Chilo partellus*)	Symptoms of damage (i.e. dead hearts, shot holes in leaves, unfilled ear-heads etc.) in 10% plants	Examine 30 plants for damage symptoms.

Contd. ...

Table 1 Contd. ...

	(iii) Sorghum midge (*Contarinia sorghicola*)	One midge/ear head at 50% flowering	Count gall midge adults on 50 random plants in the morning.
	(iv) Aphid (*Rhopalosiphum maidis*)	10-20% infested plants or 14 aphids/ central shoot	Examine 100 random plants in a field.
	(v) Painted bug (*Bagrada hilaris*)	One insect/meter row length	Count painted bug in one meter row length from 20 random sites in a field.
Chickpea	Gram pod borer (*Helicoverpa armigera*)	One larva/meter row length	Count larvae in one meter row length from 10 random sites in a field.
Okra	Leafhopper (*Amrasca biguttula biguttula*)	2-5 nymphs/leaf	Same as in the case of cotton.
Tomato	Fruit borer (*Helicoverpa armigera*)	One larva/m^2	Count larvae in 1m^2 micro plot from 10 random sites in a field.

PEST FORECASTING

One of the important objectives of the entomologists is to predict future trends of pest populations so that appropriate management practices are initiated and as when required. When chemical control is to be adopted, forecasting aims either to predict the need for spraying or to predict the most appropriate date of spraying in the light of economic threshold of a pest. Therefore, a suitable sampling technique as discussed earlier may be adopted for estimating the pest population on the host plant accurately. Cammell and Way (1987) have described various components of forecasting, which include direct measurement of pest numbers, measurement of climatic variables and measurement of changes in pest-host plant relationships. Since pest populations are greatly influenced by the environmental factors, a proper understanding of these factors may affect the pests either by directly affecting their survival, development, reproduction, emergence and behaviour, or indirectly by their action on the host plant and on natural enemies. These factors also determine the geographical limits of distribution and the time of appearance and abundance of pests. Further, climatic factors may also interfere with the management practices adopted for reducing the pest numbers. Similarly, winds may greatly influence distribution of dispersive pests affecting their direction and extent of migration. Therefore, occurrence of highly migratory pests like locusts and aphids may be strongly associated with certain wind and rainfall patterns.

Spring emergence of certain pests may vary from year to year depending upon the duration and severity of winter and onset of favourable conditions during spring. The technique of accumulating day degrees has been used for a long time to predict the appearance of the pests on the crop after winter. In the case of red hairy caterpillar, *Amsacta moorei*, and white grub, *Holotrichia* spp., emergence of adults takes place only after the first good rains after summer. *A. moorei* moths would not emerge from the soil unless the mean relative humidity after rains reached above 70 per cent and mean temperature declined to 30^{0}C or less (Saini and Ram, 2000). Thus, pest abundance can vary considerably in relation to climatic factors. Incidence of *Helicoverpa armigera* in cotton in northern India has been found to be higher in seasons with high and frequent rains (Saini *et al.*, 2004). Thus, pest abundance can vary considerably in relation to climatic factors and correct measurement of these factors to enable appropriate weather forecasts can greatly help in making predictions of pest population.

Apart from affecting the crop plants and the pests, climate may also influence natural enemies effectiveness differently which can affect pest outbreaks. Likewise, different cropping systems and cultural practices can favour the abundance of a particular pest. In northern India, inclusion of sunflower in the cropping system a few years back resulted in the increased incidence of *H. armigera* on cotton. In Haryana, with the adoption of zero tillage practice in paddy in some areas, the problem of stem borer (*Scirpophaga incertulas*) is flaring up. Further, abundance of a pest is also affected by the time of sowing as well as the variety e.g. cotton leafhopper (*Amrasca biguttula*) prefers young cotton crop in comparison to the older one.

Forecast of a pest can be made after arrival of the pest on the crop, during migration to the crop and at the source of the pest before the pest leaves for the crop. Forecasts based on crop monitoring are convenient as the field counts offer the most accurate estimates of pests. However, crop monitoring is relatively time consuming and expensive. Nevertheless, impact of various factors affecting pest populations has to be measured carefully and results interpreted properly so that reliable forecasts could be made.

SOME EXAMPLES OF PEST SURVEILLANCE AND FORECASTING IN DIFFERENT COUNTRIES

In USA, apart from private agencies, pest surveillance is a major activity for several governmental and commercial agencies. The US Department of Agriculture conducts such programmes through the Animal and Plant Health Inspection Service (APHIS). Its sub-division, Plant Protection and Quarantine Programmes (PPQ) also has responsibility in coordinating its surveillance activities with those of other agencies, including the Agricultural Research Service, the United States Forest Service, and state and private establishments (Pedigo, 1996). For Michigan, a computer-based system was developed to aid in the prediction of the occurrence of insect pests throughout Michigan. Two prediction systems were linked through computer software with sampling, weather data, temperature thresholds and biological information into a pest biological scheduling system (BIOSCHED). The first prediction system involves biological information for an array of pest events (e.g. first emergence) that are placed on computer files. The second prediction is linked by BIOSCHED in the Predictive Extension Timing Estimator (PETE), which utilizes curve fitting techniques to predict pest events. A computer software system accesses current day-degree for each of the 61 Michigan weather stations and scans each pest's historical data record (Gage *et al.*, 1982). However, in spite of all these programmes, a USDA study indicates that pest monitoring is the least widely practiced activity (Lin *et al.*, 1994).

Luttrell *et al.* (1994) have reviewed the cotton pest management programme going on in several countries. In USA and Australia, pest scouting in cotton is done by private consultants. Most fields are scouted 1-2 times per week and samples are generally based on examination of 100 terminal buds and fruiting forms per 20-40 hectares. In Australia, scouts examine upto 60 cotton plants per 100 hectares 2-3 times per week. In Brazil, government personnel are involved in scouting, but much of the cotton crop is not scouted routinely. In CIS (Commonwealth of Independent States), individual trained scouts inspect 15-20 hectares of cotton per day or 150-200 hectares during a 10 day period. The information is used by a chief agronomist to determine the need for pest control measures.

In China, information on a number of important insects is compiled in a compendium. The Chinese have a 'trilateral' organization of work teams which consists of technical persons, growers and administrators (Chiang, 1977). At the farm level, the unit of agricultural production is the 'people commune' which is made of 5000 to 50,000 members. Each commune has 5 to 15 brigades. There are plant protection teams in the commune and plant protection technicians in the brigades. The commune team monitors the start of insect activities in the spring based on catches in light traps or other sources such as plant material harbouring overwintering insects. At the appropriate time, the commune team will alert the brigade technicians to select target fields for monitoring and relay and records to the commune team to take control measures. Population density and phenology data are compiled in the commune for long term predictions.

In Europe, forecasting programmes for several pest species are being successfully implemented (Cammell and Way, 1987). Geographical information system (GIS) was found to be a powerful tool for making forecasts and for analyzing data in Finland; and AGRONET/ internet services provides a rapid and easy means of delivering the information to farmers and extension services (Tiilikkala and Brunel, 2000). GIS also has the potential to produce more sophisticated models for forecasting pest activity. Harold and Sachs (1987) have described the pest monitoring system undertaken in Germany. The system of computer-aided central pest monitoring was introduced in German Democratic Republic (GDR) in 1976. A staff of approximately 260 is responsible for pest monitoring and makes use of a central programme, in which the crops and pests to be monitored and the corresponding monitoring periods are specified. Computer aided methods are used once, twice or three times a year to determine the population densities of 70-80 of the 135 or so pest species monitored annually. The results of pest monitoring are extensively used by local and central authorities and research stations. Favrin (2000) has given an overview of pest and disease warning systems in Canada.

In Philippines, scouting is done by the Government. A potato IPM programme would have university graduates to monitor pests and advise groups of potato growers. In Indonesia, a 'crash' training programme implemented against the white rice stem borer through IPM Field Training Workshops and IPM Farmer Field Schools successfully managed the menace of this pest (Dilts, 1990). In the Netherlands, regular crop monitoring is done by farmers. Initial instructions on crop monitoring are provided by a group of experts. Data are sent to the central computer base and forecasts of whether or not treat the crop are sent to the farmers. In UK, Rothamsted Experimental Station at Hertfordshire has a network of 24 suction traps, sampling flying insects at a height of 12.2 meters, distributed over Britain. Migrant aphids, including over 20 pest species, have continuously been monitored since 1968.

IMPLEMENTING PEST SURVEILLANCE AND FORECASTING IN DEVELOPING COUNTRIES

Success of the surveillance programme in the developing countries will largely depend on the government policy, scientific package and the socio-economic condition of the farmers. Unless the government and the farmers are convinced of the importance of surveillance in adoption of IPM, no significant progress can be made in this regard. Researchers on their part will have to develop and make available the surveillance methodology which is precise, quick, convenient and cheaper for adoption by the farmers. Garforth (1993) has described several extension techniques which could help in the implementation of new technology from research institutions to the farmers. However, some of the points for strengthening pest surveillance and forecasting in the developing countries are highlighted below:

i) **Realization of the need for pest surveillance**

Owing to economic and other constraints, pest surveillance in the developing countries is not being undertaken regularly. Plant protection measures adopted by the farmers are mostly chemical intensive. However, in view of the WTO obligations, competition for selling cheaper and quality products would further increase among different countries. Farmers would be compelled to use lesser quantities of agro-chemicals to keep their produce globally competitive both in terms of price and quality. Therefore, farmers as well as research institutions in future will have to accept the challenge of using minimum amounts of agro-chemicals together with encouraging the use of other means of crop production and protection without compromising yields. This would require frequent and regular crop monitoring. It is probably due to this realization that in India the central government has shown renewed interest in educating farmers through IPM Farmers Field Schools (FFSs) to equip them with the latest techniques in crop and pest management. These schools are operating in various states through *Krishi Vigyan Kendras* (KVKs) under the mandate of Central Integrated Pest Management Centre (CIPMC), Directorate of Plant Protection, Quarantine and Storage, Faridabad (Haryana).

ii) **Working out economic threshold of key pests**

Since pest management decisions are to be based on the pest populations causing crop losses, the economic thresholds of important pests under different climatic conditions need to be worked out.

iii) **Reliable prediction of weather**

Weather conditions greatly influence pest behaviour, their development and survival. Therefore, reliable prediction of weather is most important in making accurate forecasts of pests. In India, the National Centre for Medium Range Weather Forecasting (NCMRWF) is issuing weather forecast five days in advance. Likewise, Indian Council of Agricultural Research (ICAR) has initiated NATP Mission Mode Project on Development of Weather Based Forewarning System for Crop Pests and Diseases in cotton, sugarcane, rice, pigeonpea, groundnut and mustard.

iv) **Working out relationships of pest population with the environmental factors**

Unless relationships of pest population with the weather factors, particularly temperature, relative humidity and rainfall are determined, no reliable forecasts can be made.

v) **Evaluation and strengthening of standard field skill of extension staff**

Extension personnel may be imparted training by the experts in pest monitoring. The trained staff may further educate the farmers or the scouts to undertake pest monitoring. The field staff may be provided with a practical guide containing information on standard assessment methods, diagnostic symptoms of damage by various pests, pest identification notes and photographs, standard data recording sheets, etc.

vi) **Computerization of pest monitoring programme for quicker dissemination of information**

Rapid communication is the backbone of success of pest surveillance. Even the locally validated research results remain mere academic or bureaucratic exercises if these are not passed accurately to the extension staff or the farmers. Therefore, a computer network may be established through which forecasts be sent to the user agencies quickly. Also, surveillance data can also be directly received by the central computer centre. In India, most of the data for pest surveillance is recorded manually. However, National Centre for Integrated Pest Management (NCIPM), ICAR has made some attempts to collect pest surveillance data on computerized data sheets for quicker analysis (Karel, 1996). Similarly, weather parameters are also being recorded through the automatic weather stations which would be easily transferred to computer files for analysis. NCIPM is also making efforts to use NICNET communication network for pest surveillance and forecasting.

vii) **Sharing of surveillance cost by the user agencies**

Since pest scouting is a time consuming and costly affair, it would be desirable that the user agencies share a part of the surveillance cost. Government on its part may give incentives to agricultural/entomological graduates for establishing scouting units or may offer the services of experts and other field staff to the farmers for undertaking community action. Apart from this, out of school youths and educated farmers could also be trained in pest scouting work. Alternatively, farmers can form cooperatives which may pay for the scouts.

viii) **Development of forecasting models**

The role of forecasting in insect pest management will become more wide spread in the years to come. Therefore, sampling and surveillance have key roles to play in obtaining the data needed for developing good models. In India, the National Centre for Integrated Pest Management (NCIPM) has developed forecasting models for *Helicoverpa armigera* (using trap catch data and weather parameters) and potato aphid, *Myzus persicae* (Trivedi, *et al.,* 2002).

References

Anonymous, 1994. Manual on Integrated Pest Management in Rice. Directorate of Plant Protection, Quarantine and Storage (Faridabad), Department of Agriculture and Cooperation, Ministry of Agriculture, Government of India.

Atwal, A.S. and Singh, B. 1990. Pest Population and Assessment of Crop Losses. Indian Council of Agricultural Research (ICAR), New Delhi.

Binns, M.R. and Nyrop, J.P. 1992. Sampling insect populations for the purpose of IPM decision making. *Ann. Rev. Ent.* 37: 427-453.

Cammell, M.E. and Way, M.J. 1987. Forecasting and monitoring. In: Burn, A.J., Coaker, T.H. and Jepson, P.C. (eds.), Integrated Pest Management. Academic Press, Harcourt Brace Jovanovich Publishers, London, pp. 1-26.

Chatterjee, A., Sen, D. and Gantait, V.V. 2001. Innovation of high-efficient light-trap useful for controlling agricultural pests. *Entomon* 26(Spl. Issue): 110-112.

Chiang, H.C. 1977. Pest management in People's Republic of China – monitoring and forecasting insect populations in rice, wheat, cotton and maize. *FAO Plant Prot. Bull.* 25: 1-8.

Cochran, W.G. 1977. *Sampling Techniques*, New York: Wiley 3rd edn.

Dilts, R. 1990. IPM in action: a 'crash' training programme implemented against the white rice stem borer outbreak in Indonesia. *FAO Plant Prot. Bull.* 38(2): 89-93.

Favrin, R.J. 2000. Overview of pest and disease warning systems in Canada. *Bulletin OEPP* 30(1): 49-50.

Gage, S.H., Whalon, M.E. and Miller, D.J. 1982. Pest event scheduling system for biological monitoring and pest management. *Environ. Ent.* 11(6): 1127-1133.

Garforth, C. 1993. Extension techniques for pest management. In: G.A. Norton and J.D. Mumford (eds.), Decision Tools for Pest Management, CAB International, pp. 247-264.

Ghos, G.C., Ali, M.H., Fowler, S.U. and Maslen, N.R. 1992. The development of practical and appropriate IPM methods for irrigated rice in Eastern India. *Proc. Brighton Crop Prot. Conf. Pests and Diseases* Brighton, Nov. 23-26, 1992, pp. 1021-1026.

Gowda, D.K.S., Suhas, Y., Kotikal, Y.K., Patil, B.V. and Benagi, V.I. 2002. Validation of integrated pest management of pigeonpea pod borer, *Helicoverpa armigera. International Chickpea and Pigeonpea Newsletter* 9: 46-47.

Harold, H. and Sachs, E. 1987. Ten years computer-aided central monitoring of pest occurrence in the German Democratic Republic. *Nachiichtenblatt-fur-den Pflanzenschutz-in-der DDR* 41(1): 1-4.

Karel, A.K. 1996. Insect pest sampling, surveillance and forecasting in IPM. In: Upadhyay, R.K., Mukerji, K.G. and Rajak, R.L. (eds.), *IPM System in Agriculture* Vol. 1. Principles and Perspectives. Aditya Books Private Ltd., New Delhi, pp. 133-174.

Knutson, A.E. and Wilson, L.T. 1999. The beat bucket: a rapid, reliable method for sampling predatory insects and spiders in cotton. *Proceedings of Beltwide Cotton Conference, National Cotton Council, Memphis,* TN 2: 1120-1125.

Krishna, K., Kanaujia, K.R. and Kanaujia, S. 1999. Rhythmicity and orientation of *Helicoverpa armigera* (Hubner) to pheromone and influence of trap design and distance on moth trapping. *J. Insect Sci.* 12(1): 6-8.

Kuno, E. 1991. Sampling and analysis of insect populations. *Ann. Rev. Ent.* 36: 285-304.

Lin, B.H., Vandeman, A., Fernandez, C.J. and Jans, S. 1994. Integrated pest management – how far have we come? *Agricultural Outlook* 207: 24-28.

Luttrell, R.G., Fitt, G.P., Ramalho, F.S. and Sugonyaev, E.S. 1994. Cotton pest management: Part I. A Worldwide perspective. *Ann. Rev. Ent.* 39: 517-526.

Matthews, G.A. 1984. Pest Management. Longman, London and New York, pp. 22-46.

Morris, R.F. 1955. The development of sampling techniques for forest insect defoliators with particular reference to the spruce budworm. *Can. J. Zool.* 33: 225-294.

Morris, R.F. 1960. Sampling insect populations. *Ann. Rev. Ent.* 5: 243-264.

Patel, B.S., Patel, R.C. and Patel, M.S. 1981. A note on the effectiveness of light traps for the control of Gujarat hairy caterpillar, *Amsacta* spp. *Gujarat agric. Univ. Res. J.* 7: 47-48.

Pedigo, L.P. 1996. Entomology and Pest Management (4th edn.). Prentice-Hall Inc., Upper Saddle River, New Jersey 07458, pp. 211-254.

Pedigo, L.P. and Buntin, C.D. (eds.) 1994. Handbook of Sampling Methods for Arthropods in Agriculture. Boca Raton, USA, CRC Press Inc.

Saini, R.K. and Verma, A.N. 1991. Effectiveness of light traps in suppressing populations of red hairy caterpillar, *Amsacta moorei* Butler. *Haryana agric. Univ. J. Res.* 21: 250-251.

Saini, R.K. and Ram, P. 2000. Emergence of *Amsacta moorei* moths as influenced by ambient weather parameters. *Indian J. agric. Sci.* 70(11): 765-768.

Saini, R.K., Jaglan, R.S. and Naresh, J.S. 2004. Status of American bollworm, *Helicoverpa armigera* (Hubner) on cotton in Haryana. *J. Cotton Res. Dev.* 18(1): 78-80.

Singh, S.P. and Verma, A.N. 1988. Monitoring of shoot fly, *Atherigona soccata* (Rondani) in traps and their periodic incidence in sorghum. Crop Res. 1(1): 76-83.

Southwood, T.R.E. 1978. *Ecological Methods*. London: Chapman and Hall. 2nd edn.

Thankappan, M. 2001. Access to satellite data for time-critical applications STAR and SPOTLITE. *First Australian Geospatial Information and Agriculture Conference, Sydney, Australia*, July 17-19, 2001, pp. 497-506.

Thompson, D.J. 1983. Sampling methods for pest and vector management. In: Youdeowei, A. and Service, M.W. (eds.), Pest and Vector Management in the Tropics, Longman, England, pp. 45-130.

Tiilikkala, K. and Brunel, H. 2000. Use of geographical information system (GIS) for forecasting the activities of carrot fly and cabbage root fly. *Bulletin OILB/SRDP* 22(5): 15-24.

Trivedi, T.P., Das, D.K., Dhandapani, A. and Kanojia, A.K. 2002. Development of forewarning models for insect pests and diseases of important crops. In: Harvir Singh (Ed.), *Short Course on Advances in Implementable Pest Management Technology*, Directorate of Oilseeds Research, July 8-17, 2002, Rajendranagar, Hyderabad, India.

Wang, Z.J., Zhang, A.B. and Li, D.M. 2003. Applied approaches and progress in the use of remote sensing techniques in insect ecology. *Entomological Knowledge* 40(2): 97-100.

Zhai, B.P. 1999. Tracking angels: 30 years of radar entomology. *Acta Entomologica Sinica* 42(3): 315-326.

Entomology : Novel Approaches, 2007
Eds.: *P.C. Jain and M.C. Bhargava*,(Page 43-96)
New India Publishing Agency, New Delhi
E-mail: newindiapublishingagency@gmail.com
web: www.bookfactoryindia.com

CHAPTER 3

EMERGING PEST PROBLEMS IN INDIA AND CRITICAL ISSUES IN THEIR MANAGEMENT

Paras Nath
Department of Entomology & Agricultural Zoology, Institute of Agricultural Sciences
Banaras Hindu University, Varanasi-221 005, Uttar Pradesh

INTRODUCTION

Humans have inhabited the earth for nearly 2 million years. For over 90 per cent of this period, the maximum population numbered about 200,000 and the growth rate was only about 0.01 percent per year. The use of fire and the development of tools were the first major changes affecting human population growth which increased to about 5 million individuals about 100,000 years ago. The population remained fluctuating about this mean value for the next 90,000 years depending upon the food supply. Up to this time people were hunterer-gatherers and depended exclusively on the natural ecosystem for their food. Then about 10,000 years ago, humans began to cultivate crops for food. With established agriculture came an increased and stable food supply. This contributed to an increased growth rate of the human population (Dhaliwal *et al.,* 1998).

The next increases in the population growth rate coincided with the discovery and use of fossil fuels such as coal, and gas. In particular, energy has been used to increase yields from food crops and to improve public health. Improvements in both food supply and public health have been the prime contributors to rapid population growth since then (Pimentel, 1993). The word's population did not reach one billion until about 1800. It took about 130 years more to reach two billion. About 68 years later, the world population has attained 6 billion and is increasing at an alarming rate of 1.76 per cent. The current levels of population growth are unprecedented in human history. At the current population growth rate of 2.1 per cent per year, their population is projected to be 7.2 billion, by the year 2025 (Fig.1) (Dhaliwal and Arora, 1998). A common thread that unites humanity is we all need food to survive; however only certain percentage of people is involved in agriculture. In developed countries where population growth rate is very low, very less number of people is directly dependent on agriculture while in developing and undeveloped countries situation is just reverse (Fig.2). The growing population has in turn intensified demands on the planet's ecosystem resources to provide food. The world population is likely to reach 10 billion by 2050, and most of the additional population will be living in the developing country. As a result, the demand for food in the developing countries is likely to double by 2025, necessitating a rapid increase in crop production globally to meet the needs of food security. Beirne (1967) noted that humans can not afford to feed pests in addition to themselves. Reducing the losses caused by pests is essential to continue to feed the ever-increasing population.

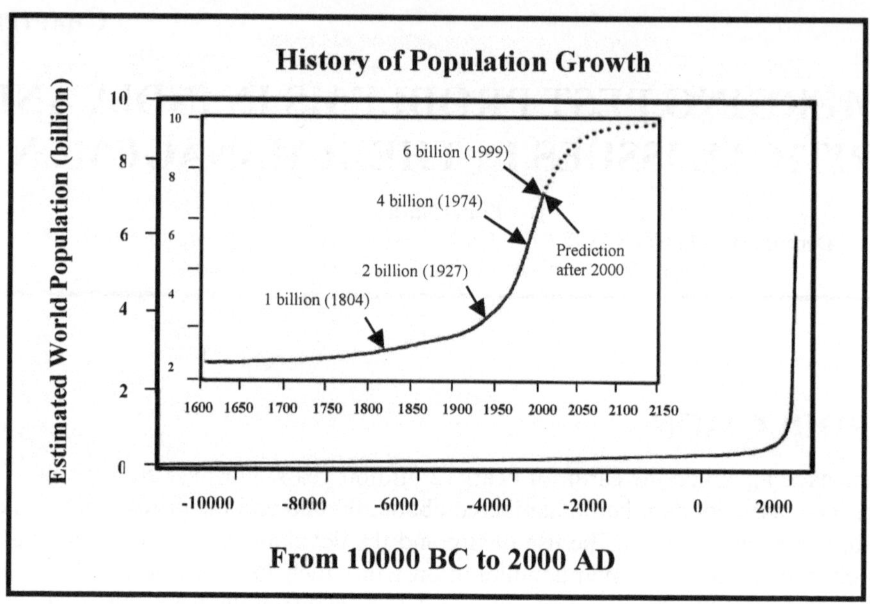

Fig. 1 Estimated world human population from 10,000 B. C. to A. D. 2150.Source: United Nations, 1999

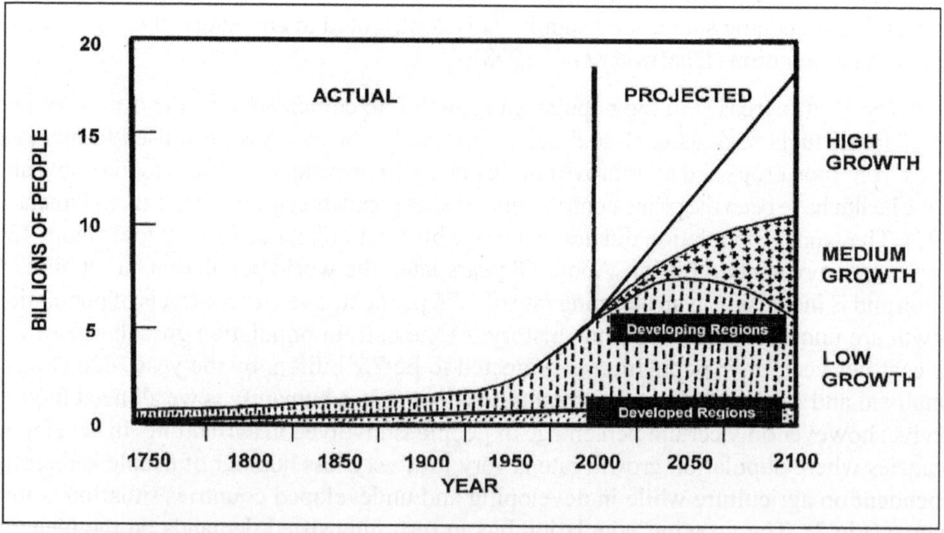

Fig. 2 Trends in world population growth (After Green, 1992), World population may stablize at 19 billion in 2100 with high growth (2.5 children/women) and at 6 billion with low growth (1.7 children/women).

The new millennium has generously acquired from the preceding century one of the unprecedented challenges to meet the demand of a growing population for food, fiber, fuel and fertilizer for improving production by increasing productivity or yield from the presently available cultivable land because fulfilling the requirement of an ever increasing population with existing crop yields is really difficult if not impossible. Food grain production has increased by 2.5 times while human population had attained a growth rate of 2-3 folds during the last 50 years (Table 1). National development is dependent on the sustained agricultural production. However, the pressure on land in our country is constantly mounting due to rapidly increasing need of the ever increasing population. Undoubtedly India has made tremendous progress in the agricultural sector over the last 50 years. It has become self sufficient in food grains and has enough stock in the storage godowns to tide over the adverse conditions. This is achieved mainly due to increase in area and adoption of modern agro-technology related to crop production and protection systems. The modern agricultural systems are input intensive and susceptible to environmental stresses. The modern developmental activities have accelerated the problem of insect pests many fold.

Table 1 Food grain production and human population of India

Year	Food grain production (Million M.T.)	Population (Millions)
1960-61	82.0	439.20
1970-71	108.4	548.20
1980-81	129.5	643.30
1990-91	176.3	846.30
2000-01	207.0	1027.00
2004-05	206.4	1080.27 (July, 2005)

Source: www.censusindia.net

The traditional Indian agriculture is heterogeneous by way of growing variety of crops and practicing polycultures by different categories of farmers. But now it is changing rapidly to fulfill the swiftly increasing demand of ever increasing population. The only way to feed the increasing population is to improve the productivity of available land. Rapidly increasing population during the last century has necessitated intensification of agriculture through introduction of high yielding varieties, expansion in irrigation facilities and application of increased rates of fertilizers and agro-chemicals. Besides, the cultural practices like spacing, crop rotation, sowing times and tillage methods were also modified to maximize productivity per unit time from the available land. The HYVs of wheat, rice, maize, and other crops have been developed under the protection umbrella provided by the broad spectrum, synthetic organic pesticides (Pradhan, 1983). While undoubtedly contributing to higher yields, this technology package has also resulted in severe outbreaks of insect pests in agricultural crops (Dhaliwal and Arora, 2001). The indiscriminate use of these pesticides has caused irreparable damage to our environment and also contributed to an aggravation of pest problems. As the agricultural productivity has to be increased manifold to meet the rising demand for food and other commodities, therefore as a corollary, pest problems will also multiply rapidly.

Therefore, the plant protection scientists have to be adequately prepared to meet the emerging challenges. An understanding of the agroecosystem including insect pests, their types, processes that have helped insects to build up their present position of dominance, the difficulties encountered in their management and other related issues may help us to face the emerging pest problems of today's and tomorrow's with greater confidence.

ECOSYSTEMS AND AGROECOSYSTEMS

The ecosystem is constituted by biotic and abiotic factors. The biotic factors include all the living organisms while abiotic factors constitute nonliving things. The species population in an ecosystem does not exist in isolation. The individuals of a species not only interact among

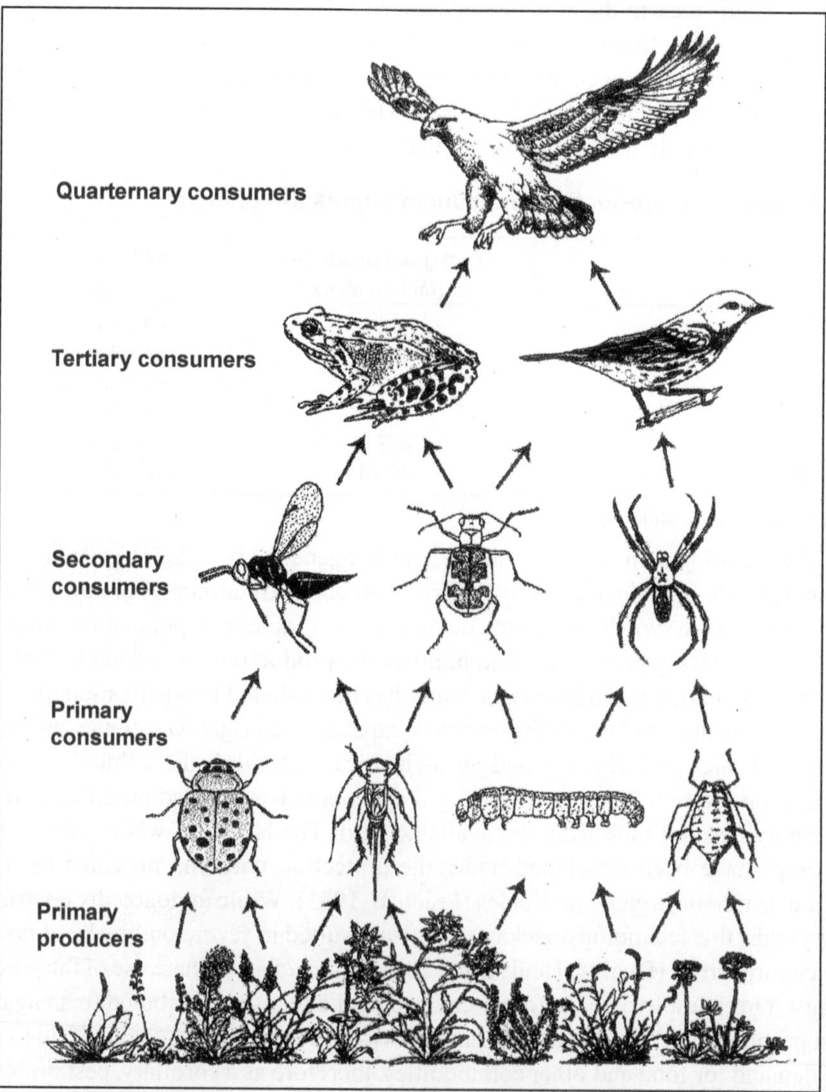

Fig. 3 A diagrammatic terrestrial food chain. Arrows indicate energy flow.

themselves for mating, feeding, and other purposes (intraspecific competition) but also interact with individuals of other animal and plant populations (interspecific competition) (Fig.3). At the very least, individuals in a population feed upon and, in turn are fed upon by individuals of other population. This interspe cific populations interacting in an area forming the food web constitute the community. Species of community vary in the size of their populations and in their niche parameters. Interactions are often expressed as a food chain, which is also a pyramid of numbers, since members of each level in the hierarchy cannot become more abundant than members of the level below without exhausting their food supply.

The physical factors of the environment strongly influence the interaction of populations in a community. These assemblages of elements, communities and physical environment are usually referred to as ecosystem. Common examples of natural ecosystems are ponds, lakes, forests etc. Ecosystems are usually the ultimate unit for study in ecology. But because of the great complexity of an entire ecosystem, it may be more convenient or desirable to consider subdivisions of it. Just as ponds, lakes, and forests are systems of interacting elements, so are agroecosystems that are largely created and maintained to satisfy a human need. Just as ecosystems are the basic unit of study of ecology, agroecosystem are the basic unit of study for insect pest management. Insects are major elements in majority of the communities. Most are primary or secondary consumers; that is, they represent links in food chain between plants (producers) and consumers of higher levels. For example, a pigeonpea plant (producer) is fed on by podborer [*Helicoverpa armigera* (Hub.)] (Primary consumer) which is attacked by hymenopteran parasitoid (*Campoletis chloridae* Uchida) (secondary consumer) (Nath and Rai, 1999). The adult parasitoid is fly about and eaten by black dragon or king crow (*Dicrurus macrocercus* Vieillot) (tertiary consumer), which in turn fall prey to hawks (*Butastur teesa*) (tetra-consumers). Similar examples are available in plenty and are given in the Fig.3. We the humans are not only the top consumers of many food chains, we also depend on many other consumers to keep our environment from being overrun by aphids, caterpillars, mice, rabbits and a host of other animals that, in the absence of check, would destroy most of our food items. The low-order consumers other than producer are the most important elements in all food chains, because without them there could be no higher levels in the food chain.

Communities change with time as a result of chance factors or of changes in physical factors. Ecological succession implies the replacement of community members in time. Many of the areas under dense forests completely denuded due to open mining have reverted to grasses and small trees. In the mean time after deforestation the faunal diversity is changed and many of the faunal populations either vanish or move to the existing forest area or settles in the near by crop fields and emerge as pest of these crops. The white grub problem is one of such several examples (Nath, 1987). If not used for factories and human settlements these denuded areas ultimately become mixed hardwood forests, with an undergrowth of plants able to flourish in the limited light the canopy permits to filter through. But there is no reversal of insect pest problems from the agroecosystem to these newly created forest cover. The change in the insect fauna through these successional changes is dramatic and complete.

Although the theory of ecosystems extends to agroecosystems, several features make plant-based agroecosystems unique:

1. Agroecosystems often lack temporal continuity. Their existence may be of limited duration and may undergo immense, abrupt changes in microclimate because of cutting, ploughing, disking, burning, chemical application, and other cultural practices.
2. Agroecosystems are, dominated by plants selected by humans, many consisting of imported genetic material. Other crop plants not from imported material have been in protective cultivation so long that they hardly resemble the parent stock from which they were derived.
3. Most agroecosystems have little species diversity, and the crop species has little intraspecific diversity; in other words, the crop tends to be genetically uniform. Usually a single species dominates an agroecosystem, and the elimination of weed species further simplifies the environment.
4. With crop plants of similar type and age in the system, the vegetative structure is uniform, and a given phenological event (for example, flowering or podding) occurs in almost all the plants at the same time.
5. Nutrients are usually added to agroecosystems, which result in crop plants with uniformly succulent, nutrient-rich tissues.
6. Agroecosystems often have frequently occurring insect, weed and disease outbreaks.

The last feature, pest outbreaks, largely results from the preceding five features. With insects, only a small fraction of the entire community finds the crop environment optimal, but those species that do are able to reproduce and survive very well, creating populations that can cause devastating losses.

DEFINITION OF PEST

The word 'Pest' is not a natural (ecological) word but a word that is coined by human being. From global perspective we were to designate as pests because pests should be defined as those biological species which impair ecosystem, productivity, diversity, and stability. According to this definition the world's major pest is man himself (earth pest) whose number and activities are threatening the stability of the biosphere. However, for better understanding of this word let us see its dictionary meaning.

The English dictionaries define pest in various ways and create confusion in understanding this word. The *Oxford English Dictionary* uses archaic definitions that state pests cause deadly epidemic or disease or plague, or anything or person that is noxious, destructive, or troublesome. *Longman's dictionary of Contemporary English*, third edition and the *Random House Webster's Unabridged Dictionary* perpetuate the belief that a pest is "an insect or small animal that harms or destroys garden plants, trees, etc" (*Random House*) or a "small animal or insect that destroys crops or food supplies" (*Longman's*). *Webster's Third New International Dictionary* (1981) and *Chamber's 21st Century Dictionary* (1996, updated in 1999) provide definitions that reflect current pest management thinking. *Webster's* states that pests are "(a) plant or animal detrimental to man or to his interests," whereas *Chamber's* states that a pest is "(a) living organism, such as an insect, fungus or weed that has damaging effect on animal livestock, crop plants, or stored products." The latter meaning of pest provided in dictionaries includes variety of organisms that may be pests and reflects current usage in pest management.

The Federal Insecticide Fungicide and Rodenticide Act (FIFRA) defines pest as "Any organism that interferes with the activities and desires of humans". According to this definition pest includes the organisms such as pathogens (including fungi, bacteria, mollicutes (mycoplasmas and spiroplasmas and viruses), weeds (all classes of vascular plants), nematodes, mollusks (slugs and snails), arthropods (including insects, mites, ticks, crustaceans and other joint legged invertebrates) and vertebrates (including amphibians, reptiles, birds, mammals).

According to Hill (1993) the definition of a pest can be very subjective, varying according to many criteria, but in the widest sense any animal (or plant) causing harm or damage to man, his animals, his crops or possessions, even if just causing annoyance, qualifies for the term pest. From an agricultural point of view an animal or plant out of contest is regarded as a pest (individually) even though it may not belong to a pest species. Thus an elephant in a shamba is a pest, but next door in a game park it is not, and is in fact a valuable national asset there. Similarly, volunteer maize plants growing in a field with cotton have to be regarded as 'weed' pests. This is further clarified that wheat plants of a variety which is out of context and have grown unwontedly in a field of wheat crop of a preferred variety and raised for seed production purpose are considered pests because if allowed to grow may deteriorate the quality of seed reducing market value.

Many organisms belong to generally accepted pest species, but individual populations are not necessarily always pests that are of course not necessarily economic pests. Therefore, here for our own interests we define pests as an organism, if it becomes sufficiently enough to harm man directly or indirectly. Thus, still we are concerned with numbers or populations of a species rather than its mere presence. This is basic to the concept of economic levels of the pest to survive and reproduce in the habitat and the way in which any factor inherent or extrinsic to the pest's physiology and behaviour violate its success or failure.

As pointed out by Norton and Conway (Cherret & Sagar, 1977), we are often somewhat over preoccupied at the present time with the state of the pest population where as probably the most important aspect of a pest species is the damage (or illness) caused by the pest and the value placed upon these consequences by human society. Thus the status of species as a pest depends on the population, level of damage and the value of damage and consequently cost of pest management.

Hence a pest is any organism whose population often increases above a certain level of economic injury and its existence conflicts with man's welfare, convenience and profits. Therefore, the outbreak of an animal pest or a disease or the spread of noxious weeds quiet often becomes a major concern of the forester, the farmer or the public health custodian. Two major classes of pest outbreaks are recognized, viz. eruptive and gradient. The eruptive outbreak spreads out from local epicentres to cover large areas. The eruptive outbreak is self-perpetuating, i.e., once initiated positive feedback processes operating at high population densities maintain the outbreak. On the other hand, gradient outbreak does not spread from local epicentres to cover large areas. It is not self driven and is completely dependent on external environmental or internal genetic conditions. This type of outbreak arises and subsides as its driving forces change in time and space.

The characteristics, biology, and role of different pest categories may be different but at the practical level, all pests have to be managed. A farmer cannot ignore one type of pest because it may destroy the crop. The relationship between crop yield and various constraints that limit crop yield is illustrated with the help of diagram by Norris *et al.* (2003) (Fig. 4). The absolute yield of crop is determined by genetic yield potential of the crop variety. The absolute yield is normally never achieved due to various constraints. The local environmental conditions of a region and the economically feasible cultural practices place limits on the attainable yield. The combined effect of all organisms affecting the crop is also a serious constraint on crop production. The effect of individual species population may not justify its pest status but in totality they may cause economic damage. Therefore a working definition of pest can be coined as all organisms within the crop environment that cause injury to the crop and are capable of reducing yield or quality.

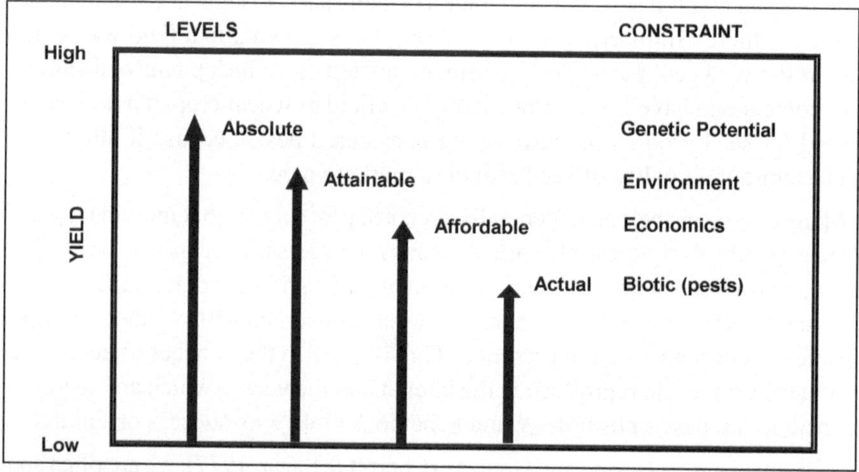

Fig.4 Relationship between crop yield and various constraints that limit crop yield.

WHAT IS INSECT PEST?

An insect whose population increases to such an extent as to cause economic losses to crops or his property (nonliving items like building, cloth, paper, wood, etc.) or a nuisance and health hazard to man and his livestock is considered as insect pest. An insect may be a pest at one place and time but may not be so at another place and time. Therefore, being a pest or not is not only a matter of insect numbers.

ANTIQUITY OF INSECT PEST PROBLEM

Insects are omnipresent in nature. They are found even in the most implausible and inimical niches on the earth. The Class - Insecta constitutes nearly 70 % of the animal species. In numerical terms, they are about one million species. They have adapted themselves to the different ecological conditions in the evolutionary process that traversed over 250-500 million years. Thus the numerical superiority and the diverse adaptability have brought them in direct association and also in confrontation with the human being and his surroundings.

While many insect species are either harmless or beneficial, a small group of insects have directly confronted and competed with the man and his needs. Some of these insects are the vectors of sorrowful human diseases, such as malaria, filaria, Japanese encephalitis, African trypanosomiasis, typhus etc; while others, mostly polyphagous insects compete with his food and shelter needs. It is this small group of harmful and destructive insects that causes loss of billions of dollars on a global scale. It is this small, yet formidable enemy, that man has been waging a war against with all his resources.

Insects have existed on this earth for at least 200 million years, while man came on the scene only half a million years ago. It is obvious; therefore, that man must have faced the problem of insect pests quite early in the history of his existence. Records of destructive insects in ancient writings support this view. Our Rig Veda, in its hymns, mentions several insects which were destroyers of grains. The Bible lists at least 11 insect pests. Dead human lice have been discovered in the hair of Egyptian mummies and many ancient men of letters and travelers have mentioned insect pests in their writings (Srivastava, 1988).

ORIGIN OF INSECT PESTS

The evolution of insects seems to have closely followed the evolution of land plants. The first land plants evolved in the Devonian and the first undoubtedly phytophagous insects are known from the Carboniferous period. The first record of winged insects almost coincided with the record of tall forest trees in the Upper Carboniferous period. Thereafter, modern orders of phytophagous insects appeared steadily throughout the fossil record. The last to appear were the Lepidoptera, coincident with the rise of angiosperms in the Cretaceous. Several other important groups evolved after the appearance of the angiosperms and a number of primitive insects became extinct (Strong *et al.*, 1984).

The dominance of angiosperms with their diversified plant surface characteristics architecture and chemical factors provided the three factors needed for insect proliferation i.e. food, shelter and oviposition sites. These factors offered an ideal habitat for insects, which consequently diversified themselves in the ecosystem (Panda and Khush, 1995). Reciprocal adaptation and counter adaptation between plants and insect phytophages has been an important mechanism driving a steady increase in plant and insect diversity over a broad sweep of fossil record.

Presently for every species of green plant there is roughly a species of phytophagous insect. Most insect orders are not phytophagous but over half of insect species are phytophagous. This pattern implies major evolutionary barriers to eating plants. Once the barriers were ruptured however the great diversity and biomass of plants have been a spectacular resource. In natural ecosystems, phytophagous insects have coexisted in a complex relationship with plant communities. Different species of plant feeding insects had to search out their own host plants from the mixed wild vegetation. In this search, they had to face the dangers of annihilation by various climatic and biotic agencies. Therefore, the damage caused by insects was probably quite limited.

The knowledge of fire and discovery of iron are the two major tools of human civilization that helped the humans in clearing the forest and starting of agriculture. Pest problems originated with the origin of agriculture. As soon as the land was cleared of natural vegetation with the help of fire and wood cutter it was replaced by a single species of food plant and humans came into conflict with phytophagous insects. These insects and other organisms feeding on the valuable crops which humans planted were called as pests. Broadly speaking a pest means any organism which is in competition with humans for some resource. In agriculture, a phytophagous insect is labeled as a pest when its population is high enough to cause significant damage in yield or quality of any of the economic plants grown by man. Not only did the insect pests originate with agriculture but the intensity of pests continued to increase with the intensification of agriculture.

What turned insects into pests initially? The following factors directly connected with man's civilization are responsible for inducing pestilence in insects (Srivastava, 1988):

1. Start of agriculture or growing of crop provided large amount of host plants to the insects that helped them to flourish easily.
2. Domestication and large-scale breeding of livestock provided ideal conditions for their multiplication (insect pest of livestock).
3. Storage of grains necessitated by large-scale cultivation encouraged insects to thrive in food stored in plenty (insect pest of stored grains).
4. The increase in human population must have attracted many insects like mosquitoes, lice, bedbugs, etc. to become human parasites. Later, as the society evolved, the creation of families and villages would have led to crowded living, favouring the spread of those parasites and with them the diseases (pathogens) they carried (insect pest of human).
5. Finally, deforestation necessitated by man's need for greater cultivation, habitation, establishment of large industries, etc. made forest insects, migrate to fields and become pests of such plants and animals which they would otherwise have not cared for.

Pest organisms: According to Norris *et al.* (2003) there are several different broad categories of pest organisms. Nematodes, mollusks, arthropods, and vertebrates feed on plants and thereby injure them. If the amount of injury results in a loss of yield or quality, then the animal is said to cause damage to the crop and thus is considered a pest. Pathogenic organisms such as fungi, molecules, bacteria, and viruses invade the host tissue(s) and in so doing disrupt the normal physiological processes. This also results in damage to the host crop and loss of yield. Weeds must interfere with crop growth sufficiently to cause yield reduction, or impact human activities in other ways. Any organism must be able to increase to damaging populations before it will be classified as a pest. The ways in which populations are naturally regulated greatly influence whether an organism will reach the status of being considered a pest, and a limitation to population increase is why most organisms never become pests.

Pest complex: Under the normal situation a crop field is attacked by a number of insects, mites, birds, mammals, nematodes, pathogens and weeds which together form a complicated interacting pest complex. The control of a pest complex is complicated and requires careful assessment, especially as to which are the key pests, and careful integration of insect control methods that may be required to avoid the related problems. This, of course, makes the process of evaluation difficult, and generally, in the past, lot of money was used on uneconomic pests for their control either through carelessness or lack of knowledge.

Pest spectrum: This is the total range of different types and species of pests recorded attacking any particular crop, and especially of concern in one particular area. The total number of insects (and mites) recorded from the major crop species are considerable; these records, incidentally, are of insects feeding or egg laying on the plant and do not include casual observations when the insect might just be resting, or when, for example, caterpillars have crawled up on to the plant just to pupate (e.g. sweet potato butterfly). Simmonds & Greathead (Cherrett & Sagar 1977) listed the numbers of pest species, on a world basis, recorded from sugarcane as 1300, cotton 1360, coffee 838, and cocoa 1400. Fortunately, for the practicing entomologist and the farmers these numbers reflect the situation globally and many of these pests are restricted geographically to one part of the world. For example, wherever cocoa is grown it will be attacked by a capsid complex (Miridae) but the actual species differ from region to region. Only a few major pests are cosmopolitan (e.g. *Myzus persicae* (Sulzer), *Agrotis ypsilon* (Hufnagel)) or pantropical (e.g. *Maruca testulalis* (Geyer)) in distribution (in point of fact, their widespread distribution is one reason for their being regarded as major pests).

Pest load: This is the actual number of different species (and numbers of individuals) of pests found on either a crop or an individual plant at any one time and space and this would usually be a pest complex but could also be a monospecific population.

Pest species accumulation: Long-term stable habitats generally exhibit extensive species diversity, both in host plants and in phytophagous arthropods. This is shown typically in old forests and also some plantation crops such as cocoa, rubber, sugarcane and tea (Banerjee, 1981). The pest species accumulation in the monoculture is in part a reflection of the area under cultivation; other important factors are the type of plant, the geographical location of the habitat (area) and its natural species richness, and the age of the actual plants and of the community (crop). It appears that in some cases the age of the plant community is not particularly important, but Banerjee observed that in tea plantations (from the Old World tropics, excluding China) age appeared to be a major factor and pest species saturation was apparently reached (in plantations in N.E. India) at the plantation age of about 35 years. After this age there was no further increase in pest species number.

Pest species recruitment: All the organisms have definite area of their origin. Like that each major crop has a more or less clearly defined area of origin in one particular part of the world and during historical times most of the crops have gradually been transported by travelers and commerce to other parts of the world where suitable climatic conditions prevail.

With some crops such as *Citrus,* it appears that most of the species in the pest spectrum in most parts of the world are allochthonous in that they have originated in the S. China /Indo-China where *Citrus* is endemic, and the pests have been gradually spread to the new areas of cultivation (Hill, 1993). Thus at the present time most of the important *Citrus* pests are pantropical in distribution.

Other crops such as rubber sugarcane and tea (Banerjee, 1981) have apparently been subjected to autochthonous pest recruitment in their new areas of cultivation. Tea, according to Banerjee has only about 3 % of its total number of pest's species common to the different areas of cultivation, the other pests being recruited locally and thus being different in each area.

As would be expected it appears that with many crops pest recruitment has been in part allochthonous and partly autochthonous. The origin of the crop pest spectrum may have an important effect so far as pest control strategy is concerned, particularly with regard to natural and biological control. It is not only the pest population which matters; there are some other parameters such as economic damage, economic injury level and economic threshold that decide the status of the pest species and likely management strategy. Therefore understanding of these important technical terms is necessary to understand the emergence of various categories of pests.

Economic damage: This is the amount of damage done to a crop that will financially justify the cost of taking artificial control measures and will clearly vary from crop to crop according to its basic value, the actual market value at the time and other factors. In practice many present farmers in the tropics engaged in subsistence farming feel that they cannot justify use of pesticides at all.

Economic injury level (EIL): This is the lowest population density that will cause economic damage, and will vary between crops, seasons and areas. But it is of basic agricultural importance that it is known for all the major crops in an area.

Economic threshold: It is desirable that control measures be taken to prevent a pest population from actually causing economic injury. So the economic threshold (Stern *et al.*, 1959) is the population density of an increasing pest population, at which control measures should be started to prevent the population from reaching the economic injury level.

Pest status: The status of an organism as a pest within an agroecosystem is not fixed. The same species can vary from being a pest and causing substantial losses, to being of no consequence. The factors that modify the status of a pest include the organism involved, the crop, the environment in which the crop grows, and time (Norris *et al.*, 2003). These interactions of host plant, potential pests, the environment, people and time can be pictured as representing a pest tetrahedron (Fig.5). This concept has been widely used in plant pathology, but it applies to all types of pests. The potential pest organism only becomes an actual pest problem when all varieties of the tetrahedron are present in an appropriate state. A suitable host has to be available, the environment must be correct for both the pest and the host, the time must be adequate for the pest to interact with the host, and an objective of human activity with respect to the host must be compromised.

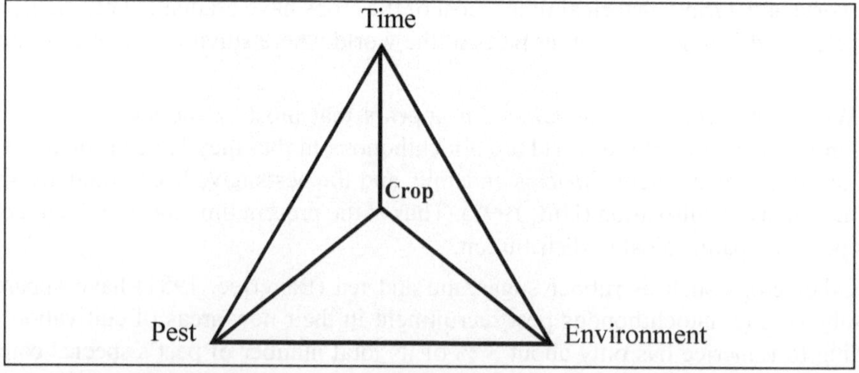

Fig. 5 Pest, crop and environment tetrahedron as determinates of the status of the organism as a pest.

CATEGORIES OF PESTS

All organisms within each category of pests are not equally important in relation to human activities. Within a pest category, organisms often are classified into various types. These types do not apply equally to all pest categories. Insect pests have been categorized as key, major, minor, sporadic and potential pests based on their population and damaging capacity. Different workers have used these terms rather loosely. In order to arrive at a definite and practicable concept different categories of pests are defined in terms of the general equilibrium position. However, the category of the pest may change according to the modification of any component of the environment.

The economic injury level (EIL) and the damage boundary is the mean value of pest density around which the pest population tends to fluctuate as changes occur in the biotic and abiotic components of the environment without permanent modification in the composition of the environment. A permanent modification of any component of the environment may alter the general equilibrium position (GEP). The lowest level of injury where the damage can be measured is called the damage boundary (DB) while the lowest number of insects that will cause economic damage is referred to as EIL.

Economic pest: On an agricultural basis we are concerned when the crop damage caused by insects leads to a loss in yield or quality resulting in a loss of profits by the farmer. When the yield loss reaches certain proportions that would justify the cost of pest management the pest can be defined as an economic pest, other wise it is **uneconomic pest**. Clearly the value of the crop is of paramount importance in this case and it is difficult to generalize but as a general guide for most crops it is agreed that most species reach pest status when there is a 5-10 % loss in yield. Obviously a loss of 10 % of the plant stand in a cereal or sugarcane field (note that this is not quite the same as a 10 % loss in yields) is not particularly serious whereas the loss of a single mature tree of *Citrus,* litchi or mango is important.

Key pest: In any one local pest complex it is usually possible to single out one or two most important pests; these are defined as key pests and are usually perennial and dominate control practices. A single crop may have one are more key pests, which may or may not vary between areas and between seasons. Key pests owe their status to several factors, including their usually high reproduction, and the type of damage. These are the most sever and damaging pests that occur in damaging levels consistently across growing seasons; they often are direct pests for which there is low tolerance. The codling moth on apples and pears is a classic example of a key pest; the larva bore into the developing fruit and a single larva in an apple either kills the fruit or makes it cull. The GEP lies well above DB and EIL. Human intervention in the form of control measures may bring the population temporarily below the EIL but it rises back rapidly and repeated interventions may be required to minimize damage. These pests present a persistent and perennial threat to our crops and are not being satisfactorily controlled with the available technology. It is of course necessary to establish economic thresholds for these key pests in order to be certain when to apply control measures, for it has been often observed that the mere presence of a few individuals of a key pest species in a crop may cause undue alarm and lead to unnecessary pesticide treatment. There is need to lower their GEP below EIL by permanent modification of one or more components of the environment. The other examples of key pests are cotton bollworms, diamond back moth, chickpea pod borer, sugarcane borer and some vectors that are frequently encountered in respective crop fields.

Major pest: These pests occur routinely and will typically influence crop yield. According to Hill (1993) the species of insects and mites are considered major pests that are either serious pests of a crop (or crops) in a restricted locality or are economic pests over a large part of the distributional range of the crop plant(s). Thus the species here regarded as major pests usually require controlling over a large part of their distributional (geographical) range, most of the time. In any one crop, in one location and at one time, there is usually only a rather small number (say 4-8) of major pests in the complex that actually require controlling. For example, although the pest spectrum for cotton world wide is 1360 species, on any one cotton crop there will probably only be about five species requiring population management. Usually for most crops in most localities the major pest species remain fairly constant from year to year, but several entomologists have commented recently that in some areas they have observed that the major pest species complex has been gradually changing over a long period of time. The brown plant hopper (BPH) of rice has risen in 10 years from obscurity to becoming the most serious pest of rice in parts of S. E. Asia. And it is reported from IRRI that there are two new major rice pests in tropical Asia (Pathak, 1980); these are the sugarcane leafhopper and rusty plum aphid. So over a period of time (say 10-50 years) it is expected that the component of major pests for a crop will change.

Minor pest: These pests occur routinely but usually only cause minor damage to crops and hence insignificant yield losses. These are the species that are recorded feeding or ovipositing on the crop plants(s) but usually do not inflict damage of economic importance; often their effect on the plant is indiscernible. They may be confined to particular crop plants or may prefer other plants as hosts. Many (but not all) pests listed as minor pests are potentially major pests (viz. BPH of rice). Many species that are major pests of one crop will occur in a minor capacity on other crops. And some times a major pest of a particular crop in one part of the world / country will be a minor pest on the same crop in different part.

Secondary pest: These are the species whose numbers are usually controlled by biotic and abiotic factors which occasionally break down, allowing the pest to exceed its economic injury threshold. These pests have the potential to cause serious damage but they are usually under adequate control by natural enemies. The term is used most commonly for arthropod pests. If the natural enemies are disrupted by agricultural practices, the secondary pest can increase and produce economic losses. An example, also from fruit crops, is spider mites that are normally under adequate control by predators. If, however, the crop is treated with nonselective insecticide that kills the predator, the spider mites will quickly increase, or flare up, in damaging populations.

Potential pest: This term is used occasionally in the literature and refers to a minor pest species that could become a major pest following some change in the agroecosystem. The potential pests normally cause negligible damage, but may become highly destructive resulting from some disturbance in the environment and the consequent increase in their number, e.g., brown plant hopper infesting paddy and armyworm on wheat crops in eastern

U.P. Only a relatively small proportion of the species listed as minor pests are really potential pests in this sense, because of their basic biology. These are similar to secondary pests in that they are typically not a problem, but if conditions change they can become a pest.

Serious pest: This is a species that is both a major pest and an economic pest of particular importance, being very damaging and causing considerable harm to the crop plants and a heavy loss in yield. It almost invariably occurs in large numbers.

Occasional pest: These pest organisms are not always present, and a close association with a particular crop is absent, but can occur as a problem from time to time. This only applies to pests that are mobile or that die out at the end of the season e.g., caseworm of rice, the castor slug caterpillar, the mango stem borer, etc. The bean leaf beetle in the USA on soybean is a good example. Conditions that favour overwintering survival may produce a larger than normal colonizing population in the spring and the species build up during the season into damaging populations. The ideal environmental conditions do not occur frequently, so the bean leaf beetle only occasionally becomes a serious pest of soybean in the Midwest, USA.

Regular pest: The regular pests are those insects that occur most frequently on a crop and such insects have close association with a particular crop, e.g., jassids [*Nephotettix nigropictus* (Stal)] on paddy, aphids [*Lipaphis erysimi* (Kalt.)] on mustard, fruit borer (*Leucinodes orbonalis* Guen.) on brinjal, thrips (*Thrips tabaci* Lind.) on tobacco, etc. These are generally found in abundance during a crop season and are expected to arrive on the crop sometime before harvest.

Sporadic pest: The sporadic pests assume pest status occasionally in certain years in a few isolated localities. It includes rice stink bug [*Leptocorisa acuta* (Thnb.)], locusts, grasshoppers, hairy caterpillars, crickets, cutworms and cotton semi-looper.

Seasonal pest: Seasonal pests are those insects which occur mostly during a particular part of the year. The incidence of these pests are largely governed by the environmental factors in a locality, e.g., red hairy caterpillar (*Amsacta albistriga* Walk.) on groundnut during April-May in certain localities of south India and the rice grasshopper (*Heiroglyphus banian* F.) during August-September in northeastern U.P.

Persistent pest: Insects which occur on a crop almost throughout the year are known as persistent pests. Scale insects and mealy bugs on sugarcane are examples.

Migrant pest: These pests are highly mobile and can infest crops for short periods of time through movement. The locust is the classic example of this type of pest.

Non-pest: This designation includes organism that never achieve status of being a problem to humans; many in fact are considered beneficial. This designation contains most of the organisms in the world.

CAUSES THAT MAKE THE INSECT AS PEST

According to Singh (2004) there are three following main causes that make the insect population as pest: (i) invasion (ii) ecological changes (iii) socio-economic changes.

Invasion: The advancement in the mode of transportation technology over the past century has increased global interaction among the nations and encouraged trade. It promoted the exchange of perishable as well as other agriculture commodities, however, the increased trade results an increased risk of new pest invasion. In spite of advanced technology used to limit the invasion, many insects used to evade inspection. Indeed, large proportions of the major pests in a modern agro-ecosystem are exotic in origin. Introduction of San Jose scale and cottony cushion scale of citrus, polyphagous serpentine leaf minor and spiraling white fly into India; gypsy moth and European corn-borer into North America; and the insect pests of *Eucalyptus* into New Zeeland and South Africa are examples.

Ecological changes: Ecological changes are another major cause of pest out breaks. The history of agriculture has been the history of constant ecological changes. Through various agro-technical practices, e.g., monoculture, selection of high yielding plant cultivars or elimination of natural enemies etc., which created conditions favourable to certain insect species and has thus induced many folds increases in their population. The high yielding new plant types of various crops are more preferred than the traditional types by the insect pest due to dense plant canopy, nutritious foliage, more number of fruiting bodies, etc. This has disrupted the population density ratios of primary consumers (phytophagous insects) and the secondary consumers (natural enemies of insect pests) in favour of the first one. Actually, these factors disrupt the interaction between phytophagous insects and their natural enemies, the essential ecological processes that contribute to the regulation of insect population. Whenever, this interaction between phytophagous and entomophagous is disrupted, the population of the phytophagous insects increased tremendously and they attain pest status because they become free from the constraints imposed by them.

Socio-economic changes: Increased urbanization has changed the life-style of human beings. For examples, the human activities have provided conditions conducive to breeding of secondary insect pests. The improved economic conditions of the urban and rural people have added new dimension to their demands and thereby long range movement of consumer items to fulfill these demands. This has helped in the wide spread distribution of insect pests and caused emergence of new pest problems. Similarly, the cotton jassid was not so serious problem on the indigenous cotton in India, but with the introduction of American cotton in the beginning of the century this insect became the dominant pest of this crop in central and south India. Introduction of susceptible and nutritious host plants into the environment of a pest species usually lead to its fast development and abundance.

There is a considerable difference between the insect population harbouring crops and non-crop plants. The insects feeding on wild plants do not occur in vast numbers, and also the plants do not appear to suffer extensively by their activities. However, contrast to this endemic situation, the insect population is extensively enormous in the cultivated fields (epidemic situation). It is, therefore, worth comparing how numbers of insects are affected in contrasting situations such as uncultivated land and crops, or comparing the same species in countries where it is a pest and in others where it does not attain pest status.

ATTAINMENT OF PEST STATUS

According to Atwal and Dhaliwal (1997) and various other workers a species may attain the status of pest by the following ways:

Entry into a new habitat: A species may become pest by entering into a previously un-colonized habitat, e.g., introduction of San Jose scale and cottony cushion scale of citrus, polyphagous serpentine leaf minor and spiraling white fly into India; gypsy moth and European corn borer into North America; and insect pests of Eucalyptus into New Zealand and South Africa.

Changes in species characteristics: A species may assume the status of a pest due to certain changes in its characteristics that did not previously compete directly with man, e.g., the sudden appearance of previously unknown and only recently involved species of cynipid wasp on chestnut plants in Japan. The pomegranate butterfly (*Virachola isocrates* (F.)) is recognized as a monophagous pest since long and now it has changed its food habit and is found infesting aonla fruits in Pratap Garh district of Uttar Pradesh. This change of food is the result of mixed planting of pomegranate in aonla orchads in large areas.

Change in habitat: The change in the habitat of the species or in the activities of man himself may increase his sensitivity to the species, i.e., increased urbanization in India has provided conditions conducive to breeding of *Culex quinquefasciauts* Say and filiariasis has thereby spread to new areas. Multiple cropping in east Uttar Pradesh has increased the problem of chickpea pod borer many fold due to continuous food supply all around the year (Nath and Rai, 1998).

Increase in abundance: A species whose interaction with man were previously negligible because of low numbers, may increase in abundance and assume the status of a pest. Such an increase may result from the greater supply of a limited resources, decreased severity or frequency of repressive environmental factors or by the simultaneous occurrence of both these changes. The introduction of a susceptible and nutritious host plant into the environment of a species may lead to its multiplication and abundance, e.g., the Colorado potato beetle attained pest status with an increase in the crop area in North America. Similarly, cotton jassid was not so serious a problem on the indigenous cotton, *Gossypium arboreum* in India, but with the introduction of American cotton, *G. hirsutum*, in the beginning of the century, this insect became the dominant pest of cotton. White grubs which were associated with pasture and forest land have shifted to agroecosystem and established as pest of various crops (Singh and Nath, 1983).

This reveals that the general evolutionary processes of nature (including man's influence) create situations for an organism to assume the status of a pest. It may not be possible for man to change the evolutionary processes but the ecological events can be altered to benefit the man kind. About 10,000 years ago, before man started living in communities and began to organize agriculture, the flora and fauna on earth remained more or less in natural equilibrium position. The supposition may be supported by the fact that the perennial forests today, untouched by man's influence, exhibit a balanced ecosystem. It can be said that the plants and the animals in a stable ecosystem live and die fitting in the energy cycle like cogs of a wheel and no one species increases in numbers out of proportion to cause a threat to the others (Atwal and Dhaliwal, 1997).

Due to geological upheavals or due to human interference the ecological conditions can change and in the process of settling to the newly created conditions, many species might be able to adjust themselves while many others might not adjust and, hence, become extinct; numerous examples are available in literature.

Another agency causing upheavals is man himself. To provide more and more food to the increasing population, man had to cut down forests and clear the *savannas* to bring more land under cultivation. In this process, he might have disturbed the balance of nature with the result that original fauna receded to whatever natural vegetation was left. On cultivated lands, a faunal vacuum was created as new niches were formed in the newly established agro-ecosystems. Perhaps, the herbivora that survived were those which could do well on the crops grown or on the original vegetation that persisted (Atwal and Dhaliwal, 1997).

In general, the removal of natural vegetation and the cultivation of soil created uncongenial conditions to those animal species, because they were exposed to weather in various developmental stages. However, some of the indigenous species that lived on weeds might have found the new plant types that are succulent and high-yielding crops more suitable and as a result they multiplied rapidly. Gradually, they increased their population to the level of a regular pest. In due course of time, they established their own less stable balance with the new environment of agro-ecosystems at a higher level of population that often rose above the economic threshold. This level might have been reached every few years and the economic threshold varied according to the nature of the injury. For example, number of insect species like aphids, thrips might go unnoticed on one type of cultivated plant, whereas a single caterpillar that bite or bore the fruits may be very much visible and make them unfit for human consumption and thereby making them unfit for marketing.

While growing crops, man either practiced monocropping (growing the same crop season after season) or multiple cropping, i.e., growing a number of crops in a locality in rotation with one another or mixed cropping i.e. growing a number of crops mixed together without any row ratio or intercropping i.e. growing number of crops in a definite row ratio. These four practices presented entirely different situations. Under monocropping, the same crop was available in a season over large areas. The favourable food was well suited to the fast multiplication of a phytophagous animal feeding on it and, year after year, its population was built up. Naturally, such a species either synchronized its life- cycle with the host crop and did not require food throughout the year or was able to sustain itself in an off-season in lower numbers, on weeds or on other vegetation of the original ecosystem that was still available in uncultivated fields, along the hedges, in forest reserves, etc. Under multiple cropping systems, the sequence of cropping provided was such that for a given pest species, food was always available in greater or smaller quantity depending upon climatic and other conditions, being favourable or un-favourable to the development and multiplication of the species.

FACTORS CAUSING PEST OUTBREAKS

Insects are designated as pests only when they are sufficiently numerous to cause economic losses. Two species of insects may be equally noxious as individuals and, yet, because of the density attained by each, one is called a pest and the other is not. This increase in the number of a species of insect is termed as a pest outbreak which could be brought about by the number of factors. These factors influence the population growth of insects by influencing birth and death rates and their movements. These factors interact in a manner that in natural system there is a balance between the phytophagous and entomophagous insects and other flora and fauna. The abiotic factors that act density-independent has little role on the pest outbreaks while the biotic factors particularly the natural enemies of the insects exert density-dependent action and put the population of phytophagous insects below the economic threshold level. Therefore, any factor that could cause the detrimental effect on the natural enemies of insect pest and create a suitable environment for the latter cause pest outbreak and knowledge of these factors is essential. For pest outbreaks numbers of reasons have been assigned by various workers (Srivastava, 1988; Atwal and Dhaliwal, 1997; Dhaliwal and Arora, 1998 and Singh, 2004). These factors are integrated together for meaningful conclusion and are given below.

Deforestation for crop cultivation: The destruction of forest for agriculture directly or indirectly influence the climate, viz., temperature, humidity, rainfall, wind velocity, etc. in that locality that create situations favouring some insects to reproduce more profoundly and assuming pest status. The insects feeding on the forest vegetations are forced to infest the cultivated crops and become new pests. Favourable weather conditions may lead to a rapid multiplication of an insect and the same conditions may turn out to be unfavourable to its natural enemies giving an added impetus to the growth of the pest to cause an outbreak.

Destruction of natural enemies: The natural enemies (parasites, parasitoids and predators) of the insects prevent the increase of their population. Their destruction either by man or other agencies (by using synthetic insecticides, burning crop wastes having harborage of dormant stages of parasitoids, e.g., burning of sugarcane leaves destroys the developmental stages of parasitoids of pyrilla) tends to increase the population of insects in an area. Sometimes the weather conditions may be favourable for the tremendous multiplication of a pest and unfavourable to its natural enemies. The phytophagous insects are more resistant to insecticides than the parasitoids and predators.

Modern agronomic practices: Modern agriculture practices, viz., mechanical tillage, timely irrigation and application of synthetic fertilizers improve the growth of the crop which reduces competition for food for insects due to which they reproduce very fast. Particularly, uses of nitrogen fertilizers increase the palatability of the crop for insect, e.g., incidence of stem borers in rice and aphids in cotton increases when high amount of urea is applied. Sometimes use of insecticides as a prophylactic measure results in reducing one of the competing species of pests while allowing the other to multiply unhindered.

Intensive and extensive cultivation of crops: Large-scale monoculture (one kind) of crops or of stock animals makes insects thrive and multiply fast. When one or more crops are cultivated in extensive areas, due to high food availability, no problem exists for the insect for food and shelter. This affords conditions that favour the proper development of the insects. The effect is more pronounced if cropping is done in more than one season for the year. If the different crops in rotation are closely related to each other or when there are alternative food plants for the insect concerned, again the population of insect is likely to increase. For example, incidence of borers and leaf hoppers is more when sugarcane crop is grown over extensive areas.

Introduction of new crops and improved strains: Plants or trees introduced into an area where these were not previously present, may serve as new hosts for some of the insects species. Most often improved cultivars of crops are susceptible to insects. Sometimes the insects which are considered of minor importance become of major importance with the introduction of such new varieties. The improved Cambodea cotton strains are highly susceptible to the spotted bollworms, *Earias* spp. and the stem weevil, *Pempherulus affinis* than the desi cotton strains. The growing of cabbage crop in the plains of Tamil Nadu as a new venture resulted in the widespread incidence of the green semilooper, *Trichoplusia ni* .

Introduction of a new pest in a new area: When an insect somehow introduced into a new area it becomes more abundant because of want of their natural enemies. For example, the woolly aphid *(Eriosoma lanigerum)* became a major pest of apple on the Nilgiri as there was no natural enemy of the pest to exercise natural control over its multiplication. Nowadays it is under control by introduction of its specific parasitoid *Aphelinus mali* collected from Punjab and Himachal Pradesh.

Accidental introduction of foreign pests: Accidental introduction of an insect by travelers or transport into a new area (country) lacking its natural enemies allows it to breed in an unrestricted manner. Modern fast transport and global trades (import and export of commodities) has increased the chances of introduction of foreign pests into areas where they were not present before. Stored grain pests, adult insects which adhere closely to the plant (e.g., scales and whiteflies), and fruit and stem borers, leaf miners, gall insects, etc., are more liable to be introduced into other countries. Some of the accidentally introduced insect pests into India from foreign countries are the diamond back moth *(Plutella xylostella)* on brassica vegetables, the San Jose scale *(Quadraspidiotus perniciosus)* on fruit trees on the hills, the green bug *(Coccus viridis)* on coffee and the potato tuber moth *(Phthorimaea operculella)* and serpentine leaf minor (*Liriomyza trifolii Burgess).*

Resurgence of sucking pests: Sometimes plants treated with systemic and contact insecticides though afford initial protection against sucking insects tend to offer physiological conditions favourable for their rapid reproduction and ultimately heavy build up of their population. For examples, in recent years in northeastern U.P. brown plant hopper *(Nilaparvata lugens)* resurge on paddy crop due to heavy use of contact position against stink bug *(Leptocorisa acuta).*

Unregulated application of insecticides: The use of insecticides leads to an outbreak in 3 ways: (i) it kills natural enemies and lifts natural control, (ii) it kills other insects (non-targets) and removes competition and struggle for existence, and (iii) it causes selective eradication of insects susceptible to insecticides so as to leave only the resistant ones to breed unaffected by the use of insecticides in future.

Large scale storage of food grains: Storage of grains in warehouses leads to increase the pest population if poorly managed.

Biotypes of pest species: Insects have great adaptation for their survival. There are several insect species that develop biotypes that are able to develop and multiply even on resistant crop cultivars. Several biotypes of aphids *(Lipaphis erysimi)* on mustard crop, gall midge and brown plant hopper *(Nilaparvata lugens)* on paddy and whitefly *(Bemisia tabaci)* on cotton, etc. have been observed in India.

Introduction of exotic crops and pests: It is well-known that, in agriculture, crops have been introduced from far-off places, even from different continents. Many times pests were introduced from the homeland along with plant materials or with the household goods. In the early stages, the introductions were due to neglect and, in more recent times since the imposition of quarantine, due to accident. These species were either known pests in the site of origin, or they were of no consequence there and became serious pests in the new country where there were few natural enemies. San Jose scale, *Quadraspidiotus perniciosus* (Comstock), is an example of the former category, being introduced from southern Asia into the United States of America. The Japanese beetle, *Popilia japonica* Newman is of the latter type. It was quite a harmless insect in Japan but when it was introduced into the United States, it became a serious pest of many economic crops and plants.

In some cases, even an indigenous insect species became a serious pest of the introduced plant, as was the case with the European corn borer, *Ostrinia nubilalis* (Hubner). This insect lived on weeds and grasses in Europe and it migrated to maize which was introduced from America. Later on, this insect was also carried to the Americas and became the most destructive pest there, too.

When a pest species was introduced into a new country, either it soon became a pest, or it took years to get adapted to the polyphagous mode of life and to cross the economic threshold. Naturally, such examples are more commonly recorded in literature from the newly colonized countries, such as the Americas, Australia, New Zealand., etc. In older places like Eurasia, where agriculture has been practiced for centuries, the early stage of the introduction or establishment of the pests has passed. In the pattern of mixed farming there, the pests have reached their rearranged balance with the environment. Presumably, the balance between the indigenous species and the environment existing at present is quite different from that which might have existed before agriculture was taken up. It is, therefore, not difficult to visualize that whether in the new countries or in the old, the dynamics of population increase of various pest species would be the outcome of the interaction of various factors, that is, the capacity to increase as modified by the type of food, climate, natural enemies or the interference of man. This may be an oversimplification, because in the last few decades we have noticed many pest species changing their population levels.

ECOLOGICAL FACTORS AND PEST OUTBREAKS

Pest populations like all animal populations are governed by their innate capacity to increase as influenced by various abiotic and biotic factors. Dhaliwal and Arora (1998) discussed the role of these factors in the pest outbreaks in the light of available information. The abiotic factors include temperature, moisture, light, etc. while biotic factors are comprised of food, intra- and inter-specific competitors as well as natural enemies of pests. The two groups of factors were termed as facultative (i.e., those causing increased percentage mortality with increase in population) and catastrophic (i.e., those killing a fixed percentage of the population irrespective of population density) by Howard and Fiske in, 1911. The same factors were called density dependent and density independent by H.S. Smith in 1935 (Clark *et al.,* 1967).

The biotic factors are greatly influenced by man's activities like agriculture and industrialization. The intensification of agriculture utilizing high yielding varieties grown under new agronomic practices with increased irrigation facilities, extensive use of agro-chemicals (Dhaliwal and Arora, 1998) and mechanization of farms has also resulted in increasing incidence and outbreaks of a number of insect pests during the period of green revolution.

***Biotic factors*:** As early as 1863, Herbert Spencer in his book *'First Principles'* observed that every species of plant and animal is perpetually undergoing rhythmical variations in number in response to the availability of food and presence/absence of natural enemies.

Food: Phytophagous insects compete with man for food supply from our agro-ecosystem and are, therefore, labelled as pests. Survival, development rate and multiplication capacity of insect pests are all determined by the quantity and quality of food. An insect population would increase in number and attain the status of a pest in case sufficient supply of a suitable host is available. The differences in varieties/species as regards their suitability for a particular insect population are governed by their recognition and acceptability as host, nutritional adequacy and absence of inhibitors/toxicants to that pest species (Atwal and Singh, 1990). The modern agriculture utilizing large scale monocultures of a few high yielding varieties of important crops provides an almost inexhaustible source of food supply for insect pests feeding on these crops and has, therefore, greatly contributed to increasing outbreaks of insect pests (Pimentel, 1977 and Jayaraj, 1988).

The introduction of early variety of pigeonpea (UPAS-120) in Varanasi region helped early start of infestation of *H. armigera* on this crop and this has also led to the development of its problem in late cultivar of pigeonpea which was not recognized earlier (Kumar and Nath, 2003). Some other changes like year round cultivation of preferred host has established this pest and now being recorded throughout the year in the agroecosystem of Varanasi region while earlier it was recorded as a seasonal migratory pest of chickpea (Nath and Rai, 1998).

Natural enemies: The role of natural enemies in the regulation of population of forest defoliators, gypsy moth, *Porthetria dispar* (Linnaeus) and brown-tail-moth, *Euproctis chrysorrhoea* (Linnaeus) was emphasized by Howard and Fiske in 1911. The importance of entomophagous insects and pathogens in regulation of pest populations was also emphasized by H.S. Smith, A.J. Nicholson and several other authors. Nicholson in a series of articles (1933 to 1958) propounded his population theory, which emphasized the role of intra-specific competition and natural enemies in regulation of pest numbers (Clark *et al.*, 1967). Populations of most insects are generally kept in check by their natural enemies. Indiscriminate use of broad spectrum conventional insecticides results in mortality of these "friends of farmers". This creates congenial conditions for resurgence of target pests or outbreaks of secondary/potential pests.

***Abiotic factors*:** Climatic factors have a dominating influence on the survival, development and reproductive capacity of insect pests and were considered to be the most important cause of fluctuations in animal populations by Elton in 1927. The role of climatic factors was further highlighted by Andrewartha and Birch (1954). On the basis of detailed investigations into populations dynamics, distribution swarm-forming grasshopper, *Austroicetes cruciata* Sauss and the apple blossom thrips, *Thrips imaginis* Bagnal, these authors found that the populations of these and other species of animals were almost entirely governed by climatic factors.

Temperature: All life survives within a certain narrow range of temperature. Departure from this optimum range on both sides is tolerated to some extent, depending upon the physiological adaptations of the concerned species or populations. Temperatures above or below these limits prove lethal. Exposure to lethal high or low temperature may result in instant killing or failure to grow and reproduce normally. Harmful effects of exposure to sub-lethal temperatures may be manifested at some later critical stage like ecdysis and pupation. Some of the insect species have developed dormancy mechanisms like diapause, hibernation and aestivation to tide over periods of unfavourable temperature in their life cycle (Atwal and Singh, 1990).

Moisture: Most terrestrial insects live in an environment, which is dry. The only source of water for phytophagous insects is the water obtained with food material from their host plants. These insects have, therefore, developed a variety of mechanisms to conserve water. In spite of these mechanisms, exceptionally dry air may prove lethal to most insects. Likewise, excessive moisture may also adversely affect many insects by encouraging disease outbreaks, affecting normal development and by lowering their capacity to withstand low temperatures. The reproductive capacity of most insects is also affected by moisture but there are great differences in the capacity of different insects to tolerate conditions ranging from extreme dryness to near saturated environments. Besides temperature and moisture, a number of other environmental factors like light, atmospheric pressure, air currents and carbon-dioxide concentration also influence insect abundance to a lesser extent.

Both biotic and abiotic factors generally act simultaneously on the pest. The populations increase due to the combined action of density independent (abiotic) and imperfectly density dependent (inter-specific competition for food, space etc. checks the increase in insect population to the point of collective suicide (Milne, 1957 and 1962). It has been reported that abiotic factors are limiting towards the margins of the distribution range of a pest species. On the other hand, in the middle where climatic factors are favourable, the population increase is checked by the biotic factors.

***Genetic factors*:** Chitty (1957 and 1960), on the basis of his studies on vole, *Microtus agrestis*, proposed that all species were capable of regulating their own population densities without destroying the renewable resources of their environment or without requiring enemies and bad weather to keep them from doing so.

Polymorphic behaviour for activity and dispersal may also be important in population dynamics. In case of western tent caterpillar, *Malacosoma californicum pluviale* (Dyar), two types of individuals were found in population. Some individuals were inactive, oviposited near their place of birth, produced mostly inactive offspring, exploited favourable habitats and were killed by harsh climatic conditions. Other individuals were more active, oviposited much further from their place of origin and always produced a higher proportion of vigorous individuals from their progeny. These active migrants tended to colonize more severe habitats or replenish the vigour of other populations. During harsh climatic conditions, individuals in marginal habitats were killed and those in favourable habitats survived to re-colonize all the vacant areas as conditions permitted (Wellington *et al.,* 1975).

Thus, pest outbreaks are the result of a combination of ecological factors that influence the capacity of insect pests to multiply. Each population has to be studied individually to identify the key factors which are playing a dominant role in its regulation.

MODERN AGRICULTURE AND INSECT PEST PROBLEMS

The problem of insect pest started with the start of agriculture. However, under traditional farming system the pest numbers were generally low as the productivity was poor. The insect pests were controlled by cultural and mechanical practices developed by the farmers largely through trial and error. The increasing human population during the 20th century has required intensive agriculture that was possible with expansion of irrigation facilities, introduction of high yielding varieties (HYVs) and application of increased amounts of agro-chemicals. Besides, the cultural practices like spacing, crop rotation, sowing times and tillage methods were also modified to achieve maximum productivity per unit time from the available land. While undoubtedly contributing to higher yields, this technology package has also resulted in severe outbreaks of insect pests in agricultural crops (Pimentel, 1977; Nath, 1987; Jayaraj, 1988; Bramble, 1989; Dhaliwal and Arora, 1993, 1994a and (Sharma and Rao, 1993). The insect pest problems in relation to various modern agriculture activities such as use of high yielding varieties, monoculture, nutrients, fertilizer, pesticides, irrigation and number of cultural practices like altered plant spacing, crop rotation etc as described by various workers are discussed here under.

High yielding varieties: The high yielding varieties are new plant types that are developed by altering the genetic makeup of the crop plant to increase yield with little or no attention to pest attack. The new plant types are susceptible to pest attack. Efforts to breed plants more palatable to human taste by elimination of factors as bitterness and hairiness may also result in development of more susceptible varieties. Moreover, the new plant types are short and heavy tillering, develop a conducive microclimate for population buildup of insect pests. The area under HYVs of paddy and wheat has been increasing since their introduction in 1966-67. This has created a number of new pests and also many minor pests have attained major status. Brown planthopper, whitebacked planthopper, green leafhopper, gall midge, leaf folder and whorl maggot have become major pests of rice since the introduction of HYVs. Even in case of wheat, where diseases are the main limiting factors in production, a number of pests like aphids, thrips, armyworm, shoot fly, cutworms, etc., are becoming increasingly serious. The problem of army work is increasing both in both rice and wheat crops (Nath, 1986 and Singh and Nath, 1998).

Development and release of new varieties/hybrids like Swarna and CSH I in sorghum pushed the status of pests such as shoot fly, *Atherigona varia* (Meigesn) and *A. soccata* Rondani and the grain midge, *Contarinia sorghicola* (Coquillett) to the forefront. The appearance of the shoot fly, *Atherigona approximata* Malloch on pearl millet has been reported after the introduction of hybrids of the crop. In sugarcane, introduction of thick and soft cane varieties resulted in severe attack of a number of borer species. The cultivation of highly susceptible varieties has also resulted in flare up of many sucking pests. The introduction of HYVs of *hirsutum* and *barbadense* species during the last two decades has escalated the severity of insect pest problems on cotton (Arora and Dhaliwal, 1996).

Monoculture: In agriculture, the natural plant community is removed, destroyed and usually replaced by a single crop species. Larger the area that is planted to a single crop, the greater the potential or pest problems. Generally, same crops are maintained in a given area year after year and this helps in survival and multiplication of the pest throughout the year. Large scale monoculturing of cotton, paddy, sugarcane and vegetable crops in India in contiguous block has resulted in rapid increase in number and intensity of pest attacks. Modern agriculture with its emphasis on high yielding varieties, further narrows down the genetic base of these crop species. Another major impact of green revolution has been that the minor crops that provided diversity in the agro-ecosystem are losing their importance. While the area under wheat has jumped from 7.4 per cent of total cultivated area in 1950-51 to 13.3 per cent in 1987-88, in case of coarse cereals it has declined from 28.6 to 21.1 per cent during the corresponding period. In case of small millets and barley, the decline has been from 3.5 and 2.4 per cent to 1.7 and 0.7 per cent, respectively. Similarly, there has been decline in percentage area under several oilseed crops like linseed, sesamum, castor, etc. The area under pulses has also been decreasing during the last three decades (Dhaliwal and Arora, 1993).

Even within the same crop, modern plant breeding with its emphasis on inbred, uniform strains has fostered a widespread trend towards genetic erosion, which is depleting the genetic base of many of our crop plants. For example, in case of wheat, about 80 per cent of the 2.5 million ha area in Haryana and Punjab is under a single variety HD 2329. In paddy, where we have the world's most diverse genetic base with thousands of traditional varieties, two-third of the area is under a few HYVs, which in turn are derived from a fewer number of parents. All these changes have made our agro-ecosystem very susceptible to insect pests. More than 80 species infest and cause damage to rice crop in India. Among which rice leaf and plant hoppers constitute one of the major groups of most important pest of rice (Nath, 1988) Misra and Israel (1968) reported 30 species of rice leaf and plant hoppers from India.

Lately, there have been many outbreaks of insect pests on the rice crop. An even greater danger comes from the continuous cultivation of rice over northeast India which has virtually eliminated the spatial and temporal hurdles that were so far operating against the movement of vectors. Five diseases of paddy that are transmitted by leaf and plant hoppers, their collateral hosts along with IPM components are described by Nath (1998). The vast rice stretch from Pakistan (Punjab) to Bangladesh provided crops of different phenology for transmission of rice tungro virus (RTV) by green leafhopper and ragged stunt virus by brown planthopper. In eastern India, rice monoculture is practiced favouring mass survival of the vectors. Along with the monsoonic air currents, the green leafhopper carrying RTV can move in large numbers from one part of the country to another (Nagarajan, 1989).

An analysis of the factors responsible for the severe outbreak of *Helicoverpa armigera* (Hubner) on cotton during 1987-88 in parts of Andhra Pradesh and Tamil Nadu revealed that the pest multiplied due to monocropping of cotton continuously over large areas, prolonging the crop growth beyond the regular duration of non-removal of crop residues before the next crop. Mono-culturing of cotton was also an important factor responsible for white fly outbreaks on this crop in south during 1980s. Mono-culture of cotton was found to harbour more insect pests than cotton inter-cropped with groundnut, cowpea and soybean. The cowpea inter-crop also attracted more coccinellid predators and resulted in increased natural parasitism of spotted bollworms. Similarly, the stable agro-ecosystem resulting from continuous monoculturing of sugarcane around sugar factories has resulted in multiplication of insect pests. Mono-cropping of vegetable crops in selected production centres especially along the periphery of the large cities, has also been an important factor in increasing outbreaks of many insect pests.

Nutrients: Altering the nutrient level in the soil affects the concentration of nutrients in the host plant, which, in return, influence the pests that are feeding on the plant. The HYVs are nitrogen responsive and require extensive use of nitrogenous fertilizers to provide maximum yield. High levels of nitrogenous fertilizers significantly increase the incidence of most insect pests. Increase in soil moisture level due to frequent irrigations also helps in build up of insect pests.

Fertilizers: Excessive use of inorganic nitrogenous fertilizers generally creates congenial conditions for rapid multiplication and subsequent outbreak of many insect pests (Dhaliwal and Arora, 1998). Consumption of fertilizers in the country has been increasing steadily since independence. The introduction of nitrogen responsive HYVs during late sixties gave further impetus to this trend. Nitrogenous fertilizers accounted for about two-third of the total fertilizer consumption of 11,571,000 tonnes during 1989-90 in India.

High levels of nitrogen fertilizers significantly increase the incidence of most of the insect pests including yellow stem borer, leaf folder, gall midge, green leafhopper, white backed planthopper, earhead bug, rice hispa, whorl maggot etc. In cotton also, it has resulted in greater incidence of jassid, whitefly and bollworms. In sugarcane, application of nitrogen fertilizers builds up stalk borer and internode borer due to softening of canes and lodging. It also results in increased incidence of pyrilla. Increase in nitrogen application from 30 to 150 kg/ha resulted in higher incidence of armyworm in wheat.

However, in few cases, application of nitrogen fertilizers reduced the incidence of insect pests. Incidence of sorghum shoot fly on sorghum cultivar CPV-86 decreased with the application of nitrogen, while in another cultivar CSH-8R, the incidence increased up to 60 kg N/ha but declined at 90 kg N/ha. Attack of *H. armigera* on tomato was minimum at the highest nitrogen level and maximum number of irrigations. Reduced nitrogen application induced proliferation of while fly in sugarcane.

On the other hand, application of potash and sometimes phosphorus also, either singly or in combination with nitrogen, results in lower incidence of many insect pests. A reduction in incidence of a number of insect pests including *Empoasca kerri* Pruthi, *Autographa nigrisigna* (Walker), *Spodoptera litura* (Fabricius), etc. following increasing application of P and K has been reported. In paddy, application of increasing dosages of K lowered the incidence of green leafhopper, brown planthopper and white backed planthopper. Aphid and thrips incidence on chillies also declined with increase in K application. Application of phosphorus and potash at higher dosages reduced the attack of *Empoasca* sp., *Therioaphis trifolii* f. *maculata* (Buckton) and *Aphis craccivora* Koch on lucerne. Application of 200 kg K/ha combined with 50 kg N/ha resulted in lowest incidence of brown planthopper in rice while 50 kg K/ha plus 100 kg N/ha had the highest incidence.

Pesticides: The discovery of synthetic organic pesticides during 1940s equiped us to gain an upper hand in the struggle against insect pests of our crops, domesticated animals and ourselves from their pestiferous attacks. However, the large scale indiscriminate use and abuse of these insecticides has resulted in increased attacks of insect pests on many crops. This may be due to development of insecticide resistance, resurgence of insect pests or outbreaks of secondary pests. The insecticides may kill natural enemies of target pest or potential pest. They may also alter crop physiology to make the plant more susceptible to attack by insects or they may even directly stimulate reproduction of surviving insects. There have been numerous reports of loss of human life due to pesticide poisoning. Pesticide residues above tolerance limits have been reported in cereals, vegetables, fruits, vegetable oils, spices and even in human milk.

According to study, only 1 per cent of the total pesticide applied is effective in controlling pests, remaining 99 per cent goes into various environmental systems. In addition to these side effects even the insect/pest problems in many crops have aggravated following continuous application of insecticides. This, in turn, has further increased the consumption of pesticides. Taking the example of cotton in Punjab, the cost of insecticides as percentage of total cost of cultivation of the crop increased from 2.1 per cent in 1974-75 to 4.6 per cent in 1979-80, 11.9 per cent in 1984-85 and 15.5 per cent in 1989-90. While the cost of cultivation has increased little over three times from 1974-75 to 1989-90, the cost of insecticides used has multiplied over 21 fold during this period. In-spite of this phenomenal increase in the cost of insecticides used, the losses due to insect pests have been increasing steadily. This is due to killing of natural enemies, resurgence of insect pests, development of insecticide resistance and outbreaks of secondary pests, etc (Dhaliwal and Arora, 1998).

Irrigation: With the expansion of surface and groundnut water irrigation facilities, the share of irrigated area in total cultivated has been increasing slowly. While only about one third of the cultivated area in the country is irrigated, the States of Punjab and Haryana have more that 80 per cent area under irrigation. The only other State with more than half of the cultivated area under irrigation is Uttar Pradesh. On the other hand, in 13 States including Assam, Karnataka, Kerala, Goa, Gujarat, Madhya Pradesh, Maharashtra, Orissa and West Bengal, less than a quarter of the cultivated area has irrigation facilities. Most of the irrigated area is under a handful of crops. Sugarcane, wheat and rapeseed-mustard have 87, 77 and 54 per cent area having irrigation, respectively. In contrast, less than 10 per cent of the area under pulses is irrigated.

The expansion of irrigation facilities has also aggravated insect pest problems in many crops. The multiplication of brown plant hopper increased as the level of the water in rice fields increased. Draining the fields and maintaining irrigation water to field capacity tend to inhibit the multiplication of brown plant hopper and gall midge. Water logging enhances the multiplication of a number of borers including stalk borer, internode borer, plassey borer and the whitefly in sugarcane crop. One of the important factors in the recent heavy build up of top borer in eastern Uttar Pradesh and Bihar has been the increased irrigation facilities. Increase in soil moisture level due to frequent irrigation also helped in build up of pyrilla population on this crop. Similarly, the population of armyworm on wheat was positively correlated with the number of irrigations (1 to 12) per plot. The multiplication of aphids on many crops increased with the increase in the moisture level of the soil (Dhaliwal and Arora, 1998).

Cultural practices: New agronomic recommendation, which are essentially meant for exploiting the full genetic yield potential of HYVs, are simultaneously favourable to most of the insect pests, diseases and weeds for their attack, multiplication and spread (Dhaliwal and Arora, 1998).

***Spacing*:** The close plant spacings currently recommended in many crops help in the build up of many insect pests. Close spacing contributed to rapid increase in population of green leaf hoppers *Nephoteltix* spp., brown plant hoppers (Nilaparvata lugens) and white backed plant hopper (*Sogatella furcifera*) in rice (Prasad and Misra, 1982). It also resulted in higher population of gall midge and leaf folder. Spacing that allowed so sunshine to reach the basal area of the plants for a part of the day resulted in smaller insect population. In autumn planted rice, infestation of stem borer and white backed planthopper increased with higher plant density. Larval population of pod borer on chickpea was 4 times as large at closest (33 plants/m^2) as at widest (3 plants/m^2) spacing. There was little advantage of more that 8 plants/m^2 in chemically protected crop. In cotton, closer spacing resulted in build up of jassid and whitefly (Sundaramurthy and Chitra, 1991). In soybean, damage by a number of insect pests including *Melanagromyza sojae* (Zehntner) and *Bemissa tabaci* (Gennadius) was greater in closely planted crop. In sugarcane, closer spacing resulted in higher population of shoot, internode and stalk borers. In contrast to all these studies, population of A*phis craccivora* Koch on chickpea was more in widely spaced crop (60x20 cm) than in closer spacing (30x10 cm). Similarly, infestation of *Hydrellia philippina* Ferino in rice decreased with increase in plant density.

***Crop rotation*:** Dhaliwal and Arora (1998) have reported that the introduction of paddy crop in the north has resulted in the practice of paddy-wheat rotation. In Punjab and Haryana alone, it is followed on 2.4 million ha, besides 4 million ha in other parts of the country. The wheat crop now serves as a host for the survival of a number of rice pests including stem borer, pink borer and shoot fly during the winter months. This has resulted in increasing attack of these pests on both crops. The large scale cultivation of sunflower in Punjab and Haryana has resulted in the practice of sunflower cotton rotation over large areas. This rotation is conducive for the multiplication of *H. armigera*, besides a number of sucking pests. New rotations like wheat - arahar, wheat - moong - rice, wheat - moong - maize, etc. have also contributed to insect pest problems in pulse crops. Modification of other practices like time and manner of planting, growing of inter-crops, trap cropping, detrashing, trenching, harvesting time and practices, which were traditionally used for minimizing damage by insect pests, has also contributed to the aggravation of insect pest problems in agriculture.

DEVELOPMENTAL ACTIVITIES AND INSECT PEST PROBLEM

The modern developmental activities of humans to meet the demand of ever increasing population have altered the natural ecosystem. As a result numbers of organisms have been displaced from forest reserve to man made ecosystems (agroecosystem). Several other activities have contributed in the transportation and settlement of these displaced species of organisms. Some of these species are now creating problem by interfering with the activities and desires of humans. Singh and Nath (1983) and Nath (1987) have discussed the impact of developmental activities on the insect pest problems. Human interventions in the forest have their impact in the forest environment and on all organisms inhabiting the forest ecosystem. The greatest human impact on forests comes from land clearance for agricultural purposes, industrial establishments, mining work and timber supply for many domestic and commercial purposes. These activities have lead to disappearance of natural forest ecosystem. The problem of white grub is one of the several examples which has shifted from the forest ecosystem to the agro-ecosystem. The establishment of power projects after 1950's; Open mining of coal and other minerals and development of urban areas near the forest have led to clearance of forest cover. Construction of roads, river bridges and high speed vehicles has helped in the successful migration of insect pests from the forest. The nocturnal phototactic insects are attracted towards the source of light. The adult insects that are phototactic and nocturnal in nature get help from the artificial source of light and distributed far and wide places.

The electrification of the nearby places of the forest and automobile vehicles attract the adult insects from the forest to the source of light and thus slowly they have reached and settled in the agro-ecosystem. It has been observed that the insects which are attracted towards the light of the trains settle down on the train and thus transported from one place to other. The agricultural crop fields by the side of railway track and roads are first infested by such type of insects. The lights on the river bridge also attract the phototactic insects that easily cross the river which usually act as a natural barrier during rainy season (Nath, 1987). The electrified pump sets and tube wells have contributed in the migration and establishment of the pest in the agroecosystem where food supply is assured due to intensive farming practices.

The assured irrigation has changed the cropping systems. The practice of intensive cropping like multiple and relay cropping have become very popular. The establishments of sugar factory and assured irrigation have not only increased the area under sugarcane cultivation in the traditional sugarcane belts, it has also been introduced in many other areas where it was never cultivated. The ratooning of sugarcane has become a popular practice in Eastern Uttar Pradesh. The groundnut crop has been introduced in the cropping system of Eastern Uttar Pradesh replacing pearl millet. This is a preferred host of white grub. These changes have accelerated the white grub problem by making assured and protected food supply. The agriculture has become a secondary business of the farmers due to their small and uneconomic land holdings. The bullock ploughing and employing the farm labourers have become a costly affair. Therefore, tractor ploughing has become these small farmers very popular during these years irrespective of land holdings. Most of the tractor owners plough the fields during night and that too only with cultivators. During night the soil inhabiting insects are neither exposed to sunlight nor being devoured by the predatory animals due to which the population of these pests are increasing rapidly and causing economic damage to the crops.

The changes brought in the forest and agro-ecosystem to fulfill the human needs have become the root cause of insect pest problems and now this require new philosophical and scientific approach to utilize these resources in a better manner.

CHANGING STATUS OF PESTS

Pest organisms are always under varied pressures. Therefore, their status is never static rather dynamic. Atwal and Dhaliwal (1997) have discussed the changing status of pests and pin pointed the prime contributors towards their changing status. According to them agriculture in India started around 2000 BC during the period of Indus valley civilization in Sindh and West Punjab, and spread to East Punjab, West U.P., Rajasthan and Madhya Pradesh. At around the same time, agriculture was also started in South and East Asia. The implements used during this period included plough, seed-drill, unwheeled cart, etc. The only references regarding pest problems during this period pertain to wild birds and animals, which were scared away by use of terracotta sling-balls. The advent of Aryans brought the iron technology but there was still no indication of any pest problems. Only around 900 AD, locusts are listed along with birds and mammalian pests as causing damage to crops. In ancient Indian literature, there are indications of sound study of useful insects but the knowledge about harmful ones in the same period was expectedly poor. Although pest problems were quite serious, the knowledge about their nature and control was very inadequate. This situation continued almost till the end of the nineteenth century.

Traditionally, the crops were grown only during the monsoon period, and winter served as a closed period for crops as well as for pests in India. With no assured irrigation and little use of chemical fertilizers, the yields obtained were also low. However, even at that time a number of insect pests were known to cause serious damage to different crops. Balfour in 1887 in his book Agricultural Pests of India and Eastern and Southern Asia, reported many groups of pests like cutworms, aphids, bugs, caterpillars and beetles damaging different field crops viz., *Leptocorisa* and greenfly on rice; sugarcane borer, *Diatraea saccharalis* (Fabricius) and sugarcane bug on sugarcane, and pink bollworm and *Helicoverpa* on cotton. Harold Maxwell Lefroy, who was appointed Entomologist to the Government of India in 1903, published a book on Indian Insect Pests in 1906, wherein he described pests injurious to cotton, rice, wheat and other crops. T.B. Fletcher, who succeeded Maxwell Lefroy as Imperial Entomologist (1913-1932), produced major treatises on cotton bollworms, sugarcane borers and stored grain pests. Fletcher (1920), at the Third Entomological Society Meeting presented a comprehensive list of several hundred species of insects associated with different crops in India. Of these, 45 species were reported as major pests on important crops. These included 8, 7,7,6,6 and 1 species causing serious damage to paddy, cotton, oilseeds, pulses, vegetables and sugarcane, respectively.

Following the introduction of high yielding varieties (HYVs) and associated technology, there has been a tremendous increase in the number of insect pests damaging various crops, viz., pigeonpea (250), rice (225), sugarcane (215), cotton (170), etc. Many minor insects have started assuming serious proportions, whereas several pests have spread to new areas.

The incidence of brown plant hopper, yellow stem borer and leaf folder has increased manifold with the cultivation of modern HYVs of rice. Gall midge, originally being confined to monsoon season, has, of late, become a pest in summer season also. The spread of rice cultivation into the non-traditional north-western states like Punjab and Haryana has brought up the problem of white backed plant hopper. The continuous cropping of maize in northern parts of India has led to the emergence of shoot fly, *Atherigona* spp., as important pest in spring sown maize. *Pyrilla perpusilla* (Walker), a serious pest of sugarcane has developed a preference of hybrids of maize and sorghum. Sorghum midge, earhead bug and aphids are gaining importance in many sorghum growing regions.

Severe outbreaks of whitefly and American bollworm have been observed on cotton during the last decade. American bollworm appeared in serious form in parts of Andhra Pradesh and Tamil Nadu during 1987-88, where it caused about 66 per cent loss in yield of seed cotton. The pest caused severe losses to cotton crop in Haryana, Punjab and Rajasthan during 1990.

A number of polyphagous pests have also attained serious status. Most dreaded of these is *Helicoverpa armigera* (Hubner), which has been recorded on more than 180 cultivated and uncultivated plant species, of which chickpea, pigeonpea, cotton, tomato and tobacco are considered to be the key source crops. Similarly, *Spodoptera litura* (Fabricius) has been recorded on more than 115 species of cultivated plants in India.

Unregulated and frequent application of insecticides for the control of certain insect pests such as *Helicoverpa armigera, Bemisia tabacci* etc in cotton crops have added new dimension to the pest problems. This has also helped in developing resistant strain or bio types of insect pest species and also helped in developing new pest problems like mite pests in the absence of natural enemies that were elimined due to insecticide application. This has resulted total failure of crop leading to suicide by the farmers due to debt trap in Andhra Pradesh, India.

Insect pest problems in agriculture are probably as old as agriculture itself. However, under subsistence agriculture, the pest numbers were generally low as the productivity was poor. The insect pests are kept under check by cultural and mechanical practices developed by farmers largely through trial and error. Rapidly increasing population during the last century has necessitated intensification of agriculture, which has also resulted in severe outbreaks of insect pests in agricultural crops.

CROP LOSSES AND CHANGING PEST SCENARIO

Dhaliwal and Arora (2001) estimated crop losses due to insect pests, diseases and weeds to the tune of Rs. 29,000 crore per annum during 1993-94. The avoidable losses caused by insect pests and diseases have been estimated to be around 60-80 % of the total crop losses in India. The present day plant protection was mainly oriented towards chemical control and in the past five decades, the uses of chemical pesticides steadily increased from 2.2 g/ha of active ingredient in 1950 to the current level of 650 g/ha (a 300 fold increase). Undoubtedly, the pesticide umbrella played the pivotal role in safeguarding the yield potential of high yielding varieties and hybrids which was practically impossible in its absence. In spite of increase in insecticide use, the losses due to insect pests have been increasing steadily, mainly due to elimination of natural enemies, resurgence of pests, development of insecticide resistance and outbreak of secondary pests. The estimated crop losses caused by insect pests under modern agriculture are enormous. From the available date of 2004-05, an effort has been made to quantity the monetary loss in all the major crops based on the figures for these crops. The estimated percent loss being the same as reported by Dhaliwal and Arora

(1998), the extent of monetary loss has increased to the tune of 74,610 crore per annum due increased production and market price (Table 2).

Table 2 Estimated crop losses caused by insect pests under modern agriculture in India

Crop	Actual production during 2004-05 (Million tonnes)*	Estimated loss in yield due to insect pests		Hypothetical production in the absence of losses due to insect pests (Million tonnes)	Monetary value of estimated losses during 2004-05 (Million Rs.)
		%**	Total (Million tonnes)		
Rice	87.80	25	29.27	117.07	1,63,912
Wheat	73.03	5	3.84	76.87	2,45,760
Maize	13.58	25	4.53	18.11	23,783
Sorghum	7.53	35	4.05	11.58	20,857
Millet	6.46	35	3.48	9.94	17,922
Pulses	13.99	30	5.86	19.53	83,915
Groundnut	6.07	15	1.07	7.14	16,050
Rapeseed -Mustard	6.40	35	3.45	9.85	58,650
Seed cotton	4.13	50	4.13	8.26	72,688
Sugarcane	234.15	20	58.54	292.69	43,613
Total					7,46,100

**Source:* Agricultural Statistics division, Directorate of Economics & Statistics, Department of Agriculture & Cooperation, Govt. of India.

***Source:* Principles of Insect Pest Management by G.S. Dhaliwal and R. Arora (1996, pp.17).

The emerging pest problems of India and critical issues in their management have been reviewed from time to time (Puri and Mote, 2003). They enumerated some important problems of insect pests in some specific crops that have been recognized as very serious and alarming either on regional or national / international basis during recent years. Some of these problems are the chickpea pod borer, *Helicoverpa armigera* on cotton, chickpea, pigeonpea, sunflower and tomato; white fly, *Bemisia tabaci* on cotton and tobacco; brown plant hopper, *Nilaparvata lugens* and green leaf hopper, *Nephotettix* spp. on rice; serpentine leaf miner, *Liriomyza trifolii* on cotton, tomato, cucurbits and several other vegetables; fruitfly, *Bactrocera* sp. on fruit and vegetable crops; mealy bugs on several field and horticultural crops; thrips on certain crops like groundnut, cotton, chilli, roses, grapes, citrus and pomegranate, etc.; gall midges on rice and fruit crops; diamond back moth specially in cabbage; hoppers on mango; *Pyrilla* in sugarcane; mites and nematodes in vegetable crops and several national, international and polyphagous pests like termites, white grubs, hairy caterpillars, *Spodoptera* sp. etc., have assumed serious magnitude in recent years in India. Some of these pests are exotic and have got entry in India either accidentally or intentionally from time to time and now they have established and posing serious threat to many crops. Some new pests like serpentine leaf miner and spiralling white fly have entered into India and they are causing severe economic losses of crops (Viraktamath *et al.,* 1993 and Singh *et al.,* 2003). Likewise, some of the minor/secondary pests such as mites in vegetables crops (Singh, 1994 and 1995) and pod bug in pigeonpea are assuming the status of major pests (Kumar and Nath, 2003 a). These pests have been successfully managed by using chemical pesticides, microbial and botanical insecticides, predatory and parasitic insects, cultural methods and resistant/tolerant crop varieties. However, there is a need to develop eco-friendly and cost effective IPM modules for sustainable agricultural production.

MENACE OF EXOTIC PESTS

Coffee berry borer: The coffee berry borer, *Hypothenemus hampei* (Ferrari) is the most serious pest of *robusta* as well as the low altitude *arabica* coffee. The larvae feed on many plants but its main host is *Coffea* sp. and the alternative hosts are various belonging to plant family Rubiaceae and Leguminoseae including *Phaseolus* and *Hibiscus* spp. This pest has been recorded from tropical Africa, Sri Lanka, South-east Asia, Indonesia, Papua New Guinea, New Caledonia, Caroline, and South America. It is native of central Africa and was detected in India during February, 1990 in Gudalur in Nilgiri District of Tamil Nadu and spread to Waganada and Idduki of Kerala, Kodagu and Satleshpur of Karnataka and Pulney hills of Tamil Nadu with an estimated area of 1, 18, 303 ha which is about 36 % of the total area under coffee in India (Prakasan *et al.*, 2001). Now, it is spread all over the major coffee growing areas of our country.

It is considered to be monophagous on coffee seeds and capable of rapid multiplication and spread causing severe crop losses. The female beetles take temporary shelter in seeds of *Crotolaria*, *Lantana*, *Maesopsis*, *Duranta*, Surinam cherry, tamarind and tea without feeding and breeding.

One or more round holes can be seen near the apex of large green or ripe berries. The damaged beans, which have a distinctive blue green staining, contain up to 20 larvae of different sizes. Infestation on tender berries results in premature fruit drop or decaying of one or both the beans. In ripe berries, the grubs make extensive feeding tunnels and cause yield loss up to 40-80 %. The tunneling also results in reduction of the crop quality. The coffee berry borer can be managed by adopting cultural and physical control methods. Heavy shade, from either shade trees or inadequately pruned coffee causes conditions unsuitable for the natural enemies of the borer and should be removed. All over-ripe or dried berries should be removed and destroyed. Old crop remains should be stripped completely. Timely harvesting of the coffee berry should be done and no berries should be left over, either on the plant or on the ground. Off season berries should be removed. Dipping of berries in boiling water for 2-3 minutes kills the pest stages inside the berries. Drying the coffee to a prescribed moisture level (arabica/robusta parchment 10 %, Arabic cherry 10.5 % and robusta cherry 11 %) helps in preventing the infestation. Mass trapping of beetles with coffee fruit extracts in 1:1 combination of ethanol and methanol helps in minimizing the beetle's population. Trap density of 27 per acre is optimum (Prakasan *et al.,* 2001).

Sapota seed borer: Sapota (*Achras sapota* Linn.) is commonly called chikoo or sapodilla. It is native to Mexico and Central America and known for its delicious fruit and for the milky latex which is the source of chicle, used to make chewing gum. It was taken to the Philippines by the Spaniards, and from there spread to Malaysia and India. In India it was confined to peninsular part and West Bengal till recently but now it is spread to drier tracts of Deccan plateau and submountane areas of Uttar Pradesh. In India, sapota trees are attacked by about 25 species of insect pests (Butani, 1979); while Hill (1993) listed 20 species out of which 5 are major remaining others are minor insect pests. But none of these reports find mention of this sapota seed borer, (*Trymalitis margirias*) Meyric as pest. The sapota seed borer is a recently introduced pest in Konkan region of Maharashtra (Puri and Mote, 2003). It was first noticed in 'Kankradi' area of Dahanu tehsil, Thane district of Maharashtra. The sapota crop suffered badly during 1999-2000 in Dahanu to the extent of 40 to 90 % during September to March (Patel, 2001). The area of 225 ha was found to be affected by this pest comprising of 19 villages in Dahanu tehsil during February, 2000. The peak infestation of the pest was reported in the month of May and December in Thane district of Konkan region. The practices like collection and destruction of fallen leaves, fruits, etc. and their destruction by burning; pruning sapota trees properly so as to allow sunlight and aeration in the orchard and installation of light traps help in reducing the infestation of the pest. Spraying the crop with endosulfan 0.05 % or dichlorvos 0.045 % or malathion 0.1 % or carbaryl 0.1 % at an interval of 15-20 days control the pest effectively (Puri and Mote, 2003).

Sugarcane woolly aphid: The sugarcane wooly aphid, C*eratovacuna lanigera* Zehntner was reported from Indonesia by Zehntner (1900) and its occurrence in the Asian continent from India through mainland S. E. Asia to China and Taiwan, and the Philippines and Java has now been reported. In India this pest has been reported from many states like Assam, Nagaland, Sikkim, Tripura, Uttar Pradesh, West Bengal, Maharashtra and Karnataka. This pest appeared for the first time on sugarcane in Maharashtra and Karnataka in July, 2003 and now it is a major pest in the country (Puri and Mote, 2003).

Sugarcane (*Saccharum officinarum*) is the main host of the woolly aphid; however, it can also feed on wild sugarcane (*S. spontaneum*) and *Miscanthus* spp. (Gramineae). Heavy infestations on the underneath of leaves of white waxy aphids cause a loss of plant sap, and at times heavy sooty mould infestation occurs on the foliage causing adverse effect on photosynthesis. This may result in the death of young shoots and affect the final yield. Within its area of occurrence it is quite a common pest but does little damage and management measures are seldom required. However, Farina (1994) reported the losses due to woolly aphids to the extent of 26 % in cane yield and 24 % in sugar content. The infesting population normally consists of winged and apterous, viviparous and parthenogenetic females and each female produces 15-35 young which are copiously covered with waxy filaments. The pest can be managed by adopting the practices like strip planting with wide spacing so as to have proper aeration and convenience of plant protection operations in the field. Normally this insect is preyed upon by many species of lady bird beetles (Coccinellidae), lacewings (Chrysopidae), harvester butterflies (Lycaenidae-Gerydinae) and syrphid flies (Syrphidae) and these together keep aphid population in check. A pyralid predator, *Conobathra aphidivora* has been reported as one of the most effective predators on *C. lanigera* (Tripathi,

1992). Malathion, methyl demeton and dimethoate are effective for the control of *C. lanigera*. Recently, brown lacewing *Micromus* sp. and syrphid flies have become abundant in endemic pockets of the pest in Kolhapur district of Maharashtra (Puri and Mote, 2003).

Spiralling white fly: The spiraling whitefly, *Aleurodicus disperses* Russell originates from Central America and has spread widely through the pacific and Southeast Asian regions since its detection in Hawaii in 1997. The pest was introduced in India from Sri Lanka during September-October, 1996 at Coimbatore (Tamil Nadu) on certain agricultural crops (Douressamy *et al.,* 1997). It is a pest of tropical and subtropical crops, and it appears as a highly polyphagous species. It has very wide host range as it attacks 27 families of plant kingdom including 38 genera and more than 100 species (Douressamy *et al.*, 1997). Srinivasa (2000) reported 481 plant species belonging to 295 genera and 90 families as host plants of this pest. Its wide host range includes many vegetable, ornamental and fruit crops, as well as numerous trees and shrubs. Among its host plants, the following crops can be mentioned: *Capsicum, Citrus, Cocos nucifera (*coconut*), Euphorbia pulcherrima* (poinsettia), *Glycine max* (soybean), *Lycopersicon esculentum* (tomato), *Mangifera indica* (mango), *Musa* spp. (banana), *Persea americana* (avocado), *Prunus* spp., *Psidium guajava* (guava), *Solanum melongena* (aubergine), etc. It is a serious pest of guava orchards in Maharashtra since January, 1977 (Sathes, 1999). The pest is spreading very rapidly in Maharashtra on several agricultural corps like banana, brinjal, mango, papaya, cotton, citrus, pomegranate, chilli, coconut, fig, sapota, groundnut, rose, cassia, guava, mulberry , cellophylleum, leucinia, jasmine, hibiscus, china rose, gladiolous, mechelia, kardale, etc. (Sathe, 1999). Damage to the crop is caused by both larvae and adult by sucking the cell sap, excreting honey dew like substance and further creating sooty mould on leaves, as a result photosynthetic activities, growth of plant and fruit setting is adversely affected. Infested leaves become yellow, dry and drop down finally. Plants are also disfigured and may be unfit for marketing. The pest is more dangerous because it remains active throughout the year. The pest population can be checked by collection and destruction of infested plant parts along with pest stages. Spraying the crop with dichlorvos 0.15 % or Malathion 0.1 % or diflufenthiuron 0.012 % controls the pest effectively. Triazophas 400 g a.i. /ha was found to be most effective on guava spiraling whitefly both on eggs and nymphs (Bhaskaran *et al*., 2003). Mani *et al.* (2004) recorded seven species of natural enemies on spiralling whitefly under Bangalore Rural district of Karnataka State. The population of this whitefly declined from 116.85 per 25 cm^2 leaf area in March 2000 to 1.10 per 25 cm^2 area in December 2001. The aphelinid parasitoid *Encarsia guadeloupe* Viggiani was the only major natural enemy encountered on the spiralling whitefly causing 20.70 % parasitism in January 2000 which had increased to 95.68 % by December 2001.

Serpentine leaf miner: Dr. K. A. Spencer, a world authority on agromyzids flies in his lectures during 1982-83 at the Indian Institute of Horticultural Research and the University of Agricultural Sciences, Bangalore on agromyzids as crop pests forewarned the Scientific community on the possibility of *Liriomyza trifolli* (Burgess) being accidentally introduced to India. True to his words between 1991-92 vegetable growers in South India noticed strange serpent like leaf miner on the leaves of many cucurbits, legumes, tomato, ornamental plants etc. Initially farmers believed it to be a curse from the snake god. But later on it was identified as the infestation of American serpentine leaf miner which started its journey from its native place Florida in the United States of America, during early 1970's. It is suspected that this was introduced to California along with chrysanthemum cuttings in the 1970's. Its first report on the incidence of *L. trifolli* was also reported from Kenya around the same period. This leaf miner later invaded Europe and was prevalent in all the European Countries by 1980. It took about ten more years to reach the Indian subcontinent. Now it has become almost cosmopolitan. This polyphagous leaf miner has increased its host range to more than 400 plants. It is probable that the leaf miner has entered India along with plant material during 1990-91 (Viraktamath *et al.*, 1993) and spread in severe form in the States of Karnataka, Andhra Pradesh, Maharashtra, Gujarat and Delhi. Srinivasan *et al.* (1995) recorded 78 host plants covering fibre crops, pulses, vegetables, ornamentals, green manuring, fodder crops, narcotins and weeds belonging to 16 families. The losses to the extent of 15-70 % in French bean, 41% in cucumber and 35 % in tomato have been reported by Krishna Kumar (1998) in Karnataka. Singh *et al.* (2003) reported number of host plants that included vegetable crops, ornamental plants, medicinal plants, weed plants etc. from Varanasi, Uttar Pradesh. Early sowing of pumpkin, sponge gourd and bottle gourd and early transplanting of tomato crops resulted reduced infestation of serpentine leaf miner (Singh *et al.,* 2004, 2004 a, 2005, 2005 a). Among the 10 tomato germplasms screened the tomato variety Flora Dade was found least infested by this leaf miner (Singh *et al.,* 2005). The pest can be managed by using yellow sticky traps in the field for monitoring the presence and mass trapping of adult flies and using the natural enemies. The chemicals abamectin has been reported to give excellent control of the larvae (Grafius and Hayden, 1988). Cyromazine, an insect growth regulator has also been found effective against larvae (Trumbles, 1985). In Hawaii, Guyana and Senegal this pest was controlled by inundative release of parasitoids viz., *Hemiptarsenus*, *Crysonotomyia* and *Chrysocharis* sp. Serpentine leaf miner is capable of developing resistance to a wide variety of insecticides in a short span of time. Insecticides resistance has been responsible for the failure to control *L. trifolli* for many years (Parrella and Keil, 1984). Insecticide resistance capability of *L. trifolli* enabled it to gradually replace other species of *Liriomyza* (Schuster and Everett, 1983).

Coconut mite: The eriophyid mite, *Aceria guerreronis* Keifer was first described in 1965 from the coconut plantations of Guerrero State, Maxico (Keifer, 1965). During the last 35 years it has spread in more than 25 countries. In the Indian subcontinent, it was first reported from Tamil Nadu in 1994. Since 1998, it has been causing very severe damage particularly in the peninsular State including Konkan region of Maharashtra and it has drawn national attention as a serious threat to the coconut plantation (Sathiamma *et al.,* 1998). The coconut mite behaves like terrorists. They live in bunkers made up of coconut bracts (polthariya); attack the tender and button nuts, causing considerable damage. The mite infestation on the nut surface under cover of the 6 bracts leads to premature nut fall and also renders the developed nuts inferior in quality reducing the size and copra yield due to sucking of cell sap. The nuts show typical triangular brown, warty skin after maturity, deformed and a 'T' shaped split may appear on the surface. Yield losses vary from place to place. Varadarajan (2000) reported 40 % losses in yield of coconut in Tamil Nadu. Muralidharan (2001) reported 30.94 % and 41.74 % losses in terms of copra and husk, respectively in Kerala. There are many scientific groups working on various aspects of this mite problem, but with little success on developing an effective control method. The mite population was lowered by 40-80 % on the nuts of palms that received boron nutrition either as root feeding with 2 % borax solution @ 250 ml/palm or as soil application of borax @ 400-600 g/palm (David *et al.*, 2000). Fenpyroximate 5 EC @ 0.75 to 1.0 ml/1 of water for spraying and 10 ml/palm for root feeding was found to provide effective control of the pest (Dey and Somchaudhary, 2001). The pesticides, wettable sulphur 0.4 %, dicofol 0.1 % and triazophas 0.02 % were also effective against this pest. As they live in bract bunkers, search and kill strategy are difficult to carry out. The safest and reasonably effective pesticide is a concoction of Neem oil, garlic and soap when applied on nuts; the pest population is reduced by about 60 %. But repeated application of pesticide is necessary to keep the infestation at levels economically non- damaging. Further, repeated application of pesticide is necessary to contain the pest whichever the chemical is used. This is of course highly cumbersome, possibly impractical and some times even uneconomic. A better solution has to be found (Fernando *et al.,* 2000). There are, however, several other alternative methods like biological, cultural and physical or a combination of two or more as an integrated pest management strategy. Injection of systemic pesticides such as monocrotophos are effective and convenient to use but can cause serious health hazards as they can rich the kernel of nuts. However, the impact of such effort is also reported to be short lived (Fernando *et al.,* 2000). The might of this 'terrorists' (mite) can be fought out with the help of trained commandos (predator mite) that can reach the bunkers (under coconut bracts). Fernando *et al.* (2000) recorded *Neoseiulus paspalivorus* and *Bdella* spp. to exert significant reduction in population of *A. guerreronis*. A pathogenic fungus (*Hirsutella thompsonii*) is also known to infect the coconut mite and in future it can be used as an effective tool of mite management.

Giant African snail: Giant African snail, *Achatina fulica* Bowditch is one of the world's largest and most damaging land snail pests. Man has been responsible for the distribution of *A. fulica* all over the world. Its native place is East Africa. It was first introduced to Mauritius before 1800. In 1847 this species was deliberately introduced from Mauritius to India and released in Kolakata. Now it is distributed most of the Indo-Pacific areas including India, Sri Lanka, Malaysia, China, Taiwan, Japan, Philippines, Hawaii, Samoa, Tahiti, New Guinea and Indonesia (Anonymous, 2000). It is also reported from Singapore, Thailand and Vietnam and throughout Africa. It was unsuccessfully attempted to be introduced in Japan during 1925. From there it was transported to Guam and near by Island and also to Okinawa. It reached New Guinea by 1946 and New Ireland and New Britain by 1949. During Second World War, it was deliberately introduced by Japanese in Andaman Island and now it is a serious menace to the local inhabitants. In India, it was not known as a serious pest till 1948, when it appeared in an epidemic form in Orissa. By 1955 it had spread to Assam, Bihar, Madhya Pradesh, and Orissa and later to Kerala and Tamil Nadu. Now it is well established as a serious pest of ornamental nurseries around Bangalore in Karnataka. It was noticed first in Aurangabad district during 1973 and in Konark region during 1989. Its sporadic outbreak was noticed in about 300 ha in Satpur during 1998 and in 150 ha area of Niphad in Nasik district of Maharashtra and caused severe damage to the crops like grapes, bitter gourd, beans, marigold, bottle gourd, groundnut, tomato, sugarcane, garlic, lucerne, chilli, okra, cauliflower. Some other recorded hosts are potato, sweet potato, carrot, French bean, cabbage, brinjal, lettuce, mulberry, banana, paddy, zinnia, money plant, areca nut plants, rubber buds, and coffee seedlings. Management of Giant African snail is a difficult task. However, there are various methods such as cultural, mechanical, chemical etc. that can be used for its proper management. Sanitation in crop fields and multiple cropping help reduce the snail population. Hand collection and destruction is quite effective as these snails undergo diapauses in congregation. The hiding places of snails must be located and the bushy vegetation sheltering these must be burnt. In rainy season, moist gunny bags or plant leaves can be heaped near the field border and the snails under these can be destroyed. 'Veg Fru Snail Kill' (Acetaldehyde Tetramer) and lime tobacco dust give significant control of snail. Metaldehyde 5 % pellets give satisfactory control of snail while admixture of Metaldehydae 5 % pellets and marygold extract increases the efficacy.

MINOR PESTS BECOMING MAJOR

American cotton bollworm (*Helicoverpa armigera* Hubner): *Helicoverpa armigera* is the most serious pest of cultivated crops. It is polyphagous in nature and is reported feeding on 181 cultivated and uncultivated plants species belonging to 45 families comprising 40 dicots and 5 monocots. It inflicts severe damage to several important crops like cotton, chickpea, pigeonpea, tobacco, sunflower, tomato, groundnut, etc. Interestingly, till sixties *Helicoverpa* was not a serious economic pest on any of the agricultural crops in India except chickpea and was, therefore, named as 'gram pod borer.' It is in the recent past that *Helicoverpa* has emerged as a dominant pest of several crops in India and of late it has attained not only the status of a National Pest but also regarded as International Pest No.1. It is a cause of serious concern in crop production, especially for cotton and pulse crops (pigeonpea and chickpea) throughout the country. This noxious pest has caused cotton crop failures during 1987-88 in Andhra Pradesh and during 1997-98 in Andhra Pradesh, Maharashtra, Punjab, Haryana and Rajasthan causing colossal crop losses which compelled many forlorn and frustrated farmers to commit suicides. The recurrence of this pest has become quite frequent on cotton, pigeonpea and chickpea crops especially in northern and central parts of India. Nath and Rai (1998) reported 21 host plants of *H. armigera* belonging to plant families Fabaceae (Leguminosae), Solanaceae, Poaiceae, (Gramineae), Cucurbitaceae, Malvaceae, Linaceae, Asteraceae (Compositae), Liliaceae, Chenopodiaceae and Convolvulaceae from Varanasi and Mirzapur districts of East Uttar Pradesh. The important crops that suffered from pod borer are chickpea, tomato, cowpea, pigeonpea, linseed, berseem and lady's finger. Rath and Nath (2001) reported to the extent of 37.50 % fruit damage in various tomato genotypes in Varanasi region. The farmers and agricultural scientists had initiated wide spread campaign against this pest causing so much damage to so many crops including commercial crops like cotton. To control the pest in a hurry the easiest tool of an array of pesticides had been applied in frenzy by the farmers. Unfortunately, serious problem of insecticidal resistance in *Helicoverpa* has arisen as an obvious result of indiscriminate use of this tool of insect management. Due to manifold resistance that has developed in various eco types of the pest, chemical pesticides are proving ineffective, often prompting farmers to use overdoses and repeated applications. Indiscriminate use of insecticides gave rise to many other problems a part from the problem of insecticidal resistance. Most part of the insecticides goes to the environment and pollutes the essential components of the ecosystem, thereby destabilizing the ecological balance of the surrounding areas of application as well as of the country. High costs of pesticides and appliances, large quantities of water required for spraying in water starved areas, side effects and health hazards due to pesticides have become the major limitations of pesticides used by the subsistence farmers, keep aside the loss of economic return from pesticide application due to the problem of resistance. Thus, farmers are embroiled in this vicious circle of using more and more pesticides with less and less economic returns. The IPM strategy composed of population monitoring, application of ecofriendly insecticides, botanical (NSKE), microbial (*B.t.* and NPV based formulations) and relatively safer chemicals) at proper doses and time is proving effective in the management of this pest.

Whitefly on cotton: Introduction of *hirsutum* cotton and later cotton hybrids replacing *Gossypium herbaceum* L. and *G. arboreum* L. accentuated the damage caused by insect pests. Cotton is assailed by multitude of pests, as it evolves through various production levels. Hargreaves (1948) recorded 1326 species of insect pests in the world. Fletcher (1920) listed 34 species of insect pest that swelled to 130 by 1964 on cotton in India (Sohi, 1964). Whitefly, *Bemisia tabaci* (Gennadius) did not find place in the list of Fletcher (1920). However, by 1964 it was recorded as minor pest of regional importance. Recently, 162 species have been reported on this crop in India. Out of these, 15 species including whitefly (*Bamisia tabaaci* Genn.) are considered as key pests in different parts of the country (Dhaliwal and Arora, 1998). Among the few sucking pests, whiteflies attained the status of pest of great concern after the introduction of varieties and hybrids with intra- and inter-specific crosses. Indiscriminate use of synthetic pyrethroids insecticides has further aggravated the problem and resulted in severe outbreaks of whitefly during the last decade. Some other factors enhancing the whitefly problem are early season spraying with broad spectrum pesticides resulting mortality of natural enemies, excessive application of nitrogenous fertilizers, close spacing, growing of highly susceptible hybrids like MECH 1, MECH 2 and DCH 32, large scale monocropping of cotton as well as other host plants like chillies, okra, and tomato round the year in the same area. Whitefly outbreak on cotton was observed in Andhra Pradesh during 1986. Thereafter, its incidence was reported from other cotton growing belts and now, it is an important pest in almost all the cotton growing areas of the country. The losses due to whitefly were estimated to be 15-20 % in Punjab (Sukhija *et al.*, 1986) and 10-15 % in Andhra Pradesh (Reddy, 1987). This is a polyphagous insect pest causing crop losses both by direct feeding and also indirectly by transmitting a number of viruses and mycoplasma diseases. The viruses transmitted by this species of whitefly are Cassava Mosaic, Cotton Leaf-curl, Tobacco Leaf-curl, and Sweet Potato Virus B (Hill, 1993). In recent years, severe incidence of cotton leaf curl virus disease (CLCV) transmitted and spread by whitefly has evoked widespread concern in Punjab, Haryana and Rajasthan. Similarly, the detection of Biotype-B of this pest in Karnataka on tomato causing severe incidence of leaf curl has sent a scare among tomato growers. However, efforts are being made to evolve suitable management strategy for B-Biotype of whitefly as a pest and vector (Puri and Mote, 2003). Earlier the problem of whitefly was not so serious but due to the introduction of pyrethroids in the late 1970's and its repeated application which protected the crop from key pests and increased the cotton produce significantly, also added new dimension to the whitefly problem. The resurgence of whitefly on cotton in Guntur and Prakasham districts of A.P. was not only witnessed but it also developed resistance to almost all synthetic insecticides. The resistance development in whitefly, *Spodoptera litura* and *Helicoverpa armigera* against almost all insecticides, which were mainly responsible for heavy yield loss, forced many farmers to commit suicide (Nageshkumar, 1996 and Gupta, 2002). Keeping these facts in view now there is clear recommendation from various cotton growing areas in the country that synthetic pyrethroids should be alternated with conventional insecticides and not more than two rounds of its application may be given in 5 spray schedule (Gupta, 2002). The insecticides such as phosphamidon 85 WSC @ 200 ml or formothion 25 EC, or Oxydemeton methyl 25 EC or dimethoate 30 EC @ 625 ml or imidacloprid 2000 SL @ 100 ml in 250 liter of water per ha can be used at its economic threshold level i.e. 20 nymphs or 8-10 adults/leaf or leaf stickiness due to deposition of honeydew by insects in 50 per cent of plants (Atwal and Dhaliwal, 1997). The spray should be directed at the undersides of the leaves as far as possible. The insecticides of different chemical groups should be used alternatively to delay the process of insecticide resistance development, for better whitefly population management and to have better crop yield and prolonged self life of insecticide that will ensure its application for a longer period of time. The whitefly is attacked by parasitoids and predators under field condition. The aphelinid parasitoid (*Eretmocerus massii* Silv.) (Aphelinidae) is associated with nymphs and pupae and incidences of parasitism reaches up to 33 % while the insect predators, antlion (*Chrysoperla* sp.) (Chrysopidae) and a lady bird beetle (*Brumus* sp.) (Coccinellidae) prey upon nymphs and adults. The problem of white fly needs sustainable management strategy for obtaining better yield and to satisfy the outcry of increasing environmentally conscious society. This can be achieved only by the integration of chemical pesticides with ecofriendly approaches that includes application of cultural and biorational (biological, botanicals etc.) inputs at proper doses, time and combinations.

Mealybugs on fruit crops: The species of pseudococcids are called mealybugs as because the females are usually elongate oval and generally covered with a mealy or filamentous waxy secretion often extended into lateral and terminal filaments; legs are well developed. Some species are subterranean while others make galls. The males are small two winged insects with long filamentous antennae and wings with reduced venation. They are often associated with ants, particularly the root inhabiling species, for the honeydew that excretes. Some species transmit viruses. The *Hibiscus* mealybugs, *Maconellicoccus hirsutus* (Green) is a polyphagous pest infesting more than 125 plant species. It is a pest of plants in the family Malvaceae, notably *Hibiscus* and *Gossypium* spp. mainly in southern and eastern states of India including S.E. Asia, up to S. China and Taiwan. Its alternative hosts are legume plants including tropical and subtropical fruits and shade trees. This is the only species of mealy bug which produces toxic saliva, whose feeding results in the stunting and sometimes death of the young infested shoots. Infestations are often very heavy and characteristic in appearance due to the shortened internodes of apical shoots, which often swell and develop a dark green colour. The leaves on these shortened shoots give the plant a 'bushy top' appearance. The adults are small (about 3 mm in lengths) and pink in body colour, but covered with a waxy secretion; the waxy filaments are very short and in general the insect resembles *Planococcus* in appearance. In India, it is a serious pest of grapevine in Andhra Pradesh, Maharashtra, Karnataka and Tamil Nadu. The damage to the extent of 80 % in grapevine has been reported in Karnataka due to *M. hirsutus* (Mani and Thontadarya, 1978). Similarly, it also occurs on custard apple, pomegranate and guava causing colossal losses annually. There is another species of pineapple mealybug (*Dysmicoccus brevipes* (Ckll.)) which is pantropical in distribution, with a few records from subtropical areas. It is common in south Indian states also attacking sugarcane, groundnut, coconut, coffee and *Pandanus*. This is a vector of the mealybug wilt virus whose first symptom usually appears in the root which ceases to grow, collapse and then rot. A flourishing crop when infested suddenly shows the symptom of slow growing poor crop. This is known as 'Quick wilt'. The mealybugs live in colonies under ground with only a small proportion living on the leaves. The occurrence of mealybug wilt is largely correlated with the subterranean colony on the roots. The aerial individuals are to be found mostly at the base of the leaves, which may have to be spread in order to make the bug's evident (Hill, 1993).

The striped mealybug (*Ferrisia virgata* (Ckll.)) is a polyphagous pest which is a vector of swollen shoot disease of cocoa. It is pantropical in distribution, but with only a few reports from Australia and South America. It is widely distributed in India and reported from almost all the states. It is an important pest of coffee. Its alternative hosts are citrus, cotton, jute, groundnut, beans, cassava, sugarcane, sweet potato, cashew, guava, tomato, and many other plants. This is a pest of new shoots, berries, and leaves and at times it attacks in very large numbers. In dry weather it may move down below ground and inhabit the roots. This mealy bug is favoured by dry weather and heavy attacks following periods of prolonged drought.

The ranges of approaches are available to manage the population of mealybugs. As the eggs are laid by wingless females in the soil during the month of April-May where it spends rest of the year in egg stage. The dead body of the female is often found sticking to the egg mass. The eggs hatch at the end of December or in January and, in a given locality, continue to hatch for about one month. Thus, nymphs appear before the fresh growth of flowers on fruit crops, (plants/trees). This is the period when nymphs ascend the trees immediately. Eventually, they congregate on the growing shoots, leaves, flowers, fruits, etc. in case of foliage feeders while incase of root feeders they congregate and feed on the roots of the plants and under severe conditions a small proportion may live on the leaves in case of certain species such as pineapple mealybug. The management strategy composed of cultural, mechanical, biological and chemical methods may be helpful in reducing the mealy bug population development. Removal of weeds and ploughing of orchards/crop fields during summer may expose the hiding stages of mealy bugs to natural enemies and extreme sun heat. The newly hatched nymphs may be prevented from crawling up the trees by applying tree bands. Srivastava and Butani (1971) found banding the tree trunks with 30 cm wide alkathene sheet (400 gauges) to be effective. The band is fixed about 30 cm above the ground level during second week of December. Before tying the band, paste a layer (40-50 cm wide) of mud or wet soil around the *tree* trunk, to make that portion smooth. After drying of mud plaster wrap the alkathene sheet around the tree on this mud plaster and tie the sheet on both sides with jute or sunhemp thread. Clean the alkathene sheet occasionally with a piece of moist cloth, to keep it smooth and slippary. The slippary band because of being smooth and slippary do not permit the mealy bug to crawl over the tree and ultimately fall down and die due to starvation. While 30 cm vide sticky band tied on the tree trunk adhere mealybug while crawling over the band to climb the tree. The sticky band may be made up of loose cotton which fibers are free that entangle the mealybugs when come in contact. Another type of sticky band may be prepared locally by mixing resin and castor oil in the ratio of 5:4 and warming for 15 minutes to mix up thoroughly. This is pasted on 30 cm wide cloth which is tied on the mud plastered tree trunk to prevent crawling over the trees. The congregated mealy bug may be cleaned and killed by spraying dimethoate, diazinon, ethion, malathion, carbaryl from time to time. Some of the highly infested trees, plants may be left without spraying insecticide for the population build up of natural enemies such as *Phygadeuon* sp. (Ichneumonidae) and *Getonides perspicax* Knal which are nymphal parasitoids while larvae of *Brinckochrysa scelestes* (Banks) (Chrysopidae) and grubs of *Rodolia fumida* (Mulsant) (Coccinellidae) are predators of mealybug (Atwal and Dhaliwal, 1997). As usual with mealybugs it is important to make sure that the insecticide reaches the body of the insect, so it is necessary to add extra wetter to the spray solution. Ants are usually associated with the mealy bugs and they feed on the honey dew excreted by the bug. The ants in return give protection to the mealybug from its natural enemies. Therefore, to enhance the activity of natural enemies, ants may be prevented by banding the trees with dieldrin-soaked cloth which will repel the attendant ants. This band may be tied at a lower height i.e. 5-10 cm below the slippery or sticky bands.

Armyworms: Armyworms are the larvae of noctuids, mostly *Spodoptera* spp., that under favourable conditions of high population density behave gregariously and its swarms 'march' off from field to field devastatingly defoliating entire crops. Most species are migratory as adults and are famous for their spectacular population fluctuations which can lead to a severe out break year resulting in catastrophic damage. The female lays eggs in batches close together and covered over by brown hair. The larvae feed gregariously for the first few days and then disperse to feed individually. In a severe year the high density of caterpillars results in a rapid defoliation of food plants. Under such conditions the caterpillars may 'march' off over the ground like an 'army' to fresh feeding locations. Because of huge populations the ground may be literally covered with the gregarious marching band of caterpillars, in many respects similar to a locust swarm.

The plants attacked are mostly Gramineae (cereals and grasses), but the genus *Spodoptera* (not all of which act as armyworms) is recorded feeding on plants from 40 different families, containing at least 87 species of economic importance (Brown & Dewhurst, 1975).

The term 'armyworm' is essentially behavioral and physiological and is not strictly taxonomic. Some species of *Spodoptera* do not act as armyworms and species belonging to other genera of Noctuidae show swarming tendencies and may be classified as armyworms, including *Achaea* spp., *Mythimna separata* (Walker) etc. Nath (1986) enumerated the causes of outbreak of paddy armyworm (*M. separate*) and forecasting its epidemic in Varanasi region of Uttar Pradesh, where it is appearing in serious proportion on dwarf paddy varieties causing enormous losses by cutting rice ear heads. It is also appearing on maize and wheat crop (Nath *et al.*, 1991 and Singh and Nath, 1998). Its spatial distribution behaviour in rice field and insecticidal management in maize field is reported by Nath and Singh (1987) and Nath *et al.* (1991), respectively. Among the 13 insecticides screened, the insecticide chlorpyriphos was the best followed by monocrotophos.

Some species of armyworms are very versatile behaviorally and may behave quite differently under different circumstances. For example, *S. litura / littoralis* in small numbers feeds on the leaves and foliage of many plants, but will sometimes act as a cutworm and at other times will swarm gregariously and act as a typical armyworm. This shows the behavioral diversity of these two almost identical species (formerly regarded as one species) (Hill, 1993).

Spodoptera spp. is a very important polyphagous insect pest on several crops. *Spodoptera litura* has assumed serious magnitude in recent years on cotton, soybean, castor, groundnut, tobacco etc. Management of armyworms is a difficult task. Handpicking and mechanical destruction of egg clusters and caterpillars during early stage of attack can reduce infestation. Fore warnings based on light trap or pheromone trap catches, or prognostications based on prevailing meteorological conditions in certain areas, followed by spraying of effective insecticides malathion 50 EC or endosulfan 35 EC (1litre) in 625 liters of water per ha reduce the pest populations. Spraying at 10 day intervals may contain the pest population. Singh and Nath (1998) have reported insecticidal management of *S. litura* infesting groundnut crop. While Singh *et al.* (1998) have found antifeedant effect of some plant extracts against *S. litara* on groundnut. However, chemical control measures have proved to be ineffective and uneconomic. Nuclear Polyhedrosis Virus (NPV) and other IPM measures are being adopted. During 2000-2001 *kharif,* the pest appeared in epidemic proportions on soybean in some districts of Rajasthan. In the areas adjoining river belts and forests, *Spodoptera mauritia* has caused crop devastation frequently in States like Assam, West Bengal, Madhya Pradesh, Chhatisgarh and Bihar (Puri and Mote, 2003).

Fruit flies: These are commonly called "fruit flies" due to their close association with fruit and vegetables. These flies are wide spread over the entire world, except the Arctic and Antarctic regions and richly predominant in the tropical and subtropical areas. This is also called "peacock flies" due to their habit of strutting about, vibrating their wings and displaying their elaborate wing and body markings. Fruit flies belonging to the family Tephritidae are one of the important pests that play havoc with horticultural industries. The adults are flower loving but their larvae are herbivorous and cause severe damage to fruit and vegetable crops. The larval hosts include a broad array of fruit, vegetables, flower heads and seeds. Some also make mines and a few form galls on different parts of the plants. In this way, they cause tremendous economic losses and great problems to fruit and vegetable growers all over the world. The species of fruit flies distributed in warmer parts of the world (subtropical and tropical countries) are polyphagous in nature and are expanding their host range while in colder parts (temperate countries) they are comparatively less mobile hence mostly monophagous in nature. Kpoor and Agrawal (1980) and Kapoor (1993) reported 159 host plants of various species of fruit flies (*Bactrocera* spp.). Recently, it has been reported that, in India, fruit flies cause losses up to Rs. 2,945 crore per annum in mango, citrus, guava and sapota in addition to 35 to 40 % losses in gourds (Anonymous, 2003). The fauna is especially diverse in hot tropical and subtropical zones and in forest areas. Oriental region is rich in this family with nearly 1,000 species so for recorded (Kapoor, 1993). Presently, 199 species of fruit flies are known from India but the host range is known only for few species and even that is in complete. Out of 199 species, only nine are known to cause economic losses to our fruits, vegetables, ornamental and oilseed plants (Kapoor,

1993). Among these, *Bactrocera dorsalis, B. cucurbitae, B. diversa, B. correcta, B. zonata* and *Dacus ciliatus* were recorded on different cultivated fruit and vegetable crops (Narayanan and Batra, 1960). If uncontrolled, these pests are capable of causing substantial losses, which can not be tolerated. They also pose major threat to global trade, since many countries have invoked restrictions to minimize the risk of establishment of exotic species. Therefore, successive cultivation and export are highly dependent on sound pest management system. The insect control methods available for fruit fly population management falls into three different categories such as control, suppression and eradication (Hill, 1993). The fruit fly control includes early harvesting of fruits, collection and destruction of all infested fruits to prevent the carry over of the pest. This will stop further multiplication. The maggots being internal feeder is inaccessible but may be killed by application of fenthion. Frequent raking of the soil under the crop plants or ploughing the infested field after the crop is harvested may help in killing the pupae after its exposure to natural enemies and solar radiation. The fruitfly spot bait traps have good promise in trapping the males and male and females both in two different traps having methyl eugenol and banana as lures, respectively (Nath and Bhushan, 2006). This may help in the suppression of fruitfly population. While applying bait spray containing 50 ml of malathion 50 EC + 0.5 kg of *gur*/ sugar in 50 liters of water per ha is helpful in reducing the infestation. Diping of fruits in 5 per cent Sodium chloride solution for 60 minutes kill the eggs, if any and also decontaminate the fruits from insecticide residue if at all present (Atwal and Dhaliwal, 1997). The parasitoids associated with *B. dorsalis* are *Opius persulcatus* Silvestri, *Biosleres arisanus* (Sonan), and *O. manii* (Braconidae); *S. afra, S. stomyoxysine* Gir. and *S. grotiuse Gir.* (Pteromalidae); and *Trybliographa daci* Weld (Eucoilidae). While the pupal of *B. cucurbital* (Coquillett) are parasitized by *Opius fletcheri* Silvestri, and (Braconidae), Mill. and *D. zonensis* Rohw. (Chalcidae).

The common parasitoids of both the species are *Opicus compensatus* Silvestri, *O. insisur* Silvestri, *Spalangia philippinensis* Mill. *Dirhinus giffardi* Silvestri. and *Pachycrepoideus debrius* Ashm. Conservation of these natural enemies may help in reducing the infestation of fruitfly.

Mites on vegetable and fruit crops: Plant mites constitute one of the important groups of crop pests causing direct as well as indirect damage to various crops. Many of the plant feeding mites recently attained pest status causing considerable loss to cultivated crop plants, particularly vegetables and fruit crops. Phytophagous mites feed on the plants, causing direct injury and loss by increasing leaf and fruit drop, decreasing fruit size, reduction in plant vigour and spoil the external appearance of vegetables and fruits. The affected vegetables and fruits have reduced market value resulting severe economic losses. Singh (1995) have listed 64 species of mites that occur on vegetable crops in India. Out of this list of plant inhabiting mites 44 species belonging to the families – Tetranychidae, Tuckerellidae, Tenuipalpidae, Eriophyidae, Tarsonemidae and Acaridae are phytophagous and remaining 20 species belonging to the family Phytoseiidae are predatory mites. The predatory mites are beneficial to the farmers as because they help in the natural biological control of phytophagous mites.

During recent years the plant feeding mites have been recognized as serious threat to the horticulturists (vegetable, fruit and flower farmers) as because these mite problem is one of the most difficult factor that limit the productivity of vegetable and fruit crops in India. The crop losses by the mite complex may be to the extent of 36.8 to 83.2 % in vegetables (Singh, 1995). They cause direct potential damage by sucking the cell sap and also indirectly by transmitting the plant diseases.

The species of mites associated with different vegetables are many, with varying intensity. *Tetranychus cinnabarinus* (Boisd.) and *T. neocaledonicus* Andre (Cucurbitae) are common mite species infesting the major vegetable crop. Red spider mite, *T. cinnabarinus* is the most common and destructive species of mite which has worldwide distribution and is highly polyphagous having a very wide range of host plants. Though, various cucurbits are its main hosts, it is also found on a number of other crops including beans, brinjal, cole crops, onion, peas, potato, sweet potato, tamarind, tomato and various ornamental plants (Butani and Jotwani, 1984). Some other species of mites causing wide spread attack on vegetable crops are *T. ludeni* Zecher on cowpea and sponge gourd; *Brevipalpus phoenicis* (Geij) on bottle gourd, bitter gourd, sponge gourd, ridge gourd, brinjal, French bean, hyacinth bean, okra, tomato, Bengal quince, apple, guava and papaya; *Aculops lycopersici* (Wolft) on brinjal and tomato and *Polyphagotarsonemus latus* (Banks) on chilli, brinjal, tomato and sponge gourd. The species of mites infesting fruit crops are *Aceria mangiferae* Sayed on mango; *A. litchi* (Keifer) on litchi; *A. granati* (Canestrini & Masalonga) on pomegranate; *Acaphylla syzygii* Channa Basavanna on jamun; and *Brevipalpus phoenicis* (Geij) on guava and papaya and *Eriophyes cernuos* Massee on *ziziphus* (Singh, 1995). The maximum activity of the pest has been observed during post monsoon period and the activity declines with the drop of temperature. The management of plant feeding mites requires close visual monitoring of its population development on regular basis as it can not be monitored through trapping devices because of being wingless. The foliage of horticultural crops (host plants of mite) should be examined regularly with a good hand lens starting from the month of February as because the mite infestation generally occurs during summer season. The road side crop fields are more prone to mite infestation (mostly spider mites) usually because of dust deposits on the foliage which provides medium for attachment of mite webs on smooth leaf surface. To avoid / reduce this mite menace non host crop should be grown. Sprinkling irrigation may further reduce the dust deposits and mite infestation by washing both of them. The acaricides such as lime sulphur, chlorobenzilate, dicofol, dimethoate, demeton –S- methyl, derris, dichlorvos, dicofol, formothion, malathion, oxydemeton-methyl, petroleum oil and phosphomidan are reported to be effective against mite pests. The foliar spray of acaricide early in the season according to the crop concerned and the conditions at the time of spraying as a full cover spray or spot spray on affected parts using as high a nozzle pressure as possible is helpfull in reducing the mite infestation. In case of severe infestations the spray should be repeated 2-3 weeks later. However, there is a need to protect the predatory mite that includes 20 species (Singh, 1995). These mites may be wrongly identified as pest species and sprayed with acaricides. Therefore, all precautions should be taken to protect the predatory mites belonging to the mite family Phytoseiidae as these are potential predator of phytophagous mite.

Nematode: The phytoparasite nematodes were unnoticed due to their small size till the beginnings of 20th century. Phytoparasitic nematodes are elongate worms generally 0.3 mm to 5.0 mm in length. Phytoparasitic nematodes are obligate parasites and feed only on the living plants. Out of the total nematode population on this earth, only 10 % are plant parasitic nematodes and cause damage to almost all the agriculturally important crops – cereals, legumes, oilseeds, ornamentals, fruits, vegetables, spices, and condiments. The average yield reduction in yield caused due to nematode pests in different crops was estimated to be 8-11 %. The yield loss, however, may be as high as 60 % in continuously cultivated crops of vegetables due to root-knot nematodes (Nagnathan, 1984).

There are approximately 20,000 named nematode species, and by some estimates, there may be as many as 1 million total nematodes species (Norris *et al.*, 2002). The more important plant parasitic nematodes along with their common name and host plants are given in Table 3.

Table 3 Important plant parasitic nematodes and their host crops

Genus	Common name	Important host crops
Meloidogyne	Root-knot nematode	Potato, cotton, field corn (maize), soybean, dry beans, wheat, rice, sorghum, tomato, sugarcane, groundnut, tobacco, alfalfa, apple, citrus, coffee, grape, cucurbits, carrots, sweet potato, onion, sugar beet, cereals.
Pratylenchus	Lesion nematode	Maize, soybean, dry beans, wheat, rice, potato, sorghum, sugarcane, peanut, alfalfa, apple, peach, strawberry, citrus, coffee, cherry.
Heterodera	Cyst nematode	Soybean, dry beans, sugar beet, broccoli, Brussels sprout, cabbage, carrots, wheat, chickpea, pigeonpea.
Ditylenchus	Stem nematode	Rice, potato, alfalfa, sugar beet, oats, flower bulbs, mushrooms.
Globodera	Cyst nematode	Potato, tobacco.
Tylenchulus	Citrus nematode	Grape, citrus.
Xyphinema	Dagger nematode	Sugarcane, grapes, apple, cherry, strawberry.
Radopholus	Burrowing nematode	Banana, citrus, tea.
Rotylenchulus	Reniform nematode	Cotton, soybean, papaya, sweet potato, pigeonpea.
Helicotylenchus	Spiral nematode	Banana, pepper, eggplant, tea.

The major nematode pests that need priority attention are *Meloidogyne incognita, M. javanica, Rotylenchulus reniformis, Heterodera avenae, H. cajani, Tylenchulus semipenetrans, Radopholus similes, Pratylenchulus zeae, M. graminicola* and *Hirschmanniella* spp. However, the root-knot (*Meloidogyne* spp.), cyst (*Heterodera* spp.) and reniform (*Rotylenchulus reniformis*) nematodes are important plant parasitic nematodes widely distributed in India.

Nematodes problems are aggravated by prevailing environmental conditions and paradigm shift in crop husbandry that favours rapid reproduction of certain species.

Infestation appears to be widespread where cultivated lands in the region are concentrated in specific ecological zones. In temperate zones, the cyst nematodes, *Globodera rostochiensis* and *H. avenae* infecting potato, and wheat, respectively, are quite prevalent. In moderately warm and subtropical zones, the root-knot nematodes, *Meloidogyne incognita, M. javanica* and *M. arenaria* commonly occur in many field and horticultural crops. *H. cajani* on cowpeas, and *H. zeae* on maize, are frequently found; and the citrus nematode, *Tylenchulus semipenetrans* is a common pest in citrus; the reniform nematode, *Rotylenchulus reniformis* is widespread in the many vegetable growing areas; and the rice nematode, *Hirschmanniella oryzae* is common in the rice growing belts (Sharma *et al.*, 2002). Although, many of the nematodes identified are capable of infecting direct severe damage to growing plants. The information on the status of nematode constraints to crop production is available from majority of the states in India. There is tremendous amount of information generated on distribution, host record, epidemiology, pathogenicity and management on few nematodes. The plant parasitic nematode management practices (cultural, physical, chemical, biological or integrated) followed have ecological base and implications. The nematode management is in fact an applied nematode ecology. However, the information on realistic / potential yield loss and package of nematode management schedule is still lacking. There is a need of developing ecofriendly approaches and hence due emphasis must be given to the cultural, physical, use of resistance, intensive bio-agents and entomophillic nematodes in integrated nematode management programme and should find a place in national level planning. Genetic manipulation in breeding crop plants resistance to plant nematodes can also play an important and crucial role in developing transgenic plants.

CRITICAL ISSUES IN MANAGEMENT

Sucking pests: Sugarcane wooly aphid, spiralling whitefly, whitefly, mealybugs and mite pests cause direct damage by sucking the cell sap from various parts of the host plants. They also cause indirect damage as they are potential vector of plant diseases. There are some special issues which are required to be looked into while implementing management strategy against these sucking pests.

1. Because of being vector of plant diseases it is difficult to fix economic threshold level.
2. Some of these sucking pests are having thick coating of waxy secretion (woolyaphid, spiralling whitefly, and mealybugs) and dust deposits on the webs (spider mites) which make the chemical pesticides ineffective; hence wax dissolving organic solvents with insecticides is required.
3. Hidden infestation makes it difficult to understand the start of infestation as incase of wooly aphid, mealybug, mite etc.
4. It is difficult to apply pesticides incase of dense plant canopy sugarcane, wooly aphid, tree (coconut mite), initial infestation confined in residential of areas on ornamental plants and fruit trees in, kitchen garden (spiralling whitefly); hence control measure be applied in such a way so that pesticide may hit the target pest at its proper site and time.
5. The management measures are required to be taken by all cultivators on campaign basis as some of the pests like wooly aphid, mite spread by wind currents.

6. Proper management of nitrogenous fertilizers and water as irrigation to avoid vulnerability of the crop and favourable climate to the pest (wooly aphid, whiteflies).
7. Wider plant spacing of pesticides is needed for easy application and better aeration which help in minimizing the humidity in the field. Inter/mix cropping also help in creating differential plant canopy that provide space for aeration, pesticide application and movement of natural enemies.
8. Conservation and augmentation of generalist predators such as lady bird beetles, chrysopa, syrphid flies, spiders etc. are necessary. The specialist predators and parasitoids of sucking pests may be identified and protected by all means so that they can multiply and have control over these sucking pests. If possible these natural enemies should be collected, reared and released in the crop fields on routine basis.
9. Farmers training in respect of identification of insect pests and natural enemies that look similar is essential to protect the natural enemies and to prevent unwanted application of pesticides.
10. Due to wide host range of some sucking pests it is very difficult to control them as because all host plants can not be sprayed.
11. Application of systemic pesticides may be effective in certain cases but not advisable due residue problem. Application of synthetic pyrethroids causes out break of white flies and aphids, hence frequent application is not advisable.
12. Polyphagous sucking pests are difficult to control because of the presence of preferred host crops throughout the years.

Internal feeders / soil pests: The insect pests like coffee berry borer, serpentine leaf miner and fruit flies are internal feeders while nematode infestation is not only internal, it is hidden, subterranean, and microscopic also; hence difficult to understand the start of infestation. These pests require carefull monitoring for assessing pest status and for implementation of management strategy for successful handling of the problem.

1. Being internal feeder it is difficult to understand the start of infestation. Mostly on appearance of external symptom, damage is already done; hence there is a need to develop some effective system of monitoring these stages which are externally visible and active outside the plant body (adult stage). These pests are difficult to control; hence all possible methods are required to apply on community basis.
2. Clean cultivation, collection and destruction of infested fruits and plant parts throughout the year on regular basis help in minimizing the incidence of coffee berry borer and fruitfly.
3. Difficult to apply pesticide on trees due to tallness and thick canopy as in case of fruit fly attacking mango fruits.
4. The information on realistic/potential yield loss and package of nematode management schedule is lacking. There is a need of developing ecofriendly approaches and hence due emphasis must be given to the development of integrated nematode management programme.
5. Nematode may transmit plant viruses or may interact with other pathogens to cause disease complexes that are worse than the disease caused by either pathogen alone.

6. Nematode infection may increase host plant susceptibility to other environmental stresses such as high temperatures or moisture. Nematode infection may nullify the host plant's resistance to other pathogens, *Fusarium* resistance in cotton.
7. Some nematode diseases may be moved from one host to another by insect pests. The pine-wilt nematode is carried from diseased to healthy pine trees by the longhorn beetle, and red-ring of coconut is carried by palm weevils.

Polyphagous, sporadic and cosmopolitan pests: American cotton bollworm and armyworm are polyphagous insect pests which are widely distributed causing economic damage to various crops. These pests are having large number of alternate hosts, hence able to complete number of generations in a year. The management strategy against these pests may be prepared after due consideration of the following points:

1. As these pests are having wide host range hence community approach should be made.
2. These pests have developed resistance against large number of insecticides; hence repeated application of an insecticide may be avoided. If need be insecticide of different chemical groups may be used alternatively to avoid the development of insecticide resistance.
3. These pests are having large number of natural enemies; hence care should be taken to protect these natural enemies. Botanical (NSKE) and microbial pesticides (B.t. and NPV) may be used to avoid the bad impact of chemical pesticides.
4. Augmentation of indigenous species of natural enemies by undertaking mix-cropping with maize, cowpea and sorghum in cotton; green gram, black gram and sorghum in pigeonpea; and mustard in chickpea help in reducing the *Helicoverpa* menace while sparse planting of castor in groundnut reduce the infestation of *Spodoptera.*

Giant African snail: It is a large sized snail which survives in high rainfall area and spread through the moss in egg stage or juvenile stage in nursery garden where moss is used for layering. Hence its spread is generally unnoticed in the beginning. These snails are generally more important in seedling crops than in mature crops. They are much more serious in no-till farming than conventionally tilled systems due to the amount of cover and food that remains on the soil surface in no-till systems. Snails have been shown to be capable of spreading tobacco mosaic virus, the fungus causing black rot of cabbages, and foot rot of *Piper* spp. among other. Clean cultivation and avoidance of crops that suffer from viral diseases help in making the cultivation profitable. Its management requires community basis approach.

SUMMARY

India has witnessed the green revolution during last three decades. This revolution has lead to quantum jump in production and productivity of the important food crops during this period. However, the growing population pressure and diminishing per capita available land demands further rise in productivity. Among the various limiting factors of higher productivity, insect pests and diseases are of prime importance. The evolution of new plant types of food and commercial crops and many other developmental activities have added new dimension to the agricultural production on one side while on the other side the agroecosystem is witnessing lot of change including several new pest problems which are causing heavy loss of crops. Among these new pests, some are indigenous while others are exotic. The exotic pests are coffee berry borer, sapota seed borer, sugarcane woolly aphid, spiralling white fly, serpentine leaf miner, coconut mite, and giant African snail while pests like American cotton bollworm, white fly on cotton, mealy bugs on fruit crops, armyworms, fruit flies, mites on vegetable and fruit crops, and nematodes hitherto considered as minor pests have changed their status to become major pests in some parts of the country. But both categories of pests are threatening to the farmers achieving the target of higher level of production by increasing the productivity of crops. Through concerted efforts, the techniques have been evolved to manage the indigenous/traditional pests but the exotic pests are posing serious threat to agricultural production and they are not responding the management measures properly due to want of natural enemies and other viable technology. Critical issues in the management of exotic as well as minor pests attaining major status are discussed. Special emphasis on pest categories/types, causes of their outbreaks, integrated management systems including cultural, biorational, ecofriendly chemicals, and other methods are discussed in the light of phytosanitary issues in the context of World Trade Organization requirements not only to reduce crop losses but also to enter export market.

References

Andrewartha, H. G. and Birch, L. C. 1954. The Distribution and Abundance of Animals. University of Chicago Press, Chicago, U.S.A.

Anonymous, 2000. DPI Notes Giant African snail. Animal and Plant Health Service, DPJ, Queensland, pp.1-9.

Anonymous, 2003. Integrated management of fruit flies in India. An Indo-UK Project launched. *Indian Horticulture 47:4.*

Arora, R. and Dhaliwal, G.S. 1996. Agroecological changes and insect pest problems in Indian agriculture. *Indian. J. Ecol.*23: 109-122.

Atwal, A. S. and Dhaliwal, G.S. 1997. Agricultural Pests of South Asia and Their Management. Kalyani Publishers, Ludhiana, Punjab, India, pp. 487.

Atwal, A.S. and Singh, B. 1990. Pest Population and Assessment of Crop Losses. Indian Council of Agricultural Research, New Delhi, India.

Bairne, B.P. 1967. Pest Management. London: L. Hill, 123.

Banerjee, B. 1981. An analysis of the effect of latitude, age and area on the number of arthropod pest species of tea. *J. Appl. Ecol.* 18: 339-342.

Bhaskaran, V., Jagdishwar Reddy, D., Subbaratnam, G.V., Amarender Reddy, S., Nath, V.V.N. and Kumar, V.D.V.N.H. 2003. Efficacy of newer insecticides against spiraling white fly, *Aleurodicus disperses* (Russell). *Pestology* 27 (4): 16-19.

Bramble, B. J. 1989. An environmentalist's view of pest management and the green revolution. *Trop. Pest Mgt.* 35 (3): 228-230.

Brown, E. S. and Dewhurst, C. F. 1975. The genus *Spodoptera* (Lepidoptera, Noctuidae) in Africa and the Near East. *Bull. Ent. Res.* 65: 221-262.

Butani, D.K. and Jotwani, M.G. 1984. Insects in Vegetables. Periodical Expert Book Agency, Delhi, pp. 65-66.

Cherret, J.M. and Sagar, G.R. 1997. *Origins of Pest, Parasite, Disease and Weed Problems, 18th Symp. Brit. Ecol.* Soc., 413 pp., Blackwell: Oxford.

Chitty, D. 1957. Self regulation in numbers through changes in viability. *Symp Quant. Biol.*22: 277-280.

Chitty, D. 1960. Population process in the *Vole* and their relevance to general Theory. *Can. J. Zool.* 38: 99-113.

Clark, L. R., Geier, P. W., Hughs, R. D. and Morris, R. F. 1967. The ecology of insect Populations in Theory and Practice. The English Language Book Society and Methuen and Co., London, U. K.

David, P.P.M., Varandarajan, M.K., Muthuswamy, M. and Kannaiyan, S. 2000. Sustainable strategies against the coconut mite, *Aceria guerreronis* Keifer. Paper presented at the *Vth National Conference of AZRA*, Dec. 27-29, 2000, Chennai, Tamil Nadu, p.213.

Dey, P.K. and Somchaudhary, A.K. 2001. Bioefficacy of Fenpyroximate against Eriophyid mite of coconut (*Aceria guerreronis* Keifer) and its impact on *Amblyseius* spp. *Pestology* 25 (10): 6-10.

Dhaliwal, G. S. and Arora, R. (eds). 1994. *Trends in Agricultural Insect Pest Management.* Commonwealth Publishers, New Delhi, India.

Dhaliwal, G. S. and Arora, R. 1993. Changing status of insect Pests and their management strategies In: K.S. Gill, G.S. Dhaliwal and B.S. Hans (eds). *Changing Scenario of Indian Agriculture.* Commonwealth Publishers, New Delhi, India, pp. 94-145.

Dhaliwal, G. S. and Arora, R. 1998. Principals of Insect Pest Management. Kalyani Publishers, Ludhiana, Punjab, India, p. 297.

Dhaliwal, G.S. and Arora, R. 1996. Principals of Insect Pest Management. National Agricultural Technology Information Centre, Ludhiana. p.17.

Dhaliwal, G.S. and Arora, R.2001. Integrated Pest Management: Concepts and Approaches. Klayani Publishers, New Delhi, p.427.

Dhaliwal, G.S. Arora, R. and Heenrichs, E.A. 1998. Insect Pest Management: From traditional to sustainable approach. In: G.S. Dhaliwal and E.A. Heenrichs (eds), *Critical Issues in Insect Pest Management.* Commonwealth Publishers, New Delhi, India, pp. 1-25.

Douressamy, S., Chandramohan N., Shivprakashan, N. Subramanian, A. and Surendra Babu, P.C., 1997. Management of spiraling whitefly. *Indian Silk,* October : 15-16.

Evans, H. E. 1984. Insect Biology, A Text Book of Entomology. Addison – Wesley Publishing Company, Inc., U. S. A.

Farina, S.R. 1994. Problems on WSCA and strategies for its control at Takakar sugar plantation. Report to the Manager of Takakar sugar Plantation. (Unpublished report), p.9.

Fernando, L.C.P., Wickramananda, I.R. and Aratchigo, N.S. 2000. Studies on coconut mite, *Aceria guerreronis* in Sri Lanka. International Workshop on Coconut Mite. CRI, Lunwila, Sri Lanka.

Fletcher, T. B. 1920. The annotated list of Indian crop pests. In: *Proc. Third Entomology Society Meeting.* Pusa, India, pp. 33-313.

Green, C. P. 1992. *The Environment and Population Growth: Decade for Action. Population Reports,* Series M, No. 10, Johns Hopkins University, Baltimore, USA.

Gupta, G. P. 2002. Management of Key pests of cotton. In: Crop Pest and Disease Management Challenges for the Milliennium, eds. D. Prasad and S. N. Puri. Jyoti Publishers, EG – 102, Inderpuri, New Delhi – 110 012.

Hargreaves, H. 1948. List of cotton insects of the world. Common wealth Institute of Ent., London, pp. 1-50.

Hill, D.S. 1993. Agricultural insect pests of the tropics and their control. First Indian Edition, Foundation books, 2/19 Ansari Road, Daryaganj, New Delhi 110 002 by arrangement with Cambridge University Press, pp.746.

Jayaraj, S. 1988. Accentuation of insect and mitee problems in the changing agriculture scenario in India: An overview. National Seminar on Changing Pest Situation in the Current Agriculture Scenario of India. July 14-16, 1988, Indian Agricultural Research Institute, New Delhi, India.

Jayaraj, S. 1988. Integrated approach vital. *The Hindu Survey of India Agriculture,* pp.155-160.

Kapoor, V.C. 1993. Indian Fruit flies (Insecta : Diptera : Tephritidae). Oxford and IBH Publishing Co. Pvt. Ltd., New Delhi. pp.228.

Kapoor, V.C. and Agrawal, M.L. 1980. Identified collection of fruitflies (Diptera: Tephritidae) in National Pusa Collection, New Delhi. *Bulletin of Entomology 17*: 1-15.

Keifer, H.H. 1965. Eriophyid Studies. Department of Agriculture, Bureau of Entomology, B-14 California, p.20.

Krishnakumar, N.K. 1998. Bio-ecology and management of the serpentine leafminer, *Liriomyza trifolii* on vegetable crops. Final Report of ICAR AD-hoc Scheme, IIHR, Bangalore.

Kumar, A. and Nath, P. 2003 a. Field efficacy of insecticides against pod bug [*Clavigralla gibbosa* (spinola) and podfly (*Melanagromyza obtuse* Malloch) infesting pigeonpea. *Ann. Pl. Protec.* Sci. 11(1): 31-34.

Kumar, A. and Nath, P. 2003 b. Pest complex and their population dynamics on an early variety of pigeonpea UPAS – 120 at Varanasi. *Indian J. Ent.* 65 (4): 453-460.

Mani, M. and Thontadarya, T.S. 1987. Record of mealybugs species on grapevine in Karnataka. *Current Science* 56 (22):1192.

Milne, A. 1957. Theories of natural control of insect populations. *Symp. Quont. Biol.* 22: 157-174.

Milne, A. 1962. On a theory of natural control of insect populatios. *J. Theor. Biol.* 3: 19-50.

Misra, B. C. and Israel, P. 1968. Leaf and plant hoppers of rice. *IRC Newsletter* 17: 7-12.

Muralidharan, K., Mathew, J. and Thomban, C. 2001. Piolot sample survey on the incidence and yield loss due to Eriophyid mite on coconut in Alappuzha district. *Indian Coconut Journal 32:* 28-32.

Nageshkumar, S. 1998. *Frontline* 15 (3): 64-67.

Nagnathan, T.G. (1984.). *South Indian Hort.*32 : 115-117.

Narayanan, E.S. and Batra, H.N. 1960. Fruit Flies and their Control. ICAR Monograph, New Delhi p.68.

Nath, P. 1986. Causes of outbreak of paddy armyworm and forecasting its epidemics. In: *Proceedings of the Training Programme of Rice Including Propogation of New Technologies on Insect Epidemics and Post Harvest Management in Paddy*, pp. 21 27, Department of Entomology and Agriculture Zoology, Banaras Hindu University, Varanasi, U.P., India.

Nath, P. 1987. Environmental stresses and the white grub problem in the agro-ecosystem of eastern Uttar Pradesh. In: Proc. Natt. Conf. Env. Impact on Biosystems (eds. T.K.R. Roa, V.A. Murthy, M. Selvanayagum and Fr. A. Alphores), pp. 223-225. P.G. and Research Department of Zoology, Loyola College, Madras, India.

Nath, P. 1988. Possible components of Integrated management of leaf and plant hoppers – vectors of viral diseases of paddy. *Bicovas.*, 2: 179-195.

Nath, P. and Rai, R. 1998. Study of the bio ecology and economic injury levels of *Helicoverpa armigera* infesting gram crop. *J. Sustain. Pest Manag.*, 1(1 & 2):45-55.

Nath, P. and Rai, R. 1999. Study of key mortality factors in the population dynamics of chickpea pod borer, *Helicoverpa armigera* (Hubner) (Noctuidal: Lepidoptera) infesting chickpea. *Tropical Ecology* 40 (2): 111-117.

Nath, P. and Singh, S.P.N. 1987. Spatial distribution behaviour of armyworm *Mythimna separata* (Walker). In: *Aspects of Behaviour* (eds. Ismail, S.A. and Alawdeen, S.S.), pp.82 91. Department of Zoology, The New College, Madras.

Nath, P., Chandra, S. and Nag, A. 1991. Insecticidal management of *Mythimna separata* (Walker) infesting maize at Varanasi. *Pestology* 15 (9): 5 9.

Norris, R. E., Caswell-Chen, E. P. and Kogan, M. 2003. Concepts in Integrated Pest Management. Prentice-Hall of India Private Limited, New Delhi, 110 001.

Panda, N. and Khush, G.S. 1995. Host Plant Resistance to Insects. CAB International, Wallingford, U.K.

Patel, Z.P. 2001. Record of seed borer in sapota, *Manikara achras* (Mill.) Forsberg. *Insect Environment 6* (4): 149.

Pathak, M. D. 1980. International Congress of Entomology (XVI). Abstracts, 480 pp., Kyoto, Japan.

Pimental, D. 1977. Ecological basis of insect pests and weed problems. In: J. M. Cherrett and G. R. Sygar (Eds.). *Origin of pest, Parasite, Disease and Weed Problems.* Blackwell Scientific Publications, Oxford, U.K., pp. 3-31.

Pimental, D. 1993. Introduction to pest management. In: D.Pimental (eds), CRC Handbook of Pest Management, Vol.1. CBS Publishers and Distributors, New Delhi, India, pp. 3-14.

Pradhan, S. 1983, Agricultural Entomology and Pest Control. Indian Council of Agricultural Research, New Delhi, India.

Prakashan, C.B., Sreedharan, K., Reddy, A.G.S. and Gokuldas, M. 2001. Mass trapping A new component in the IPM of coffee berry borer. In: *IPM in Horticultural Crop : Emerging Trends in the New Millennium* (Eds. Verghese, A. ad Parrvata Reddy P.) IIHR, Bangalore, pp. 117-118.

Prasad, J. and Misra, D. S. 1982. Effect of spacing on rice leaf and plant hoppers population. *Agric. Res. Rural Dev.* 5: 28-30.

Puri, S. N. and Mote, U. N. 2003. Emerging pest problems of India and critical issues in their management: an overview. *Proceedings of the National Symposium on Frontier Areas of Entomological Research,* Nov. 5-7, 2003, IARI, New Delhi, India, pp. 13-25.

Rath, P. C. and Nath, P. 2001. Assessment of fruit damage in various genotypes of tomato by fruit borer *Helicoverpa armigera* (Hubner). *Shashpa* 8 (1): 63 – 64.

Rath, P. C. and Nath, P. 1998. Field testing of tomato genotypes for their plant susceptibility to fruit borer *Helicoverpa armigera* (Hubner). *Bioved* 9(1, 2):1 – 3.

Reddy, A. S. 1987. Management of whitefly in cotton. In : AICCIP 11th All India Workshop, APAU, Hyderabad, January 21-24, pp.1-13.

Sathe, T. V. 1999. White fly, *Aleurodicus disperses* – A new pest of guava, *Psidium guava* in Kolhapur, Maharashtra. *Indian J. Ent., 16* (2): 195-196.

Sathiamma, B., Radhakrishnan Nair, C.P. and Koshy, P.K. 1998. Outbreak of a nut infesting Eriophyid mite in coconut plantation in India. *Indian Coconut Journal* 29: 1-3.

Sharma, H. C. and Veerbhadra Rao, M. (eds.). 1993. *Pests and Pest Management in India – The changing Scenario.* Plant Protection Association of India, Hyderabad, India.

Sharma, S. B. Sharma, H. K. and Pankaj 2002. Nematode Problems in India. In: Crop *Pest and Disease Management, Challenges for the Millennium* eds. D. Paras and S. N. Puri. Jyoti Publishers, New Delhi, India, pp. 267-275.

Singh, A. K. and Nath, P. 1998. Field screening of wheat germplasms against the infestion of insect pests. *J. Sustain. Pest Manag.* 1 (1 & 2): 71-74.

Singh, A. K., Nath, P. and Ojha, J. K. 1998. Studies of feeding response of some plant extracts against *Spodoptera litura* (Fab.) on groundnut. *Indian J. App. Ent.* 12 (1): 16 – 24.

Singh, A. K. and Nath, P. 1998. Insecticidal management of tobacco caterpillar, *Spodoptera litura* (Fab.) infesting groundnut. *Shaspa* 5 (2): 203-206.

Singh, J. and Nath, P. 1983. An analysis of the environmental stresses and white grub problem in Varanasi and Mirzapur districts of East Uttar Pradesh, India. in *The State of the Rural Environment in Developing countries, Problems, Solutions and Research Priorities* (eds. S. Ghabbour and M. A. Ayyad), Academy of Scientific Research & Technology, Cairo, Egypt, pp.83 93.

Singh, R. 2004. Elements of Entomology. First Edition, Rastogi Publications, Shivaji Road, Meerut, U. P., India.

Singh, R. K., Nath, P. and Singh, P. K. 2003. Host plant diversity of serpentine leaf miner, *Liriomyza trifolii* Burgess. Paper presented in the National Symposium on sustainable Pest Management held during January 2-4, 2003 at the Deptt. of Entomology & Agril. Zoology, Banaras Hindu University, Varanasi, U.P. pp. 70.

Singh, R. K., Nath, P. and Singh, P. K. 2004. Manipulation of sowing time of pumpkin for managing the infestation of serpentine leaf miner, *Liriomyza trifoli* Burgers. *Indian J. Environ. & Ecoplan.* 8(3): 827-831.

Singh, R. K., Nath, P. and Singh, P. K. 2005. Manipulation of transplanting time of tomato for managing the infestation of serpentine leaf miner, *Liriomyza trifoli* Burgers *J. Plant Prot. Environ.* 2(2): 40-44.

Singh, R. K., Nath, P. and Singh, P. K. 2005. Relative susceptibility of certain tomato germplasms to serpentine leaf miner, *Liriomyza trifoli* Burgess. *Indian J. Environ. & Ecoplan.* 10(2): 465-469.

Singh, R. K., Nath, P. and Singh, P.K. 2004. Effect of sowing time of sponge gourd on the population of serpentine leaf miner, *Liriomyza trifolii* Burgess *J. Appl. Zool. Res.* 15 (2): 128-130.

Singh, R. K., Nath, P. and Singh, P.K. 2005. Effect of sowing time of bottle gourd on the population of serpentine leaf miner, *Liriomyza trifolii* Burgess. *J. Exp. Zool. India* 8(1): 145-149.

Singh, R. N. 1995. Mites of deciduous fruits and vegetables of eastern part of India and their economic status. *Adv. Agric.Res. India* 3: 179-193.

Sohi, G.S. 1964. Pests of Cotton. *Entomology in India,* Entomological Society of India, New Delhi, pp. 111-114.

Srinivasa, M.V. 2000. Host plants of the spiraling whitefly, *Aleurodicus disperses* Russel. *Pest Management in Horticultural Ecosystem 6* (2): 79-105.

Srinivasan, K., Viraktmath, C.A., Gupta, M. and Tewari, G.C. 1995. Geographical distribution, host range and parasitoids of serpentine leafminer, *Liriomyza trifolii* (Burgess) in South India. *Pest Management in Horticultural Ecosystem 1* (2): 93-100.

Srivastava, K. P. 1988. A Text Book of Applied Entomology. Vol. 1, pp. 323. Kalyani Publishers, New Delhi, India.

Srivastava, R. P. and Butani, D. K. 1972. A method to prevent the mango mealy bug (*Drosicha* mangiferae Green) damage. *Entomologist's News letter* 2(5): 35.

Stern, V.M., Smith, R.F., Bosch, R. Vanden and Hagen, K.S. 1959. The Integrated Control Concept. *Hilgardia* 29: 81-101.

Strong, D.R., Lawton, J.H. and Southwood, S.R. 1984. Insects on Plants: Community Patterns and Mechanisms. Blackwell Scientific Publications, Oxford, U.K.

Sukhija, H.S., Butter, N.S. and Singh, J. 1986. Determination of economic threshold of whitefly, *Bemisia tabaci* Genn. on American cotton in Punjab. *Tropical Pest Management* 32: 134-136.

Sundaramurthy and Chitra. 1991. In: *Principles of Insect Pest Management* by Dhaliwal, G. S. and Arora, R. 1998. Kalyani Publishers, New Delhi.

Tripathi, G.M. 1992. Biology of *Conobathra aphidivora* Meyr. Pyralid predator of sugarcane woolly aphid in Nagaland. *Journal of Applied Zoological Research* 3 (1): 99-100.

United Nations. 1999. The world at six billion. http: // www.un.org/ppin/wdtrends/6billion.

Varadarajan, M.K. 2000. Ecology and management of coconut Eriophyid mite, *Aceria guerreronis* Keifer. M.Sc. (Agriculture) thesis, TNAU, Coimbatore, p. 172.

Viraktamath, C.A., Tewari, G.C., Sreenivasan, K. and Gupta, M. 1993. American serpentine leafminer is a new threat to crops. *Indian Farming* 43 (2): 10-12.

Wellington, W. G., Cameron, P. J., Thompson, W. A., Kartinsky, I. B. and Landsberg, A. S. 1974. A. stochastic model for assessing the effects of external and internal heterogeneity of an insect population. *Res. Popul. Ecol.* 17: 1-28.

Zehntner, L. 1900. Mededeelingen van het Proefstation voor suikerriet in West-Java Te Kagok Teal No.49 De Plantenluizen van het Sukkerriet of Java X. *Ceratovacuna lanigera* Zehnt. (In Dutch). Archief voor de Java Suker Industrie.

Entomology : Novel Approaches, 2007
Eds.: *P.C. Jain and M.C. Bhargava*,(Page 97-107)
New India Publishing Agency, New Delhi
E-mail: newindiapublishingagency@gmail.com
web: www.bookfactoryindia.com

CHAPTER 4

ROLE OF INSECT RESISTANT VARIETIES IN INTEGRATED PEST MANAGEMENT

AK Chakravarthy and D Rajagopal
Department of Entomology, University of Agricultural Sciences
GKVK, Bangalore 560 065, Karnataka

INTRODUCTION

The first significant economic contribution of host plant resistance in agriculture was made in 1890 when European grape vines were successfully grafted on resistant root stock to save the French Wine industry from *Viteus vilifaliac (Fitch).* Studies from the past century have led to the development of resistant varieties in about 15 crops to about 50 key insect pests. During the last 20 years alone, more than 500 insect resistant cultivars of rice, maize, cotton, sorghum, alfalfa and wheat have been developed and released world wide (Dhaliwal and Arora, 2001). The insect resistant varieties are grown on millions of ha. annually and help the farmers to save billions of dollars in insecticide costs (Dhaliwal and Arora, 2001). The role of Host Plant Resistance in Integrated Pest Management is discussed below.

VARIETAL RESISTANCE : AS A SOLE TACTIC

A spectacular success in development of improved varieties that possess resistance to as many as four insect pests and five diseases has been achieved at the International Rice Research Institute, Manila, Philippines. For example, IR 36 is resistant to brown planthopper, green leafhopper, stem borers, gall midge, blast, bacterial blight and tungro. The yield of IR 8 fluctuates widely due to the pressure of insect pests and diseases. Whereas the yield of IR 36 is alone planted in about 11 million ha in the world and yields an additional income of one million dollars annually to rice growers and processors (Khush and Brar, 1991). Rice varieties resistant to brown planthopper and green leafhopper are grown over 20 million ha of riceland in Asia. These resistant varieties can be grown with the minimum use of insecticides and are an important component of Integrated Pest Management programmes (Panda and Khush, 1995).

One of the widely distributed pests of citrus is the red scale, *Aonidiella aurantii* (Maskell). Plants with more oil glands have shown fairly high degree of resistance to the red scale (Habib *et al.,* 1972). Sandhu *et al.,* (1979) reported that red flesh, pink flesh

(Gulabi) and strawberry showed resistance/tolerance to the fruitfly, *Bactrocera dorsalis* (Hendel). They attributed this to the rough, gritty and wrinkled skin of the fruit. Probably, such a skin is deterrent to oviposition by the fruitfly. The smooth skinned Allahabad Safeda and Lucknow-49 were susceptible.

Various strategies can be adopted to prolong the useful life of the resistant varieties or to develop varieties with different genes so that farmers may have access to new varieties when the resistance of the current varieties breaks down (Panda and Khush, 1995). Several sorghum hybrids bred in South Africa exhibited greatest tolerance to stem borers damage and suffered low yield losses (Kfir *et al.,* 2002). However, there is a danger of insect developing new biotypes if highly insect resistant varieties are solely used, as it happened in wasp, *Dryocosmus kuriphilus* Yasumatsu in chestnut in Japan (Shimura, 1972).

VARIETAL RESISTANCE AND BIOLOGICAL CONTROL

By reducing pest numbers, resistant varieties help to shift the pest: predator ratios in favour of biological control (Heinrichs, 1994). Varietal resistance is usually compatible with biological control but may have an adverse effect on natural enemies. In field studies at International Rice Research Institute, Philippines, the brown planthopper, *Nilaparvata lugens* (Stal) and Spider, *Lycosa pseudoannulata* (Bosenberg and Strand) ratio increased with the level of susceptibility from ASD7 and IR 36, both highly resistant rice cultivars to IR 42 and Triveni, moderately resistant cultivars to IR 8 and TN 1, susceptible cultivars.

The host plant resistance may also enhance the predatory activity. Predation rate of the mirid bug, *Cyrtorhinus lividipennis* Reutger when feeding on the first instar brown plant hopper nymphs increased on the resistant cultivar, IR 36 as compared to the susceptible IR 8. Combinations of host plant resistance and predation by the mirid bug *C. lividipennis* have a cumulative effect on the population increase of the green leafhopper, *Nephotellix viresceus* (Distant). In case studies, the number of green leaf-hopper reached only 6 on IR 29 (resistant) with the predator and 31 without the predator, while there were 91 and 200 hoppers, respectively, on susceptible IR 22.

The compatible nature of plant resistance and biological control has been demonstrated in between resistant varieties of barley and sorghum and parasitization of the greenbug, *Schizaphis graminum* (Rondani). The movement of greenbugs on the resistant sorghum cultivars exposes them to greater parasitization.

The parasitoid, *Lysiphlebus testaceipes* (Cresson) was able to keep the biotype C greenbug population nearly static on both susceptible and resistant barley, when the initial population of the aphid was three per plant. But when 12 aphids and one female parasite were introduced per plant, the parasitoid could suppress the aphid population only on the resistant barley. Thus damage to barley was reduced by the combined effect of varietal resistance and parasitoids. There is also synergistic interaction between maize cultivars resistant to the leaf feeding fallworm, *Spodoptera frugiperda* (Smith) and nuclear polyhedrosis virus (NPV) disease of fallworm.

Some of the Soybean (*Glycine max* Linn*)* varieties have shown high levels of insect resistance and adequate agronomic characteristics under heavy stink bugs (mainly *Nezara* *viridula, Peizodorus quildinii* and *Euschistus heros)* attack in Brazil. Release of these varieties with releases of bioagents is expected to result in an enormous improvement in the Integrated Pest Management programme, since stink bugs are still, almost exclusively controlled by insecticides (Moscardi and Gomez, 1996).

Dogramaci *et al.,* (2005) studied tritrophic interactions among the parasitoid, *Lysiphlebus testaceipes,* sorghum greenbug, *Schizaphis graminum* and the Greenbug Resistant Sorghum hybrids in the northern United States. *L. testaceipes* was effective in controlling greenbugs on sorghum with antibiosis resistance (Cargill 7607E) to greenbugs. The parasitoid was also effective in controlling greenbugs on a greenbug-tolerant hybrid (Cargill 797) but not on susceptible sorghums.

VARIETAL RESISTANCE AND PHYSICAL AND MECHANICAL CONTROL

Use of resistant varieties is most useful in crops of low value where yields fluctuate due to weather and other factors. Resistant varieties are of special significance for developing countries like India where holdings are small and farmers are not equipped to take up other methods of pest control. Physical and Mechanical control methods are the oldest and offer several primitive ways of suppressing the pests. For example, *Lablab niger* Medick creeper type is generally tolerant to pod borers compared to field type *L niger.* Collection and destruction of larvae would be easier on resistant/tolerant cultivars than susceptible cultivars (Chakravarthy, 1978). Chakrabarthy *et al.,* (1970) reported Bangalore Blue and Bhokri to be resistant to the defoliator, *Adoretus drevanceli* (Blanch) than Bangalore Purple, Khalili and Anab-e-Shahi grapes and hence it was convenient to pick and destroy defoliator on resistant than susceptible grape varieties. Resistant potatoes takes care of potato tuber moth population under cold storage more effectively than susceptible potato cultivars. Use of light traps is more effective in reducing root grubs infestation on resistant/tolerant crop plants like groundnut, coconut and arecanut than root grub susceptible cultivars. There are many other physical and mechanical techniques that have useful application in Integrated Pest Management with host plant resistance as a component.

VARIETAL RESISTANCE AND CULTURAL CONTROL

The cotton bollweevil and pink bollworm have been controlled through a combination of plant resistance and short-duration cotton varieties. The damage by the cereal leaf beetle, *Dulema melanopus* (Linn.) in mixed and pure stands of resistant and susceptible wheat varieties has been compared. Biological control was more effective in the mixed cropping of beetle-resistant and beetle susceptible wheat varieties than in a pure stand of either one of these varieties. Most of the cultural practices are compatible with other pest control tactics and have long been associated with subsistence farming as an appropriate approach to low-input cropping systems. Studies by Machiappan and Bhaskaran (1983) showed that the varieties Banesham, Chinnarasum and Khander were resistant to the mango hoppers, *Idioscopus* spp. and *Amritodus atkinsoni* (Leth). The resistant trees showed less of nitrogen, but high contents of phosphorus, potassium, calcium and phenols. These results were confirmed by Khaire et al., (1987). These results indicate that mango hoppers management can be successful through soil nutrition regulation.

VARIETAL RESISTANCE AND CHEMICAL CONTROL

Host plant resistance may enhance the efficacy of insecticides. Evaluation of insecticides in rice indicated that they cause higher mortality of plant hoppers and leafhoppers feeding on resistant than on susceptible rice varieties (Heinrichs, 1994). Mortality of brown planthopper when reared on either moderately resistant ASD7 or a highly resistant cultivar Sinna Sirappu was higher than when feeding on a susceptible TN 1 cultivar. The LD_{50} of whitebacked planthopper was 9.4 on the susceptible variety TN 1 treated with ethylan but only 2.8 on moderately resistant N 22. The combination of moderate varietal resistance and low dose of pesticide resulted in effective hopper control. The integration of host plant resistance and insecticides has cumulative effect on *Nephotettix virescens* (Distant), the vector of rice tungro virus. There was no tungro infection on the resistant cultivar IR 28 even without the application of insecticides.

A lower concentration of insecticide is needed to control insects feeding on a resistant variety than those feeding on a susceptible variety. Nymphs of the wheat grain aphid, *Sitobion avenae* (Fab.) reared on resistant wheat variety. Altar possessing the antibiosis compound DIMBOA were significantly more susceptible to the insecticide deltamethrin than nymphs on the susceptible wheat variety Dollarbird. The LD_{50} was reduced by 91% for nymphs reared on the cultivar with a high DIMBOA content. Although the aphid population, *M. persicae* on the partially resistant variety of Brussels sprouts was about 85% of that on the susceptible variety, the LD_{50} of malathion was only about 55%, *i.e.* the insecticide requirement was much less on the partially resistant variety (Panda and Khush, 1995).

VARIETAL RESISTANCE AND INSECT VECTORS

Insect vectors can be very potential as they can spread disease causing organisms. For instance, some banana varieties like Musa AAB showed resistance to bunchy top. A possible explanation for this may be the resistance to the banana aphid, *Pentalonia nigronervosa* Coquered, the bunchy top vector (Verghese, 2001). Certain species of citrus like *Citrus nobilis* and *C. aurantifolia* and *C. rashmi* had significantly low population of the aphid (*Toxoptera* sp.) from among 22 species and three hybrids (Anonymous, 1984). Birch et al. (1992) showed the use of RFLP analysis, for distinguishing biotypes of the virus vector aphid, *Amphorophora idaei* thus aiding in rasberry screening (Verghese, 2001).

Interrupting vector borne diseases transmission by replacing insect vector populations with non-vector forms is an old idea that may become practicable with the use of molecular techniques and genetic engineering which could supply a population of selectively modified vectors. Because natural refractoriness may prove to be genetically unwieldy, biotechnology may provide novel ways to disrupt transmission by adding or removing specific insect functions that block or decay host/pathogen interactions. Molecular techniques are making possible to enhance classic genetic control strategies, transformation of insect symbionts and disrupt virus-host interactions (Evans, 1996).

VARIETAL RESISTANCE AND BIOTECHNOLOGY

Biotechnology offers possibilities of resistance that are not easily achievable by conventional means, as perhaps with virus resistance in some crops and insect resistance with particular pests like cereal stem borers. Current work with transgenic plants, particularly the incorporation of the *B. thuringiensis* delta endotoxin into crops for control of insects, appears to be proceeding on a vertical resistance model, based on complete resistance conferred by one or a few genes. These varieties, like those produced through conventional vertical resistance breeding, will be susceptible to the development of pest resistance. Further, they may undervalue the benefits of Integrated Pest Management approach.

Many will now agree that the risk of resistance development to *B.thuringiensis* engineered transgenic plants is considerable and probably greater than that for a *B.thuringiensis* formulation topically applied to a crop on a need based Integrated Pest Management basis, simply because of its more continuous and extensive expression. Indeed, some entomologists have come together to express their concern that use of transgenic plants may waste a valuable resource by encouraging resistance development (Anonymous, 1992).

However, field trials in USA indicated that maize plants with Cry I genes provided control of *Ostrinia nubilalis,* the European corn borer. This is an important pest of US maize and is very difficult to control with chemicals because its habit is to spend most of its life inside the maize stalk, where it is relatively inaccessible to chemical sprays. Plant expression of *B. thuringiensis* proteins is a method that can, in this case, overcome the technical barriers to insect control with other agents, and thus address an unmet need in pest management (Fischhoff, 1996).

Fischhoff (1996) outlined plans for the implementation of insect resistant transgenic cotton plants in the USA. But many of the same ideas will apply directly to the use of *B. thuringiensis* cotton in other parts of the world because many of the same or closely related pest occur on cotton in all areas. Also, in areas where other lepidopteran pests are seen, their natural history is often similar to that of some US pests. For example, in India, *H. armigera* infests many economically important crops like cotton, pigeonpea, chickpea, sunflower, corn, chillies, tomato and okra. Two years study in South and Central India demonstrated that a number of host crops of *H. armigera* support large populations at the same time that cotton is infested. Thus, these crops may act as important sources of refuge for Bt. cotton plantings in South and Central India (Ravi *et al.,* 2005). It is important to recognize that the plans for implementation of insect resistant varieties will be crop and insect specific and each example will have to be tailored to the individual cropping system.

The use of insect resistant cotton will provide benefits to the cotton grower and to society and the environment. The insect resistant cotton controls the lepidopteran pests and the crop damage that they cause with an efficacy equal to or better than that of the current control methods, while significantly reducing the application of insecticides. This substitution will have benefits to the grower. Insect resistant cotton is also likely to produce secondary benefits in pest control as an indirect result of the reduction in use of insecticides. One aspect of the use of insect resistant cotton for integrated insect management is a continued use of chemical insecticides. This is a long term strategy, may not be suited to farming systems in the tropics and subtropics.

Resistance problems involve more complex strategies of gene deployment, for instance, mixed or intercropped populations of resistant and susceptible plants or genetic methods to restrict expression of genes to certain parts of plants or certain times. The present strategy of protecting plants with engineered, single-gene defenses may offer some real opportunities for improving pest management in the short term, particularly on crops where conventional breeding can not achieve resistance (Waage, 1996).

Yencho *et al.,* (2000) reviewed how molecular markers can be used to increase our understanding of the mechanisms of plant resistance to insects and develop insect resistant crops. Molecular markers can be applied to develop more durable insect resistant crops, for example, in potato and maize plant/pest systems. Identification and characterization of Rice Gall midge R genes have played and continue to play an important role in the control of gall midge pests of agriculture. Yet concerns about gall midge adaptations are real (Harris *et al.,* 2003).

VARIETAL RESISTANCE AND INTEGRATED PEST MANAGEMENT

The effect of host plant resistance on pest population is specific, cumulative and persistent. It offers no environmental hazards and is ecofriendly. It is compatible with other pest management methods. However, occurrence or selection of pest biotypes limits the effectiveness of resistant variety. Resistance factor may be incompatible with desired economic characters. A variety resistant to one pest may be susceptible to another pest. For example, in *L niger* pod borer resistant cultivars are susceptible to the aphid, *Aphis craccivora* Koch (Chakravarthy, 1992).

Yet resistant cultivars have played a key role in the management of the major insect pests. For instance, in cotton, deep ploughing, crop rotation with non hosts, disposal of crop wastes, clean cultivation, crushing or fumigating cotton seeds, selection of pest resistant varieties, intercropping, removing plant stubbles after harvest, hand picking caterpillars, releasing *Chrysoperla* and *Trichogramma* successfully reduces pest loads on cotton. Cotton varieties showing resistance are indicated in Table 1 and of sugarcane in Table 2.

Table 1 Cotton varieties resistant to pests and diseases

Pest disease	Resistant Variety (ies)
Jassid	G.Cot 10, G. Cot 12, ABH 466 intra hirsutum hybrid
Whitefly	LPS 141, LK 861 Supriya, Kanchana
Verticillium wilt	MCW 5
Fusarium wilt	DB 312, AK 45, Sanjaya, Digvijaya, G. Cot 11, G.Cot 13, LD327, PA 32
Bollworms	LH 900, F 414
Nematode	Khandwa

Table 2 Varieties Resistant to Insect Pests of Sugarcane

Pest	Variety	Resistance Level	States recommended
Top borer	CoS 767 CoJ 67, Co 1158	R MR	Haryana Punjab and Western U.P.
Internode Borer	Co 6806, Co 621754 Co 975, CoC 771	MR MR	Tamil Nadu Tamil Nadu & Andhra Pradesh
Scale Insect	CoC 671 CoC 67d1, Co 621754 CoC 671, CoS 611 Co 1132	R R R R	Andhra Pradesh Tamil Nadu Gujarat Haryana and Uttar Pradesh
White Grub	Co 6304 Co 1158 CoS 510	MR MR MR	Tamil Nadu Haryana and U.P. Uttar Pradesh
Stalk Borer	Co 7302	R	Haryana
Redrot	Co 6806, Co 62198, Co 8021, CoSi 776 Co 6907, Co 975, Co 8014 Co 7717, Co 6914, Co 7314 CoJ 58 CoS 802, CoS 767, BO 90, BO 91 Co 7704, Co6907	R R R	Tamil Nadu Andhra Pradesh Haryana Punjab Uttar Pradesh Bihar Kerala
Smut	Co 6806	R	All states

R = Resistant; MR = Moderately Resistant; HR = Highly Resistant

Sustainable Pest Management in cotton and sugarcane in India envisages the use of resistant varieties (Raheja and Tiwari, 1996). In India, considerable progress has been made in identification and utilization of resistance for crops pests. Dhaliwal and Arora (2001) have listed varieties resistant to pests in India. Varieties possessing multiple resistance to a number of insect pests and disease are ideal in Integrated Pest Management Programme. Cultivars with multiple resistance to insects, nematodes, pathogens and tolerance to abiotic stresses (drought, soil mineral toxicity, etc.) have been developed in various crops, viz., wheat, alfalfa, cowpeas, maize, pearlmillet, sorghum, soybean and rice (Heinrichs, 1994). Use of even moderately resistant to tolerant varieties should benefit by reducing population level, cumulatively over generations (Coppel and Mertins, 1977). This makes chemical and biological control easier. Host plant resistance may serve as a means of control in unique niches where other controls are not feasible or difficult to use. For example, (a) when there is a critical timing in which the insect is exposed for only a brief period of its life cycle, (b) crop is of low economic value and (c) pest is continuously present and is a single limiting factor over a large area.

Insect resistant varieties with a yield potential comparable to commercial cultivars should be released for areas endemic to a particular pest. Greater emphasis should be placed in developing crop cultivars with multiple resistance to key insects and diseases (Dhaliwal and Arora, 2001).

HOST PLANT RESISTANCE: THE FUTURE

In recent decades substantial resistance in crops to plant pests and diseases have been produced. Buddenhagen (1991 and 1996) has argued that far more could be done in this area, and that the popularity of insecticides, and even insect-based Integrated Pest Management, has reduced efforts in resistance breeding. Most resistance breeding to date has focused on methods that result in qualitative vertical resistance. This approach is convenient because high levels of resistance can be achieved and because the method is compatible with breeding schemes used for enhancing crop performance through control of major genes. However, its gene-for-gene nature can sometimes lead to its breakdown through the evolution of resistance breaking pest genotypes. Classic examples of such resistance breakdown in insect pests include the hessian fly, *Mayetiola destructor Say* and greenbug, *Schizaphis graminum* on wheat, and the brown planthopper, *Nilaparvata lugens,* green leafhopper, *Nephoteteix virescens* and rice gall midge *Orseola oryzae*, on rice.

While plant resistance breeding is often identified as a component of Integrated Pest Management (IPM), its actual integration with other pest-control methods in IPM systems has been limited. In most cases, breeding programmes have sought single-technology solutions to pest problems (i.e. complete resistance), much like insecticides. This is often hard to achieve for some pests, e.g. stem boring moths in cereals, and while partial resistance has been possible like in field beans, *L niger* to pod borers (Chakravarthy, 1978 and 1992), it has generally not been used. Partially resistant varieties, however, may have a value when viewed in an IPM context, where the contribution of natural enemies and other factors can complement their effect (Thomas and Waage, 1993).

An IPM approach can also contribute to the durability of plant resistance to pests. The action of mortality factors, such as natural enemies, can reduce pest populations and selection pressure to overcome resistance factors. The role of natural enemies in maintaining durability in this manner has been demonstrated (Gould, 1996).

Another approach to durable resistance is to select for horizontal rather than vertical resistance. Where successful, this leads to a durable, polygenic form of resistance, resulting from many different mechanisms (Robinson, 1991). The partial nature of horizontal resistance, at present seen as a disadvantage by plant breeders, may in fact be an advantage in IPM, where it can support the continuity and contribution of natural enemies and their important density-dependent effects. Greater emphasis on horizontal resistance breeding might place on selecting with local material for local conditions is compatible with current ecological approaches in Integrated Pest Management.

REFERENCES

Ananthakrishnan, T N. 1992. Dimensions in Insect-Plant Interactions. Oxford & IBH Publishing Co. Pvt. Ltd., New Delhi, pp.90

Anonymous, 1984. Annual Report. Indian Institute of Horticultural Research (ICAR). Bangalore. pp.168

Anonymous, 1992. Call to save Bt. Biocontrol News and Information. 13(2): 19.

Birch, A.N.E., Harromar, B., Malloch, G, Catley, M. and Jones, A.T. 1992. Potential use of RFLP analysis for rapidly distinguishing biotypes of the virus vector aphid, *Amphorophoro iduci. Acta Horticulturae*. 308: 95-97.

Beck, S.D. 1965. Resistance of plants to Insects. *A. Rev. Ent*. 10: 207-232.

Buddenhagen, I.W. 1991. Better cultivars for resource poor farmers, Isle of Thorns Conference Centre, East Sussex, UK, 4-8 November 1991. Natural Resources Institute, Chatham, U K, pp.83-94.

Buddenhagen, Ivan W. 1996. Modern Plant Breeding: An overview. In: (Ed.) Persley, G.J. *Biotechnology and Integrated Pest Management*. CAB International. U.K., 205-213.

Chakravarthy, A.K. 1978. Pod borer resistance in Field beans, *Lablab niger* Medick with particular reference to the pod borer *Adisura atkinsoni* Moore. M.Sc. thesis submitted to University of Agral. Sciences, Bangalore. pp.180.

Chakravarthy, A.K. 1992. Mechanism of resistance in field beans *Lablab niger* Medick to the lepidopterous and dipterous pod borers and the black aphid, *Aphis craccivora* Koch. *Colemania* 2: 32-42.

Chakrabarthy, D.P., Ghose, G.C., Das, P.K. and Dhua, S.P. 1970. Studies on the relative susceptibility of different varieties of grape to *Adoretus devanceli* Blanch. *Technology Bihar*, 7 : 184-185.

Coppel, H.C. and Mertins, J.W. 1977. Biological Insect Pest Suppression. Springer- Verlag. Berlin. p.314.

Dent, D. 1999. Insect Pest Management CAB International, Wallingford, U.K.

Dhaliwal, G.S. and Arora, Ramesh. 2001. Integrated Pest Management. Kalyani Publishers, Ludhiana. pp.428

Dodds, K.J. and Ross, D.W. 2004. Douglas-fir beetle lipid levels in relation to tree physical characteristics. *J. Ento. Soc. British Columbia,* 101: 109-116.

Dogramaci, M., Mayo, Z.B., Wright, R.J. and Reese, J.C. 2005. Tritrophic Interaction of parasitoid *Lysiphlebus testaceipes* (Homoptera: Aphidiidae), Greenbug, *Schizaphis graminum* (Homoptera: Aphididae) and Greenbug- Resistant Sorghum Hybrids. *J.Eco.Ent*. 98 (1): 202-209.

Duan, J.J., Head, G., A. Jensen and Reed, G. 2004. Effects of transgenic *Bacillus thuriengensis* potato and conventional Insecticides for Colorado Potato Beetle Management in Oregon Potato Ecosystems. *Environ. Entomol.*33(2) : 275-281.

Evans, Elizabeth, 1996. Insect Vector Control. In: (Ed.) Persley, G.J. *Biotechnology and Integrated Pest Management*. CAB International. U.K. 323-338.

Fischoff, David, A. 1996. Insect-resistant Crop Plants. 214-217 In : (Eds.), Persley, C. J., *Biotechnology and Integrated Pest Management*, CAB International, Wallingford, Oxon, UK, pp 461.

Gould, Fred. 1996. Deploying Pesticidal Engineered Crops in Developing Countries. In: (Ed.) Persley, G.J. *Biotechnology and Integrated Pest Management*. CAB International, U.K. 264-293.

Habib, A., Salama, H.S. and Aswin, A.H. 1972. Population of *Anidiella aurantii* on citrus varieties in relation to their physical and chemical characteristics. *Entomol. Exp. Applic*., 15: 324-328.

Harris, M.O., Stuart, J.J., Mohan, M., Nair, S., Lamb, R.J., and Ro h fruitsch, 0.2003. Grasses and Gall Midges: Plant Defense and Insect Adaptation. *Ann. Rev. Entomol.* 48 : 549-577.

Heinrichs, E.A. (Ed.). 1988. Plant stress-Insect Interactions. John Wiley & Sons, New York, USA.

Heinrichs, E.A. 1994. Development of multiple pest resistant crop cultivars. *J. agric. Ent.* 11(3) : 325-353.

Horn, D.J. 1988. Ecological Approach to Pest Management. The Guilford Press, New York. pp.285.

Jyothi, C., Radhakrishnan, V.V. and others. 2004. Physical and Chemical Characteristics in Cowpea in relation to bruchid resistance. *Legume Research*, 27(3), 183-187.

Kfir, R., Overholt, W.A., Khan, Z.R. and Polarzek, A. 2002. Biology and Management of Economically Important. Lepidopteran Cereal Stem borers in Africa. *Ann. Rev. Entomol.* 47: 701-731.

Khaire, V.A., Kolhe, D.S. and Patil, J.D. 1987. Relative susceptibility of mango varieties to mango hoppers and powdery mildew. *Haryana J. Hort. Sc.,* 16: 214-217.

Khush, G.S. and Brar, D.S. 1991. Genetics of resistance to insects in crop plants. *Adv. Agron.* 45 : 223-274.

Kiplagat, O.K. 2005. The Russian wheat aphid (*Diuraphis noxia* Mord.) damage on Kenyan what cultivars and possible control through resistance breeding. Wageningen University, Wageningen, Netherlands. pp.141.

Machiappan, R.M. and Bhaskaran, P. 1983. Biochemical constitution of infloresence of certain varieties of mango in relation to mango leafhoppers. *South Indian Horticulture*, 31 :160.

Maxwell, F.G., Jenkins, J.N. and Parot, W.L. 1972. Resistance of plants to insects. *Adv. Agron.* 24: 187-265.

Moscardi, F. and Gomez Sosa, D.R. 1996. Soybean in Brazil. In: (Ed.) Persley, G.J. *Biotechnology and Integrated Pest Management*. CAB International. U.K., 98-112.

Muresanu, F. 2003. Efficient methods utilized in integrated pest management of maize crop under central west part of Trainsy/vania. *Problemade Protectioa Plantelor* 31 (1): 67-74 (Cf. RAE 93(7):1003).

Panda, N. and Khush, G.S. 1995. Host Plant Resistance to Insects. CAB International. Wallingford. UK.

Persley, G.J. 1996. *Biotechnology and Integrated Pest Management*. CAB International. Wallingford, Oxon, U.K. pp.475.

Rabindra, R.J., Palaniswamy, S., Karuppuchamy, P., Ramaraju, K. and Sridhar,R.P. (Eds.) 1999. Ecology based Pest Management. TNAU, Coimbatore, pp. 277.

Raheja, A..K. and Tewari, G.C. 1996. *Integrated Pest Management for sustainable Agriculture.* Published by Awareness programmes on Pesticides and sustainable Agriculture. New Delhi. pp.70.

Ravi, K.C., Mohan, K.S., Manjunath, T.M., Head, G., Patil, B.V. and others. 2005. Relative abundance of *Helicoverpa armigera* on different host crops in India and the role of these crops as Natural Refuge for B.t. cotton. *Envinron. Entomol.* 34(1) : 59-69

Revilla, P., Tracy, W.F. and others. 2005. Vegetative phase transition and corn borer resistance of shrunken 2 versus sugary I Sweet corn near isogenic inbred lines. *J. of the Amer. Soc. Hort. Sci.,* 130(1): 64-69.

Robinson, R.A. 1991. The genetic controversy concerning vertical and horizontal resistance. *Revista Mexicana de Fitopatologia* 9 :57-63.

Sandhu, G.S., Sohi, A.S. and Batra, R.C. 1979. Incidence of fruit boring insects in guava cultivars. *Punjab Horticultural Journal*, 19 : 171-173.

Seema Kumar and Yadav, R.P. 2005. Tritrophic interactions among Coriander genotypes, pest insect and pradaceous coccinellids. *Annals of Plant Protection Sciences*, 13(1) :136-138.

Shah, D.M., Horsch, R.B., Klee, H.J. and others. 1986. Engineering herbicide tolerance in transgenic plants. *Science*, 233: 478-481.

Shimura, I. 1972. Breeding of chesnut varieties to Chesnut gall wasp, *Dryocosmus kuriphilus* Yasumatsu, *Japanese Agricultural Research Quarterly*, 6 : 224-230.

Shirai, Y. and Takahashi, M. 2005. Effects of transgenic B.t. corn pollen on non-target lycaenid butterfly. *Applied Entomology and Zoology*, 40(1) 151-159.

Thomas, M. and Waage, J.K. 1993. Prospects for the Integration of Biological Control and Plant Resistance Breeding in IPM Systems for Resource –poor Farmers. Study commissioned by IIBC and Technical Centre for Agriculture and Rural Cooperation.

Verghese, A. 2001. Host Plant Resistance to Pests in Fruit Crops: An IPM Component. In: *Integrated Pest Management in Horticultural Ecosystems*. (Eds.): Reddy, P.P., Verghese, A. and Krishnakumar, N.K. Capital Publishing Co., N. Delhi. pp.292.

Waage, J. 1996. Integrated Pest Management and Biotechnology: An Analysis of their Potential for Integration. In : (Ed.) Persley, G.J. *Biotechnology and Integrated Pest Management.* CAB International. U.K., 37-60.

Yencho, G.C., Cohen, M.B. and Byrne, P.F. 2000. Applications of Tagging and Mapping Insect Resistance Loci in plants. *Ann. Rev. Entomol.* 45: 293-422.

Zhu. Weittong, Wang Qing Dong and others. 2004. Identification of Insect Resistance and inheritance analysis of transgenic B.t. gene in maize. *Jour. Of Henan Agricultural University,* 38(2) : 135-138.

Entomology : Novel Approaches, 2007
Eds.: *P.C. Jain and M.C. Bhargava*,(Page 109-119)
New India Publishing Agency, New Delhi
E-mail: newindiapublishingagency@gmail.com
web: www.bookfactoryindia.com

CHAPTER 5

ECOLOGICAL PERSPECTIVE OF PEST MANAGEMENT

Rakesh Gupta and Ajay Sharma
Department of Entomology and Apiculture, Dr Y S Parmar University of Horticulture and Forestry, Nauni, Solan-173 230

INTRODUCTION

Advent of agriculture in India dates back to around 2000 BC, the period of Indus Valley Civilization. South and East Asia also witnessed the start of agriculture around the same time. The crops were grown only during the monsoons. Winters served as a closed period for crops as well as for the pests. Even at that time a number of insect pests were known to cause serious damage to different crops. Agriculture has been revolutionized during the few decades since World War II (Kupatt *et al.*, 2005). Introduction of high yielding varieties and associated technology resulted in tremendous increase in the production from the same piece of land. The results of these changes include a dramatic reduction in time and resources needed to provide an unprecedented quantity of quality food and fiber. The initial introduction of plant protection agents as wonder material did not lasted long, as very soon their ill effects on human health and environment were witnessed.

The effects of pesticides and agricultural practices came to a major scrutiny with the increasing public concern for the environment in the later years of 1970's. Under this aegis an increased emphasis has been placed on finding other ways to produce food. These types of movements can be found in most of the developed and developing countries. In this alternative agriculture the focus has been on the reduction of external inputs in raising the crops. This pressure has changed the direction of much of research in plant protection. Beginning with the integrated pest management (IPM) these changes have headed towards low input sustainable agriculture (LISA). Sustainable agriculture evolves indefinitely towards greater human utility, greater efficiency of resource use and a balance with environment that is favourable both to humans and to most other species (Harwood, 1990). To deal with these emerging problems the answer is one of the oldest approaches to modify the production schemes and techniques to make crop environment less favourable for the pest. This modification of crop environment through changes in production techniques is called as cultural control. As the cultural control involves the purposeful manipulation of environment for keeping the pest population under control, hence it can also be referred to as the ecological management. It is based on thorough knowledge of pest ecology, so as to find out the weak links in the seasonal cycle of the pests and to exploit them for managing the pest population in the crops. The different approaches which can be exploited in ecological management for preventing the pest from reaching its full destructive potential involve food supply and suitability of the environment.

In our view two attributes needs to be addressed while discussing the economic perspectives of pest management. Sufficient food supply and to produce it without causing any damage to the environment and disturbing out ecology is the need of the society. The approach of pest management which should be adopted with these primary goals for sustainability are evident from this definition of IPM.

The judicious use and integration of various pest control tactics in the context of the associated environment of the pests in ways that complement and facilitate the biological and other natural controls of pests to match economic, public health and environmental goals (Anon, 2000).

Modern agriculture has largely achieved the food and fiber requirement of the burgeoning human population, but the excessive use of high yielding varieties alongwith the application of manifold increasing amount of agrochemicals have resulted in deterioration of environment and given rise to many other problems. The presence of residues of pesticides in food and other components of environment is harmful to the living organisms and to the environment.

The environmental problems are to be considered before making the critical appraisal of the role of pesticides in IPM programmes.

INSECTICIDE RESIDUES

Rachel Carson in her book 'Silent Spring' has aroused the concern about excessive use of pesticides. Hardly 0.1 per cent of agrochemicals used in pest control reach the target pests leaving the remaining 99.9 per cent to enter different components of environment, causing hazards to non target organisms (Verma, 2002). Dietary intake of DDT and HCH had exceeded the acceptable daily intake (ADI) due to ubiquitous presence of these chemicals in all food commodities (Singh and Dhaliwal, 1994). Nearly 100 per cent of human population has been found to contain some residues of pesticides. In addition to food, drinking water is also contaminated and leave the residues in humans. The soil and water samples from different parts of the country have been found to contain residues of DDT and HCH and DDT has become so widespread in our environment that its residues have been detected even in Antarctic snow (Dhaliwal and Arora, 1996).

Samples of foodstuffs have also been found to be contaminated with pesticides like endosulfan, aldrin and heptachlor etc. Keeping it in view, 20 insecticides has been banned in India and 12 has been identified for their restricted use in agriculture. The use of chemicals has accrued tremendous benefits but the excessive use of these has resulted in many adverse effects. Environmental pollution, build up of pesticide residues and contamination of food chain are the effects of exhaustive use of these chemicals. The residue levels is variable in different agro-ecosystems and is also dependent upon the weather. The presence of these residues in soil has affected the soil health and fertility. It has resulted in killing of macro and micro organisms helpful in breaking down soil litter and thus maintaining the fertility of soil. Therefore it is necessary to reduce the usage of pesticides in order to conserve the population of the soil organisms and maintain the ecological balance.

INSECTICIDE RESISTANCE

One of the consequences of the large scale use of pesticides to control insect pests is the development of resistance in target species. Resistance is the development of an ability to tolerate a dose of an insecticide, which would prove lethal to the majority of individuals in the normal population of the same species. It is a dynamic phenomenon dependent on biochemical, physiological, genetic and ecological factors. These factors depend upon the species of insect and geographical location. Resistant strains of pests develop through survival and reproduction of individuals carrying resistant genes survive after exposure to pesticides. At global level resistance to all classes of pesticides have been reported as early as in 1987 (Brown and Brogdon, 1987).

The first report of resistance to insecticides appeared in 1908 when San Jose Scale showed resistance to lime sulphur. Among the 3rd world countries, India was one of the first countries to start a large scale use of synthetic pesticides for the control of insect pests of public health and agricultural importance. India is the first country in the Asian continent to report the case of resistance. Singhara beetle *Galerucella birmanica* (Jacoby) was found resistant to DDT and HCH in 1963 (Pradhan *et al., 1963*). After this, tobacco caterpillar, *Spodoptera litura* was reported to develop resistance to HCH in 1965 since then it has become resistance to all the groups of pesticides in India. Stored grain pests have also developed resistance to different insecticides. Lesser grain borer has developed resistance against most of insecticides (Saxena *et al.,* 2003). Many other insects of public health, veterinary, agriculture and stored grain pests have developed resistance to one or more insecticides used against them.

The increase in the number of insects showing resistance to different insecticides can pose threat to the modern agriculture. In view of the resistance to the already existing pesticides, some new chemicals will be brought into use against the pests and they will further degrade our soils and water and pollute our environment. It is important to understand the dynamics of resistance to tackle this problem.

Insect Resurgence

Increase in the insect numbers as a result of the application of insecticides or the increase in the damage potential of a pest is called as insect resurgence. Maximum cases of resurgence belong to order homoptera despite the fact that homopterous pests have waxy covering on their body, which protects them from the direct contact with the insecticides. Similar is the case in lepidopterans that most of them showing resurgence are borers and they also do not come in direct contact of the insecticides. Maximum cases of resurgence are of plant hoppers. For example Brown plant hopper was initially a minor pest of rice, but the excessive use of insecticides has elevated it to be a major pests in Asia.

Increase in the population of aphids, jassids and whiteflies on different crops and especially in cotton is a result of use of excessive synthetic pyrethroids against them. Jayaraj (1987) had reported that white fly out breaks in Andhara Pradesh and Gujrat was as a result of the over-reliance on synthetic pynethroids for controlling bollworms on cotton. Arora (1996) had mentioned that different insecticides may cause resurgence of same insect by different mechanisms and even the same insecticides may produce variable effects at different sub-lethal concentrations.

Factors affecting pest populations

Abiotic as well as biotic factors greatly influence the population of different organisms. Population of insect pests is also governed by these factors. Temperature, moisture, light etc. are the different abiotic factors which affects the population of these pests. The different biotic factors are food, other organisms living in the same area which complete for food and space and their natural enemies. Both the abiotic and biotic factors act simultaneously on the pests and results in increase or decrease in the population.

Food

Food plays a major role as the limiting factor in the population of insect pests. The survival of the pest and their ability to the survival of the pest and their ability to reproduce is governed by the availability of food in the area. If the preferred food is available in plenty then the population of insect will tremendously increase to attain the status of the pest. Monoculture being practiced on a large scale provides an ample supply of food to the insect pests and results in increasing out breaks of these pests.

Temperature

Climatic factors play's a pivotal role in governing the population of different insects in an area. The rate of survival and the ability to reproduce are the most important factors which are influenced by different climatic factors of the environment. Temperature probably has the greatest affection on insect developmental rates. This is because of the insects are cold blooded; Poikilothermic All life survives within a certain narrow range of temperature. Increase or decrease beyond these limits is tolerated only upto some extent and the more diversity of temperature can result in either instant killing or inability to reproduce. Diapause, hibernation, aestivation etc. are some of the mechanisms adopted by the insects in thus life cycles to combat the adversities of the temperature.

Moisture

Phytophagous insects fulfill the need of their water requirement from the food material they consume from their host plant. Mostly the terrestrial insects inhabit dry environment. Excessive moisture and excessive dry air can result in desiccation and high moisture promotes disease incidences. Moisture conditions also affect the reproductive behaviour of certain insects. The increase in moisture also affects the normal development by lowering the capacity of insects to withstand low temperatures.

For normal growth and reproduction of the insects all the abiotic and biotic factors needs to be between the optimum range. If somehow we could target these factors and manipulate them. Then it is possible to put a check on the excessive increase on the population of the pests.

Ecological management of insect pests

The term ecological management is used to broaden the ground for cultural control of the pests. It is based on thorough understanding of ecology of pests in relation to the crop. For an organism to sustain in a particular area the major pre-requisites are food, shelter from weather conditions and enemies, preference for reproduction (mating and egg laying). Their availability in the nearby areas also supplement the need of an organism. For example grasshoppers feed on the annual crops like soyabeans etc but for egglaying the move to the so as at the edges of fields. The removal of any of these factors will render the environment unfit for that organism. These environmental requisites or the different stages of the seasonal cycle of the pests which could be targeted to disturb the population of an organism in an area are called the "weak links".

The most important aspect of ecological management is that we must understand ecological requisites for a pest, know about.

Making the ecosystem unsuitable

Ecosystem includes interacting biotic and abiotic elements in the surrounding of insects. The presence of requisites for the insect in its surrounding determines the degree of density of pest infestation and the severity of damage to the crop. To reduce the pest infestation and damage to the crop the ecosystem of the pest needs to be disturbed so as to make it unsuitable for its survival. The shelter or the habitat of the insect is only disturbed as its complete elimination is not possible.

Soil tillage

Tillage is a major cultural operation in agriculture. It is a nursery part of seed bed preparation and weed control. It also acts as a direct mean of pest control, by changing the physical environment of insect inhabitating the soil. The type and timing of tillage markedly affect the survival of insect pests and its natural enemies. The insect that diapause during winter in the soil can be managed by fall ploughing e.g. cutworms. The availability of these requisite in the surroundings and the most essential is to know about the insect behaviour in attaining these requisites. The major requisites for an insect can be grouped into availability and suitability of food and space. Interference into these two factors or their manipulation can render the area unsuitable for the pest. To fulfill the objectives of ecological management we can group the various control tactics/measures into two broad categories

a) Disturbing the food source

b) Making the ecosystem unsuitable

Disturbing the food source

Food alongwith being vital component of the life system is also the easiest fact on which could be easily targeted for manipulation. As a result of this a significant impact is created on the insect population as it disrupts both the nutrient availability as well as habitat of the insect. Crop we grow provide a suitable habitat and sufficient food for a pest to establish in that ecosystem. Manipulating the presence of that crop in time and space can eliminate the insect requisites. This will disturb the supply of food and habitat of a pest and render the space unfit for the pest, forcing it to search elsewhere for its needs.

Selection of site

It is necessary to take into consideration the crops growing around the field where we have to grow our crop. Insects have high mobility and can very easily move from one field to another as from one botanically similar crop to other. It is admissible to grow two different crops adjacent to each other. It will help in moderating the pest population. For example soybean and corn are not grown adjacent to each other. Similarly in order to avoid the potato virus transmission by aphids it is not admissible to grow seed potato adjacent to main potato plantations.

Large scale monoculture of crops like rice, cotton, sugarcane etc. have resulted in increase in insect pests of these crops e.g. outbreaks of *Heliothis armigra* and highly on cotton in the late 1980s.

Plant spacing

The objective of location of plant with in a crop is to obtain maximum yield and has always been of great interest to the agriculturists. Alongwith having an affect on the yield of crop, spacing of plants also affects the environment of insects. Plants placed close together create closed canopies that aid in insect movement, a situation that may be favourable for pests as well as natural enemies. Seeds are sown in higher rates where infested plants are removed from the crop thus reducing the insect population e.g. maize seeds as the plants having maize borers attack are removed. In rice the population of plant hoppers weevil and gall midge are found to be higher in closely spaced plants. In case of spruce weevil *Thylacites incanus* the trees growing sufficiently apart provide conjenial atmosphere for eggs to hatch as sufficient sunlight is available for the eggs. Planting of young spruce trees at wider spacing acts as a management approach against this weevil.

Crop rotation

The basic idea behind this approach is to create a gap in time in the insect food. Rotation of crop is an important method of discontinuing the pest requisite of food. In addition it also plays an important role in improving the soil fertility. The practice of crop rotation dates back to 1887, when Balfour advocated crop rotation involving cereals and pulses to minimize the insect attack. The crop rotation gives best results when pest has a narrow host range or the feeding stage is not very mobile. The root feeders have less mobility and if one crop having the food requisite of that pest is followed by a crop that does not contain the requisites of food, then these pests die as they can not move to longer distances and are not able to survive till the completion of cycle of that crop. White grubs (Caleoptera: scarcabacidae) feeding on roots are very effectively controlled by the introduction of alfa alfa in the rotation cycle.

Rotation involving botanically similar crops should be avoided e.g. introduction of paddy in north India has resulted in paddy wheat rotation. The wheat crop serves as a host for the survival of a number of insect pests including stem boner, pinkborer and shootfly daring winter months. This has resulted in increasing attack of these pests on both the crops (Nagarajan, 1989).

Time of planting

Synchronisation of the feeding stage of insect with the susceptible stage of crop is the most important factor which increases the injury to the crop. If some how we can asynchromise these stages then possibly the injury level will be decreased. Manipulation in time of planting asychromise these stages very easily and successfully. By changing the planting time we can protect the damage to crop in following ways:

a) Asynchromy between host and pest
b) Synchronising insect pests with the time when the population of natural enemies is at its peak.
c) Synchronise production of crop with alternate hosts available in the near vicinity.

The classical example is that of Hessain fly *Meyetiola destructor* (Say) control in U.S.A. The adults live for 3 to 4 days and lay eggs on upper surface of leaves. Delaying of planting the wheat till all the adults had died resulted in the ecological management of this pest (Teetes, 1993). In regions where Hesain flies are a problem, fly free dates have been established to guide autumn planting of wheat, based on seasonal occurrence of adults (Dufoux, 2001). Delaying of crop plantation also help in overcoming the damage by the migratory insects e.g. damage to sugarbeets by curnly top virus is avoided or reduced by planting the crop after migration of beet leaf hopper a vector for the disease (Norris *et al.*, 2003). The larvae of white fringed beetle (*Naupactus* spp.) feed on the sweet potato tubers in the late season. Early harvesting before the larvae reach sufficient size reduces the damage to the crop (Zehnder *et al.*, 1998). The change in planting times of Kale in Southern United States was found effective to reduce the level of infestation of cabbage looper *Trichoplusia ni* is helpful in exposing the grubs of rice root weevil to the birds and its natural enemies. Fruit fly pupates in the soil around the trees of the host plants. Deep ploughing after harvest buries infested plant parts and stubble and destroys the larvae of pests such as armyworm, *Pseudaletia unipunctela* (Harworth) (Capinera, 2001). Shallow autumn tillage may provide 90% control in sawfly population (Steffey *et al.*, 1992). Raking up and hoeing of soil around these trees helps in killing the pupa. The pests that feed on stubble after harvest may starve if the ground is tilled (Speight *et al.*, 1999).

If tillage is deliberately used for reducing insect infestation, it should be timed with insect life cycle. The immobile stage in the lifecycle of pest is to be targeted in order to expose them to the unfavourable conditions e.g. in European corn borer, ploughing of stubble results in 90% reduction of hibernating larvae (Horn, 1988).

Tillage is not always advantageous but has also certain disadvantages. It has also got affects on the natural enemies of some insects for example parasitoids of cereal leaf beetle, *Oulema melanopus* (Linn.) are severly affected by tillage whereas the pest remains unharmed (Pedig, 2002). In some cases tillage also aggravates the pest population. For e.g. in some areas the soil surface tends to form a crust. Keeping this crust intact can inhibit weed germination and prevent the plantation of soil inhabiting pests, but the tillage breaks this crust and provides congenial conditions for weeds and soils (Norris *et al.*, 2003).

Sanitation

A large number of insect pests breed or over winter in the debris in and around the fields. The removal of this debris eliminate the over wintering and egg laying sites and results in reduction of rate of survival and reproduction. Sanitation involves the removal of this debris which could provide the over wintering and breeding sites to different insect pests. This may be achieved by cloughing, shredding and chopping, burning or raking and collecting them into piles for burning. Removal and destruction of crop residues also prevents/ reduces the carry over of pests from one crop to next. Removal of damaged plants and uprooting infesting plants results in lowering down of borer, shoot fly and armyworm population in maize. Restorying of stubble in cotton after harvesting in controlling boll weevil and pink bollworm in cotton (All, 1999). Destruction of cone fresh maize stalks reduces the hibernating population of stem borers (Capinera, 2001). Collecting and destroying the dropped fruits reduce the population of this pests of codling moth *cydia pomonella* (Linn.) (Prokopy, 2001).

Probably the greatest limitation to clean residues is the cost of collection and destruction due to which the approach becomes impractical for commercial enterprises. Also the removal of plant residues play an important role in soil conservation and their removal expose the soil to wind and water erosion.

Cleaning of the tools and machinery to be used in the farms/orchards is also important before going from one field to another. Decontamination of animals, equipments and using of pest free seeds etc. are also some important operations.

Sanitation is of utmost importance for storage of grains or crops. Cleanliness, alongwith maintaining cool, dry atmosphere is necessary in preventing the stored grains from the attack of insect pests.

Alternate host management

Removing of plants from the surroundings which could act as alternate hosts when the main crop is harvested can reduce the growth rate of many pests. Many pests have to switch to other crops for fulfilling their requirements. If such alternate hosts are destroyed, insect numbers can be reduced. The plants serving as alternate hosts may occur with in the crop field or in surrounding areas. Whiteflies use many broad leaved weeds as alternative hosts in the absence of main crop (Norris *et al.*, 2003).

Many pests of annual crops leave the unprotected barren area of the field and find shelter in the nearby dense vegetation and other cover in which they spend winter. They survive the weather in a state of dormancy and acts as a source of infestation in the next season. Burning or ploughing such habitats results in reduction of such sources. For example pests of many vegetables such as stinkbugs over winter in plant cover in the edge of plantings. Elimination of such habitats reduces the infestation of these bags in cabbage and beans etc. (Capinera, 2001).

The plants that remain in the field after the crop harvest harbors large number of pests and act as a source of infestation e.g. corn plants in soybean fields promote egg laying and colonization of larvae of rootworm, *Diasrotica* species. These larvae colonize on corn in the next season.

Care must be taken when destroying the alternate hosts, as it can also disturb the habitat of some natural enemies and results in the unwanted removal of these helpful organisms.

Irrigation

Irrigation is most important component of crop production in many areas. It also plays an important role in preventing insect problems as the soil moisture has significant affect on insect survival. Flooding of fields has been recommended for reducing the attack of cutworms, armyworms, termites, white grubs etc.

Winter irrigation reduce the population of over wintering pink ball worms upto 50-70% (Beasley, 1992). Sprinkler irrigation has been effective in suppressing certain foliage feeding insects. The potato moth, *Phthorimaea operculella* is effectively suppressed by frequent overhead irrigation of potatoes in Newzealand. The diamond back moth *Plutella xylostella* (Linn.) and codling moth *C. pomonella* are effectively suppressed by frequent overhead irrigation of potatoes and overhead cabbage (McHugh and Foster, 1995).

Water management can also be used to grow more vigorous plants thereby resulting in reduction of losses. However, it is not a practice of reducing the favorability of ecosystem to the pests.

Other approaches

In many instances it is not possible to modify the food source or manipulate the environment of the pest, or it may not prove very effective in pest management. Insects with high dispersal rates are not controlled by the approaches. It is sometimes possible to take advantage of high dispersal rates and their preference to one host over another in order to manage these pests. Some of the methods which could be used are as follows.

Trap cropping or strip harvesting

Trap cropping is the practice of attracting pests to small plantings in or around a main crop. Trap crops are generally more favourable hosts for a target pests then the main crop. Strip harvesting is similar to trap cropping except that trap is created in a main crop by harvesting different areas at different times.

If a trap crop is healthy and is vigorous then the main crop is not as the insect did not leave the trap crop and incase it is begins to leave the trap crop then it can be sprayed thus preventing the main crop from the exposure to pesticides. It also helps in conserving the natural enemies in the main crop. Population of bean leaf beetle, *Cerotoma trifurcate* was suppressed in Missisipi in Southern United States by planting 5 to 10 per cent of a field with an early maturing soybean variety about 10 days to 3 week early then the main crop. Later in the season the stink bugs, *Nezara viridula* are also attracted to these traps and can be treated with insecticides. Trap cropping has been successfully employed on a large scale in four crops viz. cotton and soybean in USA potatoes in Bulgaria and Cauliflower in Finland (Dhaliwal and Arora, 1996).

The example of strip harvesting is of lygus bugs cotton and alfa alfa are grown adjacent in California. The lygus bugs move in large numbers from alfa alfa to cotton during harvesting. So in order to avoid this migration the alfa alfa is harvested in alternate strips. This approach is most applicable for forage crops, which are harvested several times in each growing season.

Both these practices results in lowering the use of chemical in controlling the pests of different crops.

Altering harvesting time

Harvest times can be modified to avoid some types of insect losses. Early or late harvesting, selective harvesting or strip harvesting can be helpful in suppressing a variety of insect pests and conserving their natural enemies. As a rule crops injured by insects should be harvested at earliest possible date. The harvesting time in cotton is very important for as pink bollworm is related. It diapause's during winter in last instar mainly in seeds of balls. In USA the crop termination is achieved by use of defoliants and desiccants so that the balls open nearly at the same time and are promptly harvested. After harvest the plants and shredded mechanically and ploughed deep in the soil.

In forestry logging is done in specific times of the year to minimize damage from wood-boring insects. Timber cut from late summer to early winter is least likely to be damaged by wood borers.

Strip harvesting of alfa alfa in order to prevent movement of lygus bug to cotton is also a type of modification of harvesting practices.

References

All. J.N. 1999. Cultural approaches to management arthropod pests In Ruberson, J.R. (ed) Handbook of pest management. Marcel Dekker, New York, pp 395-416.

Anon 2000. Definition for integrated pest management USDA CSREES, Washington, DC Available at http// www. Reeusda.gov.

Arora, R. 1996. Major insect pests of Rape seed mustard In. R.K. Upadhgay and K. G. Mukheryee (eds) IPM system in agriculture, Vol. oilseeds. Aditya Publications (p) Ltd. New Delhi India.

Beasley, C.A. 1992. Winter irrigation reduces spring emergence of pink bollworm moths. In; Proceedings Beltwide Lotton Production Research conference, Vol. 2. National cotton council of America, Memphis, Tennessee pp 943-944.

Brown, T. M. and W.G. Brogdon 1987. *Ann. Rev. Entomal* 32:145.

Capinera, J.L. 2001. Handbook of vegetable pests. Academic press. New York, 728pp.

Dhaliwal, G.S. and Arora. R. 1996. Synthetic organic insecticides. In Principles of insect pest management pp 175-221.

Dufoux. R. 2001. Biointensive Integrated Pest management. Appropriate Technology Transfer for Rural areas. (ATTRA), *Fayetteville, Arkansas* 52 pp.

Harwood, R. R. 1990. A history of sustainable agriculture. In Edwards, C.A; Lal, R. Madden P Miller, R. H. and House, G (Red) Sustainable Agricultural system. St Press, Rehvay Beach, Florida pp 3-19.

Horn, D.J. 1988. Ecological approach to pest management. Guilford press, New York, pp 285.

Jayaraj, S. (ed) 1987. Resurgence of Sucking Rests. Tamil Nadu Agricultural University Coimbtore. India.

Kupatt, C.C; Bassi, A. B and Allemann D.V. 2005. Future methods for controlling Weeds plant diseases and insects. In. Pesticides interaction in crop production pp. 545-565.

Mc Hugh, J.J. Jr and Foster, R.F. 1995. Reduction of diamond backmoth (Lepidoptera: Pultellidae) infestation in lead cabbage by overhead irrigation. *Journal of Economic Entomology* 88 162-168.

Nagaragan, S. 1989 Plant protection problems in rice wheat rotation system. A prespective. *Oryza* 26 (G) 329-333.

Norris, R. F. Caswell-chen, E.P and Kagan, M. 2003. Concepts in Integrated pest management. Prentice Hall, upper Saddle River, New Jersey. Pp 586.

Pedigo, L.P. 2002. Entomology and Pest management. Prentice Hall, Upper Saddle River, New Jersey pp 742.

Pradhan, S; Jotwani, M.G. and Sarap P. 1963. *Indian J. Entomol.* 34 :176.

Prokopy, R. 2001. Twenty years of apple production under an ecological approach to pest management Fruit Notes 66, 3-10.

Saxena, J.D; Sinha, S.R. and Sinha, S.N. 1999 *Ann Pl. Protic Sci* 7 :261.

Speight, M.R. Hunter, M.D.and Watl, A.D. 1999. Ecology of insects –Concepts and applications. Blackwell Science London 350 pp.

Srivastava, C; Agarwal, M; Sinha S.N and Devi P. 2003 Susceptibility of storage insects infesting wheat to some insecticides. In: proceedings of National Symposium on frontier Areas of Entomological Research held at IARI New Delhi from 5-7 November 533.

Steffey, K; Gray, Mand Weinzierl, R. 1992. Insect management In; Conservation tillage systems and management Midwest Rlan Services, Iowa Stalt University Ames, Iowa pp 67-74.

Teetes, G.T. 1993. The Environmental control of insects using plating times spacing. In : D. P Pimentel (ed) CRC Handbook of Pest Management in Agriculture, volume I CBS Publishers and Distributors, New Delhi, pp.169-182.

Verma, R. K. 2002. Advances in cockroach control. *Asian J. Microbial. Biotech. Environment Science* 4: 245-249.

Zehnder, G.W. Driggs, T. H. and Pittsi, J.A. 1998. Management of white fringed beetle (Coleoptera: Curcilionidae) grubs damage to sweet potato with adulticide treatments. *J. Eco Ento,* 91 708-714.

Entomology : Novel Approaches, 2007
Eds.: *P.C. Jain and M.C. Bhargava*,(Page 121-132)
New India Publishing Agency, New Delhi
E-mail: newindiapublishingagency@gmail.com
web: www.bookfactoryindia.com

CHAPTER 6

RECENT ADVANCEMENTS IN PEST MANAGEMENT: BIOTECHNOLOGY THE NEED OF HOUR

O.P. Dubey[1] and O.P. Sharma[2]
[1]Assistant Director General (OP), ICAR, Krishi Bhawan, New Delhi-110 001
[2]National Centre for Integrated Pest Management, LBS Building. Pusa , New Delhi - 110 012

INTRODUCTION

In India tremendous progress has been made in agriculture. In absence of check on population growth, it may cross 1.5 billions by 2030. As a consequence per capita availability of land 0.13 ha will further reduce to 0.1 hectare in 2020 A.D. The demand of growing population for food and fibre require country to produce higher yields from presently cultivated land. Currently an average of 2790 calories of food available each day for every human on the planet, (23 per cent more than in 1961 and enough to feed every one). The production targets and required yield to feed the ever increasing current population with present available resources will require the growth rate of 2.35 and 2.2 per cent in rice and wheat, respectively and as high as 4.45 and 3.38 per cent for pulses and oil seeds, respectively. This requires national average yield to be increased by 30-50 per cent, which is not a cakewalk. India will require producing additional 5-6 million tones annually to keep pace with ever-increasing national food and nutritional requirements. Recurrent famines, starvation death and lack of resources led to green revolution, which as of now has turned our begging bowl to agriculturally advanced cereal exporting country. Over the past four decades, worldwide food production has more than kept pace with the doubling of world population. Gains in food availability has been maximum in India, where the green revolution enabled a rise of 38 per cent between 1961 and 1998 to 2660 calories per person daily. In India the production has more than doubled since 1960, largely as a result of the introduction of new crop varieties and the intensification of agriculture - supported by increased applications of chemical fertilizers and pesticides. Plant diseases, caused by infectious fungi, bacteria, viruses, phytoplasmas and nematodes, result in economic losses to the level of 18%. These problems include reduced yields; lower quality, decreased nutritional value and contamination with toxins. Control of plant diseases is essential to provide adequate supply of food, feed, and fiber. Growers currently spend large sums to achieve partial control of pathogens that attack crops and other plants, even then, crop losses significant because of diseases. Reducing such losses has long been a high priority objective for agriculture. Besides the obvious benefits for producers and processors, plant health protection is an important mechanism for increasing food supplies without increasing the land under cultivation. The demand for certain food crops in 21st century warrants all out immediate remedial measures. Planner as well as policy makers are worried for tougher trade rules in the wake of implementation of World Trade Organization (WTO) agreement; however equally support the need and the aims of sustainable agriculture to produce sufficient food, in an environmentally and socially acceptable manner while maintaining the local natural resource base for future generations. But some scientists suggest that population and economic growth worldwide have raised the demand for food beyond levels that can be supported by extensive and environmentally friendly farming. However, like others we are equally committed to Integrated Pest Management (IPM) as defined by the FAO, as an economically viable and socially acceptable approach to crop protection. It is the top priority mission of the ICAR and Govt. of India to provide safe and effective technologies to protect against unacceptable losses caused by weeds, diseases and pests, which to an estimate works out to Rs.90,000 crores per annum at the current price and even exceeds in the time of epidemics.

Plant biotechnology ushers for plant protection workers to maintain healthy crop to optimize crop yields with minimum pesticide use. The chief objective of biotechnology is to provide nutritional security to expanding population. Genetic engineering has the potential to provide cultivars with enhanced ability to withstand or resist attack by plant pathogens. New approaches to plant disease control are particularly important for pathogens that are difficult to control by existing conventional methods. In India the average percentage of crop losses caused by plant pathogens, insect pests, and weeds, has steadily increased to 18%, while worldwide, plant diseases reduce crop productivity by 12% (Food and Agriculture Organization, 1993). Worldwide, pesticide applications costing $26 billion dollars annually are applied to manage pest losses. Genetically engineered plants resistant to plant pathogens can prevent crop losses and reduce pesticide usage. This article provides a current perspective on areas of research and application of plant genetic engineering for resistance to plant pathogens.

Biotechnology has been defined as the use of biological means to develop processes and products by studying organisms and their components. Biological means include bioreactors, immunolocalisation, biologically active cells and enzymes, cell tissues and organ cultures, genetic engineering, gene transfer, recombinant DNA technology and hybridoma techniques. These methods offer approaches and are precision tools to manage pests for sustainable yield and prevent famines by providing food security. It has already proven its effectiveness in cotton and has considerable potential to increase agricultural production through improving the yield and nutritive value of crops, extending crop production in to inhospitable habitats, unforeseen climatic stresses and improving the protection of the crops against insect pests. In view of declining acreage and food demand crop protection always emerges as a key area for research where the potential of biotechnology is discussed to prevent significant losses caused by pests. The greenness of biotechnology as an alternative to chemical insecticides makes it seem appropriate to IPM, which is the most after plant protection practices in view of Rio convention and WTO accords.

Although biotechnology is a term applicable to a range of activities, including conventional but important activities like fermentation, it is to interpret biotechnology as involving the creation and use of genetically modified organisms. Biotechnology emerged as an important tool in IPM due to one or other of the reasons viz., wide spread resistance to synthetic pesticides, inability to find replacement chemicals at same rate and unacceptable negative consequences, shift from paradigm of chemicals to that of environmentalism, pursuit of new chemicals immune to widespread resistance genes and new classes of chemicals increasingly difficult to identify, develop, register and commercialize and more importantly to revitalize the benefits of biological control .

ROLE IN PEST AND DISEASE MANAGEMENT

The proponents of each component of sustainable agriculture must recognize their proper role in the integration of scientific practice in order to ensure that their technology is a coherent element in IPM practice. Traditional technologies of IPM viz., classical biological control, augmentative releases of natural enemies, physical methods, improved pesticide application technology, pheromones, male disruptive lures, microbial pesticides provide the basis for new technology options for the role of biotechnology in IPM. Some important applications of biotechnology in IPM are as follows -

Identifying pests and diseases

The first and most important step in managing a plant disease is to correctly identify it. Although some diseases can be diagnosed quickly by visual examination, others require laboratory testing for diagnosis. These laboratory procedures may take days or even weeks to complete and are, in some cases, relatively insensitive. Delays are frustrating when a quick diagnosis is needed so that appropriate disease control measures may be taken to prevent plant injury, especially when high value cash crops are at stake. Fortunately, as the result of advances in biotechnology, new products and techniques are becoming available that will complement or replace time-consuming laboratory procedures. The characterization of organisms, biosystematics is essential to pest management. Communicating the local experience of IPM between countries requires the common language of taxonomy. Biotechnology offers more powerful tool for identification of pests and diseases. Monoclonal antibodies, polymerase chain reaction (PCR), random amplified polymorphin DNA (RAPD) and other methods hold promise.

When the organism is microscopic, polymorphic or part of a sibling species complex, or induces symptoms that cannot be easily interpreted, biotechnology can be used to find molecular characteristics that identify the relevant taxon and can use these to provide a basis for user- friendly, field-level identification. Characteristics of the organism's genome and proteins it encodes provide wealth of information, which can assist identification. Empirically chosen pre-existing genetic markers can be used, digestion of genomic de-oxyribonuleic acid (DNA) with a restriction enzyme uses the distribution of the recognition sequence as marker and the resulting fragment length pattern may be directly or indirectly used as an identifier. The technique of restricted fragment length polymorphism (RFLP) mapping can be used to reproduce and identify fragment patterns from an organism's DNA, as with use of the polymerase chain reaction (PCR) in random amplified polymorphic DNA (RAPD) and used to identify pests and diseases with complicated morphological characters.

Incorporating resistance to pests and diseases.

Although plants cannot be immunized in the traditional sense, resistance to pests and diseases can be achieved by incorporation and intracellular expression of selected foreign genes. Production of transgenic plants containing *Bacillus thuringiensis* toxin genes is currently done and a number of cotton transgenic varieties have been released for commercial cultivation throughout the country. Transformation of plants to express traits such as resistance to viruses (e.g., ring spot disease of papaya), tolerance to herbicides (e.g., herbicide tolerant soybean) and production of proteins from non-vegetative sources has been successfully demonstrated. Transgenic insects are expected to be of value in many different IPM settings. At present this transformation is limited to Drosophilids, but there is intensive effort to extend the scope to other insect groups. Genes who would be a priority target for mobilization to encode the factors, which make some strains vectors refractory for transmission of the diseases. Engineered insects is the increased efficiency of mutated strains for use in sterile male techniques. Transformed insects would be highly effective as vectors. Apart from these engineered insects would increase the fitness of beneficial species, incorporating tolerance to commonly used pesticides in to beneficial insects such as honeybees, parasitoids and predators. This would assist in maintaining population levels of these species in circumstances where chemical intervention is necessary and prevent pest resurgence subsequently.

Controlling insect pests and diseases

Biotechnology presents options for control of organisms whose abundance or location does not suit our specific purposes. These are primarily related to genetic engineering of microbial agents or entomophagus nematodes for control of insect pests and weeds. Insect viruses are desirable agents for control of pests, because of their specificity. Despite their characteristics that suit IPM, slow rate of kill prevents their wide spread use and adoption. Techniques in molecular biology allow viruses to be engineered in ways that significantly increase the rate at which they kill their hosts. This involves inserting genes that codes for an insect specific toxin or other metabolic disruptant in to the genome of the virus. Product of this can be used as knock down insecticides. Insect pathogenic bacteria Bt. are commercially available. There is intensive effort to engineer these to vary characters such as host range and persistence. Other groups include the insect pathogens such as nematodes and fungi as effective agents of pest control, which also would benefit from the use of biotechnology. Although plants have mechanisms to protect themselves against pathogen attack, in contrast to animals, there is no "immune system" per se in plants. With the advent of genetic engineering, plants can be engineered to express an antibody against a protein crucial for pathogenesis resulting in a level of immunity or resistance to the pathogen. Plants can be protected from diseases with transgenes (genes that are engineered into plants) that are derived from the pathogens themselves, a concept referred to as pathogen-derived resistance. For example, plant viral transgenes can protect plants from infection by the virus from which the transgene was derived.

Reverse genetics

This involves manipulating the genotype to induce specific phenotypic shifts and represents a powerful technology at the DNA level. These modifications are predicted on a thoroughly researched characterization of the target for instance the biochemistry of the plant cell wall structure. However, phenotypic shifts, especially for insect resistance and agronomic traits depend on the extraordinary complexities and vagaries of the environment and plant's response to it. Given the limitation of transgenics, novel options are to be explored beyond transgenics as an important tool for IPM. There is urgent need for increased strategic research on diseases/insect/plant interactions so that a much wider set of opportunities for IPM can be identified and developed e.g., cotton boll weevil's traditional host plant is not cotton but kapok. Some of the traditional natural enemies of the weevil recognize kapok as the host plant by chemical cues. Consequently cotton plants are not recognized as likely sites by natural enemies for locating weevils. Research is required to determine feasibility to transform cotton to suite and enrich weevil's natural enemies.

AREAS WHERE BIOTECHNOLOGY MADE PROGRESS AS A COMPONENT OF PEST MANAGEMENT

Biotechnological advances in crop sciences have a strong counterpart in research on insects. The Techniques of molecular biology make it possible to isolate and manipulate behaviorally, structurally and developmentally significant proteins and peptides and to determine the potential of the molecules for biotechnological approaches. Some of the areas where molecular genetics had made progress include–

Insect Control

The biotechnological tools for management of insect pests include their applications for the following areas.

Sex specific selection using chimeric genes

Application of recombinant DNA technology to isolate, manipulate and transfer genetic material to a pest insect may offer an efficient and cost effective means of achieving sex specific selection of males for use in sterile insect release programs. One general approach for genetic sexing is to impart a chemical sensitivity to the females and then select for the males by chemical treatment. The use of male limited alcohol dehydrogenase (Adh) activity is one such approach particularly in insects that have functional alcohol dehydrogenase using classical genetic techniques. Another prospective application is antisense RNA to genetic sexing. The most efficient means of introducing genes stably in to the genome of an insect is the modified P-element vector system for transformation of vectors that are phylogenetically unrestricted which is a prerequisite for the application of DNA recombinant technology for genetic modification of insect pests.

Applications on neuroendocrine systems

Advances in biotechnology on neuroendocrine system have implications for insect control. Neuroendocrine events and titers may be altered by - (i) Genetic engineering of neurohormone genes into suitable baculovirus cloning expression vectors, (ii) Disruption of neurosecretory activity by exogenous chemicals that affect aminergic neurons, (iii) Development of analogs that are antagonists or superagonists of neurohormones at receptor proteins and (iv) Inhibition of enzyme instrumental in hormone synthesis or degradation.

The variety of physiological processes affected by neurohormones suggests that disruptions in neuroendocrine-dependent processes may be used as the basis for a new approach to insect pest control. Since neurohormones such as PTTH and allatotropin and allatostatin regulate the secretion of the ecdysteroids and JH, manipulation of these neurohormones could influence molting and metamorphosis in larvae and reproduction in adults just as would manipulation of the steroids or the JHs, themselves.

The physiolocigal processes that might be affected by neuroendocrine manipulation are-

1. Genetic engineering and recombinant DNA biotechnology
2. Altered neurosecretion
3. Inhibition of enzymes for hormone synthesis or degradation and
4. Neurohormone analogs.

Insect cuticle structure and metabolism

The formation and molting of the insect exoskeleton requires a myriad of enzymatic and endocrine processes that centre principally in the epidermal cells. NMR studies identify the constituents in sclerotised and mineralized cuticles that look for a covalent cross links apart from the aminoacids and carbohydrate contents. These studies in the chemical changes provide scope for genetic manipulation and alter the chemical that may have utility in the abnormal development of the insect species.

Inhibition of juvenile hormone esterase

Juvenile hormone (JH) acting in concert with the steroid hormone ecdysone are believed to control the timing of the larval - larval molts, larval-pupal and pupal -adult transformations of the insect. The pure enzyme JH esterase is a suitable candidate for biotechnological approaches from the use of the enzyme as a probe for edocrine regulation. The use of trifluroketone affinity columns led to the purification of the pure enzyme and direct injection of the enzyme has a clear anti JH activity. The trifluroketone could be used as transition state mimics which inhibit the protease of the JH.

Insect Resistant Crop Plants

The use of Bt. Proteins and genes to produce insect resistant transgenic plants is well established. BT was discovered by Dr. Ishiwatari (a Japanese bacteriologist) in 1901 in silkworm farms as the causal agent of the sotto bacillus disease. The insecticidal activity was discovered in 1911 by Berliner (Germany) and plants are recognized to provide useful levels of insect control against many important pests. Bt is the most versatile biopesticide for use in pest management. It is used on agricultural crops, harvested crops in storage, ornamentals, bodies of water, and around the home to control various groups of insects, depending on the type of toxin produced by the specific isolate of Bt. Currently, over 180

Bt products are registered in USA. Bt is primarily a pathogen of insect larval stages. Younger larvae are more susceptible than older larvae. There are different strains or varieties of Bt available that have been selected for the control of specific insect pests. It is compatible with many pesticides and can be mixed with chemical insecticides. In determining which additional crops might be engineered to express Bt. Proteins, consideration of their utility in pest control based on local conditions and to integrate these in to insect resistance management programmes is important.

As new genes are discovered and tested, new pesticidal activities will become available from Bt. Some of these genes are likely to be useful in generating second and third generation products in the current crops, by providing genes useful for enhancing both pest management and resistance management. In further transgenic pest resistant plants of all types not just insect resistant, will be generated using biotechnology. This will depend on the discovery and development of genes with useful pesticidal properties such as protease inhibitors, lectins and enzymes and from the development of genes for fungal/ bacterial disease control.

Transgenic Plant Disease Management

Disease resistance genes could be sourced from plant pathogens themselves, as was possible with coat protein-mediated plant viral resistance and with toxin-inactivating protein-mediated bacterial resistance. Host plants can also contribute an enormous number of disease resistance genes such as those encoding pathogenesis-related (PR) proteins, which have been used against fungal diseases. Plants can be protected from diseases with transgenes (genes that are engineered into plants) that are derived from the pathogens themselves, a concept referred to as pathogen-derived resistance. For example, plant viral transgenes can protect plants from infection by the virus from which the transgene was derived. One of the most successful examples, as of date, of the use of transgenic resistance against plant diseases is that was accomplished in the management of papaya ring spot virus (PRSV) in Hawaii. Traditional breeding in bringing about resistance was of no avail, as cross ability barrier was a constraint. Under these circumstances, coat-protein-mediated resistance using coat protein gene sourced from a Hawaiian strain of PRSV was attempted resulting in resistant transgenic line. Recently, a gene silencing mechanism has been put to productive use in containing rice yellow mottle virus. An open reading frame of the virus itself is expressed in rice in order to stop the viral spread in an effective manner. Similar attempts also have been made in containing multiple viral infections (tomato spotted wilt virus and turnip mosaic virus) in plants. Xa21, a wide-spectrum bacterial blight resistance gene isolated from an African rice, *Oryza longistaminata* was back crossed into a cultivated variety at International Rice Research Institute, Philippines. The resistance gene was cloned using molecular markers. Wild fire disease of tobacco caused by *Psuedomonas syringae* pv. *tabaci* is a serious disease. A phytotoxin secreted by the pathogen drastically modifies the amino acid metabolism of the plant with the eventual accumulation of ammonia in tobacco leaves, which causes extensive blighting. Interestingly, the pathogen that synthesises the phytotoxin remains unaffected by the toxin. This formed the basis for a search of the candidate gene from the pathogen itself. A toxin-inactivating gene, 'ttr' was successfully isolated from the pathogen and cloned into tobacco cultivars, which showed excellent wildfire resistance. PR protein genes appear to be a very potential source for candidate genes for fungal resistance. These proteins may play a direct role in defense by attacking and degrading pathogen cell wall components. Typical candidate genes are that encoding chitinases and ß-1,3 glucanases. Increasing expression of individual and multiple PR-proteins in various crops have demonstrated some success in enhancing disease resistance in particular pathogens (e.g., *Rhizoctonia solani*, the sheath blight pathogen). A recent research shows a chitinase gene from an anti-fungal biocontrol fungus species (*Trichoderma viride*) confers transgenic resistance against the rice sheath blight pathogen. A rice PR-5 protein gene in wheat delays onset of symptoms caused by the wheat scab pathogen.

Biocontrol of Plant Pathogens

The environmental concern over pesticide use has been instrumental in a large upsurge of biological disease control. Among the various antagonists used for the management of plant diseases, *Trichoderma* and *Pseudomonas* play a vital role. Among the various isolates of *Trichoderma, T. viride, T. harzianum, T .virens* and *T. hamatum* are used against the management of various soil-borne pathogens causing diseases of crop plants. It has many advantages as a bio-control agent owing to its high rhizosphere competence, ability to synthesize polysaccharide–degrading enzymes, amenability for mass multiplication, broad spectrum of action against various pathogens and above all its environmental friendliness.

Fluorescent pseudomonads suppress the pathogens either directly through the secondary metabolites or indirectly by inducing plant-mediated defense reactions. The crucial factor in the success of biological control by fluorescent pseudomonads is their ability to colonize on the rhizosphere and their persistence throughout the growing season. Fluorescent pseudomonads are root colonizers because they occur in the natural habitat of rhizosphere and thus when they are reintroduced to roots through seed they colonize root surface profusely. Fluorescent pseudomonads suppress the pathogens by antibiosis through the production of various antibiotic substances such as 2,4-diacetyl phloroglucinol, phenazine-1-carboxylic acid, oomycin A, oxychlororaphine, pyoluteorin, pyrrolnitrin and pyocyanine. Siderophores are extracellular, low molecular weight substances which selectively complex iron with high affinity. Fluorescent pseudomonads produce siderophores such as pseudobactin and pyoverdine which chelate the iron available in the soil and make it unavailable to pathogen thus the pathogen dies for want of iron. In rice, seed treatment followed by root dipping and foliar spray with Pseudomonas fluorescens showed a higher induction of ISR against sheath blight pathogen, *R. solani*

Developing Virus Resistance in Transgenic Plants

Plants that express coat protein (CP) gene are resistant to infection by the strain of TMV from which the CP gene was isolated and to closely related strains. This is described as CP-mediated resistance (CP-MR). To date CP-MR has been reported to be effective against viruses in atleast 11 different classification groups and in a variety of dicotyledon and monocotyledon crop species, including potato, tomato, papaya, squash, cucumber, melon, alfaalfa, sugarbeet, rice and maize. An alternative approach to conferring resistance has been to express portions of the viral, genome that encode the viral replicase. Resistance against several classes of viruses can be engineered by expressing a gene encoding either the full viral replicase or a portion thereof. The expression of these viral sequences apparently interferes with the replication process. Attempts to interfere with virus replication through the application of antisense RNA have been shown to effective in providing resistance to viruses whose genome comprises DNA, such as the Gemini viruses. Biotechnology offers the possibility to interfere with the proteins (helper component proteins) that are necessary for acquisition of the virus by the insect vector by expressing the mutant acquisition proteins that block the involvement of the virus helper.

Approaches include preventing the spread of the virus and its accommodation in infected cells or tissues. An effective resistance could be engineered by decreasing the accumulation of virus thereby limiting the likelihood that the virus could be acquired and transmitted to adjacent plants by insect vectors.

Deployment of Pesticidal Engineered Crops

The strategies for first generation product development include - Expand sources of insecticidal genes, Identify plasmids harboring active genes, characterise insecticidal activities, construct new strains by non recombinant methodologies, optimise fermentation production and recovery process and develop formulations that maximize field efficacy. The genetic engineering of a microbial pesticide requires three components- a gene that encodes an insecticide, a plant colonizing the microbe and the technology to transfer and express the gene in the recipient colonizing microbe.

The diverse perspectives of pesticidal-engineered crops include -

1. They are one shot replacement for other pest control methods.
2. Insects may adapt to the first pesticidal plants, but by that time other toxins will be engineered in to plants as replacements.
3. Insects are capable of adapting to pesticidal plants but resistance management strategies can be used to significantly slow the rate of adaptation.
4. Insects are likely to adapt to pesticidal plants, but while the plants are still working they offer a means to limit insecticide use and to allow natural biological control agents to reestablish, making it easier for farmers to wean themselves from future reliance on insecticides.

The strategies for engineering and deploying pesticidal crops that will slow down the evolution of resistance can categorized in to the following:

1. High levels of constitutive expression of single toxins in all plants
2. High levels of constitutive expression of two or more toxins in all plants
3. Spatial or temporal mixtures of plants with high levels of constitutive expression of one or more toxins and other plants with no toxin expression.
4. Low levels of expression of single toxins interacting with the pests natural enemies.
5. Targeted gene expression.

Insect Vector Control

Effective application of any form of vector control measure must be based on an understanding of the frequency and distribution of insect vectors as well as their taxonomy, behavior and relationship to their target and the environment. Molecular tools such as restriction fragment length polymorphism (RFLP's) and random amplified polymorphic deoxyribonucleic acid (DNA)-polymerase chain reaction (RAPD-PCR) are proving useful for monitoring populations, studying their genetic and ecological relationships and detecting nuclear and mitochondria gene flow, which can be used to study migration of insects and natural enemies, movement of genes conferring resistance to insecticides or the spread of insect-vector plant diseases.

Understanding the population structure can be important for vectors, which often exist within morphologically indistinguishable sibling species complexes that can vary significantly in the potential of individuals in a vector to actually transmit disease, having an impact on the control strategy. PCR allows sensible detection of pathogens in hosts and vectors. PCR primers have been developed to detect various flaviviruses, a group of mosquito borne viruses. Manipulation of nay species generally requires a collection of genetically mapped loci. The basic size, complexity and organisation of insect genomes are being unraveled in order to understand the distribution of coding and repetitive sequences which facilitate understanding of the inheritance of insecticide resistance, the susceptibility to vector-borne pathogens, hybrid sterility, host seeking behavior, which have significance in insect control programs.

Mapping and cloning of genes provide the molecular approaches to pest control such as insecticide resistance. Strategies to employ transgenic insects for insect vector control by modifying the genetic makeup of wild populations exist. These include reducing the capacity of insects for plant pathogens; disrupt virus host interactions, transformation of insect symbionts.

Research on biological products in future need to be focused on following aspects:

1. Increased efficiency and consistency of *Bacillus thuringiensis*
2. To increase the ability of baculovirus to kill early and increase host range and virulence.
3. The time period needed to develop transgenic plants and limit the development of resistant biotypes. Production of plants with inducible and tissue specific expression systems.

CONSTRAINTS IN USING ENGINEERED PESTICIDES

The constraints include technical, implementation, training, regulatory, policy and economic. Key technical constraints include lack of understanding of crop and pest biology interactions, lack of field oriented and IPM research, inconsistent efficacy of biological agents and lack of funding. A main barrier is the perception that IPM is riskier and more expensive than conventional controls and the preference for immediate solutions. Some key regulatory constraints that have an impact on the use of genetically engineered products are diversity and uncertainty of interstate and international regulations, lack of clear EPA guidelines and unreasonable crop destruction requirements, lack of consistent national IPM policy. Successful implementation of biotechnology in IPM depends upon-

1. Demonstrated need (regulatory or economic),

2. Research: a body of knowledge on applied ecology of the pest, crop, beneficial and control alternatives,
3. Commercial development: testing via extension and compatibility with other practices and materials, and
4. Implementation: willing growers, pest control advisers, extension, marketing and sales representative. Most importantly, a demonstrated success will pave way for adoption of microbials and other new technologies such as transgenic plants in IPM.

Molecular biological advances associated with in-situ applications, portability, hybridisation imaging, non-radioactive detection and automation are now being used in entomology. PCR amplification of genes involved in insect pathogen interaction, receptor -substrate complex formation, pesticide resistance and biotype evolution, adiopokinetic hormone expression and neuropeptide production are now possible. Nucleic acid immunological probe kits are available commercially in wide array of diagnostic situations, including single-gene detection in the field. These developments will greatly expand our ability to monitor pests and predict pest outbreaks.

References

AnanthakrishnanT.N. 1992. Biotechnological Approaches to the Biological control of Phytophagous insects. Entomology Research Institute, Madras, India.

Boulter, D. 1992. Biotechnology and plant protection against insect pests. In : Plant Biotechnologies for Developing Countries. (Eds.) Sasson, A. and Costarini, V. Technical Centre for Agricultural and Rural Cooperation, Wageningen, *The Netherlands*, pp. 195.

Brown, I. 1993. Molecular detection assays for plant pathogenic fungi. *Agribiotech News and Information*, 5: 219 - 222.

Crampton, J.M. 1992. Potential application of molecular biology in entomology . pp; 3-18. In Insect Molecular Science. (Eds.). Cramptom, J.M. and Eggleston, P. Academic Press,

Kirschbaum, J.B. (1985) Potential implication of genetic engineering and other biotechnologies to insect control. *Annual Review of Entomology*, 30 : 51-70.

Luckow, V.A. and Summers, M.D. 1988. Trends in the development of baculovirus expression vectors. *Biotechnology* 6, 47-55.

Meada, S. 1989. Expression of foreign genes in insects using baculovirus vectors. *Annual Review of Entomology*, 34, 351-372.

Meredith, S.E.E.O. and James, A.A. 1990. Biotechnology as applied to vectors and vector control. *Annual Parasitology Hum. Comp*, 65 : 113-118.

Paul. A. Hedin, Julius J. Menn and Robert, M. Hollingworth . 1988. Biotechnology for Crop Protection. ACS Symposium series 379. pp. 479.

Perlak, F.J. and Fischhoff, D.A. 1990. Expression of *Bacillus thuringiensis* insect control proteins in genetically modified plants. In: Osmond, G. (ed.). Proceedings of the Vth International Colloquium on Invertebrate pathology and Microbial Control. Society for Invertebrate Pathology Adelaide, pp. 461-465

Persley, G.J. 1990. Agricultural Biotechnology : Opportunities for International Development. CAB International, Wallingford, Oxon, U.K. pp. 123-142.

Persley, G.J. 1996. Biotechnology and Integrated Pest Management. CAB International, Washington D.C., pp. 475.

Saxena, H. 1991. Microbial control of insect pests. *Farmer and Parliament*, 3:7-11.

Swaminathan, M.S. 1991. Biotechnology in Agriculture. McMillan, India.

Walker, V.K. 1989. Gene transfer in insects . Advances in Cell culture 7: 87-124.

Ward, E.R., Ryals, J.A., and Miflin, B.J. 1993. Chemical regulation of transgene expression in plants. *Plant molecular Biology*, 22 : 361 - 366.

Whitten, M.J. 1991. The relevance of molecular biology to pure and applied entomology. *Entomologia Experimentalis et Applicata* 53, 1-16.

Wyatt, G.R. 1991. Gene regulation in insect reproduction. *Invertebrate Reproduction and Development* 20 : 1-35.

Entomology : Novel Approaches, 2007
Eds.: *P.C. Jain and M.C. Bhargava*,(Page 133-144)
New India Publishing Agency, New Delhi
E-mail: newindiapublishingagency@gmail.com
web: www.bookfactoryindia.com

CHAPTER 7

GENETIC ENGINEERING OF PLANTS FOR INSECT RESISTANCE

M.C. Bhargava, R.K. Choudhary and P.C. Jain
S.K.N. College of Agriculture, Rajasthan Agricultural University, Jobner.

INTRODUCTION

While the world has been changing during the last few years both politically and economically in unexpected and remarkable way, food security remains an unfulfilled dream of billions of people. With the world population crossing six billion marks, the demand for food grains is increasing at an alarming rate. Increasing population, on the other hand, has led to decrease in cropping area. In order to meet the global food demand we have to increase the productivity. Insects have always been a major cause of yield-loss in agriculture. They account for 20-30% of crop-loss. Several approaches e.g. physical, chemical and biological have been used with varying degree of success to reduce such losses.

The insect pest management is one of the most important yield destabilizing factors in almost all the crops. While for some insects, genetic sources of resistance are available in the respective plant germplasm, for many others germplasm is almost devoid of effective resistance genes. It is because of this factor that conventional breeding strategies have not been successful in breeding insect pest's resistant varieties. The only means available at present for managing the insect is extensive application of insecticides and pesticides. Excessive usage of the pesticides causes, on one hand, environmental pollution and on the other, it forces the insects to develop resistance against these chemicals. Application of the pesticides also imposes additional economic burden on the farmers.

To increase the net productivity by preventing such losses in an ecofriendly manner is the need of the hour. Natural resistance that provides reasonable protection against insect pests is available in some crops. It means that crop or variety has a specific gene which confers the plant resistance through production of certain protein (genes are the unit of inheritance). Through breeding, the resistance can be transferred to different varieties of the same crop. But such type of transfer of resistance from one species to another is not possible through breeding programme. With the advancement of technologies in the recent past, it is now possible to identify and isolate a particular gene of one organism, and then to introduce it into another. Such techniques are referred to as genetic engineering. Such gene transfer can be done across the species boundaries. The organism which receives the gene is called transgenic. So it is now possible to introduce the insect resistance gene of one species into a desired plant which never had such type of resistance gene in its gene pool. Once introduced the new trait will be inherited generation after generation and can also be transferred to other varieties of that crop by hybridization.

In recent years, the area of genetic engineering has provided an altogether new approach for circumventing the use of chemical insecticides for controlling the insects and pests. This approach employs the techniques of molecular biology whereby genes coding for insecticidal proteins can be identified, isolated, mobilized and expressed in crop plants. Such transgenic plants thus acquire the ability, which is heritable, to protect themselves from insect pest attack.

This new technology is seen as an additional tool for the control of insect pests and could offer certain advantages over conventional use of insecticides, such as more effective targeting on specific insects, greater resilience to weather condition, no question of seasonal application, operator exposure and most importantly ecofriendly system.

Transgenic insect resistant plants can be developed through following steps

- Submission and approval of bio-safety protocols for transgenic crop development from the appropriate government and other environmental agencies along with its disposal.
- Identification of insect resistance gene.
- Isolation of the desired gene.
- Construction of recombinant DNA with a suitable vector and the desired gene or preparation of a construct.
- Transfer of the vector construct to the callus or protoplast of the desired plant.
- Regeneration of transformants.
- Study regarding expression of insect resistant character in transgenic plant.
- Examination of the stability of the transgene.
- Biosafety protocol confirmation and proper disposal of products.
- Large scale field trial at multiple locations for several generations.
- Proper notification to the biosafety agencies about the transgenic product along with failures, if any.

Genes for insecticidal proteins like δ endotoxin, protease inhibitors, α amylase inhibitors, lectins, cholesterol oxidase, isopentenyl transferase and vegetative insecticidal proteins have been identified and transgenic plants with these genes have been developed (Table 1 &2). Several other proteins with insecticidal property are being identified and characterized, and will provide valuable options for insect management using transgenic crops.

Table 1 Development of genetically engineered baculoviruses with increased speed of kill

Virus	Foreign gene	Host insect	% inc rease of kill	References
Bm NPV	Toxin from *A. australis*	*B. mori*	40	Maeda *et al.* (1991)
	Diuretic hormone from *M. sexta*	*B. mori*	LT_{50} reduced by one day	Maeda (1989)
Hz NPV	Toxin 34 from *P. tritici*	*H. zea*	47	Popham *et al.* (1995)
AcM NPV	Hv JHE with GLY for Ser201 substitution	*Trichoplusia ni,* *Heliothis virescens*	16 29	Bonning *et al.* (1995)
	EGT	*Spodoptera frugiperda*	22	O'Reilly and Miller (1991)
	Hz PBAN	*T. ni*	26	Ma *et al.* (1998)
	Toxin 34 from *Pymotes tritici* (Pt)	*H. virescens*	33	Watkins *et al.* (1997)
	Toxin 21A from Pt	*T. ni*	49	Tomalski *et al.* (1993)
	Neurotoxin from spider, *Angelenopsis aperta* (μ-Aga-IV)	*S. frugiperda*	43	Prikhodoko *et al.* (1996)
	Mag (μ-Aga-IV)+Mellitin from *Apis mellifera*	*S. frugiperda*	37	Prikhodoko *et al.* (1996)
	Bt toxin gene	*T. ni*	NS	Merryweather *et al.* (1990)
	Trunicated Cry 1A(b)	*S. exigua*	NS	Martens *et al.* (1995)
	Maize mitochondrial gene URF13	*T. ni*	40	Korth and Levings (1993)
	Chitinase gene from *M. sexta*	*S. frugiperda*	23	Gopalakrishnan *et al.* (1995)
	Neurotoxin Sh 1 from sea anemone, *Stichodactyla helianthus*	*T. ni* *S. frugiperda*	37 40	Prikhod'ko *et al.* (1996)
	Toxin from spider, *Tegenaria agrestis*	*T. ni* *S. exigua* *H. virescens*	20 18 19	Hugher *et al.* (1997)
	Latro insecto toxin from the Black widow spider	*H. virescens* *T. ni*	12 12	Watkins *et al.* (1997)

Contd. ...

Table 1 Contd. ...

	Human C mye antisense gene	*S. frugiperda*	28	Lee *et al.* (1997)
	Enhancing gene from *T. ni* (Enh 26)	*S. exigua*	10-21 fold	Hayakawa *et al.* (2000)
Ha NPV	*A. australis* toxin gene Aa 1T	*H. armigera*	40	Sun XiuLian *et al.* (2003)

Table 2 Some transgenic plants developed with insect resistance genes

Gene product	Target insect	Transformed plants
Transgenic plants with crystal protein genes of B. thuringiensis		
Cry IAb	Lepidoptera	Apple, cotton, maize, rice, potato, tomato, tobacco, white clover
Cry, IH	Lepidoptera	Maize
Cry 3A	Coleoptera	Tobacco, potato, egg plant
Cry 9C	Lepidoptera	Maize
Transgenic plants with protease inhibitor gene		
C-11 (Soybean serine protease inhibitor	Coleoptera, Lepidoptera	Potato, tobacco, poplar, oilseed
Cp TI (Cowpea trypsin inhibitor)	Coleoptera, Lepidoptera	Apple, lettuce, potato, rice Sunflower, sweetpotato, tobacco, tomato
OC-I (Rice-cysteine protease inhibitor)	Coleoptera, Homoptera	Poplar, tobacco, oilseed rape
Pot PI-II (Potato protease inhibitor II)	Lepidoptera, Orthoptera	Birch, tobacco, lettuce, rice
Transgenic plant with lectin gene		
Snowdrop lectin (GNA)	Homoptera, Lepidoptera	Tobacco, potato, rice, oilseed Rape, sweet potato, sugarcane, sunflower, tomato
Pea lectin (plec)	Homoptera, Lepidoptera	Potato, tobacco
Wheat germ agglutinin (WGA)	Lepidoptera, Coleoptera	Maize
Rice lectin	Lepidoptera, Coleoptera	Maize
Transgenic plants with a amylase inhibitors		
α amylase inhibitor of common bean (αAI-Pv)	Coleoptera	Pea, tobacco
α amylase inhibitor from cereals (WMAI-I)	Lepidoptera	Tobacco

The most common gene candidates being used for engineering plants resistant to insects pests are :

a. Delta endotoxin coding gene from *Bacillus thuringiensis* strains.
b. Protease inhibitors genes from cowpea, soybean etc.
c. Amylase inhibitors from cereals and legumes
d. Lectins, primarily from wheat germ, snowdrop plant etc.

δ ENDOTOXIN

Bacillus thuringiensis, a gram-positive soil borne facultative bacterium, was first identified by Ishiwata in 1901 in silk worm larvae. In 1909, Berliner gave detailed information about the bacterium. In adverse climatic condition, it undergoes sporulation alongwith with the formation of a bipyramidal crystal protein which is toxic to insects and known as Cry protein, Bt toxin or δ endotoxin. The spores can survive for many years but the protein (δ endotoxin) remains functional for a few days to a few months depending on the environmental conditions. Different subspecies of the bacterium produce different crystal proteins and they are very specific in their action. B. t *kurstaki* acts on lepidopterans, B. t. *israelensis* on dipterans and B. t. *tenebrionis* on coleopterans. The insecticidal crystal protein (ICP) is coded by a gene in plasmid (a self replicating extra-chromosomal (DNA). The gene first gives a proprotein of 130 kDa. Proteolytic enzyme at high pH in insect midgut breaks down the proprotein to active toxin of 65-75 kDa. This protein results in pore formation of columnar cells of midgut by osmotic lysis and subsequently leads to paralysis and death of the larvae. It is not clear whether one crystal can puncture, the cell membrane or an aggregate is needed for it. There are different types of Cry proteins, i.e. Cry l. Cry ll, Cry 111, etc. which have their own specific receptor in the columnar cells of the midgut.

Bacillus thuringiensis, a gram positive soil bacterium is omnipresent. It produces intracellular crystal proteins or delta endotoxin during its sporulation (Barton *et.al.,* 1987). These proteins often appear mocroscopically as distinctively shaped crystals and constitute about 20-30% of the dry weight of sporulated cultures (Boutler *et.al*., 1990). The Bt is being used as insecticidal spray since 1957 under different trade names like Thuricide, Agree, Dipel, Javelin, Condor, Foil and Biobit. But high cost of production and low stability in the field due to UV and sunlight have resulted in its limited use. With the advancement in the field of genetic engineering, the δ endotoxin gene of *B. thuringiensis* became an attractive candidate to be among the first gene transferred into plants. In 1987, transgenic tobacco and tomato plants with δ endotoxin gene were developed and transgenic potato plant in 1992. The crystal proteins generally exhibit a molecular weight in the range of 70 kD to 130 kD. These proteins are proteolytically cleaved and activated to a core fragment of about 60 kD inside the larval midgut. The toxic fragment of the 130 kD proteins is localized in the N-terminal half and the highs conserved C-terminal half is not essential for toxicity (Fietelson *et.al.,* 1992). The crystal proteins characterized by their insecticidal activity and spectrum are grouped in four classes. Lepidoptera-specific (Cry I). Lepidoptera and Diptera-specific (Cry II). Coleoptera-specific (Cry III) and Diptera-specific (Cry IV). Initially insect resistant transgenic plants were developed using Cry lA(b) gene with constitutive promoters. As the transformation was done using *Agrobacterium tumefaciens*, the experiments were limited to dicots only. Today more than 45 nucleotide sequences of crystal protein genes have been determined. Although they constitute a family of related genes, there are 17 distinctly different crystal protein genes. Feitelson *el.al.*, (1992) have expanded this range to 29 distinct crystal proteins based on their toxin domain structures. Identification of new strains of B.t. exhibiting toxicity to mosquitoes and beetles has stimulated extensive screening programmes for novel variants of crystal proteins. Although transgenic tobacco plants with Cry lA(b) gene became resistant to tobacco hornworm (*Manduca sexta*) and budworm (*Heliothis virescens*), use of full length native gene resulted in poor expression of ICP in transgenic plants. To overcome this problem of low expression of native δ endotoxin genes, synthetic genes were constructed. Different strategies are employed to design such synthetic genes.

- Elimination of potential intron (non-coding DNA sequence in a gene) splicing sites.
- Elimination of polyadenylation sites
- Elimination of mRNA instability signals
- Optimization of translation by altering codon usage.

Table 3 Crop plant transformed for cry gene expression

Crop	Target insect	cry gene
Tobacco	*Manduca sexta*, *Spodoptera* spp. *Heliothis* spp.	cry1A (b) cry1A (c)
Tomato	*Heliothis* spp., *Keiferia* spp.	cry1A (b) cry1A (c)
Potato	*Phthorimaea operculella*	cry1A (b)
	Laptenotarsa decemlineata	cry111A
Cotton	*Heliothis* spp.	cry1A (b) cry1A (c)
Corn	*Ostrinia nubilalis* *Chilo suppressalis* *Cnaphalocracis medinalis*	cry1A (b) cry1A (b)
Broccoli	*Plutella xylostella*	cry1A (c)
White spruce	*Chorisoneura fumiferana*	cry1A (a)

Native Cry genes are having 37% G + C. In modified Cry gene the G + C content has been increased to stimulate that of plants. Partially and fully modified genes show a 10 to 100 fold increased expression when compared with wild type gene. Transgenic cotton with modified Cry lA(c) gene have been found resistant to cotton bollworm (*Helicoverpa zea*), leaf perforator (*Bucculatrix thurberiella*), pink bollworm (*Pectinophora gossypiella*) and beet armyworm (*Spodoptera exigua*). Monsanto, an American company, has developed cotton bollworm and pink bollworm resistant transgenic cotton varieties called 'Bollgard' and 'Ingard'. Transgenic potato plants with Cry IIIA gene show resistance to Colorado potato beetle (*Leptinotarsa decemlineata*).

Designing synthetic cry genes for specific crops, keeping in view the plant's codon usage, will further improve the protein expression and insect toxicity. In addition identification of novel and more toxic crystal protein genes using PCR technology wherein primers corresponding to the toxic domain of the gene can be used to amplify the DNA from naturally occurring bacterial strains in expected to be quite useful in such an endeavour. Strategies to introduce truncates cry genes into chloroplast genome will circumvent the codon usage problems because of the prokaryote nature of the chloroplast gemone and transcriptional and translational apparatus. Such an approach will be more useful for leafy vegetables and crops in which green tissues are predated upon by insects. Chloroplast transformation, though difficult, has been achieved recently (Hilder *et. al.,* 1987). What is needed in construction of chloroplast vectors carrying a truncated version of native cry gene and its expression under a suitable chloroplast promoter. It will be of interest to investigate the level of cry gene expression in the chloroplasts and its effectiveness.

With the development of other methods of gene transfer like microinjection, gene gun and electroporation it has become easier to transform monocots. Transgenic maize with Cry IA(b) gene resistant to European corn borer (*Ostrinia numilalis*) is the first transgenic monocot. Commercially available transgenic maize varieties of USA are yieldgard, maisgard and naturegard. Transgenic rice with modified Cry IA(b) gene is resistant to striped stem borer (*Chilo suppresalis*), yellow stem borer (*Scirpophaga incertulas*) and leaf folder (*Cnaphalocrosis medinalis*).

PROTEASE INHIBITOR

Other potential candidates in engineering insect resistance in crop plants include protease inhibitors. Many protease inhibitors have been identified from many plant species, which are active against certain insect species both in *vitro* assays against insect gut proteases and *in vitro* artificial diet bioassays. The mechanism of antimetabolic action of protease inhibitors is not yet fully elucidated. Direct inhibition of digestive enzymes is unlikely to be the only effect and possibly the proteases vital in insect development, such as those involved in moulting are also affected. Many plants have natural defense mechanisms against insect based on the production of certain protease inhibitor(s). These protease inhibitors generally act on the enzymes which are responsible for the protein catabolism in insect midgut. Protease in insects includes serine, cysteine, aspartic and metallo protease that catalyse the release of amino acids from dietary protein and so provide the nutrients crucial for normal growth and development. In contrast to δ endotoxin, these proteins (inhibitors) have antimetabolic activity against a wide range of insects. Lepidopterans have serine proteases and coleopterans have cysteine proteases. Serine protease inhibitors are of 1000 Da and can simultaneously inhibit two molecules of the substrate, may be of same type or two different types. That's why they are termed as double headed or Bowman birk type.

Transgenic tobacco plants with cowpea trypsin inhibitor (CpT1) genes are resistant to larvae of tobacco budworm (*Heliothis virescens*). Transgenic tobacco with tomato or potato protease inhibitor II is resistant to tobacco hornworm (*Manduca sexta*). It inhibits trypsin and chymotrypsin activity. Although most of the works carried out to date have used serine protease inhibitors, transgenic tobacco plants with cystein protease inhibitor like oryzacystatin I are resistant to coleopteran insects. However, the major disadvantage of protease inhibitors is that it needs very high level of the inhibitor for insect mortality.

Transgenic plants with serine protease inhibitor genes from mammals and tobacco hornworm (*Manduca Sexta*) are resistant to a number of lepidoptera and homoptera. Based on in *vitro* study of inhibition of proteolysis in midgut extracts of a range of lepidopteran larvae, bovine pancreatic trypsin inhibitor (BPTI), a antitrypsin and spleen inhibitor (SI) have been identified as promising insect resistance proteins and transferred into a wide range of plants. *M. Sexta* protease inhibitor gene expressed in cotton and in tobacco has been found to reduce reproduction in *Bemisia tabaci*. Not only the genes encoding protease inhibitor isolated from cowpea have conferred resistance when expressed in tobacco but the tomato inhibitor II gene when expressed in the same model plants has also been shown to confer insect resistance (Schnepf and Whiteeley, 1981).

α AMYLASE INHIBITOR

Insecticidal activity associated with the common bean seed is due to an inhibitor of α amylase (a digestive enzyme) activity. Studies have shown that α amylase inhibitors present in seeds are active against insects and mammals but not against plant and bacterial α amylase and are thus highly specific. α-amylase inhibitor is known to be larvae of two seed feeding beetles, the cowpea weevil, *Callosobruchus maculates* (Ishimoto and Kitamura, 1988) and the azukibean weevil (*C. chinensis*). Transgenic pea plants with α amylase inhibitor gene have been found to produce 1.2% of α amylase inhibitor protein in seeds and confer resistance to both cowpea weevil and azukibean weevil.

LECTINS

Lectins are carbohydrate binding proteins found in plenty in plants, animals and bacteria. Lectins found in plants, particularly in seeds, are involved in defense against the attack of bacteria, fungi, insect and other animals. Lectin binds to midgut epithelial cells of insects and leads to disruption of critical cellular functions., Insectidial activity of lectins i.e. LC_{50} is in milligram per gram diet, whereas that of δ endotoxin is nanogram per gram diet. That's why the lectins must be expressed at a higher level in the transgenic plant. Transgenic maize with lectin genes like wheat germ agglutinin or jacalin or rice lectin gene is found to be resistant to European corn borer and southern corn rootworm. Transgenic tobacco, sweet potato and mustard with lectin gene are also resistant to various insects. Soybean lectin has been shown to be toxic to the larva of the lepidopteran, *Manduca sexta*. Similarly wheat germ agglutinin has been shown to have an inhibitory effect on European corn borer (*Ostrinia nubilalis*) and corn root worm (*Diabrotica undecimpunctella*). The disadvantage, however is using the above mentioned lectins are that they have been shown to exhibit mammalian toxicity (Vaeck *et.al.*, 1987). The pea lectin, on the other hand has been shown to be innocuous to mammals but insecticidal only in the pure form. In addition, lectin purified from snowdrop and garlic is insecticidal and non toxic to mammals. Snowdrop lectin is toxic to brown plant hopper (*Nilapervata lugens*), a pest of rice.

CHOLESTEROL OXIDASE

Screening of filtrates from microbial fermentation often reveals presence of certain proteins which attacks some insects. The protein present in two streptomyces culture filtrates which show acute toxicity to boll weevil larvae has turned out to be a cholesterol oxidase (CO). Cholesterol is required for integrity and normal function of virtually acellular membranes; therefore, any interference in its availability leads to toxicity. Histological studies on boll weevil larvae fed on diet containing CO show cellular attenuation accompanied by local cytolysis of midgut epithelial cells. CO catalyses the oxidation of cholesterol already incorporated into the membrane to produce ketosteroids and hydrogen peroxide. Midgut epithelium is the primary target for CO and thus it interferes with the process of digestion and absorption. Although newly emerged adult boll weevils exposed to CO do not exhibit acute toxicity, they display severe oostatic effects (i.e. reduction in fecundity and oviposition) due to poor development of ovaries in adult females. LC_{50} of first and second instard larvae of boll weevil is comparable to that of δ endotoxin. Enzymatic activity of CO has been detected in protoplast of transgenic tobacco with native CO gene.

ISOPENTENYL TRANSFERASE

The gene for isopentenyl transferase (*ipt*), a key enzyme in cytokinin biosynthesis, was isolated from *Agrobacterium tumefaciens*. Expression of *ipt* in transgenic tobacco and tomato plants by a wound inducible promoter has resulted in a decrease in leaf consumption by tobacco hornworm (*M. sexta*) and reduced survival of the peach potato aphid (*Myzus persicae*).

OTHER INSECTICIDAL PROTEIN FROM PLANTS

Tryptophan decarboxylase (TDC) from periwinkle expressed in tobacco is known to reduce the reproduction of whitefly (*Bemisia tabaci*) by upto 97%. TDC converts tryptophan to indole alkaloid tryptamine. Tryptamine and tryptamine derived alkaloids in plants may act as antioviposition and antifeedant agents or as inhibitors of larval and pupal development.

VEGETATIVE INSECTIDAL PROTEINS (VIPS) FROM *BACILLUS* SP.

In search of novel insecticidal proteins, it is discovered that culture supernatant fluid collected during vegetative (i.e. log phase) growth of *Bacillus* sp. is a rich source of insecticidal substances. Supernatant fluid of *Bacillus cereus* isolates possesses acute bioactivity against Northern and Western corn rootworm. Here insecticidal properties are associated with a binary system, whose two proteins are termed vip 1 and vip 2. Similarly, supernatant fluid of certain *B. thuringiensis* cultures has potent insecticidal properties against black cutworm and fall armyworm. Potency-wise it has been shown that VIPs are comparable to δ endotoxin. The cloning and characterization of the gene encoding this activity has yielded a novel insecticidal protein vip 3A which is active against lepidopteran insects particularly black cutworm, fall armyworm and beet armyworm. This protein results in complete lysis of gut epithelial cells, particularly columnar cells. Transgenic maize with vip3A gene is found to be resistant to underground pests like western and northern corn rootworm.

RESISTANCE MANAGEMENT

Insects have demonstrated a high capacity to develop resistnace to wide array of chemical insecticides. Resistance against microbial spray of δ endotoxin has already been observed. There has been some concern that use of δ endotoxin genes for insect control in transgenic plants may lead to rapid development of resistance. Such concern is however based largely on computer models.

To minimize the development of resistance to transgenic monitoring of pest population should be combined with a variety of strategies :

High dose strategy – The strategy to produce transgenic plants synthesizing very high level of insecticidal proteins has been found to be highly effective in preventing or delaying development of resistant variety of insects. This strategy may help eliminate heterozygotes (i.e. carrying only one resistant allele). Although there is a chance of decline in titre level of the protein in successive generations or with ageing, there is no report of decline of titre level below the important desired level in Bt-cotton and maize in the trials before commercial release. Again it is apprehended that high dose of toxin in transgenic plant produce may be harmful for human health. But Bt gene have been found to be species specific and are thus harmless to other non target insects, vertebrates including users and the environment.

Refuge strategy- It is adopted to provide a source of susceptible insects for mating with the resistant ones, of any. This will prevent the fixation of resistance gene. Refuges can naturally occur when a certain percentage of an agricultural acreage is planted with non-transgenic plants. Refuges may also be established by using mixture of transgenic and nontransgenic seeds, or by growing border rows of susceptible variety, or by growing susceptible variety of a little away from the crop field for high mobile insect species. However, the point of concern regarding this strategy is the size of the refuge. In India, for Bt cotton cultivation, experts have suggested a separate refuge of 30% of the land cultivated.

Multigene strategy – Transgenic plants producing additional insecticidal proteins with either different modes of action, different targets or both can solve the problem. Recent discovery of a number of new insecticidal strains of *B. thuringiensis* and new insecticidal proteins from such strains may provide more active components to be used for the purpose. New insecticidal components from sources other than *B. thuringiensis* may also be helpful for this. Several options should be available that include introduction of multiple genes with unique mode of action into the same plant, or alternatively, independent plants each with a unique insecticidal protein could be used in rotation.

REFERENCES

Barton, K.A.; Whiteley, H.R. and Yang, N.S.L. 1987. *Bacillus thuringiensis* delta endotoxin in transgenic *Nicotiana tabacum* provides resistance to Lepidoperan insects. *Plant Physiology*, 85: 1103-1109

Bonning, B.C.; Hiiver, K.; T.F.; Duffey, S. and Hammock, B.D. 1995. Development of a recombinant baculovirus a modified juvenile hormone esterase with potential for insect control. *Archives of Insect Biochemistry and Physiology* 30:177-194

Boulter, D.; Edwards, G.A.; Gatehouse, A.M.R.; Gatehouse, J.A. and Hider, V.A. 1990.Additive protection effect of incorporating two different higher plant derived insect resistance gene in transgenic tobacco plant.*Crop Protection*, 9: 351-354

Feltenlson, J.S.; Rayne, J. and Kim, L. 1992. *Bacillus thuringiensis* : Insect and beyond. *Bio. Tech.* 10: 271-275

Gopalahrishnan, B.; Muthukrishnan, S. and Kramer, K. T. 1995. Baculovirus mediated expression of a *Manduca sexta* chitinase gene: Properties of the recombinant protein. *Insect Biochemistry and Molecular Biology*, 25:255-265

Hayakawa, T.; Hashimoto, Y. and Granados, R.R. 2000. Enhancement of baculovirus infection in *Spodoptera exigua* (Lepidoptera: Noctuidae) larvae with *Autographa californica* nucleopolyhyderovirus of *Nicotiana tabacum* engineered with a granulovirus enhancing gene. *Applied Entomology and Zoology*, 35(1): 163-170

Hilder, V.A.; Gatehouse, A.M.R.; Scheerman, S.E.; Barker, R.F. and Boulter, D. 1987. A novel mechanism of insect resistance engineered into tobacco. *Nature*, 330: 160-163

Hughes, P.R.; Wood, H.A.; Breen, J.P.; Simpson, S.F.; Duggan A.G. and Dybas, J.A. 1997. Enhanced bioactivity of recombinant baculovirus expressing insect specific spider toxins in Lepidopteran crops pests. *Journal of Invertebrate Pathology*, 69: 112-118

Ishimoto, M. and Kitamura, K. 1988. Identification of the growth inhibitor on azuki bean weevil in Kidney bean (*Phaseolus vulgaris* L.). *Jap. J. Breed.*, 38: 367-370

Korth, K.L. and Levings, C.S. 1993. Baculovirus expression of the maize mitochondrial protein URF 13 confers insecticidal activity in cell culture and larvae. *Proceedings of Natural Academy of Science USA*, 90: 3388-3392

Lee, S.Y.; Qu X Chen,W.; Poloumienko, A.; MacAfee, N.; Morin, B.; Lucarotti,C. and Krause, M. 1997. Insecticidal activity of a recombinant baculovirus containing an antisense cmyce fragment. *Journal of General Virology* 78: 273-281

Ma, P.K.; Davis, T.R.; Wood, H.A.; Knipple, D.C. and Roelofs, W.L. 1998. Baculovirus expression of an insect gene that encode multiple neuropeptides. *Insect Biochemistry and Molecular Biology*, 28: 239-249

Maeda, S. 1989. Increased insecticidal effect by a recombinant baculovirus carrying a synthetic diuretic hormone gene. *Biochemical and Biophysical Research Communications*, 165: 1177-1183

Maeda, S.; Volrath, S.L.; Hanxlik, T.N.; Harper, S.A.; Majima, K.; Maddox, D.W.; Hammock, B.D. and Flower, E. 1991. Insecticidal effects of an insect specific neurotoxin expressed by a recombinant baculovirus. *Virology*, 184: 777-780

Martens, J.W.M.; Konester, M.; Weijets, F.; Groffen, S.J.A.; Hu Z. Bosch, D. and Vlak, J.M. 1995. Characterization of baculovirus insecticides expressing an insect neurotoxin potential for pest control. *Biotechnology*, 9: 848-852

Merryweather, A.T.; Weyer, U.; Harris, M.P.G.; Hirst, M.; Booth, T. and Possee, R.D. 1990. Construction of genetically engineered baculovirus insecticide containing the *Bacillus thuringiensis* subsp *kurstaki* HD-73 delta endotoxin. *Journal of General Virology*, 71: 1535-1544

O'Reilly and Miller, L.K. 1991. Improvement of a baculovirus pesticide by deletion of the egt gene. *Biotechnology*, 9:1086-1089

Popham, H.J.R.; Li, Y. and Miller, L.K. 1997. Genetic improvement of *Helicoverpa zea* nuclear polyhedrosis virus as a biopesticide. *Biological Control*,10: 83-91

Prikhod'ko, G.G.; Robson, M.; Warmke, J.W.; Cohen, C.J.; Smith, M.M.; Wang, P.; Warren, V.; Kaczorowski, G.; Van der Ploeg, L.H.T. and Miller, L.K. 1996. Properties of three baculovirus expressing genes that encode insect selective toxins mu-Aga-IV, As II and Sh I. *Biological Control*, 7: 236-244

Schnepf, H.E. and Whiteeley, H.R. 1981. Cloning and expression of the *Bacillus thuringiensis* crystal protein gene in *E. coli. Proc. Acad. Sci. USA*, 78: 2893-2897

Sun Xiulian,; Chen Xinwen,; Zhang ZhongXin,; Wang HuaLin,; Bianchi, F.J.J.H.; PengHui Yin,; Vlak, J.M.; Hu ZhiHong,; Sun, X.L.; Chen, X.W.; Zhang, Z.X.; Wang, H.L.; Peng, H.Y. and Hu Zh, 2003. Bollworm response to release of genetically modified *Helicoverpa armigera* nucleopolyhdroviruses in cotton. *Journal of Invertebrates* Pathology, 81(2): 63-69

Tomalski, M.D.; Hutchinson,K.; Todd, J. and Miller, L.K. 1993. Identification and characterization of tox 21 A: A mite cDNA encoding a paralytic neurotoxin related to TxP-1. *Toxicon*, 31: 319-326

Vaeck, M.; reynaerts, A.; Hofte, H.; Jansens, S.; De Beuckeleer, M.; Dean, C.; Zaleau,M.; Van Montagu, M. and Leemans, J. 1987. Transgenic plants protacted from insect attack. *Nature*, 328: 33-37

Watkins, M.J.; Kivatkin, N.; Hughes, D.; Beadle, D. and King, L. 1997. Study on the biological properties of a novel recombinant baculovirus. In : *Microbial Insecticides Novelty or Necessity*, BPPC Symposium Proceeding No. 68, pp. 279-284

Entomology : Novel Approaches, 2007
Eds.: *P.C. Jain and M.C. Bhargava,*(Page 145-163)
New India Publishing Agency, New Delhi
E-mail: newindiapublishingagency@gmail.com
web: www.bookfactoryindia.com

CHAPTER 8

PESTICIDES: CONTAMINATION AND ESTIMATION OF RESIDUES

V.S. Kavadia
Professor and Director of Research (Retd.), Rajasthan Agricultural University, Bikaner

INTRODUCTION

Pesticide in broad term refers to "killer of pests" but is an agent used to control, prevent, destroy, repel and mitigate a pest. Pesticides are thus poisonous substances intentionally applied in situations, which harbour so-called pests. Agriculture including poultry and veterinary (i.e. raising of crops and protection of produce and storage), health (public places-hospitals, restaurants, gardens cinema halls etc.), transportation (shipment, air rail and road transport) are some areas where pesticides are used extensively as spot application, fumigation, aerial and field sprays, soil treatments etc. Plant protection comes to be recognized as an equally important factor of agriculture production as improved seeds, fertilizers and irrigation. Pesticides over time get accumulated in the environment and contaminate all systems i.e. air, water, soil, plants, animals and micro-organisms on being transported from one system to another leading to biomagnifications.

The fate of a pesticide in the environment and the entities, which come in contact, is depicted in the Fig. 1 reflecting the pollution and contamination. Pesticides during manufacturing, transportation, storage and actual use enter the biotic and abiotic components of the environment through air, water and soil and disturb the cohesion causing great disaster sometimes. Pollution is defined as undesirable change in physical, chemical and biological characteristics of air, land and water. This change may or will harmfully affect human life or that of desirable species, industrial process, living conditions and cultural assets. It may or will waste or deteriorate our raw material resources. Pollution increases burden of human society causing loss of resources through unnecessary wasteful exploitation, the cost of pollution abatement and deterioration of human health.

Contamination of pesticides is not very simple phenomenon but involves many complex steps or stages. In order to understand this, it is pertinent to study the constitution of the pesticide, its susceptibility to temperature, humidity, light (sun and ultraviolet), decomposition chemical, biochemical and otherwise, absorption and adsorption, volatilization, its behaviour on the surface (biotic and abiotic), run off etc. During application (spraying and dusting etc.)

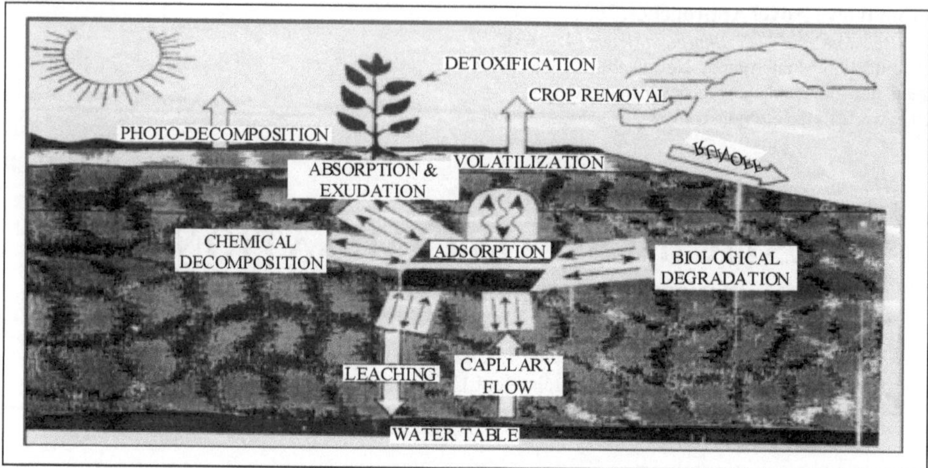

Fig. 1 Fate of pesticides in the environment

only a small portion of the pesticides is adhered to crop/plants and rest either fall down or taken up into the atmosphere by air currents or turbulence. Volatilization is a major pathway for the loss of applied pesticide, which depends upon the vapour pressure. The soil is contaminated with the pesticides either through deliberate application for controlling soil-inhabiting pests or through run off from plants or dumping of empty containers. The water sources are contaminated by surface runoff; sediment transport from treated soil, industrial and municipal wastes and directs application against aquatic pests. The plants and animals living in polluted air, water and soil environment are being contaminated or affected in one way or other abiotic or biotic routes. The contamination, thus, can be referred to as the net resultant reaction of the chemical (pesticide), substrate (surface) and the environmental factors (climatic-abiotic and biotic).

CONTAMINATION AND POLLUTION

It is difficult to make difference between contamination and pollution. It appears that these to words are synonymous to each other. In any dictionary the meaning of contaminate (v) is pollute and the pollute means contaminate, taint, infect, poison and so on. If we talk of contamination it means it also refers to pollution. In studies on fate of pesticides, the contamination and pollution go hand in hand. Persistence and residues as a result of pesticide deposit (application) on the directed surface and indirectly on living and non-living substrates (biotic and abiotic substrates) culminate in the form of contamination and pollution phenomena.

Field application of pesticides by cultivators on targeted crops not only contaminate the farm produce but also the environment, the soil, the water not only in the vicinity but also at large distances, even the biotic form and the kitchen (at home, hotels), large water bodies and so on. The contaminated substrate remains contaminated even after processing and change of forms. The studies conducted at agricultural research station, Durgapura, Jaipur (Rajasthan Agriculture University) under ICAR sponsored project reflect this(Tables 1 to 18). The farm gate and market vegetables, fruits and spices show presence of pesticides. Similarly, processed spices, milk and milk products, vegetarian and non-vegetarian diets, open and deep bore wells, green fodder and animal feeds, accumulated rain water and sewage water were not found spared from the pesticide contamination. The state of affair may be highly dangerous in places where pesticides are used indiscriminately. In the state of Rajasthan and particularly in the vicinity of large cities, the consumption of pesticides is not at a very high level (average consumption is less than 50 g) and is much less than the national average of 250 g ha^{-1}.

Table 1 Monitored pesticides residue in vegetables – Farmgate and Market at Jaipur (March and Nov.-Dec., 2002)

Vegetables	Insecticides detected	Number of samples detected			Residues mg/kg
		Analyzed	Contaminated	(%)	
Brinjal	□ HCH				0.004-0.0296
	Dimethoate				0.035-0.303
	Chlorpyriphos				0.188-0.209
	Chlorpyriphos methyl	2	2	(100)	0.067-0.068
	α Endosulfan				0.007-0.008
	α Endosulfan				0.004-0.005
	Ethion				0.018-0.027
Cabbage	□ HCH				0.024-0.043
	Dimethoate	2	2	(100)	0.028-0.366
	α Endosulfan				0.018-0.200
	PP DDE				0.004-0.005
Okra	δ HCH				0.0130.059
	Chlorpyriphos methyl				0.153-0.170
	α Endosulfan	4	4	(100)	0.025-0.166
	β Endosulfan				0.004-0.008
	Dimethoate				0.171-0.205
Onion	□ HCH				0.062-0.305
	γ HCH				0.796-1.237
	Malathion	4	4	(100)	0.024-0.028
	α Endosulfan				0.010-0.011
Tomato	α HCH				0.002-0.035
	α Endosulfan				0.094-0.098
	Dimethoate	4	4	(100)	0.052-0.064
	PP DDE				0.005-0.006

Farmgate : Tomato – Sanganer, Vatika; Onion – Sanganer; Brinjal – Sanganer; Cabbage – Sanganer

Market: Sanganer, Lalkothi and Durgapura

Table 2 Monitored pesticides residue in vegetables – Farmgate and Market at Jaipur (May and Nov.-Dec., 2003)

Vegetables	Insecticides detected	Number of samples detected			Residues mg/kg
		Analyzed	Contaminated	(%)	
Brinjal	α Endosulfan	8	2	(25)	0.090-0.120
Chillies	γ HCH α Endosulfan Chlorpyriphos Cypermethrin	10	10	(100)	0.150-0.180 0.100-0.120 0.060-0.080 0.050-0.080
Cabbage	γ HCH	6	4	(66.6)	0.080-0.100
Cauliflower	α Endosulfan β Endosulfan	6	4	(66.6)	0.005-0.008 0.004-0.007
Okra	γ HCH α Endosulfan	6	4	(66.6)	0.120-0.240 0.240-0.280

Farmgate : Okra – Sanganer; Chilies – Band and Mihanpura, Sanganer and Khejra ka vas; Brinjal – Band, Mihanpura and Sanganer

Market: Durgapura

Table 3 Monitored pesticides residue in marketed fruits (March and Sept.-Oct., 2001)

Fruits	Insecticides detected	Number of samples detected			Residues mg/kg
		Analyzed	Contaminated	(%)	
Ber	γ HCH	2	2	(100)	0.011-0.022
Grapes	γ HCH	2	2	(100)	0.039-.053
Banana	α HCH γ HCH γ HCH	10	10	(100)	0.004-0.022 0.002-0.647 0.014-0.141
Apple	α HCH γ HCH δ HCH α Endosulfan	6	6	(100)	0.003-0.012 0.567-0.963 0.019-0.057 0.005-0.032
Guava	γ HCH α HCH	2	2	(100)	0.005-0.009 1.275-1.479
Mango	γ HCH α Endosulfan	6	6	(100)	0.005-0.021 0.002-0.006

Market: Jaipur – Lalkothi, Durgapura, Gopalpura and Sanganer

Table 4 Monitored pesticides residue in marketed fruits* (June-July and Nov.-Dec., 2002)

Fruits	Insecticides detected	Number of samples detected			Residues mg/kg
		Analyzed	Contaminated	(%)	
Banana	γ HCH α HCH	4	4	(100)	0.024-0.058 0.011-0.021
Grape	α HCH α HCH	2	2	(100)	0.020-0.213 0.051-0.054
Mango	α HCH Chlorpyriphos	12	12	(100)	0.011-0.468 0.059-0.079
Pomegranate	γ HCH α HCH α Endosulfan PP DDE	2	2	(100)	0.046-0.051 0.016-0.188 0.018-0.021 0.011-0.012

Market: Jaipur – Lalkothi, Durgapura and Sanganer

Table 5 Monitored pesticides residue in ber fruits at ARS Durgapura, Jaipur and Asalpur orchard (Feb., 2002)

Ber variety	Insecticides detected	Number of samples detected			Residues mg/kg
		Analyzed	Contaminated	(%)	
Bagwari	Nil	2	-	-	
Umran	α HCH δ HCH	10	4	(40)	0.03-0.25 0.018-0.014
Sev	α HCH γ HCH	4	4	(100)	0.008-0.012 0.016-0.021
Jodhpuri	Nil	2	-	-	
Gola	α HCH	6	3	(50)	0.005-0.024
Khemli	α HCH γ HCH	2	2	(100)	0.017-0.022 0.038-0.046
Mundia	γ HCH	2	2	(100)	0.012-0.018
Chomu local	Nil	2	-	-	-

Market: Jaipur – Lalkothi, Durgapura and Sanganer

Table 6 Monitored pesticides residue in spices collected from College farm[1], home[2] and local market[3] (July, Oct., 2003)

Spice	Insecticides detected	Number of samples detected			Residues mg/kg
		Analyzed	Contaminated	(%)	
Coriander seed	α HCH δ HCH α Endosulfan	6	4	(66.6)	0.08-0.14 0.08-0.21 0.09-0.21
Cumin seed	α HCH δ HCH	6	4	(66.6)	0.014-0.018 0.011-0.031
Fennel	α Endosulfan β Endosulfan	2	2	(100)	0.011-0.017 0.005-0.008
Fenugreek seed	δ HCH	2	2	(100)	0.026-0.031
Chilli powder	α HCH	4	4	(100)	0.030-0.31 0.032-0.31
Rye	α HCH γ HCH	4	4	(100)	0.09-0.28 0.04-0.04
Pepper powder	δ HCH	4	2	(50)	0.02-0.38
Turmeric powder	δ HCH Chlorpyriphos	4	2	(50)	0.018 -0.029 -0.011-0.019

1 – Jobner, 2 – Sanganer, 3 – Shantinagar

Table 7 Monitored pesticides residue in seed spices from market* (March, 2001)

Spice	Insecticides detected	Number of samples detected			Residues mg/kg
		Analyzed	Contaminated	(%)	
Fenugreek	α HCH δ HCH Chlorothionil Chlorpyriphos	2	2	(100)	0.006-0.008 0.016-0.469 0.006-0.015 0.007-0.008
Coriander	δ HCH	2	2	(100)	0.024-0.586
Rye	α HCH δ HCH Chlorpyriphos	2	2	(100)	0.010-0.019 0.144-0.685 0.003-0.013
Fennel	α HCH δ HCH Chlorpyriphos	2	2	(100)	0.010-0.016 0.089-0.534 0.003-0.009

Contd. ...

Table 7 Contd. ...

Pepper	α HCH δ HCH Chlorpyriphos	2	2	(100)	0.004-0.011 0.0015-0.526 0.005-0.011
Cumin	α HCH δ HCH Chlorpyriphos	2	2	(100)	0.003-0.005 0.084-0.432 0.012-0.018

* Market – Jonri Bazar Jaipur

Table 8 Dissipation of insecticides in chilllies* (in green and there after drying in red chilli powder) at ARS, Durgapura, Jaipur

Dosage (g ai/ha)	Sampling interval (days)	Average residue (mg/kg)	Dissipation (%)	RL_{50} (days)
Endosulfan				
437.5	0	2.285	-	2.31
	1	2.032	11.0	
	5	0.988	56.76	
	10	0.412	81.96	
	15	0.018	99.21	
	20	BDL	100.0	
	After harvest in dry red chilli powder	BDL	100.0	
Dimethoate				
300	0	3.123	-	1.74
	1	2.615	16.26	
	3	1.381	55.70	
	5	0.515	83.50	
	10	0.066	97.80	
	15	BDL	100.0	
	After harvest in dry red chilli powder	BDL	100.0	
Cypermethrin				
87.5	0	1.423	-	0.85
	1	0.645	54.60	
	3	0.283	80.10	
	5	0.020	98.50	
	10	BDL	100.0	
	After harvest in dry red chilli powder	BDL	100.0	

BDL – Below detectable level

* Date of transplanting, 05.08.2002; date of harvesting, 04.12.2002; date of spray 1st, 24.09.2002 and 2nd, 09.12.2002 and sampling after 2nd spray; dry red chilli powder – 5 days after harvest.

Table 9 Monitored pesticides residue in milk (Feb. – Dec., 2001)

Location	Insecticides detected	Number of samples detected			Residues mg/kg
		Analyzed	Contaminated	(%)	
Vender milk Sanganer, Jaipur	α HCH Chlorothionil α Endosulfan β Endosulfan	22	22	(100)	0.004-2.329 0.002-0.009 0.001-0.281 0.001-0.038
(Saras Gold) Milk booth, Durgapura, Jaipur	α HCH Chlorothionil α Endosulfan	22	22	(100)	0.004-4.575 0.005-0.006 0.001-0.001

Table 10 Monitored pesticides residue in butter samples both branded and home made (June – Oct., 2002)

Source	Insecticides detected	Number of samples detected			Residues mg/kg
		Analyzed	Contaminated	(%)	
Branded					
Sample 1	γ HCH	2	2	(100)	0.006-0.007
Sample 2	γ HCH	2	2	(100)	0.008-0.017
Sample 3	γ HCH δ HCH	2	2	(100)	0.006-0.008 0.004-0.005
Sample 4	α HCH γ HCH	2	2	(100)	0.087 0.072
Sample 5	α HCH γ HCH	2	2	(100)	0.013-0.046 0.077-0.097
Home made*					
Sample 1	α HCH δ HCH	2	2	(100)	0.005-0.008 0.004-0.009
Sample 2	γ HCH δ HCH	2	2	(100)	0.006-0.008 0.002-0.003
Sample 3	γ HCH PP DDE	2	2	(100)	0.042-0.058 0.046-0.048
Sample 4	α HCH γ HCH	2	2	(100)	0.018-0.021 0.021-0.033
Sample 5	α HCH y HCH PP DDE	2	2	(100)	0.001-0.020 0.018-0.021 0.007-0.008

* Village Teetriya

Table 11 Monitored pesticides residue in green fodder (Sept., 2002)

Fodder sources	Insecticides detected	Number of samples detected			Residues mg/kg
		Analyzed	Contaminated	(%)	
Sample 1					
Sorghum (Teetriya)	α HCH γ HCH Chlordane	2	2	(100)	0.012-0.072 0.080-0.100 0.122-0.1 098-0.103
Cluster bean (Durgapura)	δ HCH Chlorpyriphos	2	2	(100)	0.067-0.091 0.028-0.118
Pearl millet (Durgapura)	δ HCH	2	2	(100)	0.056-0.322

Table 12 Monitored pesticides residue in honey (1st May, 2001)

Source	Insecticides detected	Number of samples detected			Residues mg/kg
		Analyzed	Contaminated	(%)	
Mansarovar market 1 {Jaipur)	□ HCH Chlorothalonil PP DDE	2	2	(100)	0.036-0.33 0.086-1.689 0.012-0.067
Mansarovar market 2 (Jaipur)	δ HCH Chlorothalonil α Edosulphan	2	2	(100)	0.043-0.046 1.897-2.503 0.011-0.015

Table 13 Monitored pesticides residue in vegetarian diet samples

Location	Insecticides detected	Number of samples detected			Residues mg/kg
		Analyzed	Contaminated	(%)	
January to December, 2001					
Dhaba & party (Durgapura) Home (Sanganer, Shantinagar and Mahaveernagar)	αHCH γ HCH Chlorothalonil αEndosulfan	24	22	(91.66)	0.003-0.007 0.001-0.335 0.002-0.004 0.001-0.04
March – Oct., 2002					
Dhaba (Anita colony and Tonk phatak) Home (Sanganer, Shantinagar, Suryanagar, Bajajnagar and Maheshnagar)	α HCH γ HCH α HCH γ Endosulfan α P DDT	16	16	(100)	0.002-0.016 0.004-0.047 0.003-0.004 0.005-0.009 0.018-0.024

Table 14 Monitored pesticides residue in non-vegetarian diet samples

Location	Insecticides detected	Number of samples detected			Residues mg/kg
		Analyzed	Contaminated	(%)	
January to December, 2001					
Dhaba and home (Sanganer) Home (Shantinagar and Gopalpura) Party (Durgapura)	α HCH α HCH Chlorothalonil	24	22	(91.66)	0.001-0.029 0.010-0.07 10.005-0.012
March – Oct., 2002					
Dhaba (Bajajnagar and Sanganer) Home and Dhaba (Tonkphatak) Hotel and Home (Durgapura, Gopalpura and Tonk road)	α HCH α HCH α Endosulfan	16	16	(100)	0.001–0.025 0.001-0.202 0.005-0.009

Table 15 Monitored pesticides residue in lake water in village Jamwa Ramgarh Jaipur (2000)

Insecticide Dec.	Jan.	Feb.	March	April	May	June	July	August	Sept.	Oct.	Nov.
α HCH -	0.70	0.66	2.14	2.98	3.72	6.0	1.65	1.12	1.1	1.07	0.8
Chlorpyriphos -	-	-	-	0.75	1.1	-	-	0.45	-	-	-

- Not detected

Table 16 Monitored pesticides residue in water (2000)

Source	Insecticides detected	Number of samples detected			Residues mg/kg
		Analyzed	Contaminated	(%)	
Open well	□ HCH α Endosulfan Chlorpyriphos Dieldrin	22 0.34	18	(81.8)	3.25 12.66 1.14 0.34
Lake	α HCH Chlorpyriphos	22	22	(100)	3.300.80

Table 17 Monitored pesticides residue in water in bore well at ARS, Durgapura, Jaipur, (2001)

Source	Insecticides detected	Number of samples detected		Residues mg/kg
		Analyzed		Contaminated
Bore well No 2	α HCH	24	16 (66.66)	0.416
Bore well No 21	α HCH Butachlor	18	14 (77.77)	0.461 0.024

Table 18 Monitored pesticides residue in open well at village Jamwa Ramgarh, Jaipur (2000)

Insecticide	Jan.	Feb.	March	April	May	June	July	Aug.	Sept.	Oct.	Nov.	Dec.
α HCH	-	-	-	-	-	-	-	-	-	0.33	-	-
β HCH	-	-	2.75	2.17	-	-	4.64	3.54	-	-	-	-
γ HCH	-	4.38	-	-	4.14	4.73	-	0.53	3.13	-	-	-
δ HCH	-	-	0.59	0.42	-	-	1.48	1.61	-	0.33	-	-
ε HCH	-	4.38	3.34	2.58	4.14	4.73	5.67	5.68	3.13	-	-	-
Endosulfan	-	-	-	-	9.48	14.16	-	-	-	-	-	-
Dieldrin	-	-	-	-	-	0.24	0.37	-	-	-	-	-
Chlorpyriphos	-	-	-	-	0.82	1.35	-	-	-	-	-	-

- Not detected

ESTIMATION OF RESIDUES

Possible risk associated with the pesticides can be in the form of (i) acute poisoning arising from the handling and application of toxic materials and,(ii) chronic risk arising from long term exposure to or ingestion of small amounts of materials in or on the treated crop on harvest. These traces, expressed as parts per million (ppm) of active ingredients and breakdown products are generally referred as residues. Both qualitative and quantitative measurements of such small amounts in or on any treated surface are known as "assay" or estimation.

A large number of techniques have been developed and perfected with high sensitivities to estimate micro-quantities. Commonly used residue estimation techniques are described below.

MICRO-BIOASSAY

Bioassay or biological assay has been defined as,"measurement of the potency of any stimulus, physical, chemical, or biological; physiological or psychological by means of the reactions which it produces in living matter (Finney, 1971). With the new insecticides for which no chemical or other method has been developed, micro-bioassay (measurement of micro-quantities) is the only method available for determining residues, decline curves or the response of the pesticides. Bioassay may be used to supplement chemical methods or for the purpose of detecting the presence of toxic metabolites or breakdown products or to speed up residue estimation. It is simple, versatile, handy, easy and highly sensitive technique for determining residues of a wide range of pesticides. It does not require sophisticated equipments or highly trained personnel, yet it gives reliable results. It is particularly useful for those pesticides for which specific methods have either not been developed or are too complicated or where metabolites have more biological activity than the parent compounds.

The main disadvantage of micro-bioassay is its general lack of specificity, although there are at least, some instances of being highly specific. The other demerits are the complexity of rearing of test organisms like houseflies, mosquito larvae, vinegar flies (Drosophila), assay methods necessary for specific organism, the susceptibility of a particular organism to plant toxicity or extractives. The result obtained by micro-bioassay in initial stages need to be verified by chemical or other methods, though this is universally true for all estimation procedures.

The micro-bioassay was first used in 1946 for the determination of DDT residues in animal tissues using housefly as the test organism.

The sensitivity of the test organism is the most crucial characteristic in its selection for micro-bioassay. Sensitivity also varies with test method, type of exposure, level of exposure period, size of exposure chamber, temperature, crop or tissue used, presence of interfering substances and toxicants.

There are several micro-bioassay methods used in estimating pesticide residues. The method used varies with the test organism, the substrate the magnitude of residues and can be in the categories of (i) direct exposure to residue containing material or (ii) extracting the residues in any solvent before exposure.

Direct exposure method: This method consists of macerating and blending the plant tissue and the exposing the test insects, vinegar flies or houseflies to it. It has advantages over extraction method in that no extraction or clean up is required. It, however, gives lower sensitivity than extraction method possibly because of masking effect of the plant and other materials.

Extraction method: (1) *Dry film method*-The test organism is exposed to a film of the pesticide in solution or deposited in crystals. Dry film technique is highly sensitive for houseflies or vinegar flies but has a demerit of difficulty and time required for extraction and cleans up and in obtaining satisfactory recovery of the toxicant. (2) *Aqueous suspension method*-This method is used for exposure of organisms of aquatic habitat such as mosquito larvae and micro-crustaceans. This method involves extraction of the pesticides, evaporating away the solvent, making suspension of residues in water and then testing it. For determining the amount of toxicant, besides mortality, physical response such as locomotary paralysis can be used. (3) *Photo-migration method*-This method is based on the habit of test organism, the mosquito larvae, to get attracted to light source. In this technique the whole organism is in constant contact with the medium and is highly sensitive. (4) *Feeding method*- It consists of evaporating the extracts to dryness over sugar, which is then dissolved in water and fed to houseflies for 24 hours. In this technique no cleanup is required. However, it is less sensitive than the dry film method and requires long period of exposure.

Enzyme inhibition method

Inhibition of cholinesterase enzyme (ChE) system in animals by pesticides, particularly organophosphates and carbamates, is the basis of this method. Most common methods to assess the cholinesterase activity for residue analysis are described here.

Potentiometric method: In this technique the enzyme is allowed to act on acetylcholine in a buffer solution for a definite period of time (usually 1-2 hours) at a constant temperature. The pH of the mixture is measured at the beginning and at the end of this time and the change in pH due to the release of acetic acid is measured which indicates the enzymatic activity. The percentage inhibition of cholinesterase can be calculated as follows:

$$\% \text{ Inhibition} = 1 - \frac{\Delta \text{ pH of sample}}{\Delta \text{ pH of sample}} \times 100$$

Where ΔpH = pH initial- pH final

The equivalent concentrations can be found out by referring to the standard calibration curve or regression equation.

Tritrimetric method: This method is based on the determination of the liberated acetic acid by titration with NaOH or KOH to a constant pH using either an acid base indicator , or a potentiometer or a conductometric titrator. It is possible to carry out titration by following changes in conductivity during the enzyme reaction.

Manometric method: The Warburg manometric cholinersterase method is based on the amount of carbon di-oxide liberated from sodium bicarbonate added to the reaction medium. The results maybe expressed as micro litres of CO_2 evolved from the bicarbonate buffer by the acetic acid, produced during the reaction. Warburg flasks are usually of standard size and shape.

Colourimetric method: Unchanged acetyl choline estimated colourimetrically at the end of completion of enzymatic reaction or cholinesterase activity requiring a substrate which after enzymatic hydrolysis yields a coloured product can be used to determine the amount of residues of a organophosphate or carbamate pesticide.

Thin layer chromatography-enzyme inhibition method: It involves use of several enzymes with cholinesterase substrates and various pH indicators to detect enzyme inhibiting pesticides or paper chromatography. Instead of paper, TLC plates can also be used for the determination of residues.

SPECTROPHOTOMETRIC METHODS

For pesticide residue analysis, three kinds of spectophotometric methods are used, depending upon the nature of electromagnetic radiation to which the sample is subjected to e.g. ultraviolet, visible and infrared. The amount of electromagnetic radiation absorbed is a measure of quantity of the sample present and can also give information regarding the chemical identity of the sample. From residue estimation point of view, the regions of interest in ultraviolet extend from 180 mμ to 380 mμ, in visible from 400 mμ to 750 mμ and in infrared from 0.75 to 25 μ.

Colourimetric method, though have some demerits, has been the backbone of pesticide residue analysis in early days. Residues of pesticides such as aldrin, BHC, DDT, diazinon, endosulfan, dimethoate, malathion, carbaryl and phorate were estimated by colourimetric methods. Lambert and Beer laws are the basis of quantitative analysis by spectrophotometry. Lambert's law states that when monochromatic light passes through a transparent medium, any layer of given thickness of the medium absorbs the same fraction of the light incident upon it. Beer's law states that the intensity of a beam of monochromatic light decreases exponentially as the concentration increases arithmetically.

Several standard solutions of known concentrations of the material under analysis are read in the spectrophotometer. A calibration curve is prepared by plotting the absorbance against the concentrations when a straight line results, if the compound follows Beer-Lambert's law. The absorption of the sample is read and the concentration computed from the curve.

An ultraviolet spectrophotometer generally consists of a source of radiation (hydrogen lamp), a manochromator (quartz prism or a grating) and a detector (photomultiplier tube). A sample can be scanned manually or automatically and the spectra can be recorded with recorders.

Visible spectophotometry was extensively used for pesticide residue analysis. In visible spectrophotometer, source of radiation is a tungsten lamp, the prism is of glass or grating and the detector is a blue photocell for wave lengths of 400 mμ to 600 mμ and red photocell for wave lengths of 600 mμ to 750 mμ.

Infrared spectrum is the most characteristic physical property of a chemical compound. It finds extensive use in "finger printing" or identifying chemical compounds.

CHROMATOGRAPHY

The term chromatography means colour writing and is used to describe such a physical method of separation. It may be defined as a method of analysis in which the flow of solvent or a gas in a sorptive medium brings about separation of substances by differential migration from a narrow initial zone.

Paper chromatography is a subdivision of liquid (solution) chromatography based on the nature of the sorptive system. The stationary phase is a sheet of paper containing water or some other polar phase and solvent is usually organic in nature. In reversed-phase paper chromatography the paper is impregnated by a non-polar solvent and developed by a polar solvent.

Separation by paper chromatography depends on the selective retention by the paper of the components of the mixture. Here two opposing forces are in operation when a solvent moves over the mixture spot. One is a driving force which acts to move the substances in the direction of solvent flow. The other is the resistive action which impedes movement of the substances by absorbing them back to the paper from the solvent. The resultant of these two opposing forces determines the distance actually traveled by any substances in any given time.

Rf values- The term used to locate a spot on the paper developed by linear flow and is calculated as:

Rf = Distance travelled by the spot (zone) / distance traveled simultaneously by solvent front

Detection of spot is made by a chemical reaction on the paper to give a colour or fluorescence. Chromogenic reagents include silver nitrate solution or bromoline-congo-red. Paper chromatography has been used extensively for pesticide residue analysis. It is very important to remember that the level of clean up of the sample has to be thorough than in case of gas chromatography.

Thin layer chromatography- Success of thin layer chromatography (TLC) depends on the preparation of uniform adsorbent layers of desired thickness on a supporting plate. Glass carrier plates are most widely used ones though aluminium, plastic or stainless steel plates have also been used. The most popular sizes of the carrier plates are 20x10 cm or 20x20 cm as these are most convenient to coat and handle. Adsorbents such as silica gel alumina, magnesium oxide, calcium hydroxide, dicalcium phosphate, bentonites, calcium and magnesium carbonate, florisil, talc and starch have been used along with 50 per cent plaster of paris as binders.

The quantitative evaluation of chromatography can be done by *(i)* separating the chromatographed compound from the adsorbent and determine it by standard micro-chemical methods like spectrophotometry, gas liquid chromatography etc., or *(ii)* measuring the spot area in relation to the weight of the material or density or colouration by a potentiometer or radioactivity. The easiest way of documenting chromatogram is photography either BW or colour. Direct use of photo-sensitive papers provides a rapid and inexpensive way of documentation.

Gas chromatography (GC) is the most commonly used analytical technique for pesticide residue studies. It provides both qualitative and quantitative analysis with speed, accuracy, precision and sensitivity in the range of nano or picograms. With this technique, not only a single pesticide but a mixture of pesticides and their metabolites can be analysed in one shot.

Separation of mixture of solutes based on the differences in physical properties is brought about by flow of gas as the moving phase; hence it is called gas chromatography. The stationary phase if a solid then it is called gas solid chromatography (GSC) and when it is a liquid then it is gas liquid chromatography (GLC).

The time lapse between the injection of sample and recording of the peak maxima is expressed on the "retention time" tR or Rt. The Rt value is a function of column temperature, carrier gas flow rate, affinity of solute for moving and stationary phase. The retention volume is the volume of carrier gas that flows through the column during the retention time at ambient temperature and pressure. Since it is a constant, it is used for calculating relative retention ratios.

A typical gas chromatograph consists of components like, source of gas i.e. gas cylinder, drying cartridge, pressure and flow controller, injection port i.e. inlet port, column, detector, amplifier and recorder. The versality of a gas chromatography lies in different types of columns, column packing, detectors, that can be utilized for the separation, identification and quantification of compounds. Selection of carrier gas depends upon the type of detectors, availability and economic feasibility. Commonly used gases are N, He and H_2. The column can be of glass, aluminium, copper or stainless steel, and its length can vary from 60 to 300 cm with internal diameter of 2 to 8 mm. Generally glass column of U shape are used. Two types of solid supports are used. Porous supports are those which are able to absorb relatively large quantities of liquid phase without loosing their flowability. These are generally made from diatomite or diatomaceous earth or Keisulguhr. The best known supports are chromosorb W or chromosorb P or similar products. The non-porous supports include micro-glass beads, used preferably for the analyses of high boiling compounds. They have low surface area and low holding capacity, 0.05 to 0.5 per cent only. Liquid stationary phase (i.e. partitioning liquids) is most important from the analysis point of view. Nonpolar liquids such as SE-30, DC-200, DC-11 or polar like fluorinated silica Q.F.1 are used widely. However mixtures of Q.F.1 with DC-200, Ov-17, SE-30 have also been used quite frequently.

The choice of detectors depends on the type of compound, sensitivity required and the economics. The most common detectors used are thermal conductivity detector (TCD), flame ionization detector (FID), alkali flame ionization detector (AFID), flame photometric detector (FPD), electron capture detector (ECD) and micro-coulometric detector (MCD). TCD gives response to all compounds but is not very sensitive. Most organic vapours have lower thermal conductivity than H_2, He or N and their presence in one of these carrier gases reduces the amount of heat conducted from the filament. Sensing filament is either a metal or thermister. FID contains two electrodes across which a potential of 500 volts is applied to capture the ions formed. The vaporized organic compounds are burnt at the jet and ionization molecule is measured. The flame jet acts as an electrode (+) while the metallic loop or cylinder surrounding the flame serves as the collector electrode (-). FID response is directly proportional to the number of C atoms in the molecule which are bound directly to hydrogen or other C atoms. Consequently the detector is insensitive to H_2O, CS_2 or inorganic compounds. In AFID, the ionization processes at flame are modified b the presence of an alkali metal which is coated over platinum coil or loop. The response could be as high as 1000 times in case of hetero atoms e.g. P. NPD, nitrogen phosphorus detector, provides maximum sensitivity to compounds containing nitrogen and phosphorus and is particularly useful in estimating organophosporus pesticides. It can detect pesticides up to levels of 10^{-11}. FID is a flame emission detector when compounds containing P or S is burnt, the excited atom emits light in variable range which is monitored with the help of a narrow band interference filters. A-526 mm maxima filters are used for P and 394 maxima for S compounds. The signal is magnified with the help of a photomultplier tube. ECD detector contains beta radiation source such as tritium or Ni63 which emits electrons and when an ionizable gas (N or argon) passes it is ionized and slow electrons are produced, A potential across the electrode is applied to collect all these free electrons that are produced as a result of ionization of gas molecules. When an electron capturing solute enters the chamber, it reacts with slow electrons to form stable negative molecule ion with a net result that electrons are removed from the system and current flowing in the chamber is reduced, which is monitored. These detectors are very sensitive to organo chlorine pesticides. It is most widely used in pesticide residue analyses because of its high sensitivity and linearity, and a pesticide having electronegative atom can be analysed upto pictogram 10^{-12} and in some cases femtogram 10^{-15} levels. MCD works on the principle that compounds containing halogen, sulphur, phosphorus and nitrogen on pyrolysis get converted into their corresponding acids which are diverted to a titration cell. The potential difference between reference and sensor electrodes is measured by this detector.

High pressure liquid chromatography or high performance liquid chromatography (HPLC) requires two essential components, the stationary and the mobile phase. Stationary phase is a solid material tightly packed in stainless steel column whereas mobile phase is a liquid. Sample mixture is dissolved in a suitable solvent and injected with syringe on to the head of column or stationary phase. The sample is carried through the column by a continuous flow of mobile phase from HPLC solvent pump. slowly than others. Those components that are more or less attracted to the stationary phase are thus separated by the time and eluted from the column and are detected by detector. Three major modes of separation are employed in HPLC depending on the choice of column and mobile phase. In Normal phase chromatography or adsorption chromatography or liquid-solid chromatography the pesticide molecule is separated on the basis of polarity. The least polar molecule elutes first from the column. However, in the Reverse phase chromatography or partition chromatography or liquid-liquid chromatography the molecules are separated on the basis of combination of solubility parameters, partition coefficients and polarity with most water soluble i.e. most polar elutes first. In the third mode the molecules are separated on the basis of their charge and are called Ion exchange chromatography.

HPLC is suitable in situations such as various metabolites isomers which were not known earlier but discovered later and pesticides which are non-volatile and thermally unstable.

Gas Chromatography: Mass Spectrometry (CMS): Mass spectrophotometer is sensitive equipment which can analyse a material which weighs les than 0.1 ì by directly determining its molecular weight and functional groups, hetro atoms and arrangement of these atoms in the molecule. With its sensitivity and usefulness, it provides an excellent tool in both qualitative and quantitative residue analyses when it is coupled with a gas chromatograph or a HPLC. It works on the principle that beam of positively charged particles are deflected by electric and magnetic field. Many of the resulting positively charged ions decompose into smaller species which are characteristic of the particular molecule. The ions are accelerated by a high potential. They then pass through an analyser region, generally a magnetic field or a quadrupole, where in effect, the ions are separated according to their masses. The number of ions representing each mass number is collected and displayed on a recorder. Mass spectrometry is more sensitive than other spectroscopic methods and gives conclusive information regarding the identity of synthetic compounds, pesticide residues and unknown metabolites.

RESIDUE ESTIMATION WITH RADIO ISOTOPES

Use of nuclear energy (radio-isotopes) for analytical purposes is one of the sophisticated, accurate and fast methods. Radio isotope is an atom of an element having the same proton number but different mass number which gives out radiation during its decay process. It is detected by its characteristic radiation. The radioactivity is measured as disintegration per second or per minute. The units of measurement are: (i) Curie (ci) is equal to 3.7 x10^{10} dps or cps (counts per second or disintegration per second). (ii) Milli Curie (mci) is equal to 3.7 x 10^{7} dps, (iii) micro curie (ìci) is equal to 3.7 x 10^{4} dps.

The use of nuclear energy in such studies is made by using radioactive atoms and then measuring it through dilution techniques. It may be used for both qualitative and quantitative measurements. Normally three methods are in vogue for residue estimations, firstly use of radio isotopes as traces of active ingredient and detecting them on a Geiger counter or proportional counters or scintillation counters, secondly measuring the penetration and translocation of radio isotopes through autoradiography and thirdly, use of stable isotopes and analyzing them through neutron activation analysis.

Radioactive materials are quite hazardous and should be handled with utmost care. The studies conducted with radioactive substances are mostly confined to the laboratories. For intense radiation which is also more penetrating the physical health care like shielding, remote control, use of rubber gloves, protective clothing and adequate ventilation in the working room are necessary.

Material samples are normally mounted on flat discs, several lengths of a milliliter (ml), within a ring of silicon greeze. Where track scattering by the mount is objectionable, a thin film of low atomic number myler, polystyrene and similar plastics may be used. Radio nuclides emitting sufficiently penetrating *beta* and *gamma* radiations are conventionally assayed in solution with dipping counter tubes or counter surrounding by hallow jackets or by insertion in the well of an ionization chamber of scintillation counter. Active gases can be introduced in to proportional counters or ionization chambers equipped with stop cock and pressure monometers.

The radioactivity can also be detected through autoradiography. This is more suitable for qualitative analysis and also penetration or translocation of radioactive atom studies.

The method of activation analyses depends upon the formulation of radionuclide from elements in the sample when they are subjected to bombardment with neutrons or charged particles. Most elements when irradiated by slow (thermal) neutrons gives rise to radioactive species of the same atomic number but one mass unit greater than its progenitor. Those radionuclide have their own characteristic radiation and modes of decay thus making radioactivation as unusual techniques for qualitative identification as well as for quantitative analysis.

The separation and analysis fo radioactive material may be performed with the help of chromatography. Radioactive material is seen through circular strip paper or thin layer chromatography and identified by calculating Rf values. The radioactive portion os cut or scratched separately and counted for radioactivity.

REFERENCES

Finney D.J. (1971). *Probit Analysis*. Cambridge University Press, pp 318

Gunther, F.A. and Blinn, R.C. 1955. Analysis of insecticides and acaricides. Interscience Publishers, London, UK, pp 696

Gupta, D.C. (Ed.) 1980. Residue analysis of insecticides (Summer institute lecture). ICAR and HAU, Hissar, pp 374

Gupta, H.C.L. 1999. Insecticides: Toxicology and Uses. Agrotech Publishing Academy, Udaipur, pp 382

Zweig, G. 1963. Analytical methods for insecticides, plant growth regulators and food additives Vol. I. Academic Press, New York

Entomology : Novel Approaches, 2007
Eds.: *P.C. Jain and M.C. Bhargava,*(Page 165-180)
New India Publishing Agency, New Delhi
E-mail: newindiapublishingagency@gmail.com
web: www.bookfactoryindia.com

CHAPTER 9

ROLE OF PARASITOIDS AND PREDATORS IN INTEGRATED PEST MANAGEMENT

Abraham Verghese and D. K. Nagaraju
Division of Entomology and Nematology, IIHR, Hessaraghatta Lake PO,
Bangalore 560 089

INTRODUCTION

The world of living organisms essentially consists of the kingdoms of plants and animals. Myriads of organisms representing these two, interact among them, ranging from symbiosis and mutualism to competition and predator-prey interactions. The last mentioned process is one the important components of ecological dynamics, that determine the balance of nature or equilibrium among the biota. To this category belong the predators, parasites (parasitoids), pathogens, and so on, all of which live at the cost of another organism, without affording any benefit to the 'victim', hence the victim is termed the 'prey'. In an agroecosystem, when an organism, explodes into an unusually high number, devouring cultivated plants it often becomes a 'pest' from a man's point of view. Sometimes the pest can be a human nuisance too. It is here, that the role of predators and parasitoids become relevant as biological agents against the pests.

A lion, which preys on herb-feeding antelopes, is also a predator. But, its roles become restricted to maintaining the balance of the antelope population. Left undeterred, antelopes can explode to levels that can threaten grasslands. Here lions play a very crucial role in maintaining (through hunting antelopes) the equilibrium of the grasslands, just sufficient for these plants to surge with every rain. Otherwise, the grasslands potentially may get denuded by the antelopes, or conversely it may override other ecosystems. This is part of the 'food web' and is found in every ecosystems; desert, aquatic, mountains, forests, agro-ecosystems, etc.

Talking of agro-ecosystems, this paper focuses on the role of predators and parasitoids in regulating pests, which mainly in the context, are insects and mites.

It is essential, here, to differentiate between the predators and parasitoids. Predators live on the prey (host), while parasitoids live in the prey. However, in the latter, all the stages are not completed in the host, unlike a typical parasite. Typical definitions of predators and parasitoids are as follows.

Predator is an animal which feeds upon other animals (prey) that are usually smaller and weaker than itself, frequently devouring them completely and rapidly. A predator most often is required to seek out and attack more than one prey to reach maturity (Coppel and Mertins, 1977).

Parasitoid is an insect parasite of an arthropod; parasitic only in its immature stages, destroying its host in the process of its development, and free-living as an adult. Generally they require only one individual host (Coppel and Mertins, 1977).

It is also essential to differentiate between a parasite and pathogen

Parasite is an animal species which lives on or in a larger animal, the host, feeding upon it, and frequently destroying it. A parasite needs only one or part of one host to reach maturity.

Pathogen is a parasitic micro-organism that disperse easily in the wind or in water.

In this paper only predator and parasitoids will be dealt with, especially their roles in man-made agro-ecosystems, where the prey is a pest affecting the welfare and/or economy of man. Again here, as mentioned before, pest in the context of this paper is an insect or mite. When a prey (pest) is devoured or parasitized by a predator or parasitoids, we say that it is effecting control over the pest. As 'life' (=bio) is controlling a pest this process is called bio-control or biological control.

Predators and parasitoids have been used since ancient days as biocontrol agents. A classical case is the use of the red ant, *Oecophylla smaragdina* (Fabricius) as a biocontrol agent against insect pests of citrus in China. Farmers of yore knew that attracting birds kept insect pests at bay. Ancient wisdom acknowledges the roles played by owls and snakes in keeping down rodents. In fact, cats were domesticated primarily to prey on rats. Thus the science of predator and prey interaction have come a long way.

ROLE OF PREDATORS

Predators are said to be generalists in that they consume an array of host spectrum, and therefore are not species or genus specific. But they can be order or group specific. Thus, coccinellids or ladybird beetles (Coleoptera: Coccinellidae) prefer homopterans as an order. Take the case of the coccinellid, *Cryptolaemus montrouzieri* Mulstant. It preys on soft-bodied insects like mealy bugs and scales, which generally come under the families Coccidae, Pseudococcidae, Margororidae, etc. all of which come under order Homoptera. This ladybird beetle rarely feeds on other soft-bodied insects like for example aphididae of homoptera and to that extent may be termed group specific. But considering that coccids, pseudococcids and margorids have hundreds of species on which *C. montrouzieri* can prey, such predators fits the bill of a "generalist".

The adults and grubs of predators such as *C. montrouzieri,* have been found to be excellent suppressors of pestilent mealy bugs and scales, which are difficult to control if flared up. Thus, release of predators contributes to their control.

Classical biological control which means importing natural enemies to control arthropod pests usually of exotic origin, has been a long practiced method of pest control. The control of cottony-cushion scale, *Icerya purchasi* (Mulstant) (Homoptera: Margarodidae) in

California using the vedalia beetle, *Rodolia cardinalis* (Mulstant) (Coleoptera: Coccinellidae) imported from Australia in 1888 provided the first dramatic success. This initial success has been repeated in other countries as the scale insect was inadvertently moved about. The vedalia beetle success stimulated further attempts at classical biological control. In India, the scale on citrus and many other crops was brought under control by *R. cardinalis* in Nilgris Hills during 1927 (Mani, 2001).

In an ecosystem, predators come in whopping numbers that even collection and identification presents a serious logistical problem. Whitcomb and Bell (1964) found over 600 species of predaceous arthropods present in Arkansas cotton fields. Smith and van den Bosch (1967) found over 1000 predaceous species associated with alfalfa in California, and more than 750 predaceous species were listed from New York alfalfa fields by Wheeler (1977). Whitcomb (1973) reported 1000 species of predaceous arthropods in Florida soybean fields.

Some ecological studies on coccinellid predators have been carried out in India. Verghese (1995) conducted a study in a guava orchard to arrive at the sampling plan for three aphidophagous predators *viz., Cheilomenes sexmaculata* (Fab.), *Pseudaspidimerus circumflexa* (Mot.) and *Scymnus castaneus* (Sicard) (Coleoptera: Coccinellidae). He found that among these, *C. sexmaculata* was spatially aggregated throughout, while *P. circumflexa* and *S. castaneus* had an initial random distribution, which later tended to aggregation with increased mean density. The sampling plans developed at 25% precision level were 18, 32 and 48 trees for respective predators. A similar study conducted by Verghese and Devaraj Urs (1995) on *C. montrouzieri* showed that, 4 – 9 trees should be sampled to get the representative samples, at 25% precision.

In grapes, the mealy bug *Maconellicoccus hirsutus* Green (Homoptera: Pseudococcidae) is being brought under a below 'injury level' by the release of *C. mantrozieri* @ 10 grubs/infested vine. Release of the grubs using small camel hair brush at the site of infestation enhances control. Similar results of good control have been obtained with respect to *Planococcus citri* (Risso) in citrus; *Planococcus lilacinus* (Cockerell) in pomegranate (all Homoptera: Pseudococcidae); *Coccus viridis* (Green) in citrus and coffee; *Chloropulvinaria psidii* (Masell) (both Homoptera: Coccidae) in guava, etc. using *C. montrouzieri* grubs. Effective control of the *Coccidohystix insolita* (Green) on brinjal by the release of *C. montrouzieri* was reported by Krishnamoorthy and Mani (1996).

In case of aphids, aphidophagous predators which are invariably not species specific like *Cheilomenes sexmaculata* (Fab.), *Scymnus castaneus* Sic, *Pseudaspidemerus circumflexa* (Mots.) (all Coleoptera: Coccinellids), and syrphids like *Peragus* sp. and *Ischiodon scutellaris* (Diptera: Syrphidae) preferentially feed on aphids. In an ecosystem the abundance of these predator may vary. Sreedevi (2003) found *C. sexmaculata* to be the most abundant species in pomegranate ecosystem followed by *Scymnus* sp.

The neuropterans like *Chrysopa* spp. (Chrysopidae) are more generalists, but again prefer soft-bodied insects. *Chrysoperla carnea* Stephen and *Mallada boninensis* Okamoto can be used for the control of whiteflies and mealybugs in chillies. These are polyphagous and are tolerant to many insecticides (Krishnamoorthy, 1985). Krishnamoorthy and Mani (1997) recommended the release of *C. carnea* at the rate of one predator for every five aphid (prey) for the effective management of the pest.

Among the above, only *C. montrouzieri* is easily amenable to rearing, followed by *Chrysopa* and so these two are the ones easily available commercially. Predators that are amenable to mass rearing are the ones therefore that can be recommended for release in the field. These would carry out the following:

a) Suppress the host population (aphids, mealy bugs, scales, etc.)

b) Enables sustainability of pest control, as they are self perpetuating.

In the long run, uses of predators are cost-effective, as they minimize the use of insecticides and mitigate yield loss due to insect pests.

There are general predators in nature that are not amenable to mass rearing and release, but are useful in keeping many insects under check. They include insectivorous birds and reptiles. The importance of birds in the management of insect pests was felt as early as 1762, when Indian Mynah (*Acridotheres tristis*) was imported to Mauritius for the control of locust.

Conserving and encouraging insectivorous birds in a particular ecosystem helps restricting insect explosions. It has been found that the black drongo, *Dicrurus adsimilis* is an excellent predator of caterpillars. By experience, farmers have found that the drongo can be attracted to an ecosystem by providing landing or perching stakes in the form of poles, which reach above the canopy height of a crop. Here the foraging behaviour of the bird is exploited. The drongos need a stake on which they perch and surveillance the field. On sighting a caterpillar they descend on to it, pick it and devour the caterpillar or bring it back to the perch and eat it. This process repeats till it is satiated. The maximum feedings are in the mornings and late evenings. When stakes are provided in paddy fields, drongos prey on *Cnaphalocrocis medinalis* (Guenee), *Nymphula depunctalis* (Guenee) (both Lepidoptera: Pyralidae), *etc.* In sunflower and cowpea they prey on *Helicoverpa armigera* (Hubner) (Lepidoptera: Noctuidae).

Another way of attracting birds is by placing water troughs in fields, especially in short-height crops like vegetables. It has been found that such water attract thirsty birds for a drink and even bath. This has been proved quite useful in cabbage fields where the caterpillars of *Plutella maculipennis* (Curtis) (Lepidoptera: Yponomeutidae) were found devoured by the wagtails, *Motacilla* spp.

Artificial nests can be provided to attract birds like tits, robins and chats, all of which are insectivorous.

It has also been found that wind-break trees around agrosystems serve as natural perches for insectivorous birds like drongos, orioles, etc. The emphasis is therefore to conserve bird habitats. The same could be told for garden lizards, frogs, toads, etc all of which are very general insect feeders.

ROLE OF PARASITOIDS

Parasitoids are more host specific than predators, while quite a few of them attack more than one host. If a parasitoid is confined to a single hosts species it is termed as *monophagous*; if on few closely related species then it is termed as *oligophagous* and if a species can develop on a number of host species it is called as *polyphagous*.

Parasitoids can attack the host at different stages of its development. For example, *Trichogramma chilonis* Ishii (Hymenoptera: Trichogrammatidae) are egg parasitoids that attacks eggs of several lepidopterans. Some of them like *Cotesia* spp. (Hymenoptera: Braconidae) attack larval stages. There are quite some which attack the larval pupal stage eg. *Exorista* sp. (Diptera: Tachinidae).

Parasitoids being target specific compared to predators, depend a lot more on cues to detect their hosts. Parasitoid adults are however not 'parasitic' and may need nectar from flower sources for survival. Almost all families of hymenopteran parasitoids-braconids, ichneumonids, platygastrids and quite some dipterans need flower visitation for adult survival.

One of the successful biological control programmes obtained in India was by using an exotic parasitoid, *Leptomastix dactylopii* How. imported from West Indies in 1984 for the control of citrus mealy bug, *Planococcus citri* (Risso) (Krishnamoorthy, 1998). Within four months of release, a complete control of *P. citri* was obtained in Karnataka and Tamil Nadu. Krishnamoorthy and Singh (1987) recommended the release of parasitoid at the rate of 5000 per hectare.

In nature, there are good number of parasitoids which can potentially bring down the pest populations. For example, the natural parasitization of epilachna beetle, *Henosepilachna vegintioctopunctata* (Fab.) on brinjal by *Pediobius foveolatus* (Cwfd) to the tune of 60-77% during June and October under Bangalore condition. Such parasitoids should be protected using less toxic insecticides. Endosulfan was found to be safer to *P. foveolatus* compared to synthetic pyrethroids (Tewari and Krishnamoorthy, 1983). The aphids, *Myzus persicae* is a major pest of chillies and this could be effectively parasitized by *Aphidius* sp. and *Aphelinus* sp. to the tune of 80 – 98.8%, if the ecosystem is free of insecticidal sprays (Mani and Krishnamoorthy, 1994). Mani (1993) found dichlorovos, dicofol and few botanicals to be safe to *L. dactylopii.*

SOME FACTORS AFFECTING NATURAL ENEMY POPULATIONS (PREDATORS AND PARASITOIDS)

Presence or absence of suitable food

This may consist of a target pest, alternate prey species, or even vegetable matter such as pollen and nectar, which serve as important sources of nutrition for many predatory mites and for adults of many natural enemies. Planting of castor in the borders of tobacco has served dual purposes. The first, it acted as repellant for aphis and the second as a pollen source for adults of *Chrysopa* adults (Singh, 1994).

Increased abundance of suitable prey is generally associated with a rise in natural enemy abundance. This may result from enhanced natural enemy reproduction, or it may derive from immigration of natural enemies from surrounding areas into the area of increased prey concentration. Therefore, habitat should have supportive alternate food resources to sustain biological control (Pimentel, 1981).

Intra/Interspecific competition and aggregation

Intra/interspecific competition in the form of territoriality and aggregated behaviour limit the distribution of predators/parasitoids. Whether directly through aggression or indirectly through competition for a limited prey or habitat source, affect the populations. Interestingly most spiders appear to avoid severe interspecific competition through territorial suppression of population densities.

Verghese and Tandon (1987) studied the interspecific association among aphid, coccinellid predator and an ant in an unsprayed guava orchard. The predator, *Cheilomenus sexmaculata* and the ant, *Camponotus compressus* were positively associated with the aphid, *Aphis gossypii,* while a negative association was recorded between the predator and the ant. For quantifying these association analyses in terms of overlapping and exclusion, spatial interspecific association analysis was carried out. Between aphid and predator, and aphid and ant, partial overlapping (γ=0.44 and γ=0.55, respectively) was obtained. Thus, although the association between aphid and predator was positive, it did not seem to be strong. Between the aphid and ant it was suggestive of mutualism, without obligate interdependence. Moreover, the predator and ant showed a tendency to independent occurrence with respect to the aphid when the spatial correlations were calculated. The predator and ant showed spatially almost complete exclusion.

A similar study was conducted by Verghese and Ramachander (1998) to know the interspecific association among *Chloropulvinaria psidii* (Maskell), the coccinellid *C. montrouzieri* and ant, *C. compressus* a prey, predator and ant, respectively. The analysis showed a positive significant association between prey-predator only at a higher predator-prey ratio. This has bearing on suppression in biocontrol release. A positive association between predator and ant was seen only when the mean prey population was lower, implying competition. At higher mean prey number, the predator and ant showed positive association implying coexistence.

"Enemity" among natural enemies

Pathogens, parasites and predators play an important role in regulating population of many predaceous/parasitoid arthropods. Predator may feed upon one another. *Polistes* spp. are important enemies of lepidopterous larvae but are intern attacked by various species of Asilidae another group of predators. More frequently, predaceous species compete for prey. A tiny dance fly (Epmididae) feeds on thrips (*Frankliniella tritici* (Fitch), *F. fuscas* (Hinds)) in Arkansas cotton blossoms, as does *Orius insidiosus* (Say) (Hemiptera: Anthocoridae). Where the dance flies have consumed all the thrips, *O. insidiosus* is forced to search further for food, thereby consuming many boll worm eggs (Pimentel, 1981). At higher food levels this is more evident. Owls for example feed on snakes which keep a check on barrowing and often less seen rodents. Shrikes hunt out lizards which are important pressure on insects.

Other entomophagous organisms affect populations of almost all the general predators. *Chrysopa* spp. are parasitized by more than 50 different species of Hymenopetra.

A number of parasitoids are parasitized by other parasitoid (hyperparasitization), a negative attribute to be avoided while selecting biocontrol agents. The potential of a parasitoid in bringing down the target prey can be best assessed in the absence of hyperparasitoids.

Stiling (1993) reported that failure of 20% of the introduced parasitoids due to hyperparasitization.

Surprisingly little is known of pathogens, which attack entomophagous arthropods. However, fungi, bacteria, virus, and protozoa attack many of the general predators. Rickettsia affect a particularly large number of predators. Buchner (1930) reviewed reckettsia affecting ants. Morel (1978) reported reckettsia attacking a wide variety of spiders. Hodeck (1973) dealt with fungal and protozoal diseases of coccenellis.

Biology of natural enemies

The biology of the natural enemies in terms of reproductive strategies, diapause, circadian rhythms and behavioral patterns have profound influence on the population of natural enemies. Most the factors have been much underrated in the past due to overemphasis on predators population fluctuations; functional and numerical responses of predators were largely neglected.

Abiotic factors

Climate and habitat characteristics exert profound influence on the dynamics of natural enemy populations. Abiotic factors affect natural enemies directly. Temperature, humidity and rain fall being especially important ones. Verghese (1992) found a positive association of the predators (*C. montrouzieri* and *C. sexmaculata*) with wind speed and relative humidity. Stiling (1993) reported that, in 23.5% of the cases the imported natural enemies failure due to climate related hazards.

Singh (1994) reported that, prolonged low temperature (5°C) for more than 7 days, drastically reduced the fecundity of *Allorhogas pyralophagus* Marsh (Hyemenoptera: Braconidae) a parasitoid of tissue borers of sugar cane, *Chilo* spp. (Lepidoptera: Pyralidae), and *Paratheresia claripalpis* (Wulp) (Diptera: Tachinidae) had poor adaptability to temperature above 30°C. Neuenschaunder (1975) found that humidity affected survival of *Hemerobius pacificus* Banks larvae stressed at extreme temperatures. Humidity can be critical in survival and development to spiders. Vitelli (1967) showed consumption of *Heliothis zea* (Boddie) eggs by the lady beetle *Coleomegilla maculata* De Geer to increase sharply with rises in temperature. It is well known that abnormally cool spring weather can precipitate aphid outbreaks by inhibiting activity of coccinellids and other general predators normally keeping aphid populations in check. Sreedevi (2003) found that more than weather, the phenology of the plant and prey density affected predators like, *C. sexmaculata, Scymnus* sp., and *Paragus serratus* (Fab.) in case of pomegranate ecosystem.

Search or host detection ability

The success of a biological control agent depends on its ability of follow cues, which may be visual, olfactory, gastatorial or tactile, to 'encounter'. Here often a strategic predator - prey interactions dynamics is evident. The prey tries to avoid being attacked, while its predator goes all out to reach its prey. A successful predator/parasitoid is one which strategises to reach the prey in the shortest time, quick to handle the prey, and efficiently devour the prey, and move on to the next prey, and ensure that its progeny hatches in the vicinity of prey abundance.

Table 1 Some major group of predators and their preys

Order	Family/Examples	Predatory stage	Prey
Odonata (dragonflies and damselflies)	All families are predators Gomphidae Lestidae	Naids and adults	Naids on the larva of mosquitoes and flies, adults on the flying insects like moths, flies, etc.
Dictyoptera (preying mantids)	Mantidae*Stagmomantis* sp.	Nymphs and adults	Flies, grass hopper, caterpillars.
Coleoptera (beetles)	Coccinellidae *Rodolia cardinalis* *Chilochorus nigritus* *Cryptolaemus montrouzieri*	Larvae and adults	Aphids, scale insects, mealy bugs
	Carabidae *Callida splendidula* *Pheropsophus hilaris*	Larvae and adults	Various soil dwelling insects and larvae of Lepidoptera
	Staphylinidae	Larvae and adults	Various soil dwelling insects
	Cicindelidae	Larvae and adults	Various insects on the ground surface
Neuroptera (lacewing bugs)	Chrysopidae *Chrsoperla cornea*	Larvae and adult	Aphids, psyllids, coccids, jassids, mites
Diptera (flies)	Syrphidae (hower flies)	Larvae	Aphids
	Asilidae (robber flies)	Adults and larvae in some cases	Various insects
	Crytochaetidae	Larvae	Coccids
	Drosophilidae *Acletoxemus indicus*		Castor and pome aleyrodids
Diptera (flies)	Chamaemyiidae *Leucopis* sp.		Aphids (*Lipaphis erysimi*)
	Tachinidae	Larvae	Various tissue borers
Hymenoptera	Vespidae (wasps) *Vespa* sp. *Philantus ventilabris* Fab.	Adults catch the prey feeds to the larvae through prophylaxis	Various insects
	Formicidae *Oecophylla smaragdina* (Fab.) *Crematogaster* sp. *Camponotus* sp.	Adults	Various insects

Contd. ...

Table 1Contd. ...

Hemiptera (bugs)	Miridae *Cyrtorhinus lividipennsi* Reuter	Nymphs and adults	Various soft bodied insects
	Lygaeidae *Geocoris bicolor*	Nymphs and adults	Eggs and nymphs of *B. tabaci.*
	Anthocoridae *Orius* sp.	Nymphs and adults	Thrips
Acari (predatory mites	Phytoseiidae Cunaxidae Stigmaeidae *Amblyseius parapelitatus* *A.cucalypticus* *Cunaxa satirostris* *Typhlodromus pyri* Scheuten	Juveniles and adults	Phytophagous mites

Source: Fenemore and Prakash, 1992; Borror *et al.,* 1992; Singh, 1994; Dhaliwal and Arora, 2000; Srivastava, 2000; David, 2001.

Table 2 Some predaceous spiders

Species	Crop/Prey
Lycosa pseudoannulata (Boesenberg and Strand)	**Rice:** Stem borer, leaf and plant hoppers **Bhindi:** *Cryptocephalus dodecospilus* Suff; *Cassida indicola* Dov. **Brinjal:** *Empoasca binotata* Pruthi; *Aphis gossypii* Glove.; *Henosepilachna vigintioctopunctata* (F.); *Myllocerus subfasciatus* Quen.
Argiope pulchella Thorell	**Bhindi:** *C. dodecospilus*; *C. indicola.* **Cucumber:** *Aulacophora foveicollis* (Lucas.); *Dacus cucurbitae* Coq.; *Mylabris pustulata* Thumb.
Zygeilla indica Tikader	**Bhindi:** *C. dodecospilus*; *C. indicola.*
Pardosa sumatrana Thorell	**Brinjal:** *E. binotata*; *A. gossypii.*; *H. vigintioctopunctata*; *M. subfasciatus.* **Sweet potato:** *Cassida indicola* Dov.; *Cryptocephalus dodecospilus* Suff; *Cylas formicarius* F.
Oxyopes lineatipes (C. L. Koch.)	**Brinjal:** *E. binotata*; *A. gossypii*; *H. vigintioctopunctata*; *M. subfasciatus.* **Potato:** *Empoasca kerri* Pruthi; *Henosepilachna vigintioctopunctata* (F.); *Myzus persicae* (Sulz.) **Sweet potato:** *C. indicola*; *C. dodecospilus*; *C. formicarius.*
Tetragnatha maxillosa Thorell	**Brinjal:** *E. binotata*; *A. gossypii*; *H. vigintioctopunctata*; *M. subfasciatus.*
Leucauge celebasiana (Wlk.)	**Brinjal:** *E. binotata*; *A. gossypii*; *H. vigintioctopunctata*; *M. subfasciatus.* **Potato:** *Empoasca kerri* Pruthi; *Henosepilachna vigintioctopunctata* (F.); *Myzus persicae* (Sulz.)

Contd. ...

Table 2 Contd. ...

	Sweet potato: *C. indicola*; *C. dodecospilus*; *C. formicarius*.
Phonognatha graffei (Keyserling)	**Apple:** *Epiphyas postvittana* (Wlk)

Source: Dondale, 1956; Nandakumar and Pramod, 1998; Satpathi, 2004.

Table 3 Some examples of parasitoids insects and their preys

Order	Family/Examples	Prey
Hymenoptera	Ichneumonidae *Gambroides javensis* *Cremastus falvoorbitalis* *Xanthopimpla punctata* Fab.	Larvae of holometabolous insects, especially Lepidoptera, Diptera, Coleoptera and Hymenoptera (hyperparasites)
	Braconidae *Cotesia flavips,* *Bracon brevicornis* Wesmael, *B. gelechidiphagus* *Apanteles taragamme* Vier	Larvae of holometabolous insects, especially Lepidoptera and Diptera
	Chalcididae *Stomatoceras sulecatiscutellum* Gir. *Brachymeria nephntidis* Gahan*Eucepsis* sp.	Larvae and pupae of Lepidoptera and pupae of Diptera
	Encrytidae *Comperiella* sp. *Metapelma* sp.	Larvae and pupae of Lepidoptera and larvae of Coleoptera
	Eulophidae *Trichospilus pupivora* *Tetrastichus schoenobii* Ferrieri	Scale insects, mealy bugs, eggs and pupae of Lepidoptera,
	Platygasteridae *Platygaster oryzae*	Rice and mango gall midge
	Scelionidae *Telenomus dignoides* Nixon *T. rowani* (Gahan)	Eggs of Lepidoptera
	Aphelinidae *Encarsia* spp.	Aphids, scale insects
	Mymaridae *Anagrus optabilis* Perk.	Eggs of BPH
	Elasmidae *Elasmus nephantids* Gahan	Larvae of Lepidoptera
	Dryinidae *Haplogonatopus* sp. Pteromolidae	Nymphs and adults BPH Larvae and pupae of Lepidoptera, Coleoptera

Contd. ...

	Trichogrammatidae *Oligaosita* sp.	Insect eggs of various orders
Diptera	Tachinidae *Stermiopsis inferens* Townsend *Diatraeophaga striatalis* Townsend *Paratheresia claripalpis* (Wulp)	Larvae of Lepidoptera, Coleoptera and Hemiptera

Source: Fenemore and Prakash, 1992; Borror *et al.*, 1992; Singh, 1994; Dhaliwal and Arora, 2000; Srivastava, 2000; David, 2001.

Table 4 Some examples of insectivorous birds

Common name	Species	Insect prey
Ashy Wren-warbler	*Prinia socialis*	*Aphis gossypii*
Common Bee-eater	*Merops orientalis*	*Aphis craccivora* and adults of *Pieris brassicae*
Black Drongo	*Dicrurus adsimilis*	Winged termites, dragonflies and exposed larvae of *Helicoverpa armigera*
Golden Oriole	*Oriolus oriolus*	*H. armigera, Diacrisia obliqua*
Hoopoe	*Upupa epops*	*D. obliqua*
House Sparrow	*Passer domesticus*	*H. armigera*
Indian Myna	*Acridotheres tristis*	*H. armigera, Spodoptera litura* Fab.
House Crow	*Corvus splendens*	*Adisura atkinsoni*
Jungle Crow	*C. macrorhynchos*	*Holotrichia* spp., *Spodoptera litura*
Wagtails	*Motacilla* spp.	*Aphis craccivora, Brevicoryne brassicae, Plutella maculipennis*
Cattle egret	*Bubulcus ibis*	*H. armigera, S. litura*
Rosy paster	*Sturnus roseus*	*H. armigera*
Tailor bird	*Orthotomus sutorius*	Aphids
Purple rumped sunbird	*Nectarinia zeylonica*	Aphids

Source: Verghese and Subramanya, 1985; Chakravarthy, 1988; Verghese, 1992; Singh, 1994.

Verghese *et al.* (2001) conducted a laboratory study to find out prey-predator interaction and to understand predator's foraging behaviour using the coccinellid predator, *C. montrouzieri* in relation to pomegranate mealy bug, *Planococcus lilacinus* (Cockrell). The correlation studies revealed that the predator's proximity to the prey and the orientation of the head of the predator towards the prey did not affect the search time. However, analyses of the variance revealed that the search time increased significantly with distance from prey (when predator was oriented with its head away from the prey). Further, proximity of prey was found to reduce the pre-feeding distance. The relationship between distance from the prey and search time was best explained with the exponential model of $Y=e^{0.55x}$, $R^2= 0.49$ for orientation towards prey. For orientation away from the prey no satisfactory model was found. Further, failed attempts of *C. montrouzieri* increased linearly with an increased in pre-feeding distance. Predator foraging tended to be random or chance encounters. However, from the results it was concluded that *C. montrouzieri* has a better detection of prey only at close vicinity of prey, as visual or olfactory cues seemed to have less role in enhancing predation. Predator not feeding on the first encounter with the prey, and subsequent feeding support a possibility of predator developing a prey image.

Pesticides

Parasitoids and predators are important components of agroecosystems. The use of plant protection chemicals has profound effect on the survivability and efficacy of these natural enemies. Use of broad spectrum insecticides leads to resurgence and outbreak of secondary pests, resulting from the destruction of natural enemies. Some examples of resurgence due to elimination of natural enemies by pesticides are given below:

Granular insecticides like carbofuran and pharate are most often used in rice ecosystems, resulting in the resurgence of *C. medinalis.* As many as 26 insecticides have been reported to cause resurgence in brown plant hopper, *Nilaparvatha ludens* (Stal.). In brinjal pyrethroids like cypermethrin and deltamethrin caused flare up in the population of *Myzus persicae* (Sulzer). Similar results are also reported in cotton ecosystems. The population of *A. gossypii* (Glover) and *Ferrisia virgata* (Ckll.) increased with the use of cypermethrin, deltamethrin and fenvalerate. In Mustard, 7 insecticides are known to cause resurgence in the aphid, *Lepaphis erysimi* (Kaltenbach) (Dhaliwal and Arora, 2000).

References

Arthur, A. P. 1961. The celptoparasitic habits and the immature stages of *Eurytoma pini* Bugbee (Hymenoptera: Chalcidae), a parasite of the European pine shoot moth, *Rhyacionia buoliana* (Schiff.) (Lepidoptera: Olethreutidae). *Canadian Entomologist.* 93: 655 – 660.

Borror, D. J. Triplehorn, C. A. and Johnson, N. F. 1992. *Study of insects* (6th edition), Harcourt Brace College Publishers, Fort Worth, Philadelphia, San Diego, New York, Orlando, Austin, San Antonio, Toronto, Montreal, London, Sydney, Tokyo.

Buchner, P. 1930. Holznahrung und Symiose, Julius Springer, Berlin.

Chakravarthy, A. K. 1988. Bird predators of pod borers of field bean (*Lablab niger* Medick). *Tropical Pest Management,* 34 (2): 395 – 398.

Coppel, H. C. and Mertins, J. W. 1977. Biological insect pest suppression (Advanced series in agricultural sciences 4). Spriner-Verlag, Berlin, Heirdelberg, New York, 314 pp.

David, B. V. 2001. *Elements of Economic Entomology.* Popular Book Depot, Chennai, 562 pp.

DeBach, P. 1964. Biological control of insect pests and weeds. New York, Reinhole.

DeBach, P. Fleschner, C. A. and Dietrick, E. J. 1951. A biological check method for evaluating the effectiveness of entomophagous insects, *Journal of Economic Entomology,* 44: 763 – 766.

Dhaliwal, G. S. and Arora, R. 2000. *Principles of insect pest management,* Kalyani Publishers, Ludhiana, New Delhi, Noida, Hyderabad, Madras, Calcutta and Cuttak, 297 pp.

Dondale, C. D. 1956. Annotated list of spiders from apple trees in Nova Scotia. *Canadian Entomologist,* 88: 697-700.

Fenemore, P. G. and Prakash, A. 1992. *Applied entomology,* Wiley Eastern Limited, New Delhi, Bangalore, Bombay, Calcutta, Madras, Hyderabad, Pune, Lucknow, Guwahati. 298 pp.

Flanders, S. E. 1930. Mass production of egg parasites of the genus *Trichogramma. Hilgardia,* 4: 465 – 501.

Flanders, S. E. 1937. Ovipositional instincts and developmental six differences in the genus *Coccophagus. University of California Publication of Entomology,* 6: 401 – 422.

Gause, G. F. 1934. The struggle for existence. Balitmore, Williams and Wilkins.

Hodek, I. 1973. Biology of Coccinellidae, W. Junk, N. V. Publishers, The Hague, Holland.

Krishnamoorthy, A. 1985. Effect of several pesticides on eggs, larvae and adults of the green lacewing, *Chrysopa scelestes* Banks, *Entomon,* 10(1): 20 – 27.

Krishnamoorthy, A. 1998. Host range, development and sex ratio of *Leptomastix dactylopii* on different stages of citrus mealy bug, *Planococcus citri. Journal of Biological control,* 2: 8-11.

Krishnamoorthy, A. and Mani, M. 1996. Suppression of brinjal mealy bug *Coccidohystrix insolita* with *Cryptolaemus montrouzieri, Insect Environment,* 2: 50.

Krishnamoorthy, A. and Mani, M. 1997. The efficacy of lacewing *Chrysoperla carnea* in controlling *Myzus persicae* on capsicum. *Proceedings of first national symposium on pest management in horticultural ecosystems: Environmental implications and its thrust.*

Krishnamoorthy, A. and Singh, S. P. 1987. Biological control of citrus mealy bug with an introduced parasite, *Leptomastix dactylopii* in India. *Entomophaga,* 32: 143-148.

Mani, M. 1993. Biological suppression of the oriental mealybug, *Planococcus lilacinus* (Homoptera: Pseudococcidae). Paper presented in National seminar on optimization of production and productivity of Acid lime at Periyakulam, Tamil Nadu, Oct. 3.

Mani, M. 2001. Biological control of fruit crop pests. In: Integrated pest management in horticultural ecosystem (Eds: Parvatha Reddy, P. Abraham Verghese and Krishna Kumar, N. K.), Capital Publishers, New Delhi. 93-107.

Mani, M. and Krishnamoorthy, A. 1994. Selectivity of pesticides to the parasitoid, *Aphelinus* sp. of the green peach aphid, *Myzus persicae* infesting sweet pepper. *Journal of Aphidology,* 32(1): 49 – 51.

Morel, G. 1978. Les maladies microbiennes des arachnids. *Symposium of Zoological Society,* London, 42: 477.

Nandakumar, C. and Pramod, M. S. 1998. Survey of natural enemies of rice ecosystem, *Insect Environment,* 4(1): 16 –17.

Neuenschaunder, P. 1975. Influence of temperature and humidity on the immatue stages of *Hemerobius pacificus, Environmental Entomology,* 4(2): 215.

Pimentel, D. 1981. *CRC Handbook of Pest Management in Agriculture*, Volume II. CRC Press, Florida.

Pimentel, D. and Wheeler, A. G. 1973. Influence of alfalfa resistance on a pea aphid population and its associated parasites, predators and competitors. *Environmental Entomology,* 2(1): 1.

Satpathi, C. R. 2004. *Predaceous spiders of crop pests.* Capital publishing company, New Delhi, Kolkata, Bangalore, 188 pp.

Singh, 1994. Fifteen years of AICRP on biological control. Project Directorate of Biological control, Post Bag No. 2491, H. A. Farm Post, Bangalore, India.

Smith, R. F. and van den Bosch, R. 1967. Integrated control, in Pest control. Biological, Physical and selected chemical methods (Eds: Kilgore, W. W. and Doutt, R. L.), Academic press, New York, 295.

Sreedevi, K. 2003. Bio-ecology, population dynamics and prey-predator interaction with special reference to pomegranate aphid, *Aphis punicae* Passerini, Ph. D. Thesis submitted to the Kuvempu University, Shimoga.

Srivastava, K. P. 2000. *A text book of applied entomology, Volume II.* Kalyani Publishers, Ludhiana, New Delhi, Noida, Hyderabad, Madras, Calcutta and Cuttak, 507 pp.

Stiling, P. 1993. Why do natural enemies fail in classical biological control programme?. *American Entomologist,* 39 (1): 31- 37.

Tewari, G. C. and Krishnamoorthy, P. N. 1983. Effectiveness of newer insecticides agasint epilachna beetle, *Henosepilachna vintioctopunctata* (Fab.) (Coleoptera: Coccinellidae) and their safety to natural enemy, *Pediobius foveoalatus* (Crawford) (Hymenoptera: Eulophidae). *Indian Journal of Agricultural Sciences,* 53: 1068 – 1070.

Verghese, A. 1992. Field ecology of coccinellid predators in guava ecosystem. Ph. D. thesis, submitted to the University of Agricultural Sciences, Bangalore. P. 129.

Verghese, A. 1995. Aggregation and sampling plan in three aphidophagous predators in a guava ecosystem. *Journal of Biologicol Control,* 9(1): 16 – 20.

Verghese, A. and Devaraj Urs, K. C. 1995. Development of sampling plan and spatial dispersion of natural population of *Cryptolaemus montrouzieri* Mulstant (Coleoptera: Coccinellidae) in guava ecosystem. *Pest Management in Horticultural Ecosystems,* 1(2):87 – 92.

Verghese, A. and Ramachander, P. R. 1998. Inter-specific association among prey, *Chloropulvinaria psidii* (Maskell) predator, *Cryptolaemus montrouzieri* Mulstant and ant, *Camponatus compressus* Fab. in a guava ecosystem. *Pest Management in Horticultural Ecosystems,* 4(1): 21 – 24.

Verghese, A. and Subramanya, S. 1985. Birds as insect suppressing agents. Non insect pests and predators. All India scientific writer's society, New Delhi 110 063, p. 145 – 151.

Verghese, A. and Tandon, P. L. 1987. Interspecific association among *Aphis gossypii, Menochilus sexmaculatus* and *Camponatus compressus* Fab. in a guava ecosystem. *Phytoparasitica,* 15(4): 289 – 297.

Verghese, A. Kamala Jayanthi, P. D. and Anil Kumar, H. R. 1998. Foraging strategies of the predator *Cryptolaemus montrouzieri* Mulstant in relation to the prey *Planococcus lilacinus* (Cockrell). *Pest Management in Horticultural Ecosystems,* 7(1): 45 – 53.

Vitelli, M. 1967. Some factors affecting the efficiency of *Celeomegilla maculata* (DeGeer) as a predator of the Bollworm, *Heliothis zea* (Boddie), M. S. thesis, University of Arkansas, Fayetteville.

Wheeler, A. G. 1977. Studies on the arthropod fauna of alfalfa. VII. Predaceous insects, *Canadian Entomologist,* 109: 423.

Whitcomb, W. H. 1973. Natural populations of entomophagous arthropods and their effects on the agroecosystem, University Press of Mississippi, Jackson, 150.

Whitcomb, W. H. and Bell, K. O. 1964. Predaceous insects, spiders and mites of Arkansas cotton fields, Arkansas Agriculture Experiment Station Bulletin, 690: 1.

Glossary of Terms used in Biological Control

Accretive release: A method of periodic introduction of biotic agents in which annual early season liberations against fairly abundant pest populations allow the beneficial organism population to increase naturally in response to rising pest densities as the season progresses.

Agroecosystem: The modified and simplified system of plants, animals and habitat used for human agricultural purposes.

Autoparasitism (adelphoparasitism): A special type of hyperparasitism in which the female develops as a primary parasitoid, but the male is a secondary parasitoid through females of its own species.

Biological check method: A method of evaluating the effects of introduced natural enemies of pest populations (particularly honeydew-producing Homoptera) in which pest-protective ants (or other species) are excluded from one plot, and encouraged in another for comparison of natural enemy effectiveness.

Biological control: Biological pest suppression in its narrow, classical sense usually restricted to the introduction by man, of parasitoids, predators, and or pathogenic microorganisms to suppress populations of plant or animal pests.

Biological insect pest suppression: The use or encouragement, by man, of living organisms or their products for the population reduction of pest insects.

Cleptoparasitism: A type of parasitism in which the adult parasitoid preferentially appropriates for its own progeny the previously paralyzed and parasitized host of another parasitoid.

Competitive displacement: An ecological concept, based on Gause's Law, which suggests that one species of insect or another organism may be capable of replacing another species over a substantial area through competition for the same niche.

Density-dependent: Refers to mortality factors or processes in the environment which destroy an increasing percentage of the subject population as the numerical population density increase, and *vice versa* (Smith, 1935).

Density-independent: Refers to mortality factors or processes in the environment which destroys a relatively constant percentage of the subject population regardless of changes in its density (Smith, 1935).

Ectoparasitoid: An insect parasite which develops externally on its arthropod host.

Endoparasitoid: An insect parasite which develops within the body of its arthropod host.

Entomophagous: Refers to the consumption of insect or their parts; insectivorous.

Entomophilic: Insect loving.

Exotic: Non native; introduced from abroad.

Extrinsic superiority: Refers to the relative competitive abilities of a beneficial organism in the environment, especially regarding efficiency in host-location and attack.

Factitious host: An unnatural, but acceptable host used in laboratory propagation of beneficial organisms.

Facultative parasitism: Here in reference to nematodes, which may either parasitize healthy insects, or develop in some other way in the environment if no insect host is encountered.

Pest resurgence (Flareback): The rapid increase of a target insect pest population subjected to an insecticidal treatment which also destroys its associated, and possibly regulative natural enemies.

Fortuitous biological control: The desirable but accidental movement of exotic beneficial organisms to new areas and /or new pests, where pest population suppression eventually results; or the successful population regulation of exotic pests by indigenous natural enemies.

Functional response: The basic behavioural response of an individual parasitoid or predator to changing densities of its host/prey; a positive response involves increased consumption with increase host/prey density and *vice versa.*

Gause's law: An ecological principle stating that different species occupying identical ecological niches (i. e. ecological homologues) cannot coexist indefinitely in the same habitat (Gause, 1934).

Gregarious parasitoid: An insect parasite which normally develops successfully at a rate of two or more individuals per arthropod host.

Harmonious use of beneficial organisms: The compatible directed use of two or more species of beneficial organisms to synergistically provide a higher level of pest suppression than that given by any one of them singly.

Hypermetamorphosis: A type of parasitic insect life cycle involving transformation through at least two distinctly different larval types, especially an active host-seeking first instar, and several more passive parasitic instars.

Hyperparasitoid: An insect parasite of another parasitoid.

Indigenous: Native to a particular region or country.

Integrated pest management: An approach to compatible utilization of all available forms of pest suppression, including mechanical, biological, chemical, and natural control, in a systematic fashion, with he primary goal of safe, effective and economical pest population reduction.

Inundative release: A method of periodic introduction of biotic agents which is analogous to insecticide treatment in that a greater amount of the liberated material is used than is actually effective, repetition may be necessary, and the effect is more or less immediate.

Monoxenous: Describes a species which requires only a single host species on which to successfully complete its annual life cycle.

Multiparasitism: A condition resulting from the simultaneous use of single host individual by two or more species of primary parasitoids.

Natural enemies: The parasitoids, predators, and pathogenic microorganisms associated naturally with a specific wild population of plants or animals, and causing mortality of debility to the individuals thereof; often used in a general sense for all parasitoids, predators and pathogen.

Parasite: An animal species which lives on or in a larger animal, the host, feeding upon it, and frequently destroying it. A parasite needs only one or part of one host to reach maturity.

Parasitism: A qualitative term referring to a kind of symbiosis in which one party (the parasite) lives at the expense of the other (the host), contributing nothing to the relationship, and frequently destroying the host in the process.

Parasitization: A quantitative term referring to the proportion of a host population attacked by parasites.

Parasitoid: An insect parasite of an arthropod: parasitic only in its immature stages, destroying its host in he process of its development, and free-living as an adult.

Period release: A method of beneficial organism introduction which involves repeated liberations to artificially maintain high population levels of indigenous biotic agents in situations where such levels are unattainable naturally.

Population dynamics: The study of numerical changes in populations of living organisms in time and space, and of the process which cause such variations.

Predator: An animal which feeds upon other animals (prey) that is usually smaller and weaker than itself, frequently devouring them completely and rapidly. A predator most often is required to seek out and attack more than one prey to reach maturity.

Primary parasitoid: An insect parasite of any arthropod which is not itself parasitic.

Protelean parasite: An insect species in which only the immature stages are parasitic.

Secondary parasitoid: An insect parasite of a primary parasitoid.

Solitary parasitoid: An insect parasite which normally develops at a rate of one individual per arthropod host.

Superparasitism: A condition resulting from the use of a single host individual by more individual parasitoids of the same species than it can successfully sustain to maturity because of nutritional limitations.

Symbiosis: The living together in close association of two or more species of organisms.

(***Source:*** Flanders, 1930; Gause, 1934; Smith, 1935; Flanders, 1937; DeBach *et al.*, 1951; Arthur, 1961; DeBach, 1964; Coppel and Mertins, 1977).

Entomology : Novel Approaches, 2007
Eds.: *P.C. Jain and M.C. Bhargava,*(Page 181-191)
New India Publishing Agency, New Delhi
E-mail: newindiapublishingagency@gmail.com
web: www.bookfactoryindia.com

CHAPTER 10

SEMIOCHEMICALS: AS POTENTIAL TOOLS IN PEST MANAGEMENT

Ashok Kumar and N.K. Bajpai
Department of Agricultural Zoology and Entomology
Rajasthan College of Agriculture, (MPUAT),Udaipur

INTRODUCTION

"An insect is like an old-fashion soldier: it does what it is ordered to do, and when it has no orders it does nothing" (a citation from Wright, 1963). The orders are very often given as chemical signals that induce certain behaviour or change the course of development in the insect. Chemical communication appears to be the primary mode of information transfer in insects. These chemicals are able to modify the behaviour of a perceiving organism at sub-micro/nanogram levels and are known as semiochemicals. The term is derived from the Greek Word *Simeone*, meaning a mark of a signal. These behaviour modifying chemicals are divided into two broad categories.

i) Intraspecific communication chemicals - These are responsible for behavioural change among the individuals of the species. *eg.* Pheromones.

ii) Interspecific communication chemicals - These are responsible for behavioural change between the individuals of different species. *eg.* Allelochemicals.

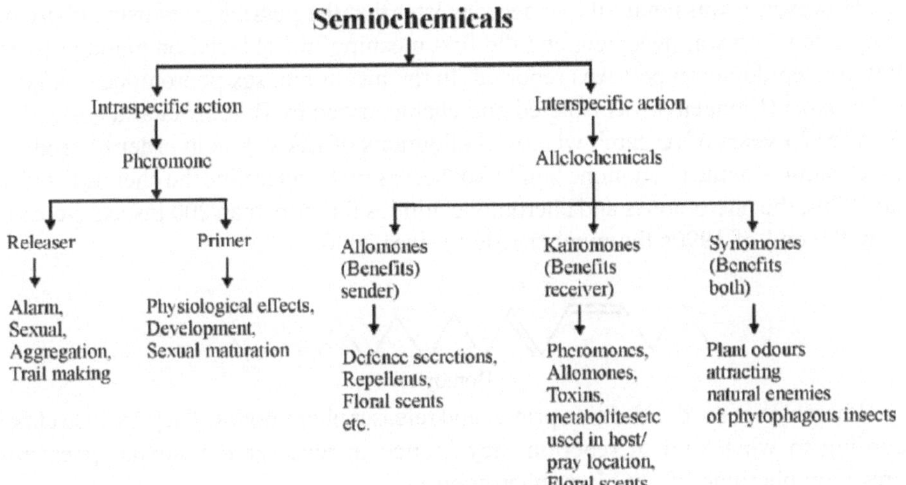

PHEROMONE

The term Pheromone was suggested by Butenandt, Karlson and Lüscher in 1959 and is derived from the Greek word p*herein*, meaning to transfer and *hormone*, meaning to excite. Thus, a pheromone can be defined as a chemical or a mixture of chemicals that is released to the exterior by an organism and that causes one or more specific reactions in a receiving organism of the same species. When a chemical not found in an insect has a pheromone like action, is called as parapheromone.

Host plants play an important role in the origin of sex pheromones of some insects through the acquisition of bioactive chemicals and required chemical precursors of pheromones via consumption, absorption or inhalation of host plant material. In some species, pheromones are passed and released with feaces, providing a strong relationship between both host plant chemistry and pheromones. In these species, there may be a temporal and spatial relationship between feeding and pheromone release.

Sex pheromone

The sex pheromones represent a diverse group of compounds. Though most commonly released by the females, they may be released by either sex. Over 150 species of insects are known in which the females produce sex pheromones and about 50 in which the males do so. The male and female sex pheromones differ in their property and action. The female sex pherome acts at a longer range, excites males to copulate and is species specific. The male sex pheromone, on the other hand, acts at a short range and only lowers (not excites) the females resistance to mating. Since at short range, other communication stimuli may come into play, the male sex pheromone is of lesser value than the female sex pheromone in insect pest control. The sex pheromone have been detected so far from Lepidoptera, Orthoptera, Dictyoptera, Diptera, Coleoptera, Hymenoptera, Hemiptera, Neuroptera and Mecoptera. In Lepidoptera, the pheromonal system has reached its evolutionary peak.

The first attempt at using pheromones for insect control was made in 1940s when crude extracts of abdominal tips of female Gypsy moths were used in monitoring traps in USA. However, it was not until two decades later that the possibility of using pheromones in crop protection was accepted, and the first meaningful field trial on mating disruption (MD) of a lepidopteron pest was reported. In the meantime, sex pheromone of silkworm, *Bombyx mori* (Linnaeus) was isolated and characterized by Butenandt and co-workers in 1959 after 25 years of research who used kilograms of silkworms in order to produce the minute amounts of the pheromone bombykol necessary to determine the chemical's identity. Upto 1970s, the pheromones and pheromone mimics for more than 200 insects were known and by the end of 1990s the number rose to over 2000.

OH

Bombykol

Pheromones are divided into primer and releaser pheromones. They are also classified according to what kind of reaction they induce in receiver *e.g.* sexual pheromones, aggregation pheromones, or alarm pheromones.

Primer pheromones influence the physiology of the recipient, initiate changes in development such as sexual maturation and the releaser pheromones influence its behaviour. Primer pheromones are used by social insects to suppress the production of sexual individuals, and by locusts to stimulate the production of migratory winged types when the population density becomes high. The releaser pheromones or analogues are most relevant as insecticides. Releaser pheromones are quite common in many insect orders. Good examples are alarms pheromones released by angry bees that tell other bees of the same hive to attack the intruder.

Pheromones chemistry ranges from the blends of aliphatic alcohols, aldehydes, esters and epoxides exploited by lepiodopterous insects to alkenoic acids and aldehydes, branched alkanones, esters, monoterpene alcohols and aldehydes, to a furanone employed by beetles (Jutsum and Gordon, 1989; Dhaliwal *et al.,* 1994). Crucial properties of releaser pheromones are their volatility, their stability, and of course the degree of specificity possible to build into a relatively small molecule. The molecular weight tends to be between 80 and 300. On the basis of evaporation and diffusion rates, it can be predicted that long distance sex pheromones would have a molecular weight between 200 and 300. A pheromone may be one chemical, but usually it is a mixture of chemicals, each of which is a component of the pheromone. Mixtures increase the specificity, a property important especially for the sex pheromones.

Alarm and Trial Pheromones

Most social insects use alarm and trial pheromones. Their specificity is not so important, but fast release and dispersal are important. They tend to be small volatile substances; for instance, (S)-4-methyl-3-heptanone is an alarm pheromone in the leaf cutting ant, *Atta textata*.

Alarm pheromone

Alarm pheromone have been reported in Homoptera, Isoptera and Hymenoptera. The pheromone is produced by a variety of organs such as a pair of cornicles or siphunculi near the tip of the abdomen in aphids, abdominal terga in *Dysdercus*, cephalic glands in termites, sting apparatus and mandibular glands in worker bees and anal, mandibular, Dufour's and poison glands in ants. Alarm pheromone are primarily an antipredator device, a warning to members of the same species about the presence or attack of an enemy (mostly a predator). This warning elicits different behaviours in different insects *e.g.* dispersion or escape in aphids and bugs, dispersion when foraging or feeding away from the nest and aggression when inside the nest in ants, attraction towards the source of emission in wasps and worker bees and aggression in soldiers of termites. Chemical nature of some of the alarm pheromones are terpenes (aphids), aldehydes (hemipterans), alkyl and alkenyl acetates (honeybee sting gland), 2-heptanone (honeybee mandibular gland), ethyl and methyl ketones and formic acid (ants) and monoterpene hydrocarbons (termite soldiers).

The alarm pheromone of aphid is identified as (E)-B-farnesene (EBF). The application of alarm pheromone alonwith host plant volatiles caused the aphids to move away from feeding sites and may be utilized to prevent colonization of the host plants by the aphids. It also increased the mobility of the aphids and this is useful for increasing the pickup of contact insecticides or pathogens by the pest. EBF is now commercially available for use against a number of aphid species including *Aphis gossypii* Glover and *Lipaphis erysimi* (Kaltenbach).

Trail pheromone

Trail pheromone are also called as terrestrial trail pheromones that are laid in the form of intermittent or continuous lines on a solid substrate that the trail followers perceive by their antennae to reach the destination which could be a mate or a source of food discovered by the trail marker. The trail pheromones are then used to find mates, or to utilise food resources more efficiently. In case of social or migrating insects, this pheromone also maintains the cohesion and social integration of the colony. In generally trail pheromones originate close to the body part that applies them to the substratum. These pheromones are multicomponent pheromones. The chemical nature of some of these pheromones are caproic acid (*Zootermopsis* spp.), 3,6,8-dodecatriemol (*Reticulitermes* spp.), hexanoic, heptanoic, decanoic, nonanic acids etc. (some formicine ants) which are all combined with other components not known in many cases. Trail pheromones are produced by variable organs *e.g.* sternal gland in termites, Dufour's gland, Pavan's gland, Poison gland, rectal gland and tibial gland in ants depending on the species and labial glands in caterpillars.

Aggregation Pheromone

These pheromones induce aggregation or congregation of insects for protection, reproduction and feeding or combinations thereof. These pheromones are known mostly in coleopterans, a pentatomid bug (*Eurydema rugosa*) and a dictyopteran (*Periplaneta americana*). The best known examples are the females of the bark beetle, *Dendroctonus frontalis*, and males of the phloem beetle, *Ips confusus.* The behaviour of *Ips* has been studied in detail. On stumbling on to a suitable host tree, the males of this insect burrow into the phloem tissue by making galleries. As they do so, they secrete the pheromone into the lumen of the hindgut where it gets incorporated into the faecal pellets voided in the galleries. The pheromone volatilises to permeate the air outside which stimulates the flying populations of both male and females. The females are stimulated to enter the galleries made by the pheromone producing males, while the males are stimulated to construct new galleries. Interestingly, after mating or continued feeding for sometime, pheromone production stops which seems to be a means of preventing overcrowding that a continuous release may cause. In some cases, this is achieved by production of antiaggregation pheromone. Some common names of aggregation pheromones are frontalin (*Dendroctonus*), ipsenol (*Ips*), periplanone (*Periplaneta*), dimethyldecanol (*Tribolium*) etc.

The pheromones have potential for use in pest management programmes. Pheromones and the behaviour associated with these semicochemicals, are crucial in the life cycle, and are, therefore, excellent targets to interfere with to obtain pest control. This could be achieved by a variety of strategies like –

- Mating disruption
- Mass trapping
- Monitoring and survey

Mating disruption (MD)

The control of insect pests by mating disruption technique is achieved by widespread application of synthetic pheromone over the treated crop. Various slow-release pheromone formulations have been developed which either permeate the air with relatively high levels of pheromones so as to achieve sensory adaptation and habituation or provide numerous discrete point sources so as to mask trail following or to create false trail.

The development of hand applied, slow release dispensers for season long control has undoubtedly contributed to make the technique effective, reliable and economical. Pink bollworm, *Pectinophora gossypiella* (Saunders) has been successfully managed in a number of countries including USA, Egypt and Pakistan by using controlled release formulations of its female sex pheromone gossyplure. In Egypt, 1,50,000 ha of cotton were treated with pheromone against *P. gossypiella.* In Pakistan, a single twist-tie formulation containing pheromones of both pink and spotted bollworms has been successfully used to obtain season-long control of bollworm complex.

Factors influencing Mating Disruption

The mating disruption is being influenced by many factors when apply in the fields. These are:-

i) Pheromone concentration
ii) Area size
iii) Weather condition
iv) Population density
v) Field conditions
vi) Plant conditions
vii) Natural enemies

Pheromone concentration

In the case of insecticide application, the dosage level has a direct relation to its effect. In mating disruption, the pheromone concentration in the field is more important than the dosage itself. The minimum concentration required for mating disruption of the target pests are not always clear, as they are changeable by the kind of target pests and the population density. The effective minimum concentrations vary in range from about 0.5 to 2 nano-gram/m^3 (Ogawa and Kobayashi, 2001). The pheromone concentration fluctuates mainly by the weather, so the dosage should be decided by these conditions. It is very important to know the pheromone concentration in the field. However, it is not so easy to measure the concentration very often. In place of the actual measurement, we should estimate the pheromone concentration in the field by considering the balance of pheromone (Table 1).

Table 1 Balance of pheromone in the field

Item	Input	Output
Dispenser	Releasing from r dispense	no
Side flow by wind	in group application	a main loss
Into sky by convection	no	a big factor at day time
Soil	desorption from soil	absorption by soil
Plant	desorption from plant	absorption by plant

Area size

The next important factor is the area size of the pheromone treated fields. A low infestation can be obtained in a centre area and a higher infestation at the border. Many researchers explain the reason of a lower efficacy at the border is because mated females from an untreated field may migrate to the border of treated field and lay eggs.

Weather conditions

The another important factor is the weather conditions. The release rate of the dispensers is dependent on the dispenser system and the weather conditions. The release rate of a specific dispenser is higher at higher temperatures but the concentration is almost the same regardless of the temperature (Table 2). The pheromone concentration has a relation to the release rate, weather conditions and with the wind velocity (Table 3). At a higher wind area, the release rate is higher but the concentration is lower. Generally speaking, the pheromone concentration is inversely proportional to the wind velocity. So the user have to adjust the dosage level according to the wind velocity, namely at windy areas a higher dosage is required. In order to reduce the side effect of wind and to keep the pheromone concentration, a wind break is useful.

Table 2 Temperature vs Release rate vs Pheromone concentration

Period of time	Concentration (ng/m^3)	Season	Release rate (gr/ha/day)	Concentration (ng/m^3)
Night	20	Summer	4.3	20
Day time	18	Spring	1.9	19

Table 3 Wind velocity, Release rate and Pheromone concentration

Area Type	Wind velocity m/sec	Release ratemg/hr/ha	Pheromone concentration (ng/m^3)
A. Windy area	2.5	235	1.2
B. Calm area	1.0	204	2.5

Population density

It is very difficult to know the precise relation between the population density and the efficacy of mating disruption, because the trap catch system can not work at pheromone treated fields. Anyway, mating disruption is usually unable to control mating in a very high population.

Field conditions

The field condition is also important. The flatness of the field and the neighboring field condition affect the pheromone concentration. At a sloping field, pheromone is moving along with a downward movement, so a higher dosage at the upper part of a slope is recommended. When the neighboring area is lower than the pheromone treated area, it is very difficult to keep enough concentration to maintain good control. Therefore, an additional higher dosage of pheromone or an additional insecticide spray, especially along the border area is required.

Plant conditions

The height of the plant is also affecting the efficacy. Usually the application of pheromone dispensers is recommended to a canopy. However, if the height of trees are even and less than 4 meters, then application of dispensers should at easy positions. Only at the border area additional dispensers should be applied to the canopy. In case of more than 4 meter tree heights, spray type formulations are better. In a tall tree when pests are living at a low position, the application at a lower position and a low dosage is also good enough. This is because there is always a higher concentration of pheromone gathering at the lower position. However, if the height of the plants are low like cabbage, a low dosage could not be recommended, because pheromone is escaping directly into the air above.

Natural enemies

Mating disruption is targeting only pests which are living inside a plant at the larval stage and so they are difficult to be controlled by natural enemies.

Key to Success

Although mating disruption is a quite simple technology, it is still a very complicated to inexperienced farmers and young researchers who are just starting out. A farmer needs 2-3 years of experience with mating disruption mainly because they cannot see the efficacy with their own eyes. In order to get a stable efficacy attention should be given to the following items :-

i) Dispenser systems	
ii) Application method →	Dosage Application timing
iii) Group application	

Dispenser systems

Life, stability and the flatness of the release rates are the important properties of a dispenser. The release rate is the most important property. In Egypt, the mating disruption system had been applied until 1997 with rather good results. But unfortunately in 1998 they suffered from high infestations and the Egyptian government decided to stop the pheromone application in order to investigate the reasons of why the efficacy was unstable. After the investigation, it was reported that the release rates are quite different according to the different types of dispensers, namely in the late season the release rates of some dispensers become too small to disturb mating. While another dispenser system, which has a constant release rate, gave good control without additional insecticide sprays. On the other hand, other dispenser systems which don't have a constant release rate, had high infestations, even if additional insecticides were sprayed. These dispensers release pheromone at a high rate in the beginning, but they cannot maintain the same rate of release later. When the rate of release decreases, insects immediately begin their mating activity, as insects are never killed by pheromone.

Application Method

a) **Dosage :** Many factors influence the efficacy of pheromone, therefore, dosage adjustment is essential according to the conditions. Usually a higher dosage is recommended at windy fields and at small fields.

b) **Application timing :** It is very important factor. The biggest reason for failure is a late application of pheromone, hence it is essential to put pheromone trap before emergence of insect population.

Group application

The application of pheromone is required at large areas even if farmers are having small fields. Hence, they should adopt-group application which has following advantages–

a) A lower dosage
b) Compensation of other factors and a stable efficacy
c) A high population of natural enemies
d) A lower border effect
e) Reduction in costs for less insecticide.

Mass trapping

Mass trapping is possible in case of adult insects with a highly developed ability to respond to pheromones. This technique is mostly restricted to protection of crops where the insecticides are not effective or environmentally acceptable. A powerful, highly specific attractant should trap a sufficiently large number of target pest individuals to reduce its population below economic threshold level. A trapping efficiency of more than 90 per cent is required in case of Lepidoptera males to achieve the desired results (Jutsum and Gordon, 1989).

A modification of mass trapping technique is the lure and kill method where instead of being trapped, the responding insects come in contact with a conventional insecticide. This technique overcomes the problem of low trapping efficiencies, since the responding insects need only to alight in the vicinity of the attractant source to pick up a lethal amount of the toxicant (Jones, 1994). This technique has been employed for the control of tephritid fruit files for several years. Recent work has shown that a microencapsulated formulation of the sex pheromone of the olive fly, *Bactrocera oleae* (Gmelin) is tank mixed with malathion or dimethoate for ground or aerial spraying.

Monitoring and Survey

Pheromone baited traps are very effective for monitoring of insect pest populations and provide a highly sensitive means of detecting the insect pests with many advantages over conventional methods such as light traps and scouting programmes. Pheromone traps can be used to detect both the presence as well as the density of pest species. Insect populations can, thus, be estimated and new areas of infestation detected at a very early stage. The trap catches may be used to forewarn regarding the outbreaks of important pests. The economic threshold level of insect pests can be determined in terms of pheromone trap catches to decide about the timing of insecticide application (Jutsum and Gordon, 1989).

Establishing correlations between trap catches and subsequent infestations is often difficult and requires a close understanding of the agroecosystem.

ALLELOCHEMICALS

The plants are equipped with various chemicals which helped them to save from damage of noxious insect pests by making them unpalatable either or adversely affect the survival, growth, development and behaviour of these pests. The defensive secretions of arthropods are another group of chemical which affect other interacting organisms. Besides some allelochemicals released by natural enemies are also important. The best known microbial allelochemical is the delta-endotoxin released by *Bacillus thuringiensis* which is toxic to a wide range of caterpillars and grubs. It is thus clear that there is a vast network of interspecifically active substances linking plants, insect pests and natural enemies into an allelochemical web (Ananthakrishnan, 1992). The exploitation of these substances offers seemingly endless opportunities for better management of insect pests.

Chemicals involved in interspecific communications were given the name of allelochemicals by R.H. Whittaker in 1970. Allelochemicals or allelochemics may be defined as non-nutrient substances originating from an organism (plant or animal) which affect the behaviour, physiological condition or ecological welfare of organisms of another species. Allelochemicals affect the survival, growth and development of herbivorous insects as well as their natural enemies. Based on the specific advantage they confer to the emitting or receiving organism, allelochemicals are classified into different categories–

a) Allomone

b) Kairomone

c) Synomone

d) Apneumone

a) **Allomone :** These are c substances produced or acquired by an organism that, when it contacts an individual of another species evokes in the receiver a behavioural or physiological reaction that is adaptively favourable to the emitter but not the receiver. The important defense of plants against insect pest is the possession of toxic or repungent allomones. These substances repel, deter or harm many potential phytophagous insects. Plant allomones might be applied to crops to act as long or short distance insect repellents or contact feeding deterrents. Therefore, allomones which reduce pest injury by rendering plants unattractive or unpalatable offer a novel approach in vector and disease management. The gossypol, a chemical allomone, in cotton plant was found toxic against *Spodoptera exigua* (Hübner), *Helicoverpa zea* (Boddie), *Heliothis virescens* (Fabricus), *Trichophisia ni* (Hübner), *Pectinophora gossypiella* (Saunders) and *Spodoptera littoralis* (Boisdual). The level of hydroxamic acid in a number of graminaceous plants including maize, wheat etc. affects the survival and development of *Ostrinia nubilalis* (Hübner), *Schizaphis graminum* (Rondani), *Sitobion avenae* (Fabricius) and other pests.

b) **Kairomone :** These are c substances produced or acquired by an organism that, when it contacts an individual of another species evokes in the receiver a behavioural or physiological reaction that is adaptively favourable to the receiver but not the emitter. Plant produced substances that serve as attractants, arrestants and oviposition and feeding stimulants for herbivores are the best known kairomones. These plant derived substances also influence organisms of the third trophic level when they are requested by herbivores and used by natural enemies to locate the herbivore. The aphid parasitoid, *Diaeretiella rapae* (M'Intosh) is attracted by the allyl isothiocyanate released by cruciferous plants harbouring various aphid species including *Brevicoryne brassicae* (Linnaeus), *Lipaphis eysimi* (Kaltenback) and *Myzus persicae* (Sulzer). *Chrysoperla carnea* (Stephens) and *Collops vittatus* (Say) are attracted to caryophyllene, a terpenoid released by damaged cotton leaves. The spined soldier bug, *Podisus maculiventris* (Say), orients to soybean plants damaged by *Trichoplusia ni* (Hübner).

c) **Synomone :** These are c substances produced or acquired by an organism, when it contacts an individual of another species evokes in the receiver a behavrioual or physiological reaction that is adaptively favourable to both emitter and receiver. Some plants respond to feeding or tissue damage by emitting synomones attractive to its natural enemies. Corn seedlings attacked by *Spodoptera exigua* (Hübner) release large amounts of terpenoid volatiles which serve as cues for females of the parasitoid wasp, *Apanteles marginiventris* (Cresson). The females of parasitoid, *Eucelatoria bryani* Sabrosky respond positively to the extracts of 19 plants which serve as food source for *Helicoverpa* spp. Corn plant contain the chemical tricosane and the corn earworm, *Helicorerpa zea* (Boddie) incorporates tricosane unchanged into its eggs. The chemical attracts egg parasitoid, *Trichogramma evanescens* Westwood, to find its host. In other cases a number of chemicals released by phytophagous insects serve as cues for their natural enemies. For example aphid odours works both as an arrestant and an oviposition stimulant for the syrphid fly, *Syrphus corollae* Fabricius and a short range attractant for *Aphidoletes aphidimyza* (Rondani). Larvae of *Chrysoperla carnea* (Stephens) respond to chemicals emanating from the scales left behind by the ovipositing females. The odour of frass or webbing from herbivorous mites is highly stimulatory to predatory mites. Some arthropod predators have evolved the ability to attract their prey chemically. The bolas spider, *Mastophora* sp. attract males of two noctuid moth species by producing an allomone similar to the sex pheromone of the female moth. The assasin bug, *Apiomerus pictipes* releases a substance attractive to the stingless bees, *Trigona fulviventris.*

d) **Apneumone :** These are a substances emitted by a non living material which evokes a behavioural or physiological reaction that is adaptively favourable to a receiving organism but detrimental to an organism of another species that may be found in or on the non living material.

References

Ananthakrishnan, T.N. (ed) 1992. Emerging Trends in Biological control of Phytophagous Insects. Oxford & IBH Publishing Co. Pvt. Ltd., New Delhi, India.

Ananthakrishnan, T.N. (ed.) 1994. Functional dynamics of phytophagous Insects. Oxford & IBH Publishing Co. Pvt. Ltd; New Delhi, India.

Ananthakrishnan, T.N. and Raman, A. (eds.) 1993. Chemical Ecology of Phytophagous Insects. Oxford & IBH Publishing Co. Pvt. Ltd; New Delhi, India.

Butenandt, A., Becknann, R., Stamm, D., and Hecker, E. 1959. Über den Sexual Lock – stoff des seidenspinners *Bombyx mori.* Reindarstellung and Konstitutionsermittlung. *Z. Naturforschung,* 14b, 283-284.

Carde, R.M. and Minks, A.L. 1977. Insect Pheromone Research. Chapman and Hall, London, 684 pp.

Dhaliwal, G.S. and Arora, R. 2001. Role of phytochemicals in integrated pest management. In: Opender Koul and G.S. Dhaliwal (eds.) *Phytochemical Biopesticides.* Harwood Academic Publishers. Amsterdam, The Netherlands, pp. 97-117.

Dhaliwal, G.S. and Bhathal, J.S. 1994. Allelochemicals : Their potential in insect management. In : S.S. Narwal and P. Tauro (eds). Allelopathy in Agriculture and Forestry. Scientific Publishers, Jodhpur, India, pp. 176-182.

Dhaliwal, G.S., Bhathal, J.S. and Bhathal, S.S. 1994. Novel approaches in insect pest management: An overview. In: G.S. Dhaliwal and Ramesh Arora (eds). Trends in Agricultural Insect Pest Management, Common Wealth Publishers, New Delhi, India, pp. 469-511.

Jones, O.T. 1994. The current and future of pesticides for semiochemicals in integrated management of insect pests. In : Brighton Crop Protection Conference. Pests and Diseases. 3. British Crop Protection Council, Farnham, U.K., pp. 1213-1222.

Jutsum, A.R. and Gordon, R.F.S. 1989. Insect Pheromones in Plant Protection. John Wiley & Sons, New York, USA.

Karlson, P. and Butenandt, A. 1959. Pheromones (ectohormones) in insects *Annu. Rev. Entomol.,* 4, 39-58.

Karlson, P. and Lüscher, M. 1959. "Pheromones" : new term for a class of biological active substances. *Nature,* 183, 55-56.

Norduland, D.A. 1981. Semiochemicals : A review of terminology. In : D.A. Norduland, R.L. Jones and W.J. Lewis (eds). Semiochemicals : Their Role in Pest control. John Wiley and Sons, New York, pp. 13-23.

Ogawa Kinya and Kobayashi Toshimi. 2001. Pheromone as a tool for IPM. In: Advances in Biological Control. Symposiumon Biocontrol based pest management for quality crop protection in the current Millennium, July-10-19, 2001, PUA, Ludhiana. pp. 47-55.

Rockstein, M. 1978. Biochemistry of Insects. Academic Press, New York. 649 pp.

Shorey, H.H. 1977. Manipulation of insect pests of agricultural crops. In: H.H. Shorey and J. Mckelvey, Jr. (eds). Chemical control of Insect Behaviour : Theory and Application. John Wiley & Sons, New York, USA, pp. 345-70.

Whittaker, R.H. 1970. The biochemical ecology of higher plants. In: E. Sondheimer and J.B. Simeone (eds). Chemical Ecology, Academic Press, New York, USA, pp. 43-70.

Wright, R.H. 1963. Chemical control of choosen insects. *New Sci.,* 5, 598-600.

Entomology : Novel Approaches, 2007
Eds.: *P.C. Jain and M.C. Bhargava*,(Page 193-210)
New India Publishing Agency, New Delhi
E-mail: newindiapublishingagency@gmail.com
web: www.bookfactoryindia.com

CHAPTER 11

UTILITY OF LIGHT TRAP IN INTEGRATED PEST MANAGEMENT

S.M. Vaishampayan
Retd. Professor & Head, Deptt. of Entomology, J.N.K.V.V. Jabalpur

INTRODUCTION

Although the concept of IPM, with emphasis on minimum use of pesticides; is well established and widely practiced in India pest population still continue to be unabated, often reaching to alarming levels. Out breaks of *Spodoptera litura* and hairy caterpillar, *Amsacta moorei* have been recorded in soybean and kharif crops recently during 2002 & 2004 in Madhya Pradesh and Rajasthan, respectively. Outbreaks of *Helicoverpa armigera* populations are well known, appearing more frequently since past few years (1987-88, 1995-96, 1997-98 & 2001) doing extensive damage in cotton and pulses. in many parts of the country. Failures in controlling these populations, even by most potent insecticides, have raised a serious doubt on the success of present day pest management practices aimed to kill the pest in immature larval / egg stage, when infestation spreads out in large areas. Insecticidal treatments require wide coverage of crop area to control them effectively. Insecticide resistance in pest species and destruction of natural enemies are obvious problems, since chemicals are applied indiscriminately to entire crop ecosystem including the non-target beneficial bicontrol species. How to minimize use of insecticide in pest control is often a question of major concern to the entomologists.

Development of non-chemical, ecofriendly methods of pest control—as a component of IPM tactics is a priority area of research today. With the reported development of multiple resistance in larval stages against recommended insecticides, spoiling of crop ecosystems with toxic chemicals and consequent health hazards the best alternative is to control the pest in adult stage before they reproduce. Use of light traps is one of such potential methods of control against the species known to be positively phototropic in adult stage.

Use of light trap in combination with pheromone traps is a part of "Adult Oriented strategy of IPM" recently proposed and discussed by Vaishampayan (2002). Poison baiting of adults using an attractant source, synchronized with peak of adult activity in trap catches is another prospective aspect of control strategy.

Adult is a most vital developmental stage in a life cycle of an insect, passing through egg, larval & pupal stages, encountering more than 98% natural mortality between egg to adult stage in a generation when population remains at a static level. Pradhan (1983) has precisely pointed out the importance of survival of adult stage in population ecology. Considering sex ratio of 1:1 the hypothetical rate of survival in a generation, from egg to adult stage shall be only 2% and 0.2% with egg laying capacity of 100 and 1000 eggs per female, respectively. In a static population, therefore, each female is needed to be replaced by only one female, at the end of a biotic circuit. Addition of every single female over this one unit shall increase the infestation level in next generation proportionately in a multiple of number of females added. Referring to *Helicoverpa*, killing of 10 female adults means controlling 10,000 to 30,000 early instar larvae emerged after oviposition and hatching, supposed to be spread out in one hectare of crop area @ one to three larvae/sq. meter. The female need multiple mating with male prior and during oviposition period and hence male and female adults both play a vital role in determining the size of future generation. Hence, it would be most strategic to attack the pest in adult stage.

Many of the insect species, mostly nocturnal, are known to be positively phototropic and attracted towards artificial light in large numbers. Entomologists have used this phenomenon since long to capture night flying insects in a device called light trap. Insects were attracted using candlelight at first, then with kerosene (Petromax lamp) or acetylene lamp and ultimately with electric lamp. In early days, prior to 1940, when chemical methods of pest control were little known, light traps were quite popular in insect control studies in India. The earliest Indian record is of Dutt (1919) reporting work done on the control of black cutworm–*Agrotis ipsilon* in Bihar. In 1921, Javeri reported successful use of light trap for the control of *Amsacta moorei* Buttler, in North Gujarat. Hussain (1930), reported results of very successful campaign of control using light traps against *Amsacta* in Hoshiarpur district of Punjab. In all 2750, light traps were set up in a total area of 11,000 acres and more than 30 lakhs of moths were captured in a season. Milne (1927) and Chopra (1928) also reported trapping of large number of moths from Hoshiarpur Distt. (Punjab). As reported by Ayyar and Ananthnarayan (1937) light trap proved very useful in minimizing infestation of rice stem borer, *Schoenobius incertulas* in Paddy crop in South India. They were able to collect 15,427 moths in 29 nights. Dina Nath (1923) recorded collection of 3194 moths of three species of cane borers (*Chilo simplex* 2180, *Scirpohaga* sp. 451 and *Emmalocera* sp. *563*) during 59 nights of trap operation in sugarcane field.

Onward 1940, however, this practice of pest control was a forgotten past till 1975, when a systematic work was initiated at JNKVV, Jabalpur in a project financed by ICAR and results of 4-5 years of study on seasonal activity and control of two major pest species i.e. *Helicoverpa armigera* and *Agrotis ipsilon* were published by Vaishampayan (1980) and (1981), respectively. Work was continued for over two decades covering important aspects of use of light trap as IPM tool.

GENERAL REVIEW OF RESEARCH WORK DONE

Monitoring Adult Activity

Light traps have been widely used in the past for survey and control of insect pest population in several crops (Otman, 1964; Kovitvadhi and Cantelo 1966; Brown *et al.*, 1969; Blair and Catting, 1974 and Betts, 1976). Beckman (1970) and Harstack *et al.* (1973)

studied the seasonal abundance of bollworms *Heliothis zea* and *H. virescesns* and emphasized the usefulness of such data for studying the population dynamics of the pest species. Ishikura (1967) in Japan used traps in the assessment of field population of rice stem borer moths. Yoshimiki (1967) studied seasonal abundance of rice stem borer moths using standardized light trap in about 6340 hectares of rice crop in Japan. *Helicoverpa armigera* (Hubner) Hardwick a new generic name of *Heliothis armigera* was adopted for the first time in India by Vaishampayan (1977).

Texas Agricultural Extension Services has used light trap catches successfully throughout Texas to forecast timing of future generations of *Heliothis* in cotton (Hartstack *et al.*, 1977). Seasonal activity of *H. armigera* adults in light trap catches was studied in India, by Vaishampayan (1980) and Vaishampayan and Verma (1987) at Jabalpur and by Nirmala Devi *et al.,* (1991) at Palampur (N.W. Himalayas). In *Agrotis ipsilon* such studies were made by Hassenein (1972), Odiya (1975) and Vaishampayan (1981).

Pawar *et al.* (1986) reported results of analysis of light trap catches of *Heliothis armigera* collected between 1977 and 1983 at ICRISAT Center Patancheru Andhra Pradesh and observed three peaks a in year i.e. August – September, November – December and March- April, respectively. Mahobe and Vaishampayan (1990a &1990b) reported two years data on seasonal activity of adults of *Plusia orichalcea* and *Plusia chalcytes,* respectively collected on light trap at Jabalpur.

Direct control

From an agricultural point of view the use of light traps is far more convenient than the indiscriminate use of insecticides, because majority of useful insects (Parasites, predators and honey bees etc.) being diurnal, are less attracted by light trap and are spared (Mazokhin, 1969).

Taylor and Deay (1950) observed 57 to 63 % reduction in the population of corn borer (*Heliothis zea*). Several workers in the United States reported significant reduction in the infestation of tobacco hornworm *Menduca sexta*, using light traps, at a density of 3 to 6 traps per square mile (Lawson *et al.,* 1963, Gentry *et al.*,1967, Lam *et al.*, 1968 and Cantelo *et al.*, 1972). Hamilton and Steiner (1939) observed significant reduction in the infestation of Codling moth (*Cydia* sp) in apples using light trap. Light trap with 15-watt black light lamp was found to be most effective against pink bollworm moths *Pectinophora gossypiella* and other insects in cotton crop (Glick and Hollingsworth 1955). In a comparative study Mercury Vapour bulb, with distinct peaks in UV and visible radiation was found to be the best light source in attracting adults of *H. armigera, A. ipsilon* and *Spodoptera litura* (Vaishampayan and Verma 1983).

THEORY, OPERATIONAL PRINCIPLES AND PHILOSOPHY MECHANISM OF INSECT ORIENATATION

For better planning of experimental work and success it is essential to have a basic knowledge of visual responses of insects and their attraction to artificial light.

An animal reacts to different objects and situations according to the information the sensory mechanism receives through light of a given wavelength and its influence on the

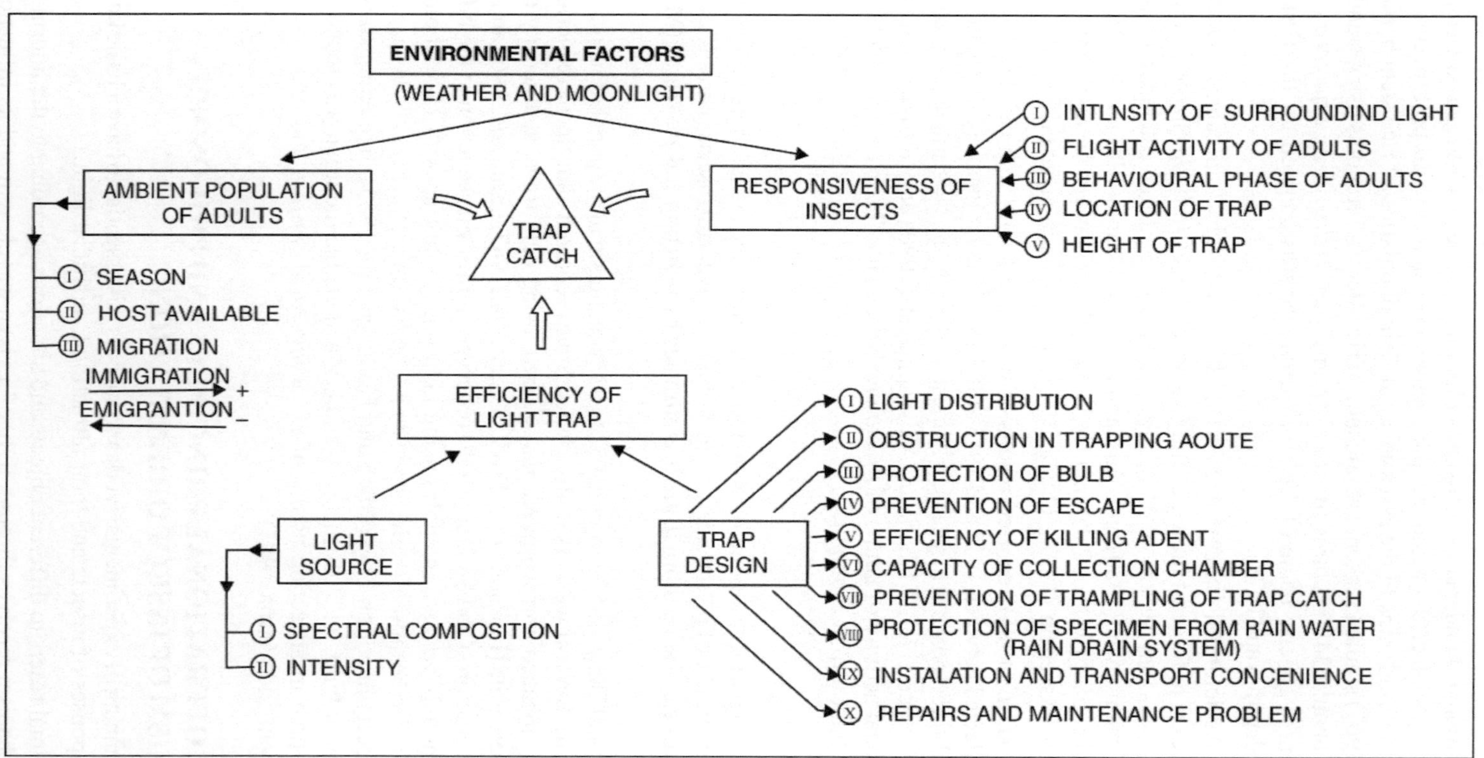

Fig. 1 Schematic diagram showing factors affecting the light trap catches of insects (Vaishampayan 1984).

effecter (motar) mechanism. Various theories have been proposed to explain when and why insect fly towards light. Marten (1956) believes that insects fly to a light source taking it as an exit from a confined space beyond whose limits they will find open space and freedom of flight. The hypothesis proposed by Mazokhin (1969) is however most logical and appropriately explained. When insects get into the region of high luminous flux, their eyes rapidly become light adopted. UV and visible radiation cause migration of the ommatidial protective pigments from nocturnal to diurnal position and visual sensitivity decreases by hundreds and thousands times and insects can not distinguish object at lower levels of illumination. The lamp is the only thing the insects are now able to see and they cannot fly to the sides of it in the darkness of the night. Thus, insects' attraction to light and their trapping is purely a function of contrast effect in illumination.

FACTORS AFFECTING LIGHT TRAP CATCHES

Success of light trap as IPM tool depends solely upon its trapping efficiency and retention of collected specimen in good condition. Design of light trap and light source used are important factors. If trapping is insufficient it would aggravate the problem by inviting more adults in the trap area. Weather extremes and bright moonlight adversely affect the size of trap catch.

The size of a total catch in light trap or net trapping efficiency (number trapped / total present) is determined by three principal factors viz.

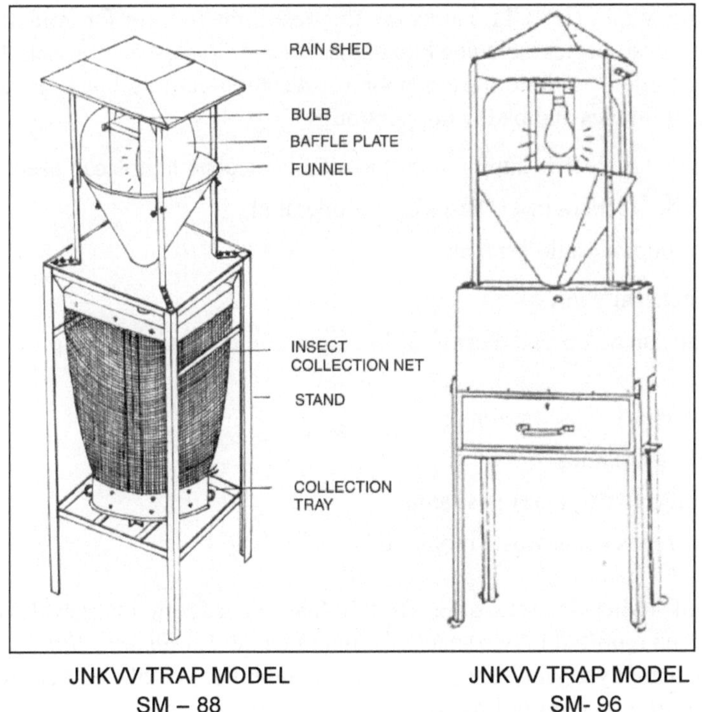

JNKVV TRAP MODEL SM – 88 JNKVV TRAP MODEL SM- 96

Fig. 2 New designs of light trap developed by Vaishampayan at Jabalpur

(1) Number of adults (population) actually present in the environment in the trap area.

(2) Behavioral response of insects towards light source influenced by:
   (i) Type of flight movement
   (ii) Bright moon light

(3) Design of trap used and light source used (spectral quality).

Factors affecting light trap catches have been discussed in detail by Vaishampayan (1985a) and are summarized in a schematic diagram in Figure 1.

Type of Light sources used

(A) Type of light sources, which can be used in light trap where electricity is available, is briefly summarized below:
   (i) Incandescent Tungsten filament lamp 100 to 150 watts.
   (ii) Latest Energy saver, Compact Fluorescent Lamps (CFL)
   (iii) High Intensity discharge Lamps M.V. lamps & Sodium Lamps
   (iv) Black light or Ultra Violet lamps 15 to 20 watt UV lamps.

Mercury vapour (MV) lamps 80 watt, 125 watt and 160 watt, emitting energy in all the wavelength bands covering insects vision spectrum ranging from 350 to 700 nm.), followed by 15 watt Ultra Violet (UV) TL lamps are the best light sources for attracting majority of noxious insect pest species including Lepidopterous and Coleopterous insects. Incandescent bulbs emitting yellow radiation are a poor source for noctuids but a good source for leafhoppers, gall midges and paddy hoppers/bugs.

(B) Where regular electric supply in not available electric lamps can used with:
   (i) Portable Generators (Diesel/Petrol operated)
   (ii) Solar photovoltaic generators

(C) Non- Electrical sources

   Light trap can be operated using following non-electrical sources producing visible light:
   (i) Kerosene lamps/ Petromax
   (ii) LPG gas lamps
   (iii) Carbide (Acetylene) gas lamps
   (iv) Oil torches using waste mobile oil

Results of studies on effect of moon light (moon phase and lunar cycle) on light trap catches have been reported in two species of noctuids i.e. *Spodoptera litura* and *Heliothis armigera* (Vaishampayan and Verma 1982) while the influence of weather factors on trap catches of *Heliothis* is reported by Verma, *et al.,* (1982).

Improved designs of light traps, most efficient in trapping insects of all orders, sturdy and easy to install, have been developed at JNKVV, Jabalpur and described earlier by Vaishampayan (1982, 1985b & 2000). Latest designs of light trap model SM 88 and SM 96 are shown in Figure 2.

MONITORING SEASONAL ACTIVITY

Use of light trap in activity study of adults is based on the assumption that changes in trap catches are mainly an expression of variation in the actual population of adults present in the environment (Persson 1976). This assumption is however not correct in toto. It holds well only if due corrections are made in trap catches for effect of certain factors.

The size of trap catch is adversely affected by weather extremes (excessive rains and very low temperature) and bright moonlight also. As such the trap catches are not always true reflection of the existing field population. The degree of correlation between trap catches and field infestation level is influenced by reproductive condition of females and the time lag between adult emergence and appearance of early instar larval stages, which is very variable in central and northern India. Due consideration, therefore, must be given to these factors to interpret the data more precisely and correctly.

A profile of weekly moving means of trap catches provides fairly good estimate of relative abundance of pest species in a season. The short-term effects of weather factors are averaged out in weekly means.

In correlation studies, comparing light trap catches and the infestation level of *H.armigera* in fields, significant correlation was established when trap catches were corrected to the effect of moon light, sex ratio and ovary condition (Table 1 and 2). Pawar *et al.* (1986) after an extensive study of light trap and pheromone trap catches conducted across the country found no significant correlation between trap catches and pest infestation in field.

Table 1 Correction factor for influence of moonlight on light trap catches of *Heliothis armigera* at Jabalpur (23"N).

Lunar group no.	Lunar days factor** group wise	Mean trap catch per lunar group (mean of 19 lunar cycles*)	Relative response (%) compared to no. Moon* days (RR)	Correction $\frac{100}{RR}$
1.	28th, 29th, 30th, 1st, 2nd, 3rd, 4th (Around "No Moon" period)	55.42	Absolute response 100.00 Suppressed response	1.000
2.	5th to 7th	32.90	59.36	1.684
3.	8th to 10th	20.90	37.71	2.651
4.	11th to 13th	21.20	38.25	2.614

Contd. ...

Table 1 Contd. ...

5.	14th, 15th, 16th, 17th, 18th (Around "Full Moon" period)	7.74	13.96	7.163
6.	19th to 21st	24.70	44.56	2.244
7.	22nd to 24th	38.60	69.64	1.435
8.	25th to 27th	59.40	Absolute response 100.00	1.000

* Observations reported by Vaishampayan and Verma (1982)

** Based on estimated reduction in trap catches due to moon-light effect.

Table 2 Relationship between light trap catch and larval population of *Heliothis armigera* (1985-86) data for various influences.

Observation Week	Total catch per week per trap phase	Catch Corrected to moon effect*	Estimated females (based on sex ratio)	Repro-ductive index (R.I.)**	Estimated fertile females NxR.I.	Early instar larvae per (end 5 m² of week)	Weekly mean minimum Temp.0°C
	X_1	X_2	N		X_3	Y	-
December							
52	16.0	118	56	1.00	56	116	10.7
January							
1	1.4	7	4	0.25	1.0	99	2.5
2	2.0	2	1	0.50	0.5	67	5.8
3	5.0	15	8	0.72	6.0	17	11.1
4	5.4	42	23	1.00	23.0	38	12.5
February							
5	50.0	127	63	0.73	46.0	43	11.6
6	83.0	83	40	0.50	20.0	27	14.1
7	29.0	70	35	0.43	15.0	22	11.6
8	22.0	256	78	0.48	37.0	32	12.9

* Corrected as per correction factor computed in Table 1.

** R. I. is based on proportion of matured ovaries

Note : Data of week number 1 & 2 excluded from correlation study due to very low temperature

Comparison of Degree of correlation:

(i) Comparison	**X_1 v/s Y**	**X_2 v/s Y**	**X_3 v/s Y**
(ii) Correlation coefficient 'r' value	**0.174**	**0.40**	**0.81**

Remark: 'r' value increase from 0.174 to 0.81

MONITORING MIGRATORY PHASE

Adults occupy an important position in population dynamics of a pest species. Its flight movement controls the distribution of pest in space. On depletion of food, increase in population pressure or on prevalence of adverse weather; the females decide to leave the area enmass, in migratory flights, in search of a favorable environment. This they achieve through physiological control over maturation of ovaries, ceasation of mating and conservation of energy (fat bodies) (Johnson 1969). On the basis of analysis of these physiological characters Vaishampayan and Verma (1987) provided first evidence on the existence of migration in *H. armigera* in India, identifying immigrant, resident and emigrant population in light trap catches. Vaishampayan *et al.* (1987) on the basis of light trap catches and other observations established a new theory on interstate migration of teak defoliator, *Hyblaea puera* from Kerala to Eastern Madhya Pradesh.

Migratory movement has significant impact on population dynamics of the pest species. Outbreaks of *H. armigera* in Rabi season on pulses occurred in 1987-88 in M.P. (Central India) were found to be influenced by migration of *Heliothis* adults from southern states including Tamil Nadu, Andhra Pradesh and Maharashtra (Vaishampayan 1995). How these migrant adults influence the build up of larval population of *H. armigera* on major host crops early in the season has been discussed in detail by Vaishampayan (2000). The IPM strategies therefore, should take into consideration such migratory movements on regional basis.

DIRECT CONTROL

1. Philosophy is to kill the noxious insects at time and place when they are most Concentrated, Immobile and Accessible (C.I.A. factors). Light trap concentrates the adult population present in the surrounding environment within a trap zone, arrest their movement by trapping (make them immobile) and are killed easily using a drop of insecticide.
2. Concept is to control pest population in its nucleus egg stage prior to oviposition, when it is concentrated and confined within the ovaries of female adults.
3. Light trap works as pest management tool indirectly though monitoring the adult activity, which decides the need of pesticide application, and it's timing if necessary and directly by mass trapping of adults (male and female both) and disrupting reproduction and oviposition. Mass trapping minimizes infestation in field significantly without any toxic hazards.

The concept is to detect the potential danger of population explosion well in advance by monitoring the peak periods of activity of adults with their reproductive potential, and decide the proper time of application of insecticides synchronizing with these peaks. Forecasting of pest population level in field shall be more accurate and reliable if adult activity is also taken into account.

PRACTICAL UTILITY

In general, use of light trap would be more successful in species

(i) Having annual life cycle or limited number of generations in a year,

(ii) Adults have more egg laying capacity and

(iii) Breeding is confined to the local area where adults are non-migratory. Mercury Vapour 80 watt and Ultra Violet 15 watt are the best light sources at Present. CFL above

20 watts, in violet-blue colour, is the potential light source of the future especially for noctuid Lepidoptera.

(iv) There is a certain time lag between adult emergence and oviposition (Pre-oviposional period). More this period, greater are the chances of trapping adults before they lay the eggs

Light trap works as a selective control device killing the pest in adult stage only with no hazards of toxic chemicals. In Organic Farming where use of pesticides is prohibited and in crop ecosystems like cotton, badly spoiled by excessive use of pesticides, use of light trap is inevitable as a component of IPM.

It has a tremendous potential as IPM tool in the management of bollworms, cutworms, hairy caterpillars, stem borers (sugarcane & paddy), *Spodoptera*, diamond back moth, fruit sucking moths, codling moth, white grub beetles, field crickets, grasshoppers, and against paddy pests including plant hoppers, gallmidges, gundhi bugs etc. It has a specific significance in IPM of Sal heart woodborer in Sal forest and Woodborers in forest plantations of Teak and Eucalyptus. For the first time, light traps are now available in India in sturdy plastic moulds for large-scale use by the farmers.

DATA ON IMPORTANT SPECIES OF INSECT PEST COLLECTED IN LIGHT TRAP AT JABALPUR

Systematic data were collected on trap catches of seven species of Lepidoptera and three species of Orthoptera operating the trap every night round the year using 125 W/ 160W mercury vapour bulb. Abstract data of yearly collection for the period 1982-1987 are presented in Table 3.

Table 3 Yearly distribution of major insect pest of pulses collected on light trap at Jabalpur during 1982-83 to 1986-87.

Name of Insect pest species	Total annual catches per trap					Major host crops
	82-83	83-84	84-85	85-86	86-87	
LEPIDOPTERA: Gram pod borer *Heliothis armigera*	2080	14192	2505	4683	16291	Cotton, sorghum, sunflower, tomatoes
Tobacco caterpillar *Spodoptera litura*	2630	2691	3120	2290	2046	Cotton, soybean, sunflower, tobacco, groundnut, crucifers, castor
Green semilooper *Plusia chalcytes*	4406	4723	6637	8611	6720	Cotton, soybean, sunflower, cruciferous, vegetables.
Cabbage semilooper *Plusia orichalcea*	2682	6514	7172	4134	6682	Cotton, sunflower, crucifers, soybean, vegetables.

Contd. ...

Table 3 Contd. ...

Black cutworm *Agrotis ipsilon*	912	1099	1201	1922	1610	Rabi crops (Wheat, gram, potatoes, crucifers, linseed
ORTHOPTERA: Field crickets ii) *Gymnogryllus humeralis*	15754	26577	31030	12460	22884	Soil insect - damage germinating seeds, young seedlings, roots (Kharif and Rabi both)
Mole crickets *Gryllotalpa grylliotalpa*	2773	4464	2405	1268	3411	Soil insect - damage germinating seeds, young seedlings, roots (Kharif and Rabi both)
Grass hoppers i) *Trilophidia cristata* ii) *Gastrimargus transversus*	1226	852	2690	309	1666	Polyphagous (Kharif Crops)

Five years data (1982 to 1986) of monthly distribution if light trap catches of *Heliothis armigera* and *Spodoptera litura* are presented in Table 4.

Table 4 Monthly distribution of light trap catches of *Heliothis armigera* and *Spodoptera litura* recorded in different years at Jabalpur

Period	Total monthly catch/ trap (Corrected to 30 days) in different									
	Heliothis armigera					*Spodoptera litura*				
	85-86	86-87	87-88	94-95	95-96	82-83	83-84	84-85	85-86	86-87
June	2	-	2	-	-	145	04	31	00	—
July	3	8	7	-	-	05	45	43	22	30
August	13	8	33	-	-	102	373	898	112	126
September	12	113	102	9	1	480	345	707	679	546
October	2	47	'26	32	6	965	343	713	224	515
November	2	5	22	10	17	211	183	69	118	94
December	115	55	72	66	423	234	30	34	34	19
January	162	145	887	71	297	106	95	28	273	35
February	660	1189	955	245	1452	91	298	58	359	106
March	2579	1882	3101	6493	4189	183	458	225	320	134
April	1133	12574	450	5658	3370	78	371	141	149	297
May	-	265	90	1340	82	30	146	172	-	144

OBSERVATIONS ON BENEFICIAL PARASITIC AND PREDATORY SPECIES OF INSECTS COLLECTED ON LIGHT TRAP AT JABALPUR

During 1983 and 1984 crop season's observations were made on beneficial parasitic and predacious insects collected on light trap. In all 21 predacious and 8 parasitic species were recorded to appear in significant numbers. Their proportion, compared to the catch of harmful pest species was however, very low below two percent. Highest trap collection was recorded in families of Carabidae & Cicindellidae, among predacious Coleoptera and Reduvidae among the Hemiptera. In parasitic species Ichneumonids and Braconids of the order Hymonoptera were predominent with maximum activity observed in March, & April when Rabi crops are harvested. *Campoletis chloridae* (Ichneumonidae) an important larval parasite of *Heliothis* was of rare occurrence only. Results reveal that light trap operations will have no significant adverse effect on existing natural biocontrol activity in a crop ecosystem.

OBSERVATIONS ON CONTROL OF *AGROTIS IPSILON* AND *HELIOTHIS ARMIGERA* USING LIGHT TRAPS

Excellent results were obtained in controlling population of cutworms *Agrotis ipsilon* in about 20 hectares of chickpea crop in village Kathonda. Four traps, using 160-watt mercury vapour lamp, were operated from 15th of October till maturity of crop. Comparative data of incidence of cutworms in the endemic area, recorded for 9 years since 1974-75, presented in Table 5 demonstrates very significant reduction in cutworms damage, requiring no insecticidal treatment.

Table 5 Relationship between rainfall, light trap catches and the incidence of black cutworm, *Agrotis ipsilon* at Jabalpur

Year of Observation	Total rainfall (mm)		Total trap Catch/trap in non Endemic in area (Total of Nov. & Dec)	Cumulative light trap catch in Octo. to Dec. Endemic area (Maximum mean)	Incidence of pest Dec./Jan. population per meter row weekly
	Mansoon season Larval July to Oct.	Winter season Nov. & Dec.			
1974-75	747	2.0	265+	Nil	3.5
1975-76	1238	0.0	35+	Nil	3.8
1976-77	918	8.2	78	Nil	7.0
1977-78	1424	30.5	64	Nil	9.6
1978-79	921	0.0	57+	Nil	4.8
1981-82	465	42.8	55	Nil	0.4
1982-83	1221	17.1	21.3	883	3.0**
1983-84	1252	8.4	65	465*	3.3**
1984-85	1284	0.0	174	5096	1.0**

*December only * Late installation of light trap. 100% losses in February.

** 1.Anticipated very high incidence of the pest (6 to 8 larvae/meter)

2. Light trap operation reduced the incidence very significantly.

In case of *Heliothis armigera*, 5 years data presented in Table 6-revealed 50 to 60% reduction in pest population in chickpea crop around light trap within a radius of about 100 meters.

Table 6 Larval population of *H. armigera* on gram crop at various distances from light trap at Jabalpur (mean of four years, 1982 to 1986)

Month	Larval population/10 samples (1/2 x 1/2 m crop area)			
	50 m	100 m	150 m	200 m
December				
I	5.0	5.0	5.0	21.0
II	-	-	-	-
III	-	-	-	-
IV	1.6	5.6	1.7	0.5
January				
I	2.5	7.5	6.2	11.2
II	2.6	4.2	4.5	5.8
III	2.9	10.2	7.1	4.5
IV	7.8	9.8	7.3	6.9
February				
I	14.2	12.8	18.1	15.3
II	19.9	25.6	20.2	16.6
III	8.5	21.5	14.5	13.5
IV	14.7	21.3	22.4	17.6
March				
I	12.2	16.5	27.2	14.0
II	2.0	13.0	7.0	9.0
III	22.0	40.0	7.5	85.0
Mean	**8.90**	**14.85**	**16.63**	**18.9**

OBSERVATIONS ON USE OF LIGHT TRAP IN THE MANAGEMENT OF RED HAIRY CATERPILLAR *AMSACTA MOOREI*

Extensive work was done during the years 1982-1986 to test the use of light trap in survey and control of *Amsacta moorei*, a polyphagous pest of Kharif crops in the endemic area of district Jhabua (Madhya Pradesh).

Light traps of JNKVV model with 160-Watt Mercury Vapour lamps were operated every night during the active season and trap caches were examined every day. Observations on larval population on major host and trap crops were taken following standard sampling procedures. Data are presented in Tables 7 & 8 . Results of significant importance are summarized below.

1. Pest is active from mid June to August
2. Moth emergence and collection of moths in the traps starts in significant numbers on 3rd and 4th day after the first monsoon showers with precipitation above 50 mm within 2 days. Egg laying is followed immediately. Relative size of trap catches will help to predict areas of low as well as high infestation (Table 7).

Table 7 Weekly distribution of rainfall and light trap catches of *Amsacta moorei* (Arctiidae) during different years.

Period/		Total trap catches / trap / week (JNKVV trap)							
(Weekly)		1983		1984		1985		1986	
		Rain fall (mm)	Trap Catch	Rain fall (mm)	Trap Catch	Rain fall (mm)	Trap Catch	Rain fall (mm)	Trap Catch
Location : Jhabua									
June	25	53.0	-	6.2	7	-	-	52.2	326
	26	31.0	-	0.0	2	8.0	4	10.7	6092
July	27	41.2	192	70.9	476	0.0	2	-	377
	28	40.5	322	1.7	100	16.0	1	8.2	1
	29	79.2	1107	63.0	21	7.0	165	35.5	2255
	30	53.2	567	6.0	77	27.0	16	-	-
August	31	23.0	613	-	-	50.0	37	-	-
	32	119.2	2282	-	-	19.0	18	-	-
	33	66.0	1050	-	-	15.0	930	-	-
	34	12.7	507	-	-	11.0	418	0	0
	35	0.0	-	-	-	-	16	-	-
Total		**519.2**	**4660**	**147.7**	**683**	**153.0**	**1607**	**85.0**	**9051**
Location : Alirajpur (Jhabua District)									
June	25	6.2	11	-	-	65.6	1215	0.2	2068
	26	1.6	42	9.1	50	10.0	8702	17.8	1779
July	27	63.2	4920	2.0	83	3.8	2433	0.8	202
	28	27.6	245	4.0	61	5.0	0	-	291
	29	-	14	73.0	3856	32.0	337	8.6	436
	30	-	10	170.0	1568	68.2	75	0.1	53
August	31	-	-	38.0	11	-	-	-	22
	32	-	-	35.0	89	-	-	-	-
	33	-	-	0.0	171	-	-	-	-
	34	0	0	2.0	137	-	-	-	-
	35	-	-	0.0	30	-	-	-	-
Total		**88.6**	**5242**	**333.0**	**6056**	**184.6**	**12762**	**27.5**	**4851**

3. Second generation larvae emerged in August or September, pupate in soil and undergo long diapause period of about 10 months in pupal stage. Trapping of these adults

during August/September will check the carryover of the pest and reduce its activity in the following year.

4. Over 80% reduction in larval population was observed in field crops close to light trap, covering around 50 hectares of crop area. Reduction was absolute 100% on Jobat farm in about 20 hectares, where light trap was in operation consecutively for 3 years since 1982 (Table-8).

Table 8 Survey of the incidence and control of Red hairy caterpillar *Amsacta moorei* in Jhabua Distt. July 1984

Location	Crop observed	Number of larvae per 10 samples*	Remarks
Jhabua Block			
(i) Gopalpur	Maize	20 larvae	Control**
(ii) Dhekal	Maize	5 larvae 2 larvae	Light trap installed One year operation
(iii) Dhekal	Maize		
Bhabra Blocks			
Kakarbari	Maize & Urid mix.	14 larvae	Control**
Jobat Blocks			
(i) Kilajobat	Cowpea, Maize Maize Cowpea, Moong Cowpea, Moong	30 larvae 19 larvae 49 larvae 34 larvae	Control**
(ii) Kalikhaetar	Sunhemp Maize Cowpea	1 larvae 1 larvae 0 larvae	Light trap installed in 1984(One year operation)
(iii) Jobat farm	Moong Urid	1 larvae 0 larvae	Light trap installed in 1982 (Three years operation)

* Sample Unit- ½ X ½ meter (1/4th Sq. meter)

** Control means "No light trap operation"

References

Ayyar, T.V.R., and Anantanarayana, K.P. 1937. The stem borer pest of the rice *Schoenobius incertulas* in South India. *Agric. and Livestock in India*, 7 (2): 171-179.

Beckman, C.M. 1970. Seasonal abundance of *Heliothis* spp In Georgia piedmont, *J. Ga. Entomolo Soc*. 5 (3) : 138-142.

Betts, E. 1976. Forecasting infestations of tropical migrant pest : the desert locust and the African army worm. Pp. 113-114 in Rainey, R.C. (Ed.) *Insect Flight* – 287 pp. Oxford, Blackwell Scientific.

Blair, B.W. and Catling, H.D. 1974. Outbreaks of African army worm, *Spodoptera exempta* (Walker) (Lepidoptera : Noctuidae) in Rhodesia, South Africa, Botswana and South West Africa from February to April 1972. Rhod. *J. Agric. Res*. 12, 57-67.

Brown, E.S., Betts, E. and Rainey, R.C. 1969. Seasonal changes in distribution of African army worm, Spodoptera exempta (Wik.) (Lep. : Nocutidae) with special reference to Eastern Africa. *Bull Ent. Res.* 58, 661-721.

Cantelo, W.W., Smith, J., Baumhover, A.H., Stanley, J.M. and Henneberry, T.J. 1972. Suppression of an isolated population of the tobacco horn worm with black light traps, unbaited or baited with virgin female moths, *Environ. Entomol.* 1(2) : 252-258.

Chopra, R.K. 1928. Annual report of the Entomologist to the Govt., Punjab Lyallpur for the year 1926-27. *Rep. Dept. Agric. Punjab* 1(2) : 43-49.

Dina Nath 1923. Prelimainary observations on the attraction to light of moths of sugarcane borers. *Rep. Pro. V Entomo. Mtgs Pusa*: 65-67.

Dutt, H.L. 1919. The methods of control of *Agrotis ipsilon* in Bihar. Rep. *Proc. 3rd Entomol. Meet.* Pusa II : 622-625, Calcutta.

Gentry, C.R., Lawson F.R., Knott C.M., Stanley J.M. and Lam J.J. Jr. 1967. Control of hornworms by trapping with black light and stalk cutting in North Carlina. *J. econ. Entomol.* 60 : 1437-42.

Glick, P.A. and Hollingsworth, J.P. 1955 Response of the pink bollworm and other cotton insects to certain ultraviolet and visible radiation. *Tech. Bull USDA* 48 : 173-177.

Hamilton, D.W. and Steiner, L.F. 1939. Light trap and codling moth. *J econ. Ent.* 32(6): 867-72.

Hartstack, A.W., Jr., Hollingsworth, J.P., Ridgway, R.L. and Coppedge, J.R. 1973. A population dynamics study of the bollworm and the tobacco budworm with light trap. *Environ. Entomol.* 2, 244-252.

Hartstack, A.W., Jr., Henson J.L., Witz. J.A., Jackman J.A., Hollingsworth, J.P. and Frisbie R.E. 1977. The Texas program for forecasting *Heliothis* spp. Infestations on cotton Proceedings 1977. Beltwide Cotton Production Research Conference, Atlanta, G.A.

Hassanein, M.H. 1972. Abundance and population density of three lepidoterous insects in upper Egypte. (Lepidoptera : Noctuidae). *Bull. Soc. Entomol.* Egypt 55: 79-83.

Hussain, M.A. 1930. *Entomology Rep. Dept. Agril. Punjab*. 1928-29, (1) : 53-63.

Ishikura, H. 1967. Assessment of the field population of rice stem borer moth by light trap. Proceedings of a symposium on the major insect pests of rice plant 14-18 Sept. 1964. Los. Banos Laguna, Philippines. Johns Hopkins Press : Baltimore M.D. 169-179.

Javeri, T.N. 1921 Notes on 'Katra' (Hairy caterpillars) and their controlling measures. *Rep. Prod. 4th Ent. Mtgs.* Pusa : 98-100.

Johnson, C.G. 1969. Migration and dispersal of insects by flights. Methuen and Co. Ltd., London, pp. 763.

Koitvadhi, K. and Cantelo, W.W. 1966. The use of light traps to measure relative population densities of the rice gall midge, pachydiplosis oryzae and other insect pests of rice. *Int. Rice Comm. News Letter* 15(2) : 16-19.

Lam, J.J.Jr., Stanley, J.M. Knott, C.M. and Baumbover, A.H. 1968 Suppression of nocturnal tobacco in sects population with balck light traps. Trans. Asae. *(Amer. Soc. Agr.Eng.)* 11 (5) : 611-12.

Lawson, F.R., Gentruy C.R. and Stanley, J.M. 1963 Effects of light traps on hornworm populations in large areas, *USDA, ARS*. (Ser). 31-91, 18p.

Mahobe, J. and Vaishampayan, S.M. 1990a. Studies on the seasonal variation in the reproductive potential and ovary development of soybean semilooper *Plusia chalcytes* Fabr. *Geobios* (17): 237-241.

Mahobe, J. and Vaishampayan, S.M. 1990b. Seasonal variation in the reproductive potential of female of Plusia *orichalcea.* Collected from light trap. *Proceed. Nat. Acad. Sci. India.* 60(B): IV: 373-76.

Marten (1956). *Entomol. Z.* 66: 121-133.

Mazokhin Parshnyakov 1969. *Insect Vision*. Plenum Press, New York.

Milne, D. 1927. *Entomology Rep. Dept. Agric*. Punjab, 1925-26(1) : 49-54.

Nirmala Devi, Deshraj and Mahavir Singh 1991. Seasonal abudance of two noctuid pests (*Plusia orichalcea and Heliothis armigera*), in northwest Himalayas (India). *J. Ent. Res.* 15(2) : 120-124.

Odiyo, Onyango P. 1975. Seasonal distribution and migration of *Agrotis ipsilon* (Hufnagel) (Lepidoptera : Noctuidae) Cent. Overseas Res. Trop. *Pest. Bull* 4: pp. 26.

Otman, E.R. 1964. Orchard insects surveys with black light traps. *J. econ. Entomol.* 57(1) : 6-8.

Pawar, C.C. Shrivastava C.P. & Reed,W. 1986. Some aspects of population dynamics of *Heliothis armigera* at ICRISAT Center, III Oriental Entomology Symposium, Februray 21-24, 1984, Trivandrum. Proceedings March 1986, pp. 79 to 85.

Persson, B. 1976. Influence of weather and nocturnal illumination on the activity and abundance of population of Noctuids (Lepidoptera) in South Coastal Queensland. *Bull. Ent. Res.* 66 : 33-63.

Pradhan, S. 1983. Origin of insect pests Chapter 4 In : *Agricultural Entomology and pest control*. ICAR Publication, New Delhi.

Taylor, J.G. and Deay, H.O. 1950. Electric lamps and traps in corn borer control. *J. Amer. Soc. Agric. Engin.* 31: 503-5.

Vaishampayan, S.M. 1977. A note on Helicoverpa armigera (Hubner) Hardwick, a new generic combination of gram pod borer Heliothis armigera. *Current Sci.*46(8): 271.

Vaishampayan, S.M. 1980. Seasonal abundance and activity of gram pod borer moths *Helicoverpa (Heliothis) armigera* (Hubner) on light trap equipped with mercury vapour lamp at Jabalpur. *Indian J. Ecol.* 7 : 147-154.

Vaishampayan, S.M. 1981. Light trap studies on Black cutworm *Agrotis ipsilon* (Hfn.) In : *Progress in soil biology and ecology in India*. Edt. Veeresh G.K. : 192-200. *UAS Tech. Sem.* No. 37, Univ. Agril. Sci. Bangalore.

Vaishampayan, S.M. 1982. New design of light trap for survey and management of insect pest population in Agro and Forestry ecosystems. *Indian J. Ent.* 44 (3) : 201-205.

Vaishampayan, S.M. 1985a. JNKKV Trap SM 84 Model Light trap for survey of insect pests and vectors. Pp : 152-155. In : Use of traps for pest / vector research and control. Eds. : Mukhopadhyay, S. and Ghosh, M.R., Bidhan Chandra Krishi Vishwa Vidyalaya, Kalyani, West Bengal.

Vaishampayan, S.M. 1985b. Factors affecting the light trap catches of insects with emphasis on design aspects. pp:62-67. In : "Use of traps for pest / vector research and control". Eds. : Mukhopadhyay, West Bangal.

Vaishampayan, S.M. 1995. Analysis of factors for the outbreak of polyphagous crop pests creating national problems *Helicoverpa* (*Heliothis*) *armigera*, In: Proceed. *National Seminar on changing pest situation in the current agriculture scenario of India,* ICAR Publication, New Delhi, pp. 261-270.

Vaishampayan, S.M. 2000. Ecology and management systems of *Heliothis (=Helicoverpa) armigera* Hubner. In: *IPM System in agriculture.* 7: 31-54, Aditya Book Pvt. Ltd., New Delhi, India.

Vaishampayan, S.M. 2002. Use of light trap as a component of adult oriented strategy of pest management. In: *Resources management in Plant Protection.* II: 139-144. Publ. Plant Prot. Asso. India, Hyderabad.

Vaishampayan, S.M. and Verma, R. 1982. Influence of moon light and lunar periodicity on the light trap catches of gram pod borer *Heliothis armigera* (Hubner) moths. *Indian J. Ent.* 44 : 206-212.

Vaishampayan, S.M. and Veda, O.P. 1980. Population dynamics of gram pod borer *Heliocoverpa armigera* (Hubner) and its outbreak situation on gram Cicer arietinum L. at Jabalpur. *Indian J. Ent.* 42(3) : 453-459.

Vaishampayan, S.M. and Verma, R. 1983. Comparative efficiency of various light sources in trapping adults of *Heliothis armigera* (Hubn.) *Spodoptera litura* (Boisd) and *Agrotis ipsilon* (Hufn.) (Lepidoptera : Noctuidae). *Indian J. Agric. Sci.* 53 (3) : 163-167.

Vaishampayan, S.M. and Verma, R. 1987. Seasonal change in reproductive potential of female moths of *Heliothis armigera* (Hubner) (Lepidoptera : Noctuidae) collected on light trap at Jabalpur. *Indian J. Agric. Sci.* 57 (1)

Vaishampayan, S.M., Verma, R. and Bhoumik, A.K. 1987. Possible migration of teak defoliater *Hyblaea puera* (Lepidoptera: Hyblaeidae) in relation to the movement of the south-west monsoon as indicated by light trap catches. *Indian J. Agric. Sci.* 57 : 41-46.

Verma, R., Vaishampayan, S.M. and Rawat, R.R. 1982. Influence of weather factors on the light trap catches of Heliothis armigera (Hubn.) *Indion J. Ent.* 44 (3):213-218.

Yoshimeki, M. (1967). A summary of the forecasting programme for rice stem borer in Japan. The major insect pests of rice plant. Baltimore, Maryland. The John Hopkins Press, pp.729.

Entomology : Novel Approaches, 2007
Eds.: *P.C. Jain and M.C. Bhargava,*(Page 211-244)
New India Publishing Agency, New Delhi
E-mail: newindiapublishingagency@gmail.com
web: www.bookfactoryindia.com

CHAPTER **12**

INSECTICIDE RESISTANCE IN IPM

Chitra Srivastava and Berin Pathrose
Division of Entomology, Indian Agricultural Research Institute, New Delhi-110 012

INTRODUCTION

Insecticide resistance poses a severe threat to agricultural productivity in India and in other countries of the world. Humans have had the ingenuity to develop thousands of pesticides to protect foodstuffs, livestock, and health and the use of these poisons has triggered a rapid evolutionary response in many target species and some nontarget species of invertebrates, microorganisms, higher plants and vertebrates. Insecticide exposure led to nullify their toxic effects through genetic selection for individuals with the biochemistry or behaviour. Continual development of new compounds and its greater use and increase in input insecticide cost is mainly due increase in resistance. Reduction in insecticide efficacy from insect resistance has major economic, environmental and human health implications. Integrated Pest Management (IPM) is a convenient descriptor for wide range of still-evaluating pest control program that emerged from chemical prophylaxis practiced. IPM theory has fostered a transition from primarily chemical control to programs tempered with biological information and input tailored to each agro ecosystem, crop and pest situation in production.

Definition, History and Status of Insecticide Resistance

Resistance is the microevolutionary process of genetic adaptation through the selection of biocides (Whalon and McGaughey, 1998). One consequence of resistance is the failure of a plant protection tool, tactic, or strategy to control a pest where such failure is due to a genetic adaptation in the pest. This definition has traditionally been applied to insect populations that escape the effects of a chemical insecticide. However, nearly all classes of organisms provide an example of resistance to pest manage-ment measures, chemical or otherwise. (Mota-Sanchez *et al.*, 2002).

Just as resistance evolves over time, the definition of resistance has been developed and refined. A panel of World Health Organization (WHO) experts defined resistance as "the development of an ability in a strain of insects to toler-ate doses of toxicants which would prove lethal to the majority of individuals in a normal population of the same species" (WHO, 1957). This definition was the operational definition for years. After more than 60 years of synthetic insecticide applications, insect populations all over the world have been exposed to, and selected by, one or more pesticides, making it very difficult to find a normal population. In addi-tion, the WHO definition is for populations rather than individuals, a distinction with more significance today because new biochemical and physiological tech-niques facilitate the detection of resistance in single individuals (Mota-Sanchez *et al.*, 2002).

In 1960, J. F. Crow presented a more inclusive definition of resistance that considers single individuals as well as populations. He proposed "resistance marks a genetic change in response to selection" (Crow, 1960). This definition is not re-stricted to high resistance levels or dependent upon the failure of an insecticide in the field. Incipient resistance is included in this definition as well. However, perhaps the most significant consequence of pesticide resistance is missing: field failure. In 1987, R. M. Sawicki improved upon Crow's definition by adding the significance of field failure to the definition as follows: "Resistance marks a ge-netic change in response to selection by toxicants that may impair control in the field" (Sawicki, 1987).

The exponential increase in the worldwide cases of resistance during the first three-quarters of the twentieth century, combined with scientific and public pressure, led the pesticide industry to form various "resistance action committees" including ones for insecticides (IRAC), fungicides (FRAC), and herbicides (HRAC). These action committees worked in various aspects of resistance management, specifically monitoring pro-grams. The criteria developed by IRAC for defining resistance include the follow-ing circumstances (Tomlin, 1997):

An insect should be viewed as resistant only when

1) The product for which resistance is being claimed carries a use recommen-dation against the particular pest mentioned and has a history of success-ful performance.
2) Product failure is not a consequence of incorrect storage, dilution, or appli-cation and is not due to unusual climatic or environmental conditions.
3) The recommended dosages fail to suppress the pest populations below the level of economic threshold.
4) Failure to control is due to a heritable change in the susceptibility of the pest populations of the product.

Resistance to insecticides develops mainly because of their intensive use in situations that would otherwise favour rapid pest build up. There are three phases in the development of insecticide resistance in a population: *First* when the frequency of resistant insects is very low (one in a million or more), and the insecticide gives satisfactory control. *Second,* with the continuous use of the same or related insecticide the frequency of resistant individual increases, leading to occasional crop failures, and *third* with the continuous pressure of insecticides there is a large increase in the number of resistant individuals in the population and the insecticide becomes ineffective (Mehrotra and Phokela, 2000). A result of the large scale use of pesticides to control insect pests is the development of resistance in the target species, whose number has increased to 540 (Whalon *et al.*, 2004).

Resistance was first recognized for what is in 1908 by Melander, who noted an unusual degree of survival of the San Jose scale following attempts at control with customary lime-sulfur treatments in the Clarkston Valley of Washington (Melander, 1914). Of partic-ular interest is his comment that, if the problem became worse, it may be necessary "to introduce a weak strain of the San Jose scale to cross with the immunes and thus return to the normal susceptible population".

During the next 38 years, only 11 addi-tional species of arthropods had developed resistance to pesticides. However, not long after the advent of the synthetic organic insecticides, marked by the introduction of DDT, there was a sudden and marked increase in the rate of appearance of new cases of resistant pop-ulations (Forgash, 1984). Numbers of species alone do not reveal the full story, however, nor the gravity of the situation that we find ourselves in today. We must also recognize that great increases have occurred, and continue to occur, in geographical distribution of resis-tant populations, in the frequency of resis-tance genes, and in the numbers of different pesticides to which resistance has devel-oped. It is also instructive to view the re-sistance picture over time in terms of in-secticide types, numbers of orders, and im-portance of the pests involved (Forgash, 1984).

The first ever report of pesticide resistance in an agricultural pest in India was when the reported failure to control Singhara beetle, *Galerucella birmanica* was traced to the development of resistance (Pradhan *et al.,* 1963). Another insect pest of agricultural importance to have become resistant to pesticides is the tobacco caterpillar, *Spodoptera litura.* Srivastava and Joshi (1965) reported that failure of HCH Lo control *S. litura* on cauliflower by the growers around Ajmer and Jaipur was due to the development of resistance. The real development of pesticide resistance in this pest took place during 10 year period from 1971 to 1980. Reddy and Reddy (1984) working on the populations of *Spodoptera* collected from Kurnool and Guntur districts in Andhra Pradesh observed that there was distinct development of resistance in Guntur when he compared the LC_{50} of various pesticides against population of *Spodoptera* originating from Kurnool. A study was also conducted in 1983 by Ramakrishnan *et al.* (1984) when they collected *S. litura* populations from the agricultural fields of Tenali (Andhra Pradesh) and showed that the *S. litura* larvae were 5.73-fold resistant to malathion, 14.73-fold to pyrethrum, 16.25-fold to lindane and 85.91-fold to endosulfan. Armes *et al.* (1997) working on twenty two strains of *S. litura* collected from groundnut crops of eight locations in Andhra Pradesh between 1991 and 1996 reported resistance levels ranging from 0.2 to 197-fold to cypermethrin, 8 to 121-fold to fenvalerate, 1 to 13-fold to endosulfan, 1-29-fold to quinalphos, 2 to 362-fold to monocrotophos and 0.7 to 19-fold to methomyl.

Diamond back moth (DBM), *Plutella xylostella* is a serious pest of crucifers throughout the world. It is notorious for developing resistance very easily and has developed resistance to nearly all classes of insecticides used against it all over the world (Sun, 1990). In India, the first report of *P. xylostella* developing resistance to pesticides was that of Verma and Sandhu (1968) when they observed that DDT and parathion failed to control this pest around Ludhiana. A high degree of resistance to pyrethroids, cypermethrin, fenvalerate, deltamethrin and an organophosphate, quinalphos has been reported in this pest from various parts of the country (Saxena *et al.,* 1989).

Mustard aphid, *Lipaphis erysimi,* is a major constraint in the production of mustard and rape seed in the country. At the Indian Agricultural Research Institute, New Delhi, Dhingra (1991) using the data generated by Sarup *et al.,* (1969) as the baseline data concludes that *L. erysimi* has developed 20-, 12-, 5-, and 4-fold resistance to methyl parathion, dimethoate, malathion and lindane, respectively over twenty three years from 1969-1991.

Appearance of pesticide resistance to cotton bollworm, *Helicoverpa armigera*, to pyrethroids is now well established. It had been suggested as early as 1982 that the *Helicoverpa* obtained from various regions of the country showed a differential response to pesticides indicating that perhaps these- populations were subjected to different degrees of pesticide pressure in different parts of the country (Reed and Pawar, 1982). Purely on the basis of biochemical parameters it was also suggested that *Helicoverpa* populations in various parts of the country differ significantly and that resistance to pesticides in this important pest may be just around the corner (Mehrotra and Phokela, 1986), Indeed, the failure to control *Helicoverpa* in Andhra Pradesh during cotton season 1987-88 and in Punjab in 1990-91 cotton season was traced to the development of high degree of resistance to synthetic pyrethroids (Dhingra *et al.*, 1988; McCaffery *et al.*, 1989; Phokela and Mehrotra, 1989; Mehrotra and Phokela, 1992; Sidhu *et al.*, 1994).

Among the stored grain insect pests, the pesticide resistance was first reported in flour beetle, *Tribolium castaneum* in 1971 against DDT and malathion by Bhatia and Pradhan (1970), and Bhatia et *al.* (1971). Later, lindane resistance in *T. castaneum* was reported from Bombay, Rajasthan, Uttar Pradesh and various locations in Punjab (Champ and Dyte, 1976; Kalra *et al.*, 1975). Malathion resistance varying from 16 to 38-fold, DDT resistance (1.4 to 2.77-fold), dichlorvos resistance (2.2 to 13.3-fold) and phosphine resistance (2.08 to 5.84-fold) was also detected in this species from various locations (Bhatia et *al.*, 1990; Saxena *et al.*, 1991a, b; 1992; Rajendran *et al.*, 2000).

Similarly, lesser grain borer, *Rhyzopertha dominica* is also known to be resistant to malathion, lindane and phosphine (Champ and Dyte, 1976). Khapra beetle, *Trogoderma granarium* population with 40-fold resistance to phosphine was reported from Punjab by Borah and Chahal (1979)

Origin and development of resistance

The development of pesticide resistance, being preadaptive in nature, depends on the presence within, the population of genes necessary for resistance to the chemical ap-plied. The evolution of resistance in a pop-ulation is a step-wise process. During early exposure to the pesticide, there is a latent period during which the resistance genes become segregated and associated with other genes that govern factors favorable to survival of the pest; this may be accompa-nied by some noticeable increase in toler-ance to the chemical (Forgash, 1984).

Next comes a period in which resistance gradually develops, followed by a period of rapid increase, during which multiple fac-tors may contribute to the resistance of the population. It is probably the inheritance of these associated resistance features during the latent period that enables rapid devel-opment of resistance to unrelated types of chemicals that are subsequently used after failure of the initial pesticide (Forgash, 1984).

It is widely accepted that the development of insecticide resistance by insects is due to the selection of variants in the population carrying pre adaptive genes for resistance. The rate at which resistance develops in a population depends on:

1. The frequency of resistant genes present in the population
2. The nature of these genes (either single or multiple, dominant or recessive)

3. The intensity of selection pressure, that is, the magnitude of the population exposed to the chemical and the proportion killed and
4. The rate at which the species breeds (number of generations per year)

Out of these four the latter two are found to be more important. In some instances, geographical isolation, confined populations, or highly inbred strains of insects may not contain the prerequisite resistant genes; therefore, they may fail to develop resistance (Perry *et al.*, 1998).

Toxicological observations indicate that there exists a dynamic variation in tolerance to a poison among individuals of a given population. However, truly sublethal doses of insecticides do not induce resistance in susceptible populations, or increase the level of enzymes responsible for the metabolism of the chemical. On the contrary, the cumulative effect of daily sublethal doses of insecticide may render the insect more susceptible to the poison (Perry *et al.*, 1998).

Mechanisms of resistance

There are three major classes of mechanisms of resistance to insecticides in insects (Jeschke and Nauen, 2005). They are as follows

a) Allelic variation in the expression of target proteins with modified insecticide binding sites, *e.g..*, acetylcholinesterase insensitivity towards organophosphates and carbamates, voltage – gated sodium channel mutations responsible for knockdown resistance to pyrethroids, and a serine to alanine point mutation (*rdl* gene) in the ã-aminobutyric acid (GABA) – gated chloride channel at the endosulfan/fipronil/dieldrin binding site.
b) The second and often most important class of resistance mechanisms in insect pest species is metabolic degradation involving detoxification enzymes such as microsomal cytochrome P-450 dependent monooxygenases, esterases and glutathione S-transferases. This increased enzyme activity can be attributed to gene amplification, increased gene expression and/or gene splicing.
c) The third least important mechanism is an altered composition of cuticular waxes which affects penetration of toxicants. Reduced penetration of toxicants through the insect cuticle has often been described as a contributing factor, in combination with target site insensitivity or metabolic detoxification (or both), rather than functioning as a major mechanism of its own (Oppenoorth, 1985).

Altered cholinesterase

OP and carbamate insecticides exert their neurotoxic effects by inhibiting the enzyme acetylcholinesterase (AChE), thereby prolonging the residence time of acetylcholine at cholinergic synapses and producing hyperexcitation of cholinergic pathways (Soderlund and Bloomquist, 1990). This was first observed in organophosphorous (OP) resistant spider mites, *Tetranychus urticae* (Smissaert, 1964). Insecticide insensitive AChE has been widely documented biochemically from different insect species. Linkage mapping of *Ace* gene of *Culex pipiens* suggests that it is at a sex linked locus whereas it is not sex linked in other insect species (ffrench-Constant *et al.* 1999).

Resistance of AChE to OP and carbamate insecticides is due to modifications of the active site (amino-acid replacements) and these modifications also appear to alter the catalytic activity of the enzyme towards insecticides and substrates. The replacements can be phenylalanine to serine, isoleucine to valine, glycine to alanine, phenyalanine to tyrosine, glycine to valine, phenylalanine to tyrosine, glycine to serine, serine to phenylalanine, phenylalanine to leucine, alanine to serine or phenylalanine to tryptophan at different positions in various insect species (Li and Han, 2004). These mutations may affect the orientation of active-site serine, which is phosphorylated by OPs and carbamylated by carbamates, affecting the binding of insecticide.

Traditionally, insects and invertebrates were thought to have only one *Ace* locus. However recent works revealed that some insects had at least two AChEs. Phylogenetic analysis by Li and Han, 2004 showed that insect AChEs are divided into two subgroups. Of which, *Ace*1 is the paralogous gene whereas *Ace*2 is the orthologous gene of *Drosophila* AChE. Whether only one or both of the two AChE genes in these species are functional or contribute to resistance requires further work.

Resistance due to kdr gene

Knockdown resistance is a generic term that is applied to one of the two major types of resistance to pyrethroids. Unlike resistance due to enhanced metabolic detoxication, knockdown resistance is unaffected by synergists that inhibit insect esterases and monooxygenases. Instead, knockdown resistance is caused by a reduction in the sensitivity of the insect nervous system to pyrethroids. Because, knockdown resistance usually limits the effectiveness of all pyrethroids, its presence as a mechanism in the field has severe consequences for sustained use of pyrethroids in insect control. The early detection and characterization of knockdown resistance are therefore critical to the development of strategies for resistance management (Soderlund and Knipple, 2003). The *kdr* trait confers resistance to both the rapid paralytic and lethal actions of all known pyrethroids as well as the pyrethrins and DDT, but does not diminish the efficacy of other insecticide classes. A second resistance trait in the housefly (designated *super-kdr*) that confers much greater resistance to DDT and pyrethroids than that found in *kdr* strains has also been isolated genetically and mapped to chromosome 3. The *kdr* and *super-kdr* traits are widely presumed to represent allelic variants at a single resistance locus on the basis of their similar spectra of resistance and their common localization to chromosome 3 in housefly.

The *para*-orthologous sodium channel gene (designated *Vssc 1*) and the *kdr* and the *super-kdr* resistance trait have tight genetic linkage (within ~1 map unit) (Williamson *et al.*, 1993; Knipple *et al.*, 1994). Comparison of partial and complete sequences from 15 housefly strains representing multiple examples of susceptible, *kdr* and *super-kdr* phenotypes consistently identified two point mutations that were associated with resistant phenotypes: mutation of leucine to phenylalanine at amino acid residue 1014 (L1014F) in all *kdr* and *super-kdr* strains, and the additional mutation of methionine to threonine at residue 918 (M918T) only in *super-kdr* strains. So far, 26 unique sodium channel amino acid sequence polymorphisms are associated wit pyrethroid resistance (Soderlund, 2005). These can be at S5 and S6 transmembrane segments that form the homology domains I, II, and III; at the pore forming region lying between the S5 and S6 segments of the homology domain III;

at the short intracellular loops connecting S4 and S5 transmembrane helices of homology domains I, II and III and those occurring near the beginning or the end of the intracellular linkers that separate the homology domains (Soderlund and Knipple, 2003).

Point mutation at the GABA-chloride channels

In the *rdl* (resistant to dieldrin) *Drosophila* strain, insecticide resistance arises form the substitution of a single amino acid (A302S), which contributes to the M2 pore forming region of the ion channel. A similar mutation (A285S) confers dieldrin insensitivity to a clone *H. virescens* RDL equivalent (Wolff and Wingate, 1998). Similarly, cyclodiene resistance in the aphid, *Nasonovia ribisnigri*, a main pest of salad crops, is also due to an A302S mutation (Rufingier *et al.*, 1999).

Cytochrome P450 monooxygenases

The P450 monooxygenases are ubiquitous enzymes found from bacteria to mammals. They are involved in endogenous metabolism as well as in the metabolism of xenobiotics. P450 monooxygenase activities can be involved in the metabolism of virtually all insecticides, leading to an activation of the molecule or, more generally to a detoxification. For some insects, this detoxification is so active that the insecticide does not reach its molecular target before being metabolized and degraded by these enzymes: such individuals are resistant to insecticides (Berge *et al.*, 1999)

P450 enzymes bind molecular oxygen and receive electrons from NADPH to introduce an oxygen atom into the substrate and to form water with the other oxygen atom according to the reaction:

$$\text{Substrate (S)} + (\text{NADPH}+\text{H}^+) + \text{O}_2 \rightarrow \text{S(O)} + \text{NADP}^+ + \text{H}_2\text{O}$$

The electrons necessary for this reaction are transferred from NADPH on the 'substrate P450' complex by an NADPH cytochrome P450 reductase. The stability of initial product [S(O)] can vary, leading to final overall reactions as diverse as hydroxylation, epoxidation, O-, N-, and S- dealkylations, N- and S- oxidations and to such various chemical reactions, and products that these enzymes have been called 'diversozymes' (Coon *et al.*, 1996).

Traditionally, the first line of evidence for a role of a P450 enzyme in resistance has been the use of an insecticide synergist (e.g., piperonyl butoxide), a suppression or decrease in the level of resistance by treatment with the synergist being diagnostic. But this can be misleading in the absence of subsequent biochemical confirmation as piperonyl butoxide can be a poor inhibitor of the P450s responsible for resistance, the synergist used as P450 inhibitor can decrease the activation of a proinsecticide and in some cases resistance is enhanced by piperonyl butoxide (Ringo *et al.*, 1995). An independent and additional line of evidence is the measurement of total P450 levels or metabolism of selected model substrates. An increase in either or both being viewed as diagnostic (Feyereisen, 2005). An increase in the metabolism of the insecticide itself in the resistant strain is more conclusive of the above methods. Cytochrome P450 monooxygenases are a major metabolic mechanism responsible for pyrethroid resistance in India, Pakistan and China (Yang *et al.*, 2004).

The Learn Pyrethroid Resistant (LPR) strain of house fly showed a 10-fold higher level of CYP6D1 mRNA when compared to a susceptible strain. In contrast to resistance mediated by esterases gene amplification does not appear to be involved in overexpression of P450s (Scott, 1999). Cytochrome *c* oxidase subunit I gene, *COXI*, were found to be overexpressed in pyrethroid German cockroaches (Pridgeon and Liu, 2003).

Nauen *et al.,* 2002 has suggested the possible involvement of cytochrome P450 dependant monooxygenases in neonicotinoid resistance in whiteflies. Rauch and Nauen (2003) biochemically confirmed that whiteflies resistant to neonicotinoid insecticides showed a high microsomal 7-ethoxycoumarin O-deethylase activity.

Esterases

Esterase based resistance to organophosphorous and carbamate insecticides are common in a range of different insect pests. The esterases either produce broad spectrum insecticide resistance through rapid binding and slow turnover of insecticide, *i.e.* sequestration, or narrow spectrum resistance through metabolism of a very restricted range of insecticides containing ester bond. The majority of esterases which function by sequestration are elevated through gene amplification (Hemingway, 2000). There was a 5.1 fold higher level of esterase mRNA in the resistant strain of *Lygus lineolaris* due to the enhanced gene expression (Zhu *et al*., 2004). Esterases which produce resistance by increased metabolism are thought to occur by single point mutations in the structural genes, although few have been characterized at the nucleotide level. These mechanisms often involve resistance to the organophosphorous insecticide malathion. Resistance to malathion is caused by a single W251L substitution within the blow fly E3esterase. A second G139D in E3 confers broad spectrum cross-resistance to a range of organophosphates, excluding malathion (Campbell *et al*., 1998). In aphids there are two common amplified esterase variants E4 and FE4, which appear to have had single independent origins (Devonshire *et al*., 1999).

Glutathion-S-transferases

Glutathion-S-transferases (GSH transferases) catalyze the nucleophilic attack of the endogenous tripeptide glutathione (GSH) on a variety of reactive substrates. These enzymes are of particular importance in the metabolism of OP insecticides, in which either the alkyl or aryl substituents on phosphorous can be cleaved. The role of GSH-transferases in OP metabolism can be demonstrated by the formation of GSH conjugates of the cleaved alkyl or aryl moieties using appropriately radiolabeled insecticides (Soderlund and Bloomquist, 1990). Detoxification of OPs occurs via an O-dealkylation or O-dearylation reaction.

Elevated GST activity has been implicated in resistance to DDT and pyrethroids also. DDT dehydrochlorination is a major route of detoxification in insects. The enzyme catalyzing this reaction was not initially recognized as a member of the GST family partly due to the inability to detect even a transient GST conjugate and also because of the lack of DDT dehydrochlorinase activity in some purified GSTs. Evidence that DDT dehydrochlorinase is a GST was finally provided in the housefly (Clark and Shamaan, 1984). GSTs can also catalyse the secondary metabolism of OP insecticides. This insecticide class is usually applied in its non-insecticidal phosphorothionate form and activated to the insecticidal form by the action of cytochrome P450s in the insect (Hemingway *et al*., 2004)

GSTs have not yet been shown to be involved in the direct metabolism of pyrethroid insecticides. Nevertheless, they play a very important role in conferring resistance to this insecticide class by detoxifying lipid peroxidation products induced by pyrethroids (Vontas *et al*., 2001). A delta class GsT present in elevated quantities in a pyrethroid resistant strain of *Nilaparvatha lugens* was expressed *in vitro*. The recombinant protein had high peroxidase activity and a role for this enzyme in the prevention or repair of oxidative damage was proposed (Vontas *et al*., 2002). GSTs have also been suggested to protect against pyrethroid toxicity in insects by sequestering the insecticide.

Ecological genetics of resistance

Theory predicts that the number of resistance genes selected to confer resistance depends on whether selection acts within or outside of the phenotypic distribution of the susceptible population. Selection from within this distribution selects preferentially for polygenic resistance, by combining common pre-existing resistance factors that have a minor effect, such as body size and developmental rate, whereas selection outside of this distribution selects for rare mutations in single genes that have a major effect-a monogenic response. Examination of insecticide target-site mutants, isolated in field studies, supports the importance of single genes that have a major effect (ffrench-constant *et al*., 2004). While laboratory selection studies usually produce polygenic resistance (Oppenoorth, 1984).

The dichotomy between laboratory and field selection in the evolution of resistance shows that the genetic basis of character variation is not intrinsic to the character, but rather to the selection regime applied. At very high or very low doses, there is no discrimination among genotypes in favor of resistance, so no selection can occur. Selection can occur only at those doses that permit discrimination between susceptible and resistant genotypes. The selecting dose in laboratory programs can be very closely controlled and precise, permitting discrimination even among very similar genotypes within the normal physiological distribution of phenotypes. Because of uneven coverage, field doses are unlikely to be nearly so precise, but may still discriminate between susceptible and rare resistant genotypes when major alleles are involved. Laboratory selection programs are designed to allow at least some survivors in each generation. Thus they are well suited for the accumulation of many genes, each of small effect, over several generations, which results in polygenically inherited resistance (Roush and Daly, 1990). Another factor contributing to the rarity of polygenic resistance in the field is that monogenic resistance is more likely to spread than polygenic resistance, particularly when pesticide exposure varies across time and space.

The contrasts between resistance selection under laboratory and field conditions imply that experimental studies of resistance-management strategies should be based on resistances that are genetically similar to those that commonly develop in the field. Such studies should involve strains that include major resistance genes isolated from field populations, rather than strains selected only in the laboratory (Roush and McKenzie, 1987).

The rate of change in allele frequency at any given locus in a closed population is a function of allele frequency, dominance and the relative fitness of the various genotypes. Measurement of initial frequencies is difficult. The resistant phenotype, and therefore an efficient method to detect it, cannot be determined until resistance develops, which generally occurs only after the pesticide has been in widespread use (Roush and Miller, 1986). In the absence of data, researchers have assumed that resistance allele frequencies may range from 10^{-2} to 10^{-13} based on mutation-selection equilibrium theory (Roush and McKenzie, 1987).

Dominance refers to the phenotypic expression of a character in heterozygotes relative to expression in the homozygous parents. High doses of insecticides can make the toxicological phenotype of resistance effectively or functionally recessive by killing the heterozygotes. A lower dose can make resistance effectively dominant. However, dominance is a property of phenotypic characters, not of alleles. Nerve insensitivity mechanisms for pyrethroids and DDT, similar to *kdr*, tend to be recessive or semirecessive in their inheritance. Resistance by altered acetylcholinesterases and changes in the cyclodiene target site tend to be codominant to dominant in their inheritance. Most metabolic mechanisms of resistance tend to be dominant (Roush and Daly, 1990)

Resistant strains of arthropods studied in the laboratory often show reproductive disadvantages in developmental time, fecundity, and fertility. In the field, frequencies of resistant individuals usually decline over time in the absence of pesticide use. However, it is often difficult to associate fitness disadvantages specifically with resistance (Roush and Croft, 1986). The most serious and consistent disadvantages are usually associated with esterases. This may be because a relatively large portion of body protein in these strains is devoted to esterases. Resistant *Myzus persicae* may be more susceptible to predation than susceptible aphids because of decreased responsiveness to alarm pheromone (Dawson *et al.*, 1983). In contrast, little or no reproductive disadvantages have been associated with the *kdr*-like mechanisms, the altered acetylcholinesterase mechanism, malathion specific carboxylesterases or with induced microsomal monooxygenases.

While some populations of major insects develop resistance to most insecticides to which they are repeatedly exposed, many other populations of the same or closely related species remains relatively susceptible, even though exposed to similar spray rates and number of applications. Variation in population structure probably contribute to these differences. Population structure describes the ways in which populations are subdivided into finite breeding units as a result of behavior, sex ratios, discrete distribution of resources, population size, the degree to which population deviates from random mating and restrictions on dispersal that limit the flow of genes from one location to another. Pest control regularly reduces the population to small size. Evolution can occur more rapidly in small, subdivided populations than in large, panmictic ones. In Australia *Helicoverpa punctigera* and *H armigera* are both sprayed with similar insecticides several insecticides several times each year and have similar generation times. However, only *H. armigera* has developed resistance to a number of insecticides. This may be because a higher proportion of the *H. punctigera* than of the *H. armigera* population resides in refugia (Roush and Daly, 1990).

Monitoring and detection of insecticide resistance

Resistance management requires more effective techniques for detecting resistance in its early stages of development. Much of the new diversity in resistance monitoring methods has been the result of improved knowledge of resistance mechanisms and expression, as well as a change in the goals and philosophy of monitoring. Monitoring can be (a) merely to document the efficacy of an insecticide, (b) to determine the frequency of resistant genotypes with accuracy, or (c) to have a high chance of detecting a given frequency of resistant genotypes (ffrench-Constant and Roush, 1990). One of the most promising aspects of resistance management from the perspective of reducing pesticide use is the development of rapid methods that allow users to choose the most effective, safe, or inexpensive pesticides from among those affected by resistance. Developing a resistance profile immediately before choosing the pesticide to be applied reduces the frequency of ineffective, wasted applications. Further, resistance detection efforts should be initiated when resistance is still rare, thus providing an early warning or some indication of the genetic potential for resistance to occur.

Brent (1986) has given the aims of resistance monitoring as follows.

1) Documentation of resistance
2) To measure and identify resistant genotypes accurately
3) To provide early warning of an impending resistance problem
4) To determine changes in the distribution or severity of resistance
5) To make recommendations for pesticides least affected by resistance
6) To measure the biological characteristics of genotypes under field conditions and
7) To test the effectiveness of resistance management tactics.

Detection techniques can be broadly divided into two categories (a) *in vivo* assays on intact individuals, which usually involve pesticide exposure, and (b) *in vitro* biochemical and molecular techniques that assess enzyme activity or the nature or quantity of DNA coding for specific resistance genes in insect preparations. Use of both approaches will be more accurate and mutually confirmatory.

Pesticide bioassay

Traditionally, monitoring has involved comparison of LD_{50}s, LD_{90}s, or the slopes of dose response curves between field and laboratory populations. Although adequate for documenting resistance at high frequencies, both LD and slope estimates are insensitive to small changes in resistance frequency, particularly when resistance is first appearing in the population (Roush and Miller, 1986). Diagnostic tests (such as fixed dose and exposure period) are more efficient for detecting low frequencies of resistance because all individuals are tested at an appropriate dose and none are wasted on lower doses, where percentage mortality is no informative (ffrench-Constant and Roush, 1990). Test techniques that accurately discriminate between resistant and susceptible genotypes (*i.e.,* kill >99.9% of susceptibles but < 0.15 of resistant individuals) are significantly more efficient for detecting resistance than techniques that misclassify more frequently.

In most practical situations a perfectly discriminating test is unknown either because resistance has not yet been documented or because the expression of resistance has not yet been examined in sufficient detail to facilitate the choice of an appropriate method, particularly if resistance intensity is low. In such circumstances, the best monitoring method for resistance detection will be the use of a dose that kills about 99% of susceptible individuals (a susceptible LC_{99}). The ideal test technique should be fast and efficient and would correlate closely with field control, even if the stage tested or mode of exposure differs from that under field conditions (ffrench-Constant and Roush, 1990).

Different monitoring methods can be classifies as follows,

(a) Immersion, whereby appropriate life stages are dipped or washed in solutions of known concentration or alternatively, reared in diet containing insecticide
(b) Residue or surface contact, whereby individuals are exposed to a dry residue of pesticide on a natural (*e.g.,* leaf) or artificial (*e.g.*, glass) substrate, and
(c) Topical, where a known dose of insecticide is applied directly to individual insects commonly *via* a microsyringe.

Exposure methods that give steeper dose-response curves (with lower variation and standard deviation) tend to give better discrimination between genotypes because fewer susceptible individuals will survive relatively high doses. Thus, if resistant genotypes are not yet available, a general rule may be to choose exposure methods that give steeper dose-mortality lines (ffrench-Constant and Roush, 1990).

Diagnostic doses for some important crop pests of India have been given by Ayyasamy and Regupathy, 2004a; which is given in table 1.

Table 1 Discriminating doses for major pests of India

Chemical	Method	Diagnostic dose	Reference
***Helicoverpa armigera* (3rd instar larvae)**			
Fenvalerate	Topical	0.2 µg/larvae	Forrester and Cahill, 1987
Cypermethrin	Topical	0.1 µg/larvae	Gunning *et al.,*1984
Deltamethrin		0.0125 µg/larvae	Ayyasamy and Regupathy, 2004
λ – cyhalothrin		0.025 µg/larvae	Forrester *et al.*, 1993
β - cyfluthrin		0.2 µg/larvae	Ayyasamy and Regupathy, 2004
Endosulfan	Topical	10 µg/larvae	
Quinalphos	Topical	0.75 µg/larvae	
Profenophos	Topical	2 µg/larvae	
Methomyl	Topical	1.2 µg/larvae	
Chlorpyriphos		1 µg/larvae	Gouthaman and Regupathy, 2003
Spinosad		10 µg/larvae	Ayyasamy and Regupathy, 2004
Piperonyl butoxide		50 µg/larvae	
Propargyloxythalimide		25 µg/larvae	
Pongamia oil		50 µg/larvae	Gavi Gowda and Regupathy, 1995
DEF		20 µg/larvae	Ayyasamy and Regupathy, 2004
Emamectin benzoate		0.80 ppm	Stanley, 2004

Contd. ...

Table 1 Contd. ...

Plutella xylostella **(3rd instar larvae)**			
Fenvalerate	Vial	115 ppm	Chandrasekaran and Reghupathy, 1996
	Spray tower	170 ppm	
	Larval dip	130 ppm	
Quinalphos	Vial	3 ppm	
	Spray tower	10 ppm	
	Larval dip	10 ppm	
Monocrotophos	Vial	140 ppm	
	Spray tower	170 pp	
	Larval dip	150 ppm	
Cartaphydrochloride	Vial Spray tower	5 ppm 10 ppm	
	Larval dip	10 ppm	
Carbosulfan	Vial	4 ppm	
	Spray tower	20 ppm	
	Larval dip	15 ppm	
Emamectin	Leaf dip	2 ppm	Lavanya, 2004
Spinosad		12 ppm	
Spodoptera litura **(3rd instar, 30-40mg weight, 8 day old)**			
Endosulfan	Topical	2.0 ppm	Niranjankumar and Regupathy, 2001
Profenofos		3.0 ppm	
Chlorpyriphos		0.15 ppm	
Fenvalerate		0.25 ppm	
Emamectin	Leaf dip	0.40 ppm	Stanley, 2004
Spinosad		125 ppm	
Cnaphalocrocis medinalis **(4th instar larvae, 20-30 mg)**			
Chlorpyriphos	Topical	1.00 μ	Anandan, 1997; Anbalagan, 2001
Monocrotophos		0.35 μ	
Phosalone		1.90 μ	
Phosphamidon		5.50 μ	
Quinalphos		0.40 μ	
Aphis gossypii **(wingless adult aphids of ca 1.45 mm size and weighing ca 0.19mg)**			
Thiamethoxam	IRAC	10 ppm	Praveen, 2003; Praveen

Contd. ...

Table 1 Contd. ...

Imidacloprid	Method	20 ppm	and Regupathy, 2003a
Dimethoate	No.8	50 ppm	
Methyl demeton		400 ppm	
Acephate		40 ppm	
Monocrotophos		20 ppm	
***Thrips tabaci* (2nd instar nymphs)**			
Imidacloprid Thiamethoxam	Leaf dip	0.7 ppm 1.0 ppm	Praveen, 2003; Praveen and Regupathy, 2003b
Dimethoate		1000 ppm	
***Amrasca devastans* (3rd instar nymph of ca 1.3mm size and weighing ca 0.14mg)**			
Thiamethoxam	IRAC	0.008 ppm	Jeya Pradeepa, 2000; Jeya Pradeepa and Regupathy, 2002
Monocrotophos	Method	0.55 ppm	
Imidacloprid	No.8	0.005 ppm	
Dimethoate		400 ppm	
Methyl demeton		800 ppm	
Acephate		850 ppm	
Coccus viridis			
Imidacloprid	Leaf dip	66 ppm	Senthilkumar, 2003; Senthilkumar and Regupathy, 2004
Thiamethoxam		35 ppm	
Dimethoate		500 ppm	

Biochemical assay

Currently available biochemical assays include (a) detection of enzyme activity in unprocessed insect homogenates using model substrates, (b) use of enzyme-specific antisera to isolate the activity of resistance-conferring enzymes, and (c) detection of specific DNA sequences (ffrench-Constant and Roush, 1990).

Increased hydrolysis of 1-naphthyl acetate has been widely used for the detection of elevated esterase activity, often on convenient microtiter plates (Brogdon *et al*., 1988, Zhu *et al*., 2004) or on filter papers (Pasteur and Georghiou, 1989). Zhu *et al*., (2004) employed microtiter plate assays using PNPA (*p*-nitrophenyl acetate), α- and β-naphthyl acetate for determination of elevated esterase activity in tarnished plant bug. Absorbance of enzyme assay is monitored at 450 nm. Elevated hydrolysis of a substrate alone is not sufficient evidence of resistance, unless it is proved that altered enzyme responsible also confers resistance. Rossiter *et al.,* (2001) has evaluated the effectiveness of polyacrylamide gel electrophoresis (PAGE) for the rapid detection of esterase mediated OP resistance in *Oryzaephilus surinamensis.* Resistant strains with enhanced esterase activity produced thicker and darker esterase bands. It was also useful in distinguishing between two highly resistant strains where total esterase activity was similar.

Acetylthiocholine iodide has been widely used as a model substrate for monitoring acetylcholinesterase activity in the presence or absence of insecticides, in a variety of individual insects. Kinetic microplate readers have led to refinement of this technique, allowing the rapid and accurate identification of genotypes in a number of insects (Li and Han, 2004).

Enhanced monooxygenase mediated metabolism of aldrin in propaxur resistant strain of housefly was detected by the measurement of conversion of aldrin to dieldrin (Plapp and Cassida. 1969). Subsequently, monooxygenase substrates like methoxyresorufin, ethoxyresorufin, ethoxycoumarin (Lee and Scott, 1989) *p*-nitroanisole (Yang *et al.*, 2004) and many others are used. Pattern of activity toward model substrates varied between strains suggesting there may be no single marker enzyme to detect monooxygenases-mediated resistance, aside from the actual insecticide(s) to which the strain is resistant. Although model substrates can become useful tools after mechanism of resistance has been elucidated in a strain, they are not, by themselves, indicative of the mechanism of resistance (Wilkinson, 1983).

For assays quantifying glutathione *S*-transferase (GST) activity, the substrates 1-chloro-2,4-dinitrobenzene (CDNB) and 3,4-dichloronitrobenzene (DCNB) are used.

Changes in DNA sequences associated with resistance can be used for determination of specific causes of resistance. Various PCR (Polymerase chain reaction) based techniques like PCR/RFLP (restriction fragment length polymorphism), RAPD-PCR (randomly amplified polymorphic DNA-PCR), PASA (PCR-mediated amplification of specific alleles), SSCP (Single-stranded conformational polymorphism) and minisequencing are used for this.

PCR/RFLP was used to detect A302S point mutation at GABA receptor associated with cyclodiene insecticide resistance. The resistance associated mutations cause a loss of a *Hae* II site and helps to determine the resistance frequencies in field populations. (ffrench-Constant *et al.*, 1993). This technique was also used to identify alleles involved in resistance at the *Ester* locus associated with OP resistance in mosquitoes. The restriction enzyme *Hae*III generated distinct and repeatable profiles for the *Ester* super locus (Berticat *et al.*, 2000). Two resistance-associated point mutations of insensitive AChE-mediated OP resistance in *Bactrocera oleae* was detected by PCR-RFLP utilizing an *Acc*I restriction site created by I214V and a *Bss*HII restriction site destroyed by a second G488S mutation (Hawkes *et al.*, 2005).

Amplification of genomic DNA by RAPD-PCR was used to differentiate between deltamethrin resistant and susceptible *Culex pipiens pallans* (Zhu *et al.*, 1998). Sharma *et al.*, (2003) have identified RAPD markers for imidacloprid resistance in whitefly. PASA relies upon the specific amplification of one allele in preference to others at a given magnesium concentration within the PCR reaction. A forward allele-independent primer and an allele-specific reverse primer are use here. The alanine to serine mutation associated endosulfan resistance in *Hypothenemus hampei* was detected by PASA (ffrench-Constant *et al.*, 1994). PASA was also used to monitor the frequency of *kdr* mutation in natural populations of *Cydia pomonella* (Brun-Barale *et al.*, 2005) and *Leptinotarsa decemlineata* (Clark *et al.,* 1999). Point mutation associated with AChE insensitivity in azinphosmethyl resistant *L. decemlineata* was also determined by PASA (Zhang *et al.*, 1999).

An improvement to PASA named bi-PASA (bi-directional PASA) was used by Bass *et al.,* (2004) to detect *kdr* mutation in cat flea. A bi-PASA reaction contains four separate primers (three in PASA, allele non specific, susceptible allele-specific and resistant allele-specific). The two inner primers are allele-specific at the mutation site, and the two outer primers are allele non-specific. In bi-PASA, one allele is amplified in one direction whereas the other allele is amplified in the opposite direction, resulting in two allele specific fragments that differ in size.

Single strand conformational polymorphism was used to detect cyclodiene-resistance associated mutations (Coustau and ffrench-Constant, 1995). SSCP is capable of detecting all mutations in a given stretch of DNA. It can overcome some of the limitations of the previous PCR-based techniques in that it (a) can detect novel resistance-associated mutations of previously unknown sequence (b) can identify polymorphisms closely linked to resistance associated mutations and thus identify different resistant haplotypes and (c) it requires only the analysis of a single PCR reaction.

In minisequencing, a DNA polymerase is used specifically to extend a primer that anneals immediately adjacent to the nucleotide position to be analysed with a single labeled nucleoside triphosphate complementary to the nucleotide at the variant site. It is adaptable to high through put analysis. Zhang *et al*., (1999) has used this method to validate point mutation associated with acetylcholinesterase sensitivity to azinphos-methyl in Colarado potato beetle.

Clark *et al*., (2001) has developed and compared bi-PASA, SSCP and minisequencing for the detection of insensitive acetylcholinesterase and insensitive sodium channel point mutations associated with point mutations associated with azinphos-methyl and permethrin resistance, respectively, in *L. decemlineata*. They have suggested that DNA-based genotyping is best employed in combinations, with the bi-PASA as the primary method and the SSCP and minisequencing as the secondary validating methods.

Resistance Management Tactics

Resistance management seeks to slow, prevent or reverse the evolution of resistance in pests. A secondary goal of resistance management is to promote evolution of resistance in beneficial species, such as natural enemies (Tabashnik, 1990).

Mathematical models can be useful tools for working toward the goals of resistance management. Models have helped to build a conceptual framework for resistance management and generated hypotheses for experimental tests. Models can provide a relatively fast, safe and inexpensive way to project the consequences of different assumptions about resistance and to weigh the merits of various management options. Models can help to organize data and identify gaps where more research is needed (Tabashnik, 1990). The models can be classified by the basic assumptions employed, the modelling approach taken, the variables considered, and the problem addressed. Four modeling approaches are analytical, simulation, optimization and empirical models (Hoy, 1999).

Analytical models aim to analyze general trends by providing a simple mathematical description. Such models aim to analyze general trends by providing a simple mathematical description. Such models do not attempt to capture realistic details; instead they seek to define fundamental principles. Analytical models of the evolution of insecticide resistance usually assume simple population dynamics with discrete generations and no age structure. Population growth is usually determined by some form of the logistic equation. But, few arthropods have discrete generations; most of those that are prone to developing resistance are multivoltine, and have overlapping generations (Hoy, 1999).

Simulation models are more complex and realistic than analytical models because they attempt to incorporate details of the biology, behavior and ecology of the population and often contain complex population dynamics, including age structure, overlapping generations and temporal and spatial variation in pesticide dose. The modeling of complex details typically involves series of coupled equations that cannot readily be solved analytically, so computer programs are used to determine numerical solutions (Tabashnik 1990).

Optimization models evaluate, using dynamic programming techniques, which management strategy will maximize profit when pest susceptibility to a pesticide is considered a non-renewable natural resource. The goal is to balance the future cost of reduction in pest susceptibility with the present losses in crop yield due to the effects of the target pest. The details of the biology of the target species are usually simplified and considered a constraint to the model, which focuses on an economic analysis (Tabashnik 1990).

Empirical models are based on actual relationships among variables, with no assumptions made about the causal mechanisms. The models are derived from data and probably are appropriate only to the specific conditions of the observed populations (Tabashnik 1990). Thus, empirical models are least useful for developing a strategy for delaying resistance in an unknown situation if we assume that the important variables can vary between populations (Hoy, 1999).

Georghiou (1983) divided resistance management tactics into three categories: management by moderation (lower the mortality caused by any given chemical, by saturation (saturation of the resistance mechanisms of the pest by higher doses or synergists), or by multiple attack (mixing or alternating pesticides). Two or more tactics must often be used together in an integrated resistance management (IRM) approach (Tabashnik, 1990).

Selection can only occur at concentrations that kill (or reduce the reproduction) of at least some susceptible homozygotes but allow at least some resistant homozygotes to survive. Tactics that reduce this discrimination, particularly the discrimination between susceptible homozygotes and heterozygotes, will generally slow the rate of resistance evolution. From a genetic perspective Roush and Daly (1990) has suggested the following techniques for manipulating selection pressure in IRM strategies.

A. Reduce fitnesses of resistant individuals when insecticide is applied

1. Increase the dose to kill heterozygotes or resistant homozygotes
2. Use compounds that confer lower levels of resistance
3. Treat the most vulnerable life stage
4. Use synergists to suppress detoxification mechanisms
5. Mix pesticides of differing modes of action and metabolism

B. Reduce the total amount of selection pressure applied

1. Decrease the concentration of insecticide so that some susceptible individuals exposed to the pesticide can survive

2. Reduce the number of pesticide applications
3. Use pesticides with short residual activity and avoid slow release formulations
4. Do not spray all habitats of the pest; use spot treatments
5. Rotate pesticides so that not all generations are exposed to the same one, but avoid spatial mosaic treatments.

Increase the dose of insecticide to kill heterozygotes or resistant homozygotes

In theory if the concentration of insecticide is so high that the heterozygotes are killed (*i.e.,* resistance is recessive), if the frequency of resistance is sufficiently low that very few RR individuals can be found in the population, and if relatively high frequencies of susceptible individuals escape exposure altogether and mate with surviving resistant genotypes to assure that they will not produce RR off-spring, increases in the frequency of resistance may be very slow. If the concentration can be increased even higher to kill all resistant homozygotes, the strategy might work even if the gene frequency is high and all the population is treated. In short, the strategy discriminates poorly between resistant and susceptible genotypes; the R allele is neutral, and selection should not lead to any increases in frequency (Roush and Daly, 1990).

However, there are many practical problems for such a strategy. For those species that tend to mate in the same habitats where they are controlled, pesticide residues may be so persistent that they kill or repel susceptible individuals, thereby inhibiting the mating of susceptible with resistant individuals. In addition, the environmental and economic costs of using a dose high enough to kill the heterozygotes may be great (Tabashnik and Croft 1982). Consequently, a tactic of increasing the dose applied may be useful only in populations in which resistance is not yet a problem, where doses applied can be tightly controlled and uniform, and where resistance levels in heterozygotes are likely to be low (Roush and Daly, 1990). Nevertheless, this strategy was employed to contain pyrethroid resistance in cattle ticks (Reid, 1989).

Use compounds that confer lower levels of resistance

Within any group of pesticides that show cross-resistance there are usually some that show lower resistance than others. In delaying the evolution of resistance the compound that discriminates most poorly would be favored. Among the cyclodienes and related compounds, for example, dieldrin usually shows high resistance, whereas lindane shows considerably lower resistance (Oppenoorth 1985). In at least some aphids, pirimicarb selects less strongly for resistance than does demeton--S-methyl or a deltamethrin-heptenophos mixture (ffrench-Constant *et al.*, 1987).

Treat the most vulnerable lifestage

In a similar fashion, treating the lifestage that shows the lowest levels of resistance can also weaken the discrimination between genotypes. Relative fitnesses vary with the lifestages treated because some resistance mechanisms are more poorly expressed in some life stages than others. In other cases, some life stages are so sensitive that they are killed even if resistance is expressed. In the Lepidoptera, genotypically resistant neonate larvae of both the spotted tentiform leafminer (Pree, 1990) and *H. armigera* can be killed by fresh deposits of pesticides, even though larger resistant larvae are not. Both pyrethroid-resistant and susceptible *H. armigera* are killed by pyrethroids until the larvae are at least three days of age, suggesting that no selection for resistance occurs in small larvae (Daly *et al*., 1988). However, in many lepidopterous pests, the generations are not closely synchronized. Even though the target of insecticide applications may be eggs or small larvae, larger larvae may also be exposed. Strategies aimed at moderation in selection of small larvae may therefore fail if selection occurs against these other life stages. Careful scouting of crops on a regular basis is necessary in these situations (Roush and Daly, 1990).

Use synergists to suppress detoxification mechanisms

Synergists, when applied with an insecticide, increase the potency of the insecti-cide, by blocking the action of the resistance mechanism. This reduces the effect of the mechanism, thereby reducing the discrimination between resistant and susceptible genotypes, much as if the insecticide alone conferred only a low level of resistance. Synergists can only be used effectively if there is a single mechanism for resistance, if resistance does not subsequently develop to the combination of synergist and insecticide, if the synergist is stable under field conditions and has low mammalian toxicity, and if it is cost-effective.

Mixtures

Of all the strategies proposed to manage resistance, mixtures may be the most controversial. If resistant alleles at two separate loci exist in the population at very low frequency, it is extremely unlikely that any one individual will carry both alleles. If unexposed susceptible individuals are relatively abundant and mate randomly with the resistant survivors, mixtures must delay resistance (Curtis 1985). Thus, the use of a second insecticide negates much of the advantage of carrying a resistance for the first pesticide, and worsens the discrimination be-tween genotypes. Recent models suggest that equal resistance of both pesticides is extremely important to the successful use of mixtures. Although a mixture strategy is most effective if imple-mented while resistance to both pesticides are still rare (Roush 1989).

Decreasing the dose of application to that which kills most, but not all of the susceptibles

Curtis (1985) has shown that this is not a very viable strategy. From a genetic perspective, any dose that allows some susceptibles to survive will always allow even greater survival of the heterozygotes (unless resistance is completely recessive) and an increase in the frequency of resistance. However, this tactic can be very useful when alternative controls, are available. This approach is especially feasible where pesticides are selective (or natural enemies are resistant) and reduced rates can enhance predation by stabilizing prey densities (Roush 1989).

Apply insecticides less frequently so that susceptibles have a chance to reproduce and to dilute resistance by interbreeding with resistant individuals

This can be achieved by using a higher pest population threshold for insecticide application as part of an integrated pest management strategy. Conflicts can arise, however, between the need, to control pest numbers in crops and the need to manage insecticide resistance (Roush and Daly, 1990).

Use chemicals with only short residual and avoid slow release formulations

Persistent pesticides or formulations are functionally equivalent to frequent applications of less persistent materials and may also deter the successful immigration of susceptible individuals. The decay of residues in long residual pesticide can allow survival of resistant individuals that would have been killed by the initial concentration (Roush and Daly, 1990).

Allow refuges for escape of susceptibles

As noted earlier, the recruitment of susceptible genes into a population is impor-tant to the delay of resistance; hence, the preservation of refuges for susceptibles to breed is an important approach to the management of resistance. This tactic may often be applied, although inadvertently, in most agricultural crops because 100% kill of susceptibles with registered doses of insecticides is not normally observed (Roush and Croft 1986). The only way to assure that a true refuge has been provided is to leave some habitats completely untreated. One practica1 way to achieve this is by treating only those parts of a field that has damaging populations. Permanent refuges can be created if certain habitats are always left untreated, such as native vegetation. In many insect pests, maintenance of refuges may be difficult, as little is known about the size of the population outside treated areas or the magnitude of migration between areas. Although many insect pests are highly mobile, migration may not be sufficient to retard the spread and increase in resistance. Another way to provide refuges for susceptible genes is to direct control against adults after they have mated. Resistant females that mate before the spray may carry refuges for susceptible genes in the form of sperm, whereas females that mate after application will be mating with resistant males that survived. This approach is most useful for those species that generally only mate once (Roush and Daly, 1990).

Rotate pesticides so that not all generations are exposed to the same one, but avoid spatial mosaic treatments.

Given two or more insecticides with unique modes of action and metabolism, one could alternate their use either in the short term (rotations) or long term (sequential introduction), or across space (mosaics). All three approaches limit selection pressure on any given locus, but simple genetic simulation models and experiments suggest that rotations limit selection more effectively than mosaics (Curtis, 1985, Roush, 1989).

Integrated Resistance Management and Integrated Pest Management

It is difficult to "manage" resistance once resistant individuals make up more than 5 to 10% of the population. Generally, the best that can be achieved is to delay the development of high levels of resistance for a few years, most often by using the product less often. A more effective resistance management strategy will combine a variety of effective pest management tactics along with a reduction in numbers and rates of pesticides applied. IRM strategies must complement good intergrated pest management (IPM) practices and only when IRM is properly incorporated into acceptable IPM programmes will there be any hope of successful resistance management (Forrester and Fitt, 1992) and it has been suggested by Hoy (1999) that IRM is synonymous with IPM. Effective pest management tactics include monitoring, evaluating economic injury levels so that pesticides are applied only when needed, biological control, host plant resistance, cultural controls, biorational pest controls, and genetic control methods (Hoy, 1995).

Developing a resistance management program can take several years; studies typically are conducted to develop an appropriate monitoring method, estimate the frequency of resistant individuals in populations, detect cross resistances, and evaluate mode of inheritance and stability of the resistance (Roush & Daly 1990). Meanwhile, unless pesticide applications are discontinued, selection for resistance continues. Initial detection of resistance usually requires that resistant individuals comprise at least 5% of the population (Brent 1986). Thus, by the time resistant individuals are detected, selection by additional pesticide applications is likely to increase their frequency in the population. This scenario is particularly familiar with multivoltine and highly fecund pests such as aphids, spider mites, whiteflies, and leafminers (Hoy, 1995).

It has not been widely acknowledged is that resistance management programs should include: 1) Altering the way pesticides are developed and registered, and 2) Recognizing that resistance management must be a broad-based, multitactic endeavor. It seems appropriate to assume that the development of resistance is nearly inevitable and the issue is not whether resistance will develop, but when. With this assumption, resistance management programs have the goal of delaying rather than preventing resistance (Hoy, 1992).

Registration requirements

Assessing resistance risk is a fundamental part of the development process for new molecules and is increasingly becoming a requirement of registration alongside toxicological and environmental risk data. Laboratory studies, including elucidation of target sites and metabolic degradation pathways, mutagenesis, computer models and cross-resistance tests, and field studies, including establishment of baseline sensitivities and evaluation of anti-resistance strategies, all play an important part in such assessment. The challenge is then to devise management strategies which are relevant to local practice and actually reduce selection pressure to a point where product life is preserved. A preventive strategy should be in place at time of launch of pesticides. Monitoring resistance levels in field populations after product launch enables any fine tuning of tactics over time (Jutsum *et al.*, 1998).

Pesticide resistant natural enemies

Pesticide-resistant natural enemies are a special category of pesticide selectivity that can help to delay resistance in a multi-tactic IPM program. Relatively few natural enemies have developed resistance to pesticides through natural selection, but several have been deployed in IPM programs (Croft 1990; Hoy 1990). Artificial selection of phytoseiid predators for pesticide resistance can be a practical and cost-effective tactic for the biological control of spider mites (Hoy 1990).However, development of pesticide-resistant natural enemies should not be considered before exploring other, less expensive options for IPM and pesticide resistance mitigation (Hoy, 1999).

In India, efforts were being carried out towards the development of insecticide resistant strains of bioagents like *Trichogramma* and *Chrysoperla* (Patel and Yadav, 1997). Project Directorate of Biological Control (PDBC, Bangalore) has developed a strain of *T. chilonis* tolerant to endosulfan and this strain has been named as 'ENDOGRAM' (PDBC, 1998).

Multiple insecticide tolerant strain of *T. chilonis* utilising the endogram strain has also resulted in tolerance to monocrotophos and fenvalerate. This strain also showed cross-resistance to dimethoate, acephate, phosphamidon and methyl demeton (PDBC, 1999).

Negatively correlated insecticides

It has been showed that in certain circumstances resistant insects develop higher sensitivities to certain insecticides than their susceptible counterparts. The green leafhopper, *Nephotettix cincticeps*, has developed cross-resistance to whole aryl N-methylcarbamates due to alteration of the AChE. The altered AChE became insensitive to the N-methylcarbamates, while it became more sensitive to N-propylcarbamates. For the field resistant strain, the combination of N-methyl- and N-propylcarbamates was effective, while N-propylcarbamate was powerful in controlling the strain highly resistant to N-methylcarbamates (Yamamoto *et al.*, 1983). Insecticidal dihydropyrazoles have been shown to exhibit negative cross-resistance to a *super-kdr* strain of houseflies insensitive to pyrethroids (Khambay *et al.*, 2001). A field strain of *Spodoptera littoralis* resistant to conventional insecticides was proved to be more susceptible to spinosad (Temerak, 2003).

Microbials

The reduction in enzyme activity caused by the virus infection decreased the ability of pest to degrade insecticides. This kind of declining activity was reported in enzymes like cytochrome P-450 (Brattsten, 1987) in *H. virescens* and *Manduca sexta* (Linnaeus) by cytoplasmic polyhedrosis virus, acetyl cholinesterase (AchE) and mixed function oxidases (Huang and Dai, 1991) in *Pieris rapae* by PrGV, carboxylesterase (Geetha *et al.*, 1999; Song *et al.*, 2000), glutathione-S transferase (Geetha *et al.*, 1999), MFO and AchE (Song *et al.*, 2000) in *H. armigera* by HaNPV and carboxylesterase, MFO and GST (Rajagopalbabu, 2001) in *P. xylostella* by PxGV. Combination of NPV, GV and *Bacillus thuringiensis* with insecticides resulted in induction of susceptibility in resistant insects (Ayyasamy and Regupathy, 2004b).

IRM of *Helicoverpa armigera* in India

The cotton bollworm, *Helicoverpa armigera* Hubner (Lepidoptera: Noctuidae), which was never traced as a major bollworm of cotton in any part of the country before 1986 has become number one agricultural pest in India. The crop loss due to this pest was estimated at 47-90% (Fakrudin *et al*, 2002), the monetary value of damage was more than $500 million (Fakrudin *et al.*, 2003). To combat the unprecedented pressure from *H. armigera*, farmers in South India had applied over 30 sprays as against the recommended 6-10 sprays (Fakrudin *et al*, 2002). More than 75% of insecticides sprayed over the cotton crop are being targeted towards *H. armigera* (Banerjee *et al.*, 2000) of which, synthetic pyrethroids constituted 50-70% (Jayaswal, 1989). Frequent outbreaks of this pest on cotton crop led to severe social disturbances, with several reports of suicide by farmers (Patil *et al.*, 1996). The failure to control the *H. armigera* was traced to the development of resistance to insecticides (Kranthi *et al.*, 1997). Resistance in the American bollworm, *Helicoverpa armigera* to almost every class of insecticides has been documented (Singh *et al.*, 1994).

The IRM against *H. armigera* strategy relies on IPM to avoid unnecessary selection pressure by chemical groups. It incorporates soft chemistries to minimize adverse effects on beneficial fauna and sets up a three window rotational approach for insecticide use. The strategy restrict the use of pyrethroids to the later part of the cotton season to coincide either with the 2nd or 3rd window (Kranthi, 2004). Since LD_{50} slopes of the field strains in India consistently indicated a high level of heterogeneity in population response to pyrethroids, it is anticipated that the frequency of resistant individuals would increase rapidly in field populations after only a few pyrethroid applications. Thus avoidance of pyrethroids on the first few generations of *H. armigera* in cotton, and restricting the use to later generations of bollworms may help in preventing the resistance problem in India (Russell *et al.*, 1998). The application of pyrethroids as late season sprays during mid October-November would be preferable not just against slightly less resistant *H. armigera* but because these would also simultaneously target the first thresholds of the pink bollworm infestation during this period (Kranthi, 2004).

The resistance management strategies are being proposed for the management of *H. armigera*.

1. Grow early maturing crops to avoid dominant *H. armigera* populations late in the season (Fakrudin *et al.*, 2003).
2. Avoid growing alternative host crops near main crop, as they serve as early season nursery crops for resistant insects (Fakrudin *et al.*, 2003).
3. Destruction of crop residues to prevent carry over of pest populations and summer ploughing to destroy resting stage insect populations (Kranthi, 2004).
4. Utilization of host plant resistance (Fakrudin *et al.*, 2003). Cultivation of sucking pest tolerant genotypes to help in delaying the first spray, thereby conserving the initial build-up of natural enemies (Kranthi, 2004).
5. Inter-cropping with cowpea, soybean and blackgram to encourage natural enemies (Rao *et al.*, 1994).
6. Chemical seed treatment to help delay the first spray (imidacloprid 70 WS or thiamethoxam 70 WS @ 5-7 g/Kg for protecting the crop against leafhoppers up to 30-60 days (Kairon and Kranthi, 1998; Surulivelu *et al.*, 2000).
7. Use of a rational and sensible sequence of insecticides that are effective on the target species, cause less disturbance to beneficial fauna and minimize selection pressure (Kranthi, 2004).
8. Check crops frequently and thoroughly and spray on threshold. This can minimize the need for sprays (Fakrudin *et al.*, 2003).
9. Hand picking and destruction of grown up larvae may be suggested as one of the component of IRM where the human resource is abundant (Ramasubramanian, 2004).
10. Avoidance of broad spectrum organophosphates such ass monocrotophos, methyl demeton, phosphamidon, acephate etc. especially as early sprays as these sprays strongly disrupt the natural enemy populations (Kranthi, 2004).
11. Use of endosulfan during the initial cropping phase representing the first window for bollworms. The use of soft chemistries such as *Bacillus thuringiensis* or HaNPV (nuclear polyhedrosis virus of *H. armigera*) or neem based insecticides can be used

as initial sprays to help conservation of natural enemies. For north Indian conditions wherein, the spotted bollworm occurs initially in the season, spinosad will be the preferred insecticide for the first window (Kranthi, 2004).

12. In the second window organophosphates or carbamates can be used as effective larvicides during 90-110 DAS based on ETLs. Resistance levels against certain organophosphate group of insecticides (quinalphos, chlopyriphos, and profenophos) and carbamates (thiodicarb) have been found to be low (Kranthi, 2004).
13. The application of pyrethroids as late season sprays during mi October-November constitutes the third bollworm window (Kranthi, 2004).
14. Avoid consecutive spraying of pyrethroids (Fakrudin *et al.*, 2003; Ramasubramanian, 2004).
15. Use of indoxacarb against pyrethroid resistant strains of *H. armigera* as it exhibits negative cross resistance to pyrethroid resistant strains (Ramasubramanian, 2004).

CONCLUSION

For managing resistance, altering pesti-cide use patterns, and reducing pesticide use are ef-fective resistance management tactics (Tabashnik, 1990). Resistance management pro-grams should include: I) Altering the way pesticides are developed and registered, and 2) Recognizing that resistance management must be a broad-based, multitactic endeavor (Hoy, 1992).

Nearly all major insect and mite pests will eventually become resistant to all classes of pes-ticides given sufficient selection pressure over sufficient time. There may be some ex-ceptions, but this generalization is reasonable given the documented record of resistance development in arthropod pests during the past 40 years. Resistance to stress is a fundamental and natural response by living organisms. On an evolutionary time scale, it is apparent that insects will develop multiple and di-verse mechanisms to survive extreme temperatures, allelochemicals, and other envi-ronmental stresses. Thus, most insects develop resistance to most pesticides, if subjected to appropriate and sustained selection. New pesticide classes have been proclaimed to be potential. It seems appropriate to assume that the development of resistance is nearly inevitable and resistance manage-ment programs have the goal of *delaying* rather than preventing resistance (Hoy, 1992).

There are increas-ing social, economic, and ecological pressures to reduce pesticide use and to increase the use of nonchemical control tactics such as host plant resistance, biorational meth-ods, cultural controls, and biological controls (National Research Council, 1989). There is an increasing priority by research scientists, regulatory agencies, legislators, and the public on using pesticides that are non-toxic to biological control agents and have minimal impact on the environment. Compatibility of pesticides with natural enemies and other nonchemical tactics is critical for improving pest management and environmental quality, and for managing resistance to pesticides. Enhancing the compatibility of pesticides and biological control agents is complex and sometimes difficult (Hoy, 1985; 1990), but can end in improved pest control (Metcalf, 1994) and pes-ticide resistance management (Tabashnik & Croft, 1985).

High rates of insecticides disrupt effec-tive biological control, leading to additional pesticide applications, which exert unnecessary selection for resistance in the pest. Under these circumstances, it may be appropriate for the label to contain two different directions for use; one rate could be recommended for the traditional strategy of relying on pesticides to provide control (although this is becoming a less and less viable option). A lower rate could be recom-mended for use in an IPM program that employs effective natural enemies. This ap-proach to labeling could reduce the number of pesticide applications and reduce rates, resulting in reduced selection for resistance in both target and non-target pests (Hoy, 1995).

Another innovation in pesticide registration would require that the toxicity of the pesticide to a selected list of biological control agents be determined for each cropping system. This information should be provided, either on the label or in readily available computerized data bases, perhaps via the internet. Without such information, use of these pesticides could disrupt benefits derived from effective biological control agent. This often results in unnecessary use of pesticides leading to more rapid evolution of pesticide resistance. Enhancing biological control not only leads to improved pest man-agement, but is also an essential tool in managing pesticide resistance (Hoy, 1995).

It is often difficult to predict the impact of pesticides under field conditions based on laboratory assays (Hoy 1990, Robertson & Preisler, 1992). Increased in-ternational consultation and cooperation between scientists, regulatory agencies, and pesticide companies could resolve many of the questions related to development of resistance. There is general agreement that reduced pesticide use is one of the essential elements of any resistance management program (Tabashnik, 1990; Metcalf, 1994).

Attempts to manage pesticide resistance generally has involved making relatively minor tactical shifts in use patterns. Application rates and the numbers of applications per growing season should be made with the un-derstanding that they affect the speed with which resistance will develop. In some cases new products may not be developed because they are toxic to biological control agents and thus could disrupt effective IPM programs already in place. This would only speed up the development of resistance in specific pests (Hoy, 1995). Thus, the compatibility of pesticides and biological control agents is a crucial issue in pesticide resistance management.

Some people suggest that financial incentives may have to be provided to induce pesticide companies to develop and register products that are harmless to biological control agents (= physiological selectivity). Others argue that many pesticides can be applied in a manner that affords substantial selectivity (= ecological selectivity) if the timing, location, rates, and methods of application are altered (Hull & Beers, 1985). Relying on ecological selectivity is more likely to be cost effective than developing large numbers of special use products with physiological selectivity. However, a com-plete financial analysis may indicate that, over the long term, selective pesticides (based both on physiological and ecological selectivity) are among the most cost effec-tive approaches to managing resistance to pesticides (Hoy, 1995).

References

Anandan GK (1997) Studies on insecticide resistance in rice leaf folder, *Cnapholocrocis medinalis* Guenee. M.Sc. Thesis, Tamil Nadu Agricultural University, Coimbatore.161p.

Anbalagan G (2001) Studies on insecticide resistance in rice leaf folder, *Cnaphalocrocis medinalis* Guenee. Ph.D. Thesis, Tamil Nadu Agricultural University, Coimbatore.161p.

Armes NJ Wightman JA Yadhav DR and Rao GVR (1997) Status of insecticide resistance in *Spodoptera litura* in Andhra Pradesh, India. *Pesticide Science*. 50: 240-248.

Ayyasamy R and Regupathy A (2004a) Discriminating dose technique for monitoring insecticide resistance in India. *Resistant Pest Management Newsletter*. 14(1): 36-40.

Ayyasamy R and Regupathy A (2004b) Role of microbials for the management of insecticide resistant *Helicoverpa Armigera* Hub. and *Plutella Xylostella L.* in India. *Resistant Pest Management Newsletter*. 14(1):26-32.

Banerjee SK Turkar KS and Wanjari RR (2000) Evaluation of newer insecticides for the control of bollworms in cotton. *Pestology*. 24: 14-16.

Bass C Schroeder I Turberg A Field LM and Williamson MS (2004) Identification of mutations associated with pyrethroid resistance in the *para*-type sodium channel of the cat flea, *Ctenocephalides felis*. *Insect Biochemistry and Molecular Biology*. 34: 1305-1313.

Berge JB Feyereisen R and Amichot M (1999) Cytochrome P450 monooxygenases and insecticide resistance in insects. In: Denholm I Pickett JA and Devonshire AL (Eds.), Insecticide Resistance: From Mechanisms to Management. CABI Publishing, Oxon, pp. 25-30.

Berticat C Dubois M Marquine M Chevillon C and Raymond M (2000) A molecular test to identify resistant alleles at the amplified esterase locus in the mosquito *Culex pipiens*. *Pest Management Science*. 56: 727-731.

Bhatia SK and Pradhan S (1970) Studies on resistance to insecticides in *Tribolium castaneum* (Herbst) II. Cross resistance characteristics of the p,p' DDT -resistant strain. *Indian Journal of Entomology*. 32: 32-38.

Bhatia SK Saxena JD and Sinha SR (1990) Status of insecticide resistance in *Tribolium castaneum* in India. I. Resistance to malathion. *Bulletin of Grain Technology*.28: 250-254.

Bhatia SK Yadav TD and Mookherjee PB (1971) Malathion resistance in *Tribolium castaneum* (Herbst) in India. *Journal of Stored Product Research*. 8: 227-230.

Borah B and Chahal BS (1979) Development of resistance in *Trogoderma granarium* Everts to phosphine in the Punjab. *FAO Plant Protection. Bulletin*. 27: 77-80.

Brattsten LB (1987) Sub-lethal virus infection depresses cytochrome P-450 in an insect. *Experientia* 43: 251-454.

Brent KJ (1986) Detection and monitoring of resistant forms: an overview. In: Pesticide Resistance: Strategies and Tactics for Management. National Academy Press, Washington, D.C. pp.298-312

Brogdon WG Beach RF Stewart JT and Castanaza L (1988) Microplate assay analysis of organophosphate and carbamate resistance distribution in Guatemalan *Anopheles albimanus*. *Bulletin WHO* 66: 339-346.

Brun-Barale A Bouvier J Pauron D Berge J and Sauphanor B (2005) Involvement of a sodium channel mutation in pyrethroid resistance in *Cydia pomonella* L, and development of a diagnostic test. *Pest Management Science*. 61: 549-554.

Campbell PM Newcomb RD Russell RJ and Oakeshott JG (1998) Two different amino acid substitutions in the ali-esterase E3, confer alternative types of organophosphorous insecticide resistance in the sheep blowfly *Lucilia cuprina*. *Insect Biochemistry and Molecular Biology*. 28: 139-150.

Champ BR and Dyte CE (1976) Report of FAO global survey of pesticide susceptibility of stored grain pests. *FAO Plant Production Protection Series No.5,* FAO Rome, pp. 1-297.

Chandrasekaran J and Regupathy A (1996) Insecticide resistance in DBM, *Plutella xylostella* Lin. in India. Discriminating dose studies. In: 20th International Congress of Entomology, Firenze, Italy, Aug. 25-31, Abst. 19-067p.602

Clark AG and Shamaan NA (1984) Evidence that DDT dehydrochlorinase from the housefly is a glutathione S-transferase. *Pesticide Biochemistry and Physiology*. 22: 249-261.

Clark JM Lee SH Kim HJ Yoon KS and Zhang A (2001) DNA-based genotyping techniques for the detection of point mutations associated with insecticide resistance in Colorado potato beetle *Leptinotarsa decemlineata. Pest Management Science*. 57: 968-974.

Clark JM Zhang A Dunn J and Yoon K (1999) Molecular detection of insecticide resistant alleles. *Pesticide science*. 55: 606-608.

Coon MJ Vaz AD and Bestervelt LL (1996) Peroxidative reactions of diversozymes. *FASEB Journal*. 10: 428-434.

Coustau C and ffrench-Constant (1995) Detection of cyclodiene insecticide resistance-associated mutations by single-stranded conformations polymorphism analysis. *Pesticide Science*. 43: 267-271.

Croft BA (1990) Arthropod Biological Control Agents and Pesticides. Wiley-Interscience, New York.

Crow JF (1960) Genetics of insecticide resistance: general considerations. *Misc Pub Entomol Soc Am*. 2:69-74.

Curtis CF (1985) Theoretical models of the use of insecticide mixtures for the management of resistance. *Bulletin of Entomological Research*. 75: 259-265.

Daly JC Fisk H and Forrester NW (1988) Selective mortality in field trials between strains of *Helicoverpa armigera* (Lepidoptera: Noctuidae) resistant and susceptible to pyrethroids: Functional dominance of resistance and age class. *Journal of Economic Entomology*. 81: 1000-1007.

Dawson GW Griffiths DC Pickett JA and Woodcock CM (1983) Decreased response to alarm pheromone by insecticide-resistant aphids. *Naturwissenschaften*. 70: 254-255.

Devonshire AL Field LM Foster SP Moores GD Williamson MS and Blackman RL (1999) The evolution of insecticide resistance in the peach potato aphid *Myzus persicae*. In: Denholm I Pickett JA and Devonshire AL (Eds.), Insecticide Resistance: From Mechanisms to Management. CABI Publishing, Oxon, pp. 1-8.

Dhingra S (1991) Periodic checking to detect development of resistance in mustard aphid, *Lipaphis erysimi* Kalt. to recommended insecticides used for more than two decades. *Journal of Entomological Research*. 15: 88-92.

Dhingra S Phokela A and Mehrotra KN (1988) Cypermethrin resistance in the populations of *Heliothis armigera* Hubner. *National Academy of Science Letters* 11: 123--125.

Fakrudin B Patil BV Badari Prasad PR and Prakash SH (2002). Insecticide usage patterns in South Indian cotton ecosystems to control cotton bollworm *Helicoverpa armigera*. *Resistant Pest Management Newsletter*. 12: 35-38.

Fakrudin B Vijaykimar Krishnareddy KB Badariprasad Patil BV and Kuruvinashetti MS (2003) Insecticide resistance in *Helicoverpa armigera* (Hubner): A mini review. *Resistant Pest Management Newsletter*. 13(1): 5-8.

Feyereisen R (2005) Insect Cytochrome P450. In: Gilbert LI Iatrou K and Gill SS (Eds.) Comprehensive Molecular Insect Science Vol.4. Elsevier, Oxford. pp. 1-77.

ffrench-Constant RH and Roush RT (1990) Resistance detection and documentation. In: Roush RT and Tabashnik BE (Eds.) Pesticide Resistance in Arthropods. Chapman and Hall, New York. pp. 4-38.

ffrench-Constant RH Devonshire AL and Clark SJ (1987) Differential rate of selection for resistance by carbamate, organophosphorous and combined pyrethroid and organophosphorous insecticide in *Myzus persicae* (Sulzer) (Hemiptera: Aphididae). *Bulletin of Entomological Research*. 77: 227-238.

ffrench-Constant RH Phillip JD and Golf GL (2004) The genetics and genomics of insecticide resistance. *Trends in Genetics*. 20: 163-170.

ffrench-Constant RH Pittendrigh B Vaughan A and Anthony N (1999) Why there are so few resistance associated mutations in insecticide target genes? In: Denholm I Pickett JA and Devonshire AL (Eds.), Insecticide Resistance: From Mechanisms to Management. CABI Publishing, Oxon, pp. 9-17.

ffrench-Constant RH Steichen JC and Brun LO (1994) A molecular diagnostic for endosulfan insecticide resistance in the coffee berry borer *Hypothenemus hampei* (Coleoptera: Scolytidae). *Bulletin of Entomological Research*. 84: 11-16.

ffrench-Constant RH Steichen JC Rocheleau TA Aronstein K and Roush RT (1993) A single-amino acid substitution in a ã-aminobutyric acid subtype A receptor locus is associated with cyclodiene insecticide resistance in *Drosophila* populations. *Proceedings of National Academy of Sciences USA*. 90: 1957-1961.

Forgash AJ (1984) History, evolution and consequences of insecticide resistance. *Pesticide Biochemistry and Physiology*. 22: 178-186.

Forrester NW and Cahill M (1987) Management of insecticide resistance in *Helicoverpa armigera* (Hubner) in Australia. In: Ford M Holloman DW Khambay BPS and Sawicki RM (Eds.) Combating Resistance to Xenobiotics: Biological and Chemical Approaches. Ellis Horwood, U.K., pp. 127-137.

Forrester NW and Fitt GP (1992) Management of insecticide resistance in *Heliothis armigera* in Australia ecological and chemical countermeasures. In; Denholm I Devonshire AL and Hollomon DW (Eds.) Resistance '91: Achievement and Developments in Combating Pesticide Resistance. Elsevier Applied Science, London and New York. pp. 88-99.

Forrester NW Cahill M Bird LJ and Layland JK (1993) Management of pyrethroid and endosulfan resistance in *Helicoverpa armigera* (Lepidoptera: Noctuidae) in Australia. *Bulletin of Entomological Research* Suppl., No.1: 132 p.

Gavi Gowda and Regupathy A (1995) Pyrethroid synergism with vegetable oils. *Pod Borer Management Newsletter*, 6: 7.

Gavi Gowda and Regupathy A (1996) Potential use for Pongamia oil as an adjuvant with pyrethroid insecticides. *Pod Borer Management Newsletter* 7: 7.

Geetha N Rabindra RJ Muthukumar B Thayumanavan B Regupathy A and Sundara Babu PC (1999) Effect of nuclear polyhedrosis virus infection on the activity of carboxylesterase and Glutathione S-transferase in *Helicoverpa armigera* (Hbn.). *Tropical Agricultural Research* 11: 408-424.

Georghiou GP (1983) Management of resistance in arthropods. In: Georghiou GP and Saito T (Eds.) Pest Resistance to Pesticides, Plenum Press, New York. pp. 769-792.

Gouthaman S and Regupathy A (2003) Monitoring Techniques for Detecting Insecticide Resistance in *Helicoverpa armigera* (Hubner). *Resistant Pest Management Newsletter* 13(1.) 38-41.

Gunning RV Easton CS. Greenup LR and Edge VE (1984) Pyrethroid resistance in *Heliothis armigera* (Hubner) (Lepidoptera: Noctuidae) in Australia. *Journal of Economic Entomology*, 77: 1283-1287.

Hawkes NJ Janes RW Hemingway J and Vontas J (2005) Detectioin of resistance-associated point mutations of organophosphate-insensitive acetylcholinesterase in the olive fruit fly, *Bactrocera oleae* (Gmelin). *Pesticide Biochemistry and Physiology*. 81: 154-163.

Hemingway J (2000) The molecular basis of two contrasting metabolic mechanisms of insecticide resistance. *Insect Biochemistry and Molecular Biology*. 30: 1009-1015.

Hemingway J Hawkes NJ McCarroll L and Ranson H (2004) The molecular basis of insecticide resistance in mosquitoes. *Insect Biochemistry and Molecular Biology*. 34: 653-665.

Hoy M A (1985) Almonds: integrated mite management for California almond orchards. *In*: Helle W and Sabelis MW [Eds.]. Spider Mites, Their Biology, Natural Enemies, and Control, Vol. 1B. Elsevier, Amsterdam. pp. 299-310.

Hoy MA (1990) Pesticide resistance in arthropod natural enemies: variability and selection responses. In Roush RT and Tabashnik BE [Eds.] Pesticide Resistance in Arthropods, Chapman and Hall, New York. pp. 203-236.

Hoy MA (1992) Proactive management of pesticide resistance in agricultural pests. *Phytoparasitica* 20: 93-97.

Hoy MA (1995) Multitactic resistance management: An approach that is long overdue. *Florida Entomologist*. 78: 443-451.

Hoy MA (1999) Myths, models and mitigation of resistance to pesticides. In: Denholm I Pickett JA and Devonshire AL (Eds.), Insecticide Resistance: From Mechanisms to Management. CABI Publishing, Oxon, pp. 111-119.

Huang H and Dai GQ (1991) Studies on the synergism of *Pieris rapae* granulosis virus and insecticides. *Journal of South China Agricultural University* 12: 96-103.

Hull LA and Beers EH (1985) Ecological selectivity: Modifying chemical control practices to preserve natural enemies. *In* Hoy MA and Herzog DC (Eds.) Biological Control in Agricultural IPM Systems. Academic, Orlando. pp. 103-122.

Jayapradeepa S (2000) Studies on insecticide resistance in cotton leaf hopper, *Amrasca devastans* D. M.Sc. Thesis, Tamil Nadu Agricultural University, Coimbatore.161p.

Jayaswal AP (1989) Management of American bollworm on cotton in Andhra Pradesh. *Indian Farming*.17: 6-7.

Jeschke P and Nauen R (2005) Neonicotinoid insecticides. In: Gilbert LI Iatrou K and Gill SS (Eds.) Comprehensive Molecular Insect Science Vol.5. Elsevier, Oxford. pp. 53-105.

Jeyapradeepa S and Regupathy A (2002) Generating base line data for insecticide resistance monitoring in cotton leafhopper, *Amrasca devastans* (Distant). *Resistant Pest Management Newsletter* 11(2):1-6.

Jutsum AR Heaney SP Perrin BM and Wege PJ (1998) Pesticide resistance: Assessment of risk and the development and implementation of effective management strategies. *Pesticide Science*. 54: 435-446.

Kairon MS and Kranthi KR (1998) Non-insecticidal methods in cotton pest management – A critical re-appraisal. In: Chari MS Qayum MA Sanghi NK and Shastri MV (Eds.) Workshop Proceedings: Non-pesticidal Management of Cotton and Pigeonpea pests. April 10-11, 1998, MANAGE, Hyderabad, India pp. 33-36.

Kalra RL Udeann AS and Bindra OS (1975) Status of insecticide resistance in stored grain insects in the Punjab. *Indian Journal of Plant Protection*. 3: 186-196.

Khambay BPS Denholm I Carlson GR Jacobson RM and Dhadialla TS (2001) Negative cross-resistance between dihydropyrazole insecticides and pyrethroids in houseflies *Musca domestica*. *Pest Management Science*. 57: 761-763.

Knipple DC Doyle KE Marsella-Herrck PA and Soderlund DM (1994) Tight genetic linkage between the *kdr* insecticide resistance trait and a voltage sensitive sodium channel gene in the house fly. *Proceedings of National Academy of Sciences USA*. 91: 2483-2487.

Kostaropoulos I Papadopoulos AI Metaxakis A Boukouvala E and Papadopoulou-Mourkidou E (2001) Glutathione S-transferase in the defense against pyrethroids in insects. *Insect Biochemistry and Molecular Biology*. 31: 313-319.

Kranthi KR (2004) Managing *Helicoverpa armigera* resistance to insecticides. In: Subrahmanyam B Ramamurthy VV and Singh VS (Eds.) Frontier Areas of Entomological Research. Proceedings of the National Symposium on Frontier Areas of Entomological Research, IARI, New Delhi, India, Nov. 5-7, 2003 (Invited lectures), pp.355-375.

Kranthi KR Armes NJ Nagarjun GV Rao RS and Sundaramurthy VT (1997) Seasonal dynamics of metabolic mechanisms mediating pyrethroid resistance in *Helicoverpa armigera* in central India. *Pesticide Science*. 50: 91-98.

Lavanya D (2004) Studies on the susceptibility of *Plutella xylostella* (Lepidoptera: Plutellidae) on cabbage the insecticides of new chemistry (special reference to Avermectins). M.Sc. Thesis, Tamil Nadu Agricultural University, Coimbatore, India, 59p.

Lee SST and Scott JG (1989) Microsomal cytochrome P450 monooxygenases in the house fly (*Musca domestica* L.): biochemical changes associated with insecticide resistance and phenobarbital induction. *Pesticide Biochemistry and Physiology*. 35: 1-10.

Li F and Han Z (2004) Mutations in acetylcholinesterase associated with insecticide resistance in the cotton aphid, *Aphis gossypii* Glover. *Insect Biochemistry and Molecular Biology*. 34: 397-405.

McCaffery AR King ABS Walker AJ and El Nayir H (1989) Resistance to synthetic pyrethroids in the bollworm, *Heliothis armigera* from Andhra Pradesh, India. *Pesticide Science*.27: 65-76.

Mehrotra KN and Phokela A (1986) Variations in carboxylic ester hydrolase (Ec. 3 1.1.1) activity in *Heliothis armigera* Hubner larvae. *Current Science*. 55: 508-509.

Mehrotra KN and Phokela A (1992) Pyrethroid resistance in *Helicoverpa armigera* Hubner V. Response of populations in Punjab in cotton. *Pesticide Research Journal*. 4: 59-61.

Mehrotra KN and Phokela A (2000) Insecticide resistance in insect pests: Current status and future strategies. In: Dhaliwal GS and Singh B (Eds.) Pesticides and Environment. Commonwealth Publishers, New Delhi. pp. 39-83.

Melander AL (1914) Can insects become resistant to sprays? *Journal of Economic Entomology*. 7:167-173

Metcalf RL (1994) Insecticides in pest management. *In* Metcalf RL and Luckmann WH (Eds.) Introduction to Insect Pest Management. John Wiley and Sons, New York. pp. 245-284

Mota-Sanchez D Bills PS Whalon ME (2002) Arthropod resistance to pesticides: status and overview. In: Pesticides-in-agriculture-and-the-environment. pp. 241-272.

National Research Council (1989) Alternative Agriculture, National Academy Press, Washington, D. C.

Nauen R Stumpf N and Elbert A (2002) Toxicological and mechanistic studies on neonicotinoid cross resistance in Q-type *Bemesia tabaci* (Hemiptera: Aleyrodidae). *Pest Management Science*. 58: 868-875.

Niranjankumar BV and Regupathy A (2001). Status of insecticide resistance in tobacco caterpillar *Spodoptera litura* (Fab). in Tamil Nadu. *Pesticide Research Journal* 13(1): 86-89.

Niranjankumar BV and Regupathy A (2001b). Resistance levels of tobacco cutworm, *Spodoptera litura* (F.) to different insecticides in south India. In: International Conference on Pesticides, Environment, and Food security. 19-23, November, 2001. Society of Pesticide Science India, IARI, New Delhi. p 271-272.

Oppenoorth FJ (1984) Biochemistry of insecticide resistance. *Pesticide Biochemistry and Physiology*. 22: 187-193.

Oppenoorth FJ (1985) Biochemistry and genetics of insect resistance. In: Kerkut GS and Gilbert LI (Eds.) Comprehensive Insect Physiology, Biochemistry and Pharmacology. Pergamon Press, New York, pp 731-773.

Pasteur N and Georghiou GP (1989) Filter paper test for rapid determination of phenotypes with high esterase activity in organophosphate resistant mosquitoes. *Mosquito News*. 41: 181-183.

Patel IS and Yadav DN (1997) Cross resistance of monocrotophos resistant strain of green lacewing, *Chrysopa scelestes* Banks. *India Journal of Agricultural Sciences*, 67: 33-34.

Patil BV Lingappa S Srinivasa AG and Bheemanna M (1996). Integrated Pest Management strategies for cotton. Paper presented at XX International Congress of Entomology, 25th-31st August, Firenze, Italy.

PDBC (1998) Annual report. Project Directorate of Biological Control, Bangalore. pp.172.

PDBC (1999) Annual report. Project Directorate of Biological Control, Bangalore.pp.232.

Perry AS Yamamoto I Ishaaya I and Perry R (1998) Insecticides in agriculture and environment. Narosa Publishing House, New Delhi. 261pp.

Phokela A and Mehrotra KN (1989) Pyrethroid resistance in *Heliothis armigera* Hubner. II. Permeability and metabolism of cypermethrin. *Proceedings of Indian National Science Academy* B55: 235-238.

Plapp FW Jr and Casida JE (1969) Genetic control of house fly NADPH-dependent oxidases: relation to insecticide chemical metabolism and resistance. *Journal of Economic Entomology*. 62: 1174-1179.

Pradhan S Jotwani MG and Sarup P (1963) Failure of BHC and DDT to control Singhara beetle, *Galerucella birmanica* Jacoby. *Indian Journal of Entomology* 34: 176-179.

Praveen PM (2003) Studies on insecticide resistance in early season sucking pests of cotton in Tamil Nadu. Ph.D. Thesis, Tamil Nadu Agricultural University, Coimbatore.161p

Praveen PM and Regupathy A (2003a) Generating baseline data for insecticide resistance monitoring in cotton aphid, *Aphis gossypii* Glover *Resistant Pest Management Newsletter*.12 (2) 26-27.

Praveen PM and Regupathy A (2003b) Insecticide resistance in field Populations of cotton thrips, *Thrips tabaci* Lindeman in major cotton growing regions of Tamil Nadu. *Pesticide Research Journal* 15: 197-200.

Pree DJ (1990) Resistance management in multiple-pest apple orchard ecosystems in Eastern North America. In: Roush RT and Tabashnik BE (Eds.) Pesticide Resistance in Arthropods. Chapman and Hall, New York pp. 261-276.

Pridgeon JW and Liu N (2003) Overexpression of the cytochrome *c* oxidase subunit I gene associated with a pyrethroid resistant strain of German cockroaches, *Blatella germanica* (L.). *Insect Biochemistry and Molecular Biology*. 33: 1043-1048.

Rajagopalbabu S (2001) Interaction of granulosis virus of *Plutella xylostella* (L.), (Yponomeutidae: Lepidoptera) with some insecticides. M.Sc. (Agri.) Thesis, Tamil Nadu Agric. Univ., Coimbatore. 109 p.

Rajendran S Shrilatha B Rakha Savanth TL and Nayak KR (2000) Response of field strains of stored grain insect pest to phosphine fumigant. *Pesticide Research Journal*. 12: 63-67.

Ramakrishnan N Saxena VS and Dhingra S (1984) Insecticide resistance in the population of *Spodoptera litura* (F.) in Andhra Pradesh. *Pesticides*. 18: 22--27.

Ramasubramanian T (2004) Management, mechanism and management of pyrethroid resistance in *Helicoverpa armigera* Hubner in India. *Journal of Entomology*. 1: 6-11.

Rao NV Rajasekhar P Venkataiah M and Rama Rao B (1994) Cotton pest control problems in Andhra Pradesh, India: Optimizing pest management options for a more sustainable approach to cotton cultivation. In: Constable GA and Forrester NW (Eds.) Proceedings, Challenging the Future, World Cotton Research Conference 1 pp. 563-568.

Rauch and Nauen R (2003) Biochemical markers linked to neonicotinoid cross-resistance in *Bemesia tabaci* (Hemiptera: Aleyrodidae). *Archives of Insect Biochemistry and Physiology*. 54: 165-176.

Reddy CNVS and Reddy GPV (1984) Preliminary assessment of resistance in tobacco caterpillar, *Spodoptera litura* Fab. to insecticides. *Andhra Agricultural Journal*. 31: 85-91.

Reed W and Pawar CS (1982) *Heliothis:* a global problem. Proceedings of International Workshop on *Heliothis* Management. ICRISAT, Patancheru, India, p.9.

Reid TJ (1989) Acaricide resistance in Queensland. Information series Q189011, Department of Primary Industries, Queensland Government, Brisbane.

Ringo J Jona G Rockwell R Segal D and Cohen E (1995) Genetic variation of resistance to chlorpyriphos in *Drosophila melanogaster* (Diptera: Drosophilidae) infesting grapes in Israel. *Journal of Economic Entomology*. 88: 1158-1163.

Robertson JL and Preisler HK (1992) Pesticide Bioassays with Arthropods. CRC Press, Boca Raton, Florida.

Rossiter LC Gunning RV and Rose HA (2001) The use of polyacrylamide gel electrophoresis for the investigation and detection of fenitrothion and chlorpyriphos-methyl resistance in *Oryzaephilus surinamensis* (Coleoptera: Sylvanidae). *Pesticide Biochemistry and Physiology*. 69: 27-34.

Roush RT (1989) Designing resistance management programs: How can you choose? *Pesticide Science*. 26: 423-441.

Roush RT and Croft BA (1986) Experimental population genetics and ecological studies of pesticide resistance in insects and mites. In: National Research Council (Eds.) Pesticide resistance: strategies and tactics for management. National Academy Press, Washington DC pp. 257-270.

Roush RT and Daly JC (1990) The role of population genetics in resistance research and management. In: Pesticide Resistance in Arthropods. Chapman and Hall, New York pp. 97-152.

Roush RT and McKenzie JA (1987) Ecological genetics of insecticide and acaricide resistance. *Annual Review of Entomology*. 32: 361-380.

Roush RT and Miller GL (1986) Considerations for design of insecticide resistance monitoring programs. *Journal of Economic Entomology*. 79: 293-298.

Rufingier C Pasteur N Lagnel J Martin C and Navajas M (1999) Mechanisms of insecticide resistance in the aphid *Nasonovia ribisnigri* (Mosley) (Homoptera: Aphididae) from France. *Insect Biochemistry and Molecular Biology*. 29: 385-391.

Russell D Singh J Jadhav D Surulivelu T Reghupathy A and Kranthi KR (1998) Management of insecticide resistance in *Helicoverpa armigera* in cotton in India. In: *Proceedings of the World Cotton Research Conference*. Greece, 6-12 September 1998. pp. 679-688.

Sarup P Singh DS and Lal R (1969) Testing of different insecticides as contact poisons against the adults of mustard aphid, *Lipaphis erysimi* Kalt. (Homoptera: Aphididae). *Indian Journal of Entomology*. 31: 21-25.

Sawicki R (1987) Definition, detection and documentation of insecticide resistance. In: Ford MG Holloman DW Khambay BPS Sawicki RM (Eds.) Combating Resistance to Xenobiotics: Biological and Chemical Approaches. Ellis Horwood, London. pp. 105-I17.

Saxena JD Rai S Srivastava KM and Sinha SR (1989) Resistance in the field population of the diamond back moth to some commonly used synthetic pyrethroids. *Indian Journal of Entomology*. 51: 265-268.

Saxena JD Sinha SR and Bhatia SK (1991a) Status of insecticide resistance in *Tribolium castaneum* (Herbst) in India V-Resistance to DDT. *Bulletin of Grain Technology*. 29: 89-92.

Saxena JD Sinha SR and Bhatia SK (1991b) Status of insecticide resistance in *Tribolium castaneum* (Herbst) in India IV-Resistance to phosphine. *Bulletin of Grain Technology*. 29: 148-151.

Saxena JD Sinha SR and Bhatia SK (1992) Status of insecticide resistance in *Tribolium castaneum* (Herbst) in India III - Resistance to dichlorvos. *Bulletin of Grain Technology*. 30: 28-31.

Scott JG (1999) Cytochromes P450 and insecticide resistance. *Insect Biochemistry and Molecular Biology*. 29: 757-777.

Senthil Kumar CM (2003) Diversity of *Lecanicillium (=Verticillium) lecanii* (Zimmermann) Viégas and spiders in coffee ecosystem applied with thiamethoxam. Ph.D. Thesis, Tamil Nadu Agricultural University, Coimbatore.161p.

Senthil Kumar CM and Regupathy A (2004) Generating base line data for insecticide resistance monitoring in coffee green scale, *Coccus viridis* (Green). *Resistant Pest Management Newsletter*. 13 (2) 5-7.

Sharma S Dilawari VK and Gupta VK (2003) RAPD-PCR analysis to monitor imdacloprid resistance in cotton whitefly. *Resistant Pest Management Newsletter*. 13(1): 77-81.

Sidhu AS Singh J Arora R Brar DS Dhaliwal GS Bal RS and Singh J (1994) Adoption of plant protection measures *vis-a-vis* effectiveness of recommended insecticides against *Helicoverpa armigera* (Hubner) during 1990-a year of cotton crop failure in Punjab. *Journal of Insect Science*. 7: 10-13.

Singh HN Jose Rose and Sing HK (1994) *Helicoverpa armigera* insecticide resistance management at Varanasi. *Pod borer Management Newsletter*, No. 4:3.

Smissaert HR (1964) Cholinesterase inhibition in red spider mites *Tetranychus urticae* susceptible and resistant to organophosphate. *Science*. 143: 129-131.

Soderlund DM and Bloomquist JR (1990) Molecular mechanisms of insecticide resistance. In Roush RT and Tabashnik BE [Eds.] Pesticide Resistance in Arthropods, Chapman and Hall, New York. pp. 58-96.

Soderlund DM and Knipple DC (2003) The molecular biology of knockdown resistance to pyrethroid insecticides. *Insect Biochemistry and Molecular Biology*. 33: 563-577.

Soderlund DM (2005) Sodium channels. In: Gilbert LI Iatrou K and Gill SS (Eds.) Comprehensive Molecular Insect Science Vol.5. Elsevier, Oxford. pp. 1-24.

Song WC Lian SX and Yu ZG (2000) Synergism of *Helicoverpa armigera* nucleopolyhedrovirus in combinations with chemical insecticides and lecithin. *Acta Entomologica Sinica*. 43: 346-355.

Srivastava BK and Joshi HC (1965) Occurrence of resistance to BHC in *Prodenia litura* Fab. (Lepidoptera: Noctuidae). *Indian Journal of Entomology*. 27: 102-104.

Stanley J (2004) Studies on base line toxicity of emamectin and spinosad to *Helicoverpa armigera* (Hubner) and *Spodoptera litura* (Fab.) and their bio-efficacy on brinjal fruit borers. M.Sc. Thesis, Tamil Nadu Agricultural University, Coimbatore, India

Sun CN (1990) Insecticide resistance in diamondback moth what can we do with existing information? In: Second international workshop on management of diamondback moth and other crucifer pests, Abstract volume, Asian Vegetable Research and Development Center, Shanhua, Taiwan.

Surulivelu T Venogopal K Kannan R and Pandi V (2000) Imidacloprid seed treatment effect on sucking pests, predators, plant growth and productivity in cotton. In: Proceedings of World Cotton Conference 2 Athens, Greece, 6-12 September 1998, pp.874-877.

Tabashnik BE (1990) Modeling and evaluation of resistance management tactics. In Roush RT and Tabashnik BE [Eds.] Pesticide Resistance in Arthropods, Chapman and Hall, New York. pp. 153-182.

Tabashnik BE and Croft BA (1982) Managing pesticide resistance in crop-arthropod complexes: interactions between biological and operational factors. *Environmental Entomology*. 11: 1137-1144.

Tabashnik BE and Croft BA (1985) Evolution of pesticide resistance in apple pests and their natural enemies. *Entomophaga* 30: 37-49.

Temerak SA (2003) Negative cross-resistance to spinosad: An interesting observation in the field population of cotton leaf worm larvae, *Spodoptera littoralis* in Egypt. *Resistant Pest Management Newsletter*. 13(1): 33-36.

Tomlin CDS (1997) The Pesticide Manual. Farnham, UK: Br Crop Protection Council.

Verma AN and Sandhu GS (1968) Chemical control of diamond back moth, *Plutella xylostella* (Curtis). *J. Res. Punjab Agric. Univ.* 5: 420-423.

Vontas JG Hejazi MJ Hawkes NJ Cosmidis N Loukas M Hemingway J (2002) Resistance-associated point mutations of organophosphate insensitive acetylcholinesterase, in the olive fruit fly *Bactrocera oleae*. *Insect molecular Biology*. 11: 329-336.

Vontas JG Small GJ and Hemingway J (2001) Glutathione S-transferases as antioxidant defence agents confer pyrethroid resistance in *Nilaparvatha lugens*. *Biochemical Journal* 357: 65-72.

Whalon M Mota-Sanchez D and Duynslager L (2004) www. Pesticideresistance.org/DB/species.php.

Whalon ME McGaughey WH (1998) *Bacillus thuringiensis:* Use and resistance management. In: Ishaaya I Degheele D (Eds.) Insecticides with Novel Modes of Action: Mechanism and Application. Springer-Verlag, New Delhi. India. pp 106-137.

50Wilkinson CF (1983) Role of mixed function oxidases in insecticide resistance. In Georghiou GP Saito Y (Eds.) Pest Resistance to Pesticides. Plenum Press, New York, pp. 175-206.

Williamson MS Denholm I Bell CA and Devonshire AL (1993) Knockdown resistance (*kdr*) to DDT and pyrethroid insecticides maps to a sodium channel gene locus in the housefly (*Musca domestica*). *Mol. Gen. Genet*. 240: 17-22.

Wolff MA and Wingate VP (1998) Characterisation and comparative pharmacological studies of a functional gamma-aminobutyric acid (GABA) receptor cloned from the tobacco bud worm *Heliothis virescens* (Noctuidae: Lepidoptera). *Invertebrate Neuroscience*. 3: 305-315.

World Health Organization (1957) Expert Committee on malaria. 7th Report. WHO Tech Rep Ser 125.

Yamamoto I Takahashi Y and Kyomura N (1983) Suppression of altered acetylcholinesterase of the green rice leafhopper by N-propyl and N-methyl carbamate combinations. In: Georghiou GP and Saito T (Eds.) Pest Resistance to Pesticides, Plenum Press, New York pp.579-594.

Yang Y Wu Y Chen S Devine GJ Denholm I Jewess P and Moores GD (2004) The involvement of microsomal oxidases in pyrethroid resistance in *Helicoverpa armigera* from Asia. *Insect Biochemistry and Molecular Biology*. 34: 763-773.

Zhang A Dunn JB and Clark JM (1999) An efficient strategy for validation of point mutation associated with acetylcholinesterase sensitivity to azinphosmethyl in Colorado potato beetle. *Pesticide Biochemistry and Physiology*. 65: 25-35.

Zhu YC, Snodgrass GL and Chen MS (2004) Enhanced esterase gene expression and activity in a malathion resistant strain of the tarnished plant bug, *Lygus lineolaris*. *Insect Biochemistry and Molecular Biology*. 34: 1175-1186.

Zhu, CL Tian HS Li JM Yu XZ and Ye BH (1998) Identification of deltamethrin resistant mosquito, *Culex pipiens pallens*, with RAPD-PCR. *Chin. J. Parasitic Dis. Cont.* 11: 214-216.

Entomology : Novel Approaches, 2007
Eds.: *P.C. Jain and M.C. Bhargava,*(Page 245-266)
New India Publishing Agency, New Delhi
E-mail: newindiapublishingagency@gmail.com
web: www.bookfactoryindia.com

CHAPTER 13

PEST MANAGEMENT IN CEREALS

Prem Kishore and R.K. Sharma
Division of Entomology, Indian Agricultural Research Institute, New Delhi- 110 012

INTRODUCTION

The introduction of high yielding varieties and hybrids in late sixties led to accumulation of insect pest problems in cereals. The varieties and hybrids developed during green revolution era with the sole objective to enhance production to feed increasing population were better in quality parameters than traditional locals but were highly susceptible to insect pests. Over 200 pests attack different cereals and insects that were not known or were of minor importance have now become major pests and several new insect pests have appeared on cereals. Out breaks are frequent due to several obvious reasons. Major pests are serious, perennially occurring and persistent species which cause considerable loss in yields every year in large areas require control measures. Occasional pests attack the crops sporadically but cause potential damage and sufficient loss, though in some localities or only in some seasons, so as to warrant human intervention.

CONCEPT OF IPM

Concept of integrated pest management implies the rational integration of various methods of insect pest control viz. cultural, chemical, biological etc. to suppress pest populations to a level below economic injury level. Eco-friendly approaches like light traps for attracting and mass killing of moths and beetles can also be used to monitor the field population of insects. The trapped moths and beetles can be killed in kerosene water to reduce the insect population in the field. Light traps also helps to correlate seasonal incidence and abundance of pests with climatic factors. Manual removal and destruction of dead-hearts and infested plants has been found successful in reducing the stem borer damage.

The purpose of cultural control practices is to make the environment less favourable for the pest and more favourable to its natural enemies. It includes decisions on varieties to be grown, time of planting, crop rotation, tillage, plant spacing, sanitation etc.

Varieties: The use of resistant varieties is the most effective tool for minimizing the pest attack. Grow only recommended hybrids and composites, which are relatively less, susceptible to insects, particularly the borers.

Sowing date: Early sown crop usually suffers more borer attack then a crop sown later. While reverse is true for shoot fly species. Through population monitoring of pests, the sowing dates can be suitable adjusted to minimize losses due to insect pests.

Crop rotation: Crop rotations can check insect populations build up by removing hosts of the pest for extended periods. Cereals can be rotated with pulses, groundnuts, cotton etc.

Plant density: Spacing affects the relative rate of development of a plant and its pest population. It has been observed that with an increase in plant density, there is higher incidence of stem borer. Recommended and optimum plant density can play an important role in reducing pest incidence. Using higher initial plant density, then subsequently thinning of dead- hearts helps in reducing shoot fly damage and maintains required plant population.

Farm sanitation: Collection and burning stubble and stalks of plougyhing and destroying crop residue, removal of infested plant parts or infested plants at hoeing can prove effective in minimizing pest population, if carried out by farmers over large area on community basis at village level.

Tillage: Deep ploughing after harvest and before sowing reduces the potential for carryover of pest from one season to another.

Balance use of fertilizer and irrigation: This helps in checking the plant population

Biological control: Efforts should be made to release the known parasites and predators of borers or other insect pests and the use of insecticides should be restricted to protect the biotic agents. The egg parasite *Trichogramma* spp. can be utilized for suppression of stem borers. The egg parasite @ 1.0 lacs per hectare should be released on 7 days old crop followed by second application on 12 days old crop to minimize the population of stem borers. Many species of spiders have been reported to feed on eggs and early larval stages of stem borers in plant whorls. The adults of coccinellid beetles, *Menochilus sexmaculata* Fab., and *Coccinella septumpunctata* L. are predacious on the freshly emerged larvae of stem borer. Information on potent parasites and predators must be utilized for their use against key pests in pest management programmes.

Chemical Control: Several insecticides have been tested as seed treatment, foliar and other methods of application against maize pests. The insecticides, at least in foreseeable future, will remain as a major tool in pest management system. However, they must be used judiciously. Pesticide application based on economic threshold levels will reduce the number of insecticidal applications.

Pest management in cereals is a viable option and should be implemented. Efforts have been made to present IPM programme for different cereals in following paragraphs.

IPM IN WHEAT

The last three decades has witnessed several outbreaks of sporadic pests like armyworm, cutworms, flea beetles, sugarcane leafhopper etc. The major insect pest problems in India are termites, which are very serious in dry areas, and also aphids, shootfly and brown wheat mite have become regular occurring pests. The gujhia weevil, a serious pest of yesteryears, has assumed a non-pest status. Wheat was considered to be a pest free crop before green revolution as only termites and gujhia weevil were regarded as pests. However, the post green-revolution period witnessed the rise in pest problems, both in the number of insect species as well as their severity. This all occurred due to complete replacement of older tall varieties with dwarf high yielding ones possessing more succulent foliage and wide ranging changes in wheat agro-ecosystem. The awareness of changes in field pest complex of wheat led to regular involvement of entomologists in the project during seventies. Due to shift in sowing time of wheat, availability of relatively photo-insensitive varieties, temperature tolerant genotypes and also due to considerable changes in agro-techniques involving higher fertilizer inputs and irrigation led to changes in pest complex of wheat.

Termites, *Odontotermes obesus* Ramb and *Microtermes obesi* Holm

Termites are traditional pests of wheat and several others crops mainly under unirrigated areas having light soils. About 16 species of these polyphagous insects attack wheat crop in India. Among these, the most important and widely distributed species are *Odontotermes obesus* Ramb and *Microtermes obesi* Holm. Termite incidence on wheat crop starts with sowing and continues till harvest. The damage is caused by feeding of workers on roots and underground portion of stems resulting in drying up of the infested plants/tillers. Subsoil existence of the termites, their polyphagous nature and the sources of infestation (termnitaria) usually situated away from the fields, pose difficulty in crop protection.

Effective crop protection in termite prone areas is feasible only through prophylactic chemical treatments. Earlier recommendations included pre-sowing soil treatment with insecticidal dusts but later studies showed that more effective and economical control could be achieved through seed treatment. The treatment is done by spraying required quantity of insecticide diluted in 5 litres of water over one quintal of seed 1-2 days before sowing. Previously aldrin was used but since its ban now, endosulfan or chlorphyriphos @ 240 and 90 grams a.i. per quintal of seed, respectively, are recommended. This prophylactic seed treatment is most effective method of crop protection in termite prone areas and the method is compatible with fungicidal seed dressing. In situations where no prophylactic seed treatment is undertaken and termite incidence is noticed in standing wheat crop, some check of pest damage could be achieved through broadcast of insecticide treated soil. For this treatment chlorpyriphos or lindane @ 800 and 600 g a.i. respectively are diluted with 5 litres of water, mixed with 50 kg loose soil and broadcast in one hectare field. If feasible, a light irrigation may be given thereafter (Singh, 1986 and 1998).

Shootfly, *Atherigona naquii* and *A. oryzae*

Shootfly has emerged as a pest of wheat crop in India after the advent of high yielding dwarf varieties. Since initial reports of shootlfy infestation from Rajasthan and Madhya Pradesh during 1965-66, the pest has been found widely distributed in the country and has been recognized as a regular pest in major wheat growing states like Rajasthan, Gujarat, Madhya Pradesh, Uttar Pradesh, Delhi, Haryana and Punjab. A taxonomic review of the genus *Atherigona* by Pont, in 1972, revealed that the wheat shootflies in India comprised of *A. naqvii* and *A. oryzae*. Out of these, *A. naqvii* is more prevalent and widely distributed and was first time reported infesting wheat in early seventies. Shootfly damage to plants is characterized by formation of 'dead hearts', which result in drying of the central portion of tillers. The infestation occurs throughout the plant growth but damage to young seedlings and tillers are most important. Studies conducted at Delhi have shown that the most vulnerable period of crop is during the first 45 days after germination and after that damage to newer tillers is not of much economic importance. With the death of the tiller, the infested plant throws up newer tillers, which in turn are further infested and in cases of severe infestation the plant gives a bushy appearance.

Critical observations showed that shootfly infestation was greatly dependent on time of sowing and that the pest has two clear-cut periods of activity during crop season. It was inferred form observations taken form different fields with varied sowing time that under Delhi conditions the wheat crop sown early (before middle of October) or late (after December) was infested, while the normal (November and early December) sowing escaped shootlfy damage. While relative humidity and rainfall did not affect pest incidence, the low temperature during mid-December to mid-February was found to arrest the pest activity.

The most effective management strategy against this pest is to strictly adhere to normal sowing time and avoid late sowings. However, due to the late vacation of field from the previous crop, late sowing of wheat has become common. Under such situations, increased seed rate of 120-130 kg/ha may be used and the infested shoots (dead hearts) removed through a campaign after 10 and 20 days of germination.

The protection to the crop can be given when the tiller damage exceeds 10 per cent. Two foliar spray of cypermethrin (@ 50 g a.i./ha), first within a week of germination and the second two weeks later has been found effective against this pest. The finding at I.A.R.I. revealed that yield loss due to shootfly is correlated with the reduction in number of effective tillers. Also that yield losses are significant only when shoofly infests more than 20% tiller in the crop.

A number of good sources of resistance to this pest have been identified through multilocation screening nurseries under AICWIP which includes UPD8, UP85042, HD2307, PDW 215, PDW 206, Raj 3190, Raj 3214, HUW 206, PBW 225, WH841, WH283, HI 8381, K 9025, PBW 299, HP 2189, Lok 1,PDW 232, K 9000, WH 147, HP 1731, PBW 175, PBW 377, PDW 243, WH 589, WH 610, WH 629, CPAN 2092, ISW 38, Hyb 65, UP 2377, UP 2378, UP 2379, WH 611, Raj 3826 and WH 627.

Mites, *Petrobia latens* Muller

Mite is becoming serious pest on wheat in recent years. Conventional acaricides like kelthane are not effective against the pest, but it is effectively controlled with foliar sprays of formation, phosphamidon, oxydemeton methyl or monocrotophos @ 160, 250, 160 and 200 ml a.i./ha, respectively. The treatment needs repletion after 15 days interval depending on the pest incidence (Bhatia *et al.,* 1974).

Deep ploughing during summers reduces the mite incidence to some extent. Field evaluations at Delhi, Jaipur and Ludhiana have led to identification of resistant lines like C 306, A9-30-1, DT18, HDR 132, HPW 42, WH 589, WH 610, WH 629, Hyb 65, ISW 38, CPAN 092, UP 2377, UP 2378 and UP 2379. It is interesting to note that C 306, the most prevalent rainfed wheat variety, has tolerance mechanism of resistance as it harbours considerable mite population yet yields well (Singh 1986 and 1998).

Army worms, cutworms, aphids and semiloopers

Sometimes armyworms outbreaks have been observed on wheat. Sowing time has been found to affect armyworm incidence. Mid-October to mid-November sowing escape pest attack while low incidence is evident in mid-November to early December sowings and further delay in sowing causes higher pest attack. Also higher nitrogen application and irrigation increases armyworm infestation. No resistance varieties are known but dwarf varieties having broad leaves, thick foliage provided more conductive environment. Chemical control of armyworm is important as often the time lag between initiation of infestation and crop devastation is very short due to voracious feeding habit of the pest. The present recommendations are foliar sprays with fenitrothion or dichloros @ 500 ml a.i./ha) or carbaryl (1 kg/ha) when the incidence reaches 1 caterpillar per m^2 area (Singh 1986 and 1998).

Three species of cutworm viz. *Cirphis loreyi* Dup., *Euxoa spinifera* Hub. and *Agrotis ypsilon* Rott and two species of semiloopers viz., *Plusia orichalcea* Fabr and *Plusia nigrisigna* Wlk. Infest wheat-crop. These polyphagous pests are distributed throughout the country and have gained importance as sporadic pests during post-green revolution period. Among these, cutworms are more important and their caterpillars, like those of armyworm, remain hiding under clods, crop debris and within soil crevices during day time and feed voraciously on leaves and cut shoots during night. The control measures recommended against armyworm are also effective against these lepidopterous pests (Singh 1986, 1996 and 1998).

Aphid, *Rhopalosiphum padi* usually appears during later part of the crop season starting from January till crop maturity, with its peak during heading stage of the crop. Delayed sowings and cloudy whether are conductive but winter rains are detrimental for aphid multiplication. Sucking of plant sap by adult and nymphs from leaves, stem and earheads results in retardation of plant growth and earheads develop shriveled and fewer grains. The insecticides recommended are oxydemeton methyl and minocrotophos. Foliar sprpays may be given when 11 aphids/tillers or 4 aphids per earhead are noticed. Varietal differences have been noticed but so far no resistant wheat lines have been identified in India (Singh 1998 and Vidya Sekhar and Singh 2000). Another aphid, *Rhopalosiphum rufiabdominalis* (Sabaki) infests roots of wheat and several other cereals under light soils. The pest is known as sporadic pest from Delhi, Madhya Pradesh, Utrtar Pradesh, Haryana, Punjab and Rajasthan, but infestation is more severe in Madhya Pradesh. This pest is active during January to March. Lindane applied with irrigation water is effective in checking the pest attack (Singh 1996 and 1998). Application of chlorpyriphos 20 EC @ 4 litres per hectare with irrigation is recommended in place of lindane.

Pink stem borer, *Sesamia inferens* Wlk.

It is found all over the country but is common on wheat in Madhya Pradesh, Rajasthan, Gujarat, Mahahrashthra and Karnataka. Of late its incidence in Punjab, Haryana, Uttar Pradesh and Bihar is also increasing. The caterpillars bore into stems causing characteristic 'White ears' having fewer and chaffy grains. The incidence occurs only during earing stage. At times up to 8% earheads have been reported damaged but usually it is less and noticed late, making its control difficult. Staggered sowing and uneven flowering causes more incidence in late flowering crop. Hand picking of infested tillers and their destruction reduces borer attack. Also removal and burning of stubbles after harvest is helpful in reducing its incidence in next crop (Singh , 1986, 1996 and 1998).

IPM IN RICE

Table 1 Insect pests scenario in different states of India

Insect pest	Intensity	States
Major Pests		
Yellow stem borer, *Scripophaga incertulas*	Severe	Andhra Praesh, Kerala, Maharashtra, Orissa, Punjab
	Moderate to severe	Haryana, Madhya Pradesh, Uttar Pradesh
	Low to moderate	Andaman and Nicobar Islands, Assam, Bihar, Gujarat, Himachal Pradesh, Jammu and Kashmir, Karnataka, Meghalaya, Manipur, Pondicherry, Tamil Nadu, West Bengal.
Leaf folder, *Cnaphalocrosis medianalis*	Moderate – severe	Andhra Pradesh, Haryana, Kerala, Maharashtra, Punjab, Tamil Nadu
	Light-moderate	Assam, Bihar, Goa, Gujarat, Himachal Pradesh, Jammu and Kashmir, Karnataka, Meghalaya, Manipur, Pondicherry, Tamil Nadu, West Bengal
Brown Planthopper, *Nilaparvata lugenms*	Severe	Andhra Pradesh, Kerala
	Moderate-severe	Karnataka, Madhya Pradesh, Meghalaya, Orissa, Pondicherry, Tamil Nadu
	Low-moderate	Assam, Bihar, Goa, Gujarat, Maharashtra, Punjab, Rajasthan, Uttar Pradesh, West Bengal
White backed plant-Hopper, *Sogatella furcifera*	Moderate-severe	Haryana, Madhya Pradesh, Punjab, Uttar Pradesh
	Low-moderate	Andhra Pradesh, Assam, Bihar, Gujarat, Himachal Pradesh, Jammu and Kashmir, Maharashtra, Orissa, Tamil Nadu, Uttar Pradesh, West Bengal
Green leaf hopper, *Nephotettix virescens*	Moderate-severe	Andhra Pradesh, Madhya Pradesh, Pondicherry, Tamil Nadu.
	Low-moderate	Assam, Bihar, Goa, Gujarat, Haryana, Himachal Pradesh, Kerala, Jammu and Kashmir, Maharashtra, Orissa, Tamil Nadu, Uttar Pradesh, West Bengal

Contd. ...

Table 1 Contd. ...

Gall midge, *Orseolia oryzae*	Severe Moderate-severe Low-moderate	Andhra Pradesh, Madhya Pradesh, Manipur, Orissa. Assam, Bihar, Goa, Karnataka, Kerala, Meghalaya. Goa, Maharashtra, Tamil Nadu, Uttar Pradesh, West Bengal
Minor pests		
Rice-hispa, *Dicladispa armigera*	Moderate-severe Low-moderate	Assam, Gujarat, Himachal Pradesh, Punjab, West Bengal Andhra Pradesh, Bihar, Haryana, Madhya Pradesh, Orissa, Meghalaya, Uttar Pradesh, West Bengal
Gundhi bug, *Leptocorisa acuta*	Moderate-severe Low-moderate	Andhra Pradesh (Telangana region), Uttar Pradesh Assam, Bihar, Gujarat, Haryana, Orissa, Tamil Nadu, Uttar Pradesh, West Bengal, Pondicherry.
Case worn, *Nymphula depunctalis*	Low-moderate	Andhra Pradesh, Goa, Karnataka, Kerala Madhya Pradesh, Maharashtra, Orissa, Tamil Nadu, Uttar Pradesh, West Bengal
Whorl maggot, *Hydrellia philippina*	Low-moderate	Andhra Pradesh, Assam, Bihar, Haryana, Karnatakka,. Goa, Madhya Pradesh, Maharashtra, Orissa, Tamil Nadu, Uttar Pradesh.

(*Source:* Garg *et al.,* 2004)

Table 2 Some multiple resistant varieties to rice pests

Variety	State in which released	Resistant against
Bhuban	Orissa	GM, BLB
Daya	Orissa	GM, BPH, GLH, BLB
Kunti	West Bengal	GM, BL
Lalat	Orissa	GM, BPH GLH, BL
Rasmi	Kerala	GM, BPH, BL
Samalei	Orissa, MP	GM, BPH, GLH, BL
Shaktiman	Orissa, West Bengal	GM, BPH, WBPH, BL
Suraksha	AP, Orissa, West Bengal	GM, BPH, WBPH, BL
Vikramarya	AP	GM GLH, RTD

(GM- Gall midge, BPH- Brown plant hopper, WBPH- White backed plant hopper, GLH- Green leaf hopper, RTD- Rice tungro disease, BLB- Bacterial leaf blight)
(*Source:* Garg *et al.,* 2004)

Table 3 Promising insecticides for management of insect pests

Insects	Insecticides
Green Leaf Hopper	1. Cartap 750 g a.i./ha* 2. Monocrotophos 400 g a.i./ha** 3. Ethofenprox 75 g a.i./ha** 4. Fipronil 50 g a.i./ha**
Brown Plant Hopper	1. Carbofuran 750 g a.i./ha* 2. Fipronil 75 g a.i./ha* 3. Phosphamidon 500 g a.i./ha**
Hispa	1. Carbofuran 750 g a.i./ha* 2. Fipronil 75 g a.i./ha*
Leaf Folder	1. Cartap 750 g a.i./ha* 2. Chlorpyriphos 500 g a.i./ha** 3. Fipronil 50 g a.i./ha** 4. Quinalphos 500 g a.i./ha**
Whorl Maggot	1. Carbofuran 750 g a.i./ha*
Gall Midge	1. Fipronil 75 g a.i./ha* 2. Phorate1250 g a.i./ha*
Stem Borer	1. Fipronil 75 g a.i./ha* 2. Phorate1250 g a.i./ha* 3. Cartap 750 g a.i./ha* 4. Chlorpyriphos 500 g a.i./ha**
Cut Worms	1. Monocrotophos 400 g a.i./ha**

*Granular formulations ** Spray formulations

Early spraying of insecticides should be avoided so as to conserve natural enemies. The need based and judicious use of insecticides should be undertaken. The releases of egg parasitoid, *Trichogramma chilonis* and *T. japonicum* are beneficial against the leaf folder and stem borer complex. The fungal pathogens *Beauveria bassiana* has been reported useful against rice hispa and BPH.

IPM IN MAIZE

With the presently available technology, maize (*Zea mays* Linn.) can be profitably grown in kharif, rabi and zaid season (spring and summer). In fact, the success of winter maize cultivation has become one promising area to contribute towards crop diversification, a necessity for sustainability of production systems in developing countries like India, where agricultural requirements also necessitates serious efforts in enhancing per unit production. Thus, maize is available at one stage or more stages throughout the year. This has added new dimensions to the pestilence front. About 220 insects and mite species have been reported on maize as pest from sowing to harvest. Of these the stalk borers, *Chilo partellus* Swinhoe) and *Sesamia inferens* Walker are predominant, while other pests including shootfly, *Atherigona* spp.; *thrips, Anaphothrips, sudanensis Trybom; climbing cutworm, Rhycia herculea* Corti Draudt; army worm, *Mythimna* spp.; sugarcane pyrilla, *Pyrilla perpusilla*

Walker; termites), *Microtermes obesi* Holmgren; cut worms, *Agrotis* spp.; have become serious during certain seasons/locations.

Table 4 Important field pests of maize and their classification

S. No.	Pest species	Status	Season	Location	Nature of damage
1.	*Chilo partellus*, Spotted stem borer (Crambidae Lepidoptera)	Major	Kharif	Across the country	Pinholes, shoot holes, Streads and midrib breakage in the young plants due to larval feeding and dead-hearts later
2.	*Sesamia inferen,* pink borer (Noctuidae Lepidoptera)	Major	Rabi	Peninsular India	-do-
3.	*Atherigona naqvi A. soccata* shoot flies (Anthomyiidae Diptera)	Major	Spring	A.P., U.P., Karnataka, Delhi, Mahrastra, Rajasthan	Dead of central shoot result in the dead-heart
4.	*Agrotis ipsilon,* greasy cutworm (Noctuidae: Lepidoptera)	Serious (localized)	Throughout the year	Bihar, Delhi, Haryana, M.P., H.P., Orissa, W.B., Southern states	Caterpillar feed on leaves at dusk and cut the tender stems of young and growing plants
5.	*Rhyacia herculea,* Climbing cutworm (Noctuidae Lepidoptera)	Serious (localized)	Rabi	North Bihar	Larvae cut the leaves from the top or from eges. Larvae feed only on the leaves.
6.	*Rhopalosiphum maydis* Corn aphid (Aphidadae Homoptera)	Minor	Round the year	Across the country	Nymphs and adults suck the sap of growing leaves. Leaves show yellow and brown patches.
8	*Mythima separata* Army worm (Crambidae Lepidoptera)	Serious (localized)	Round the year	Rajajsthan, Delhi, Assam, Bihar, M.P., Orissa, Punjab, U.P., Haryana	Infested plant show completely wilting, root become weak earthen galleries appears on the outer surface of stem and on all parts of plant including cob.

Source: (Panwar and Sharma 1998, Panwar *et al* 1995, 1996 and 1999)

Maize Stalk borers: It has been estimated that the average avoidable losses in yield of maize due to *C. partellus* and *S. inferens* ranged from 27.6 to 80.4 per cent and from 25.7 to 78.9 per cent, respectively. The larvae of *C. partellus* and *S. inferens,* after hatching feed on leaves where they form small, irregular holes and finally bore into the stem to form tunnel. As a result of such feeding the main shoot dies in young plants to form dead-heart and the grown up plants suffer in vigour. Both the borer attack all parts of plant except root.

Use of light traps for adult collection, ploughing after harvest and buying stubbles, use of cut fodder, increase in seed rate, sowing of maize in first week of July, roughing the infested plants and destruction of dead-hearts help in reducing the borer population to a great extent.

Spray endosulfan 35 EC, 0.1% on 10-15 days after sowing followed by granular application of 1% granules of lindane or 4% granules of endosulfan @ 15 kg/ha in plant whorls on 25 days old plants, if required. Alternatively, the spray of 0.05% lindane 20 EC in 15 days old crop has been recommended to control the borers.

Shootflies: Spring sown maize is heavily damaged by shootfly, *Atherigona* spp., in northern India. Three leaf stage is normally preferred by the adult flies for oviposition. The maggots enter the pant tissue at the ground level by puncturing it from lateral side. As a result of tissue penetration the growing leaves turn brown and the central axis dries to form dead-heart. The symptoms are apparent within 5-7 days after egg laying.

At the time of sowing in furrows apply phorate 10G or carbofurn 3G @ 1.0g/m row or quinalphos 1.5G@12.5 kg/ha.

Thrips: Maize grown during winter and spring seasons and also occasionally, during dry spell of kharif season is infested with thrips at seedling stage. Both nymphs and adults appear on leaves, such the sap of growing tips and gradually congregate within the leaf sheath to form colonies. The feeding on oozed out sap results into small silvery specks on the leaves. Severely damaged leaves show lacerations and the laminae form variable loops.

Flooding the fields results into natural resistance of plants to thrips. Carbofuran 3G @ 150g a.i/ha and 75g a.i/ha as seed dressing in sandy loam and clay soil, respectively give control up to 30 days after sowing or Endosulfan 35EC @ 0.1% spray in 10 days old crop followed by 4% granules @ 7.5 kg/ha15 days after the first application.

Climbing cutworm: The caterpillars attack on 20-30 days old plants during winter season in Bihar. They feed voraciously on the apical portions and the margin of the leaves, consuming all but the mid-rib. The losses caused by the pest vary from 18 to 34 per cent.

Two application of carbofuran 3G (0.5 kg a.i/ha) in 30 days old plant whorls followed by second application 20 days after the first or two spray of endosulfan @ 0.5% on 15 days and 35 days after sowing proved effective for the control of the pest.

Army worms: The caterpillars of *Mythimna* spp. cause heavy losses to maize at two stages of its growth: (1) immediately after germination up to development of nodes and internodes and (2) one month after sowing when spacious whorls are formed. Early infestation gives the appearance of a grazed crop while late infestation results in complete defoliation due to larval feeding from leaf edge towards the mid-rib. Later instars are usually gregarious.

The larvae excrete faecal matter in the form of pallets which are seen in the plant whorls. Spray the crop with endosulfan 35 EC @ 0.1 per cent.

Sugarcane pyrilla: The pest is prevalent in Uttar Pradesh, Punjab and Delhi. Both nymps and adults gather on the ventral surface to tender leaves along mid-rib or on stem. In severe outbreaks, these may be seen spread over all parts of the plants. The leaf sap is sucked by the insects and the infested leaves turn yellow, brown and wilted. Under severe infestation the photosynthetic activity is adversely affected.

Bagging the adult nymph or their collection by net and destroying them in kerosinised water. A number of natural enemies including *Ooencyrtus pyrillae* Crawford, *tetrastichus pyrillae* Crwford, *Epipyopes melanolenca* Fletcher, *Metarrhizium anisopliae* (Metsoh) and *Hirsutella* spp. kill the insect in different stages of its growth. Use of monocrotophos 40 EC, endosulfan 35 EC prove effective up to 5 days after application to restruct the population of *Pyrilla* infesting maize.

Termites. The termites remain quite active in sandy and sandy loam soils during dry season. Maize is attacked by *Microtermes obesi* Holmegren and Oden soon after germination and also at any stage of crop growth. The early infestation is evident by the presence of dead plants in the row. Such plants are easily pulled out along with the chewed roots. The grown up plants infested with termites show complete wilting, root loose strength and usually the plants fall over the ground. Earthen galleries appear on the outer surface of the infested stem and the inner hollow space of the stem is also filled with the soil.

Plough the field after harvesting the crop to expose the hidden pupae and larvae to surface. Use of carbofuran 3G (25 kg/ha) at the time of sowing will be improve.

OVERALL IPM STRATEGY IN MAIZE

(Panwar and Sharma, 1998; Sharma and Sharma, 2001 and Sharma and Panwar 1998)

1. Grow only recommended hybrids and composites, which are relatively less susceptible to insects, particularly the borers.
2. Early sown maize crop during kharif season is severely attacked by borers. Thus, in maize should be sown with first rain to avoid borer attack.
3. Crop rotation with non-host crops reduces the pest infestation by interrupting the continuity of food chain, available for insects.
4. It has been observed that plant population plays an important role in reducing the insect infestation. Insect incidence increases with the increase in plant density. Recommended plant population should be maintained in the fields to reduce the insect incidence.
5. In areas where termite is a problem, dip maize seeds in 0.1 per cent or 20 ml/kg seed chlropyrifos solution or apply lindane dust (a.i. 1.3) @ 25-30 kg/ha before sowing. Destroy termitaria in the vicinity of fields on community basis at village level.
6. In areas where seed maggot, wire worms, white grubs, termites, cutworm and field crickets are usually prevalent every year, apply and mix with soil to a depth of 12 cm, phorate 10 per cent granules @ 10-15 kg/ha at the time of spreading of basic manure. Spray quinalphos 25 EC, 0.05 per cent on the standing crop.

7. Use high seed rate and removal and destruction of infested plants and dead-hearts has been found to be successful practice in reducing the pest incidence.
8. Place phorate 10 per cent granules @ 10-15 kg/ha below seeds or apply carbofuran 3G around the shootfly infested plants.
9. Spray endosulfan 35 EC (0.1 per cent) or lindane 20 EC (0.05 per cent), 10-15 days after germination, followed by granular application of 1 per cent granules of lindane or 4 per cent dust of endosulfan @ 15 kg/ha in plant whorls a fortnight later, if required for the control of borers.
10. Light traps should be used to monitor the field population of insects. The trapped moths can be killed in kerosene water to reduce the insect population in the field.
11. Efforts should be made to release the known parasites and predators of borers or other insects and the use of insecticides should be restricted to protect the biotic agents. The egg parasite *Trichogramma* spp. can be utilized for suppression of stem borers. The egg parasite @ 1.0 lacs per hectare should be released on 7 days old crop followed by second application on 12 days old crop to minimize the population of stem borers. Many species of spiders have been reported to feed on eggs and early larval stages of stem borers in plant whorls. The adults of coccinellid beetles, *Menochilus sexmaculata* Fab.,. and *Coccinella septumpunctata* L. are predacious on the freshly emerged larvae of stem borer.

Removal and destruction of stubbles after harvest to reduce pest populations, destruction of crop residue and other alternate host plants, and chopping of stem harbouring diapausing larvae could be very effective in reducing borer population.

IPM IN SORGHUM

Insect pests greatly affect the quality and productivity of sorghum. Their pest problems are almost identical. The infestation of various pests varies in time and space. Whatever the magnitude of damage, it is the loss in yield of grain and fodder resulting through restricted flow of vital nutrients drained by borers. It has direct bearing on the per unit area production per unit time and *in vitro* dry matter digestibility per cent. Also, shoot fly is becoming a problem in single and multicut management system. The second cut in multicut genotypes generally falls during late July, which is the most favourable period of activity of shoot fly in the northern tract. Of late, the changes in cropping pattern has led to the emergence of *Pyrilla* and grey weevil as serious pests of sorghum. These pests need very careful watch as some of the promising genotypes entered in coordinated trials for initial and advanced trials have shown vulnerability to these pests. Further, it is needless to emphasize that for grain yield, the management of earhead pest like midge, head bug and head worms is essential (Table 5). Obviously, the felt need is to provide viable options to contain insect pests for evolving an effective management strategy based on eco-friendly control methods (Prem Kishore, 1987 and Prem Kishore and Gahukar, 1992).

Table 5 Nature of damage of major insect pests of sorghum along with their economic thresholds

Insect Pest	Nature of damage/ incidence/ extent of losses	Economic thresholds (as reported by various workers
Atherigona soccata Rond. (sorghum shoot fly)	Larvae cut the growing point resulting in the production of dead-hearts. 22.4-39.5 kg/ha grain yield loss. Causes an average loss of 5%. Shoot fly incidence was > 30% at several locations over the past one decade.	4-10, 3-9 and 6-15% shoot fly infestation in CSH-1, CSH-5, and Swarna, respectively (Rai *et al.*, 1978) 1 % increase in infestation reduced grain yield by 89.10 kg/ha in CSH-5 and 30.49 kg/ha in CSH 8R (Mote, 1986a) 1 % increase in dead hearts reduced the grain and fodder yield by 55.2 and 42.3 kg/ha, respectively (Mote, 1983)
Chilo partellus Swin. (spotted stem borer)	Larvae feed on leaves, damage the growing point resulting in dead heart formation and also tunnel the stem. 55-83% grain loss in north India. 28% loss in forage yield.	1 % increase in stem tunnelling reduced the grain yield by 1.01, 0.39 and 0.62 g/plant in CSH 8R, Spy 86 and M 35-1, respectively (Mote, 1986b). A unit increase in stem tunnelling resulted in 0.59 unit loss in grain yield. A unit increase in leaf injury reduced the yield by 0.02 units. (Prem Kishore, 1990).
Stenodiplosis sorghicola Coq. (sorghum midge)	Larvae feed on developing grain. 20-40% grain yield loss.	0.67-2.2 midges/panicle on CSH-5 (Mote *et al.*, 1991) 0.1-0.2 midge/panicle on ICSV I, ICSV 112, and CSH-I (Sharma *et al.*, 1993).
Calocoris angustatus Leth. (earhead bug)	Nymphs and adults feed on developing grain which shrivel and become red brown. 5-84% yield loss. females ovipositing for 3 days, caused a loss of 31.5 g/panicle. 55-88% loss in commercial cultivars CSH-I, CSV-I and CSH-5.	7.9-19.1 nymphs, 5.4-10.5 adults or 0.06-0.12 ovipositing adults/panicle (Natarajan and Sundra Babu, 1988). 1.3-1.4 bugs/panicle on CSH-I, 0.4 on ICSV I, and 0.2 on CSH-5 at flowering stage (Sharma and Lopez, 1989).

VIABLE OPTIONS FOR FORMULATING PEST MANAGEMENT STRATEGY

Field sanitation and cultural practices: Fields should be ploughed immediately after crop harvest and shortly before sowing. Crop residues should be collected and destroyed before the onset of monsoon to reduce the carryover of pests from one season to another. Based on the population dynamics of the major pests, sowing dates need to be adjusted to keep the pest population under check. Early and uniform sowing of the same cultivar over large areas reduced the damage by shoot fly, midge, and head bugs. Cultural operations, *viz.,* high seed rate, balanced fertilizer application, field sanitation, weeding, and intercropping with legumes reduced the damage by shootfly, stem borer, armyworm, and sorghum midge.

Host plant resistance : The use of pest resistance cultivars should form the backbone of the pest management programme in sorghum. Efforts to develop, release and popularise varieties that are resistant to key pests should be intensified. P 311, SPV 1015, M-35-1, Swati, and CSV 14R are less susceptible to shoot fly, and suitable for cultivation in the rainy and post rainy seasons, where shoot fly is a major problem (Prem Kishore, 1996a). ICSV 197, ICSV 88032 are highly resistant to sorghum midge, and their yield potential is comparable to commercial cultivars. They can also be utilized as a component of IPM in sorghum. Cultivars having loose panicles such as ICSV 88032 may be grown in areas endemic to head bugs and caterpillars. Many new sources of resistance have been identified against major insect pests of sorghum and the promising lines are presented in Table 6.

Table 6 Location of sources of pest resistance in various sorghum germplasms

Pest	Resistant/multiple resistant germplasm
Varieties	
Shoot fly	SPV 1413, SPV 1450, SPV 1467, Sel. 3
Stem borer	SPV 1155, SPV 1359, SPV 1466
Delphacids	SPV 1413, SPV 1155, SPV 1359, SPV 1451, SPV 1461, SPV 1462, SPV 1463, SPV 1464
Aphids	SPV 1450, SPV 1359, SPV 1155, SPV 1380, SPV 1413. SPV 1452, SPV 1453, SPV 1462, SPV 1465, Sel. 3
Hybrids	
Shoot fly	SPH 1174, SPH 1077, SPH 1166
Stem borer	SPH 1162, SPH 1078, SPH 1079, SPH 1165
Delphacids	SPH 733, SPH 1010, SPH 1026, SPH 1075, SPH 1159, SPH 1164, SPH 1168, SPH 1171
Aphids	SPH 1162, SPH 1174, SPH 1020, SPH 1026, SPH 1079, SPH 1161
Parents	
Shoot fly	RR 9807
Stem borer	RR 9807, SPV 570, SPV 839, RR 9808
Delphacids	SPV 570,117 A, 2219 B
Aphids	RS 29, R 354, 296 B

Chemical control: On the basis of economic thresholds, control schedules should be formulated to restrict use of insecticides. This will facilitate implementation of IPM more effectively. In areas where shoot fly is known to be a serious problem or the crop is sown late, it is desirable that freshly treated sorghum seed with carbofuran (5 parts active ingredient per 100 parts of seed) may be used. The cost of seed treatment is reduced substantially i.e., calculated as 1/3 of the cost of granules of the same insecticide. However, if sowing is done by dibbling, spot application with carbofuran granules may be done at a reduced rate (1.0 kg a.i.lha). Stem borer can be effectively controlled by the application of endosulfan (4%), carbaryl (5%) or lindane (1.3%) dust twice during the crop growth. The first application @ 5.0 kg/ha should be given 25 days after germination of the crop and the second @ 7.5 kg/ha, 10 days after the first application. Application of dust formulation should be directed towards the 'leaf whorls'. For controlling the earhead pests, *viz.,* head bug, spray the earheads at the time of 90% panicle emergence (50% flowering) with endosulfan (35EC) @ 1 litre in 500-600 litres of water or dust the earheads with endosulfan (4%) or fenvalerate (0.4%) @ 12-15 kg/ha (Prem Kishore, 1987 and Prem Kishore and Gahukar, 1992).

Possibility of including insecticides of plant origin such as those obtained from *Azadirachta indica (neem), Annona squamosa, Acrorus calamus, Catharanthus roseus, Bluemea eriantha,* etc., should be explored in IPM.

Biocontrol: Several natural enemies have been recorded on shoot fly eggs. *Trichogrammatoidea bactrae* Nagaraja and *T. simmondsi* West; *Trichogramma evanescens* Sch., *T. kalkae* Sch. & Feij, *T. chilonis* Ishii and *T. australicum* Gir. have been reported to parasitize eggs of sorghum shoot fly (Zongo, 1995; Rao *et al.,* 1987 and Taley and Thakare, 1979). Among the shoot fly larval parasitoids, *Aprostocetus* spp. and *Neotrichoporoides (Tetrastichus) nyemitawus* Rohwer are the most important. Parasitism due to *Aprostocetus* spp. reached 15% in Sep-tember 1975 and 35% in August 1977 (Prem Kishore, 1987),.

Various natural enemies, *viz., Trichogramma chilonis* Ishii, *Bracon chinensis* Szepl. and *Cotesiaflavipes* Cam. for stem borer; NPV, and *Steinernema feltiae* Filipjev for armyworm; *Aprostocetus diplosidis* Crawford and *Orius maxidentex* Ghauri for midge and write, *T. chilonis* Ishii and *Campoletis chlorideae* Uchida for head caterpillars may be effectively utilized (Prem Kishore, 1987).

Bioinoculants: Seed inoculation with *Azospirillum* significantly reduced the formation of dead-hearts due to shoot fly. Also, *Azotobacter* with fermented residue to sorghum was effective in reducing shoot fly dead-hearts. New promising variety PC-121 responded well to *Azospirillum* and *Azotobactor* (Prem Kishore, 1998) and showed significantly low shoot fly dead-hearts.

Breeding for resistance: After taking the overall view and stock of situation about breeding for resistance to major insect pests of sorghum in the country, it can be said that temperate x tropical crosses provided the basis for combining resistance in more productive backgrounds. However, the derivatives obtained from these crosses showed initial promise but later lost the desired level of resistance. Low levels of resistance, lack of persistence on stability of resistance over the years and low inheritance of resistance resulted in the lack luster performance within 2 to 3 years after the release of most of the released sorghum varieties developed from such crosses. Moreover, the variability amongst the most widely used sources of resistance, *viz.,* IS 1054 (Maldandi or M 35-1), IS 1055 (BP 53), IS 1122 (Karad local) and IS 1151 (Aispuri) was limited besides quantitative nature of resistance to different insect pests made the breeding task difficult.

Evidently, such a situation involving varietal approach to increase production and combat insect attack was an exercise in futility. Thus, the focus shifted to hybrid development with the sole aim to exploit hybrid vigour, but there too situation was not very encouraging, except a few hybrids such as CSH 9, CSH 12R, etc. showing some promise. Others failed badly. From the, entomological point of view, the level of resistance transferred in these genotypes was unable to cope up the intensity of insect attack, specially when the plant is attacked by a number of key pests at different growth stages. The situation became even more worse due to overlapping generations of different pests. These overlapping generations may have different biological relationship with the plant. In addition, resistant x resistant, and resistant x susceptible crosses did not exhibit an improvement over parents indicating no diversity among resistant lines. All these shortcomings indicated the futility of single cross (temperate x tropical) approach and needed to adopt different strategy to overcome this intricate problem.

A new strategy was thus evolved, where intermediate derivative was crossed with another intermediate derivative or resistant x intermediate crosses were made with the objective of pooling the genes for obtaining sorghum varieties and hybrids resistant to various pests. Intermediate derivatives represent high yielding derivatives obtained from resistant x susceptible crosses. This approach was found better than the single cross approach. Selections based on least injury by different pests and quality parameters were practiced from F_2 to F_6 generations. This also showed gradual improvement in the levels of resistance to different pests. Initially, this strategy involved two major pests, *viz.,* shoot fly and stem borer.

By adopting this strategy, nine new sources of resistance to stem borer, *Chilo parte/ lus, viz.,* P-84, P-217, P-291, P-296, P-297, P-467, P-471 and P-500, were evolved. Such germplasms performed well under both the natural and artificial conditions of infes-tation. Three cultivars, *viz.,* P-217, P-297 and P-500 were outstanding under artificial infes-tation conditions during kharif i.e., rainy season.

OVERALL IPM STRATEGY IN SORGHUM

Insect pests, particularly the stem borer, *Chilo partellus* (Swinhoe) and the shoot fly species greatly affect the quality and productivity of sorghum. Their pest problems are almost identical. Irrespective of the magnitude of damage, the loss in yield of grain and fodder resulting through restricted flow of essential nutrients drained by the pests is of vital concern. Because of the adaptability of sorghum to drier and low fertility conditions, both these crops form an important component of food security in the country. Obviously, there is a great need to provide viable options to contain insect pests and do away with insecticides due to their adverse affects to humans and environment and prohibitory costs. The identification of multiple pest resistant sources, *viz.*, CSH 9, CSV 14 R, SPV 1015, ICSV 197, ICSV 88032, P 311, SPV 1413, SPV 1450 etc. have to a certain extent eased the pest situation. Growing of such cultivars or other pest tolerant sources soon after the onset of monsoon with increased seed rate by 20 to 30 per cent and subsequently removing the dead-hearts due to shoot fly/ stem borer and maintaining good plant stand would result in good quality yield and fodder. Use of biopesticides like *neem;* natural enemies, *viz.*, *Trichogramma* spp. *Aprostocetus* spp., *Cotesia flavipes,* N.P.V. and *Bacillus thuringiensis,* and application of bioinoculants like *Az.ospirillum* and *Azotobactor* as seed treatment are newly emerging trends being utilized in integrated manner for suppressing insect pests of sorghum and pearl millet. Breeding for resistance to major pests of sorghum has shown great promise. This seed borne technology has great relevance to these subsistence crops.

IPM IN PEARL MILLET

Pearl millet *(Pennisetum typhoides)* is the fourth important food crop after rice, wheat and sorghum, grown in varying ecological environment of low to high productivity levels. Because of its adaptability to drier and low fertility conditions, pearl millet has an edge over other coarse cereals in these environments and, thus, is an important component of food security in the country. Pearl millet is not only a staple food of millions of rural people inhabiting in these regions, but is also valued as a source of fodder to cattle population.

Recent studies have shown that the serious economic losses by potential pests (Table 7) have been overlooked because the pearl millet grower can hardly afford any input on a crop of low economic return. Evidently, pearl millet is an ideal crop for implementing the concept of pest management more effectively (Prem Kishore, 1996b). Implementation of insect pest management programme needs basic information on various aspects like biology, bionomics, economic thresholds and data on applied ecology. Very few of the various control methods have been tested under actual field conditions against the major pests. The current research programmes under the All-India Coordinated Pearl Millet Improvement Project are oriented to collect data on all these aspects. However, till such information is made available there exists sufficient knowledge for initiating control measures against a few major pests of pearl millet based cropping system which will entirely obviate the use of insecticides.

HOST PLANT RESISTANCE APPROACH IN PEARL MILLET

The most practicable long term solution for insect management in pearl millet lies in developing high yielding resistant varieties and hybrids. Thus, the identification of sources of resistance to major insect pests has to be another major strategy for future. It will provide materials for the breeders to breed resistant varieties and hybrids (Prem Kishore, 1995 and 1997).

Table 7 Potential insect pest of pearl millet based cropping system

Pest	Remarks
Atherigona approximata Malloch	The fly maggots feed on the seedlings and produce dead-heart. Sometimes, the shoot is not killed because of quick growth of the plant. The fly also causes damage to earheads in later stages of crop growth.
White grubs, *Hololrichia* spp. *Anomala* spp.	The grubs feed on roots in the soil. Most severe injury has been reported from Rajasthan and Gujarat on light soils.
Cotton grey weevil, *Myllocerus* spp.	Adults cause severe damage to leaves leaving only the midribs. Beetle grubs feed on the roots. In some cases, the damage due to grubs in seedling stage becomes very serious.
Grasshoppers, *Hieroglyphu snigroreplelus* (Bol.), *Colemania sphenarioides* FJol.	These insects cause severe defoliation of the crop. The first species occurs throughout the country while the remaining is common in southern states.
Hairy caterpillars, *Esligmene lactinea* Cram (Black), *Amsacta moorei* Butl. (Red)	These caterpillars frequently cause severe defoliation of the crop.
Armyworms, *Mythmina separala* (Walk.) *Leucania loreyi* (Dup.)	These caterpillars are nocturnal in habit. They defoliate the crop and occur throughout the country.
Cutworms, *Agrotis ypsilon* (Hufn.), *A. flammalra* R.	These caterpillars cut the seedlings at the soil level.
Stem borer, *Chilo partellus* (Swinhoe), *Sesamia inferens* Walk.	Borer larvae feed in plant whorls and make tunnel in stem.
Leaf roller, *Marasmia trapezalis* Guen	Larvae feed on the tips of leaves causing leaf rolling.
Blister beetles, *Myllabris pustulata* Thunb., *Lytta lenuicollis* Cast.	Adults feed on pollen and flowers of *bajra* earheads. Most severe in Gujarat and southern states.
Aphid, *Aphis ufoabdoninalla,*	Both species suck sap from the plant whorls.

Contd. ...

Table 7 Contd. ...

Rhopalaaiphum maidis, Fitch	
Earhead caterpillars *Eublemma sillicula* Swinh., *Heliothis armigera* Hb.	These caterpillars feed on maturing grain in the earheads.
Sugarcane leaf hopper, *Pyrilla perpusilla* Walk.	Adults and nymphs both suck the sap from tender leaves.
Midge, *Geromyia penniseti* Felt	Midge larvae destroy seeds developing in panicles. Severe in southern states.
Shoot bug, *Peregrinus maidis* (Ashm.)	Both adults and nymphs suck sap from plant tissues in leaves and leaf whorls.

Some of the promising sources of resistance selected for different pests were identified for shoot fly, resistant cultivars included MP-16, MP.-19, MP-53, MP-67, MH-49, MH-52, MH-9, MH-82, MH-99 and MH-I05. For stem borer, resistant sources were located in MP-19, MP-21, MP-31, MP-39, MP-47, MP-53, MP-56, MP-60, MP-63, MH-31, MH-91, lCMS-7703, lCMS-7704, ICH-241, WCC-75 and IVPS-77. Other resistant entries comprised 29 MO, RSK, and 268 for earhead caterpillar; 26 J, 98, 103 and 107 TD for shoot bug; 26 0; 29 MD and 146 for leaf roller; 36 D, IP Nos. 44, 79, 214, 263, 1266, 1301, 1307, 1345, 1395, 1402 and 2708 for *Pyrilla;* 23A x KS30, RSK Nos. 1086,213,315,1826,833 and 23A x K 230 for whitegrubs; MP-38, GHB-45, MH-85 and ICH-241 for midge; and MP-17, MP-38, MH-34, MH-36, MH-62, MH-82, MH-88, MH-90, MH-93 and MH-94 for grey weevil (Prem Kishore, 1989 and 1993).

OVERALL IPM STRATEGY IN PEARL MILLET

Shoot fly: The damage by shoot fly is severe on late sown kharif crops. Hence, it is recommended that sowing should commence after the onset of monsoon or latest within 10-15 days after the first monsoon shower. Staggered sowing in the same area is to be avoided since it results in the buildup of shot fly population. Also, the increase in seed tate to 3.3 kg and later resorting to removing or destroying affected seedlings would help in checking the shoot fly damage. In case of heavy attack in endemic areas, spraying the crop with 0.07% endosulfan can be undertaken. In areas confronting water scarcity, dusting the crop with 4% endosulfan dust @ 15 kg/ha may be used.

White grubs: In the endemic areas, immediately after the first shower, collection and destruction of adult beetles visiting trees like *neem* or *Acacia* for mating is advised. Spraying with 0.05% chlorpyriphos or 0.2% carbaryl on the host trees, with the onset of monsoon can also be done, However, all the host trees should be sprayed within 2 to 3 days after the receipt of the rain. Application of FYM and lindane (2:3) @ 100 kg/ha in furrows at the time of sowing is quite effective to control white grubs damage. Also, application of carbofuran @ 25 kg/ha before sowing would provide the desired control.

Armyworm, hairy caterpillars, grasshoppers, Pyrilla, grey weevil and stem borer: Clean cultivation along with deep ploughing after harvesting of the crop would expose the pupae of armyworms and hairy caterpillars and egg-pods of grasshoppers in soil to weathering agencies and bird predators. This will help in reducing infestation of these pests in the next season. Scraping and dusting the bunds with 1.3% lindane immediately after the outbreak of the monsoon is advised since initial infestation of the grasshopper is restricted to grasses on the bunds. Further, dusting the crop with 1.3% lindane or 4% endosulfan @ 25 kg/ha or spraying with 0.05% endosulfan is also effective in checking these pests.

REFERENCES

Garg, D.K., Jeswani, M.D. and Pathak, Mahesh. 2004. IPM in rice ecosystem. In: Advances in plant protection sciences. Eds., D. Prasad and America Singh, Akansha Publishing House, New Delhi- 110059.pp 404-419.

Kumar, P., Sharma, R.K., Midha, R.K., Siwach, P, Raghav, D.K. and Sharma, A. 2003. IPM in chickpea and maize in Punjab. *In: Proc. National Symposium on Frontier Areas of Entomological Research.*, November 5-7, 2003, Division of Entomology, Indian Agricultural Research Institute, New Delhi 110 012. Extended Abs.: pp- 250-251.

Kumar, P., Sharma. R.K. ,Siwach, P, Raghav, D. and Sharma, A. 2004. Validation of IPM in Maize in Punjab. *In: National Symposium on Biotechnological Approaches for the Management of Plant Diseases,* January 30-31, 2004, Agrawal

Mote, U.N. 1983. Relation between shoot fly damage and sorghum yields during rainy season. *Indian J. Pl. Prot.,* 11: 145-147.

Mote, U.N. 1986b. Correlation between degree of damage due to stem borer, *Chilo partellus* Swin. and the yield of sorghum grain. *Indian 1. Ent.,* 48: 357-358.

Mote, U.N. and Jadhav, S.S. 1990. Incidence and losses caused by sorghum head bugs. *1. Maharashtra agric. Univ.,* 15: 121-122.

Mote, U.N., Ghule, B.D., and Jadhav, S.S. 1991. Studies on economic threshold level of sorghum midge on sorghum earheads. *1. Maharashtra agric. Univ.,* 16: 349-352.

Mote,. U.N. 1986a. Correlation between the deadhearts caused by shoot fly, *Atherigona soccata* Rond. and the yield of sorghum hybrids. *Indian 1. Ent.,* 48: 356-357.

Natarajan, N. and Sundara Babu, P.C. 1988. Ecmomic injury level of sorghum earhead bug, *Calocoris angustatus* Leth. in southern India. *Insect Sci. Applic.,* 9: 95-98. P.G. College, Jaipur, Rajasthan. Abs.: pp 49-50.

Panwar V.P.S. and Sharma, R.K. 1998. Evaluation of maize germplasms for locating multiple resistant sources to tissue borers, *Chilo partellus* (Swinhoe) and *Atherigona* spp., *Indian J. Ent.* 60 (2) : 171-176.

Panwar, V. P. S., Sharma, R.K. and Singh, N.N. 1995. Maize stalk borer, *Chilo partellus*- a serious threat to maize cultivation in Asia., F. A. O., *Tropical Asian Maize Network Newsletter*. 2 (1) : 7.

Panwar, V. P. S., Singh N. N. and Sharma, R. K. 1996. The effect of cropping system on the incidence of *Chilo partellus* (Swinhoe) in maize. *Abst., 2nd International Crop Science Congress* held in New Delhi from 17-24 November, 1996., p-77

Panwar, V.P.S., Singh, N.N. and Sharma, R.K. 1999. Impact of previousseason farming practices and fertilization regimes on *Chilo partellus* (Swinhoe) infestation in two high yielding maize cultivars. *Ann. Agric. Res.,* 20(2): 220-226

Prem Kishore and Gahukar, R.T. 1992. Operational integrated pest management strategy involving cultural cum host plant resistance in sorghum. *J. ent. Res.* 16(2) : 155-162.

Prem Kishore, 1987. Key pests of sorghum, pearl millet and smaller millets and their management, pp. 243-259.In Y.K. Mathur, A.K. Bhatnagar', N.D. Pandey and J.P. Srivastava, [eds.], *Recent Advances in Entomology,* Gopala Prakashan, Parade, Kanpur (India).

Prem Kishore, 1989. Insect plant relationship studies in pearl millet: Resistance to shoot fly, stem borer and grey weevil. *J. ent. Res.,* 13(I) : 64-66.

Prem Kishore, 1990. Strategies for pest management in sorghum in India. *J. ent. Res.,* 14: 16-20.

Prem Kishore, 1992. Evaluation of twin resistance sources amongst advance generation derivatives in sorghum to shootfly, *Atherigona soccata* Rondani and stem borer, *Chilo partellus* (Swinhoe). *J. ent. Res.,* 16(3): 236-241.

Prem Kishore, 1993. Insect resistance in newly developed hybrids and populations of pearl millet. *J. ent. Res.,* 17(4) : 215-218.

Prem Kishore, 1995. Search for new sources of resistance in newly developed genotypes of pearl millet. *J. ent. Res.,* 19(2) : 187-190.

Prem Kishore, 1996a. A new approach to develop shootfly and stem borer resistant cultivars for insect pest management in sorghum through crossing of intermediate in resistant derivatives. *J. ent. Res.,* 20(2): 173-175.

Prem Kishore, 1996b. Evolving management strategies for pests of millets in India. *J. ent. Res.,* 20(4) : 287-297.

Prem Kishore, 1997. Selection of resistance sources in breeding material of pearl millet against shootfly, *Atherigona approximata* Malloch. *J ent. Res.,* 21(4) : 381-383.

Prem Kishore, 1998. Response of sorghum variety Pusa Chari-121 to carrier based irioculants *(Azotobacter* and *Azospirillum),* fermented residue and shootfly under filed conditions. *J. ent. Res.* 22(2) : 101-105.

Rai, S., Jotwani, M.G. and Jha, D. 1978. Economic injury level for sorghum shoot fly. *Atherigona soccata* Rond. on sorghum. *Indian 1. Ent.;* 40: 126-133.

Rao, KJ., Thontadarya, T.S. and Suhas, Y. 1987. *Trichogrammatoidea bactrae* Nagaraja - A new egg parasitoid of sorghum shoot fly, *Altherigona soccata* Rond. *Curr Sci.* 56: 283.

Sekhar, J.C., Sharma, R.K., Venkatesh, S. and Reddy, M.L.K. 2003. Avoidable losses due to *Spodoptera exigua* in rabi maize. *In*: *Proc. National Symposium on Frontier Areas of Entomological Research.*, November 5-7, 2003, Division of Entomology, Indian Agricultural Research Institute, New Delhi 110 012. Extended Abs.: pp- 36

Sekhar, J.C., Venkatesh, S., Sharma, R.K. and Reddy, M.L.K. 2003. *Spodoptera exigua* (Hubner) as a pest of rabi sweet corn. *Indian J. Ent.* 65 (1) : 581- 583.

Sharma, H.C. and Lopez, V.F. 1989. Assessment of avoidable losses and economic injury level for the sorghum head bug, *Calocoris angustatus* Leth. (Hemiptera: Miridae). *Crop Prot.,* 8: 429-435.

Sharma, H.C., Vidyasagar, P. and Nwanze, K.F., 1993. Effect of host plant resistance on economic injury level for sorghum midge, *Contarinia sorghicola. Int. J. Pest Mgt.* 39: 435-444.

Sharma, R.K. and Panwar V.P.S. 1998. Search of source for resistance amongst maize germplasms against *Atherogona soccata* and *A. naqvii* during spring season. *Ann. Agric. Res.*, 19(3): 359-360.

Sharma, R.K. and Sharma, K. 2001. Effect of seed treatment with different insecticides on damage and yield of maize by termites. *In: National Symposium on Soil Arthropods and their Management.,* September 27-28, 2001, Agricultural Research Station, Durgapura, Jaipur, Rajasthan. *Abst*, p 46.

Sharma, R.K., Sharma, Kirti and Shekhar, J.C. 2002. Evaluation of plant protectants on damage and yield of rainfed maize by termites. *Pesticide Res. J.,* 15: 36-39

Singh, V.S. 1986. Management of insect and mite pests, In: Twenty five years of coordinated Wheat Research (1960-1986) Eds: Tandon, J.P. and Sethi, A.P., Wheat Project Directorate, IARI, New Delhi: 158-188.

Singh, V.S. 1996. Pest Management in cereal crops, In: *Recent Advances in Indian Entomology*. Ed: O.P. Lal, A.P.C. Publication, New Delhi: 311-316.

Singh, V.S. 1998. Pest Management in Wheat, Indian Farming, Special issue- India CIMMYT Partnership. 48 (1) 47-50 and 59-63.

Taley, Y.M. and Thakare, K.R. 1979. Biology of seven new hymenopterous parasitoids of *Atherigona soccata* Rond. *Indian J. agric.* Sci.. 49: 344-354.

Zongo, J.O. 1995. Biological control of the shoot fly. *Atherigona soccata* :71 Yrs. of Research. *In Proc.* X *Mngmt., & Scient. Conf. African Assoc. insect Scientists. 5-10 Sept.* 1993. *Mombasa. Kenya.,* 1995: 31-36.

Entomology : Novel Approaches, 2007
Eds.: *P.C. Jain and M.C. Bhargava,*(Page 267-286)
New India Publishing Agency, New Delhi
E-mail: newindiapublishingagency@gmail.com
web: www.bookfactoryindia.com

CHAPTER 14

INTEGRATED PEST MANAGEMENT IN RICE

H. P. Misra and B. C. Jena
Department of Entomology, College of Agriculture, Bhubaneswar – 751 003, Orissa

INTRODUCTION

Rice is the staple food of about 50% of the world's population that resides in Asia, where 90 % of the world's rice is grown and consumed. In the International year of Rice celebrated in 2004 the theme was "Rice is life" which reflects the importance of rice as a primary food source of human beings. In the world, annually rice is cultivated in about 151.54 million hectares with an annual production of 593 metric tonnes and average productivity of 3.91 tonnes/ha (Rai, 2004). From its Asian homeland, it is now cultivated in 113 countries and plays a variety of roles related to important aspects of food security as well as rural and economic development. In Asia this crop is grown in 136.07 million ha followed by 7.67 million ha in Africa and 5.09 million ha in Latin America with a production of 539.84, 16.97 and 19.54 metric tonnes and productivity of 2.97, 2.21 and 2.84 tonnes, respectively (Rai, 2004). In India, rice is the staple food for more than 65 % people (Pasalu and Katti, 2004). Rice is grown under various geographical, climatic and cultural conditions. The current level of rice production in the country is around 90 million tonnes. Based on the present rate of population growth @ 1.5 % and per capita consumption of 250 g of rice/day in the country, the demand for rice production is expected to be about 100-105 million tonnes by 2010 and 130-140 million tonnes by 2025. To achieve this target, it requires increasing productivity per unit area by the adoption of modern and intensive agricultural practices, such as use of high yielding varieties and improved cultural and management practices. Experience of post green revolution period in last four decades indicates that these practices often aggravate other biotic constraints such as insects, diseases and weeds. However, insect pests are considered to be the major limiting factor in enhancing the productivity of rice.

INSECT PEST SCENARIO IN RICE

More than 200 species of insect pests attack the rice crop in India, but about 20 insect pests cause severe damage in rice (Heinrichs, 1994). Cultivation of high yielding rice cultivars in mid sixties with the onset of green revolution has necessitated incorporation of high input like excess use of chemical fertilizers (particularly nitrogen) in rice cultivation that has altered the microclimate and accentuated the insect pest problem. Number of insect pests

(Table 1) which were considered important have risen from 3 to 13 during 1965-1995 (Krishnaiah *et al*., 1999).

Table 1 Changing scenario of insect pests in rice

Stem borer(SB) Gall midge (GM) Green leafhopper (GLH)	Brown Plant hopper (BPH) Stem borer, Gall midge, GLH, White backed plant hopper (WBPH)	BPH stem borer, Gall midge (3 biotypes), GLH, WBPH, Cutworm, Gundhibug, Hispa	Stem borer, Gall midge (Biotype 4), Leaf folder (LF), BPH, WBPH, GLH, Cut worm, Gundhibug, Hispa	Stem borer, Gall midge, Leaf folder, GLH,BPH, WBPH, Cutwork, Gundhi bug, Hispa, Thrips, Mites	Stem borer, Gall midge, (Biotype 5), Leaf folder, BPH, GLH, WBPH, Cut worm, Gundhibug, Hispa, Thrips, Mites, Root weevil, Black bug
1965	1975	1980	1985	1990	1995

Stem borer are the major pests infesting rice. Out of five species of stem borers, *viz.*, yellow stem borer, *Scirpophaga incertulas* Walker, pink borer, *Sesamia inferens* Walk.), stripped borer, *Chilo suppressalis* (Walk.); dark headed borer, *Chilotraea polychrysa* and white borer, *Scirpophaga innotata* Wlk. The yellow stem borer (YSB) assumes major status in medium and lowland rice. It was earlier less known in areas like Punjab and Haryana but recently it has appeared as one of the most disastrous pest. The pest inflicts severe losses particularly on scented basmati varieties cultivated for export purpose. Rice gall midge (*Orseolia oryzae* Wood-Mason) which was sporadically occurring is now a wide-spread pest causing serious losses in many new areas like Uttar Pradesh, Bihar and North-Eastern state of Manipur in addition to its traditional areas like Orissa, Andhra Pradesh and Kerala. Leafhoppers, planthoppers and leaf folders which were of minor importance have assumed major pest status. The brown planthopper, *Nilaparvata lugens* Stal and whitebacked planthopper, *Sogatella furcifera* Horvath hitherto minor pests in Punjab, Haryana and parts of Uttar Pradesh are now of serious concern. Leaf folder, *Cnaphalocrocis medinalis* L. has assumed major pest status in the entire country both in wet season and dry season rice. These insects are considered to be national pests because they are responsible for significant decline in rice yield across the country.

Some other pests of regional significance in India are (i) rice hispa, *Dicladispa armigera* Olivier occurring in Andhra Pradesh, Himachal Pradesh, Bihar, West Bengal, Orissa and North Eastern region, (ii) gundhi bug *Leptocorisa* spp. in upland rice in Uttar Pradesh, West Bengal, Bihar, Orissa, Madhya Pradesh, Manipur and parts of Andhra Pradesh, (iii) green leaf hopper, *Nephotettix virescens* Distant prevalent in Bihar, West Bengal, Assam, Orissa, MP, AP and Tamil Nadu, (iv) climbing cut worm, *Mythmna separata* Walker in upland rice growing areas and swarming caterpillar, *Spodoptera mauritia* Boisduval in low lying rice areas of Bihar, West Bengal, Assam and Orissa are important. Recently, rice thrips (*Sternochaetothrips biformis* Bagnall) in southern region, mealy bug

(*Brevennia oryzae* Lindinger) and case worm (*Nymphula depunctalis* Guenee) in eastern region have emerged as potential pests. Besides insect pests, leaf and panicle mite, *Oligonychus oryzae* Hirst, *Steneotarsonemus spinki* Smiley as well as nemotodes are becoming serious in southern states of India.

Some minor pests of regional importance occurring inrice include (i) termites, *Odontotermis obesus* Rambur in uplands (2) Whorl maggot, *Hydrellia philippina* Ferino, (3) green horned caterpillar, *Melanitis leda ismene* , (4) green semilooper, Naranga aenescens, (5) rice skipper, Pelopidas mathieas, (5) Crickets, *Gryllotalpa concinnus*, (6) short horned grasshopper, *Oxya* spp., *Hieroglyphus banian* Fab. , (7) mole cricket, *Gryllotalpa africana*, (8) black bug, *Scotinophora* spp., and (9) root weevil, *Echinocnemus oryzae* Marshall. The intensity of insect pests problems of rice in different states and union territories in India have been presented in Table 2 (Krishnaiah *et al*., 1999).

FACTORS RESPONSIBLE FOR CHANGE IN INSECT PEST SCENARIO

Of the several factors contributing towards change in insect pest scenario in rice over years, the major ones are (1) cultivation of high yielding varieties (HYV's) over extensive areas which have narrow genetic base, are photoinsensitive and could be cultivated in non-traditional areas. These varieties respond to high fertilizer input and possess plant architecture congenial for multiplication

Table 2 Intensity of insect pest problems of rice in India

State/ Union Territory	Stem borer	Gall midge	Brown plant hopper	Green leaf hopper	Leaf folder	White backed plant hopper	Cut worms	Hispa	Gundhi bug
Andhra Pradesh	S	M	S	S	S	M	M	M	L
Assam	M	-	L	L	M	-	M	M	M
Bihar (undivided)	M	M	-	M	L	L	-	L	L
Gujarat	M	-	-	M	M	L	L	L	L
Haryana	VS	-	-	L	S	S	L	L	L
Himachal Pradesh	M	-	-	L	M	M	-	L	-
Jammu and Kashmir	L	-	-	M	M	L	-	-	L
Karnataka	M	M	M	L	M	M	-	L	L
Kerala	S	S	S	M	M	L	L	L	L
Madhya Pradesh (undivided)	M	S	M	S	M	M	L	L	L
Maharastra	S	S	M	L	M	L	L	L	L
Manipur	L	S	-	L	M	-	-	-	L
Meghalaya	L	L	L	L	L	-	-	L	L
Nagaland	L	-	-	L	L	L	-	L	-
Orissa	S	S	M	S	M	L	M	L	L

Contd. ...

Table 2 Contd. ...

Pondichery	S	M	S	M	S	M	L	L	L
Punjab	VS	-	-	L	M	M	-	M	-
Rajasthan	L	-	-	L	L	L	L	-	-
Tamil Nadu	S	M	M	M	S	M	L	M	L
Tripura	L	L	-	L	L	L	L	L	L
Uttar Pradesh (undivided)	M	L	M	M	M	M	M	L	M
West Bengal	S	M	S	M	M	M	-	M	M

L = Low M = Moderate S = Severe VS = Very severe

of insect pests (2) Growing of varieties which have high yield potential that do not possess resistance to major pests , (3) cultivating rice throughout the year (monoculturing) providing permanent food source to the pests, (4) changes in the nutrient use pattern. For example, traditionally rice was grown with organic manures and green manures till mid- 20th century. Later on with the introduction of HYV's, use of chemical fertilizer in higher quantities in stead of balanced dose, particularly application of high levels of nitrogen lead to more of free amino acid concentration in the plant system creating better physiological conditions for multiplication of pests, (5) Indiscriminate use of synthetic chemical pesticides has been considered as one of the important reasons for upsurge of many pests leading to pest resistance, resurgence, secondary pest outbreaks and other detrimental side effects.

Extent of loss

Yield loss due to insects has been estimated as 5-30 % (Cramer, 1967). Plots protected from insects yielded an average of 1000 kg more rice per hectare than did the plots that had undergone similar agronomic practices but were not protected (Pathak and Dyck, 1973). Yield loss due to insect pests ranging from 21-51 % were estimated because of moderate to serious incidence of stem borer, gall midge, planthoppers and other sporadic pests in extensively rice growing areas of India. Among different stages of crop growth, the period between 30-60 days after transplanting was found to be most vulnerable where maximum yield loss (20-68 %) occurred in majority of the areas mainly due to stem borer, gall midge, leaf folder and brown planthopper. During first 30 days after transplanting, significant yield losses were recorded mainly due to stem borer and whorl maggot only in 10-15 % of the locations in India. During the period beyond 60 days after transplanting, the crop was inflicted by stem borer and leaf folder causing 10-48 % damage to rice crop (AICRIP, 1980). Multi location studies conducted under the All India Coordinated Rice Improvement Project (AICRIP) have shown that the avoidable losses due to insect pests averaged 28.8 % in rice crop. India has also been a witness to many pest outbreaks resulting in extensive loss in yield. In the recent past, gall midge outbreak in Kuttanad area of Kerala resulted in a yield loss of Rs.60 million. The total monetary loss in rice due to insect pests has been estimated to vary between Rs.17, 100 to 94,680 million rupees in the country. At the global level, the monetary loss in yield for rice was the highest among crops and annually amounted to US $ 113 billion (Pasalu and Katti, 2004).

Importance of IPM

In view of the fact that hitherto farmers have been mostly relying on single tactic of chemical control as first line of defense for managing the insect pest problem in rice, but in the last few years the ill effects resulting from over reliance on the harmful chemicals has made them to look for better and safer alternatives such as use of resistant varieties, adopting suitable cultural practices and other non-insecticidal practices which are eco-friendly, economically viable and socially acceptable to the farmers. Though there are several alternatives to chemical control, individually each component can not tackle the pest problems on its own. Hence, it is felt essential to bring all the viable pest control tactics including the insecticides under a single umbrella and develop a holistic system of tackling pests to make it more effective, low cost and safe for the farmers. Such a system of integrating pest control options is popularly referred to as Integrated Pest Management (IPM) which is probably the best alternative to manage rice insect pests at present.

Components of IPM in rice

The IPM in rice involves the following components.

***Host plant resistance*:** Host plant resistance is the most effective, economical, practical and easiest means of encountering the pest problems and it is compatible with all other methods of pest control. Most of the modern varieties released are high yielders and grown widely in pest prone areas, possess resistance to at least one insect pest or sometimes multiple resistances to many insect pests. In the last three decades 632 varieties were developed and released for commercial cultivation in India by Central and State Variety Release Committees for different ecosystems (Rai, 2004). Of the 632 varieties, 51 are resistant to gall midge, 25 to brown plant hopper, 3 each to stem borer and green leaf hopper and 2 to white backed plant hopper (Table 3).

Table 3 Sources of resistance and released varieties against important insect pests of rice in India

Insect pest	Donors	Released varieties
1. Gall midge	CR 143, Eswarakora, Leuang 152, 0b 677, Ptb 10, Ptb 18, Ptb 21, Siam 29	Sneha, Pathara, Kakatiya Erramallelu, Kavya, Rajendradhan 202, Karna, Ruchi, Samridhi, Usha, Asha, NDU 3, Bhuban, Samalei, Orugallu, Abhaya, Shakti, Suraksha, Daya, Pratap, Udaya, IR 36, Shaktiman, Tara, Kshira, Sarasa, Neela, Lalat, Phalguna, Mahaveer, Vibhava, Divya, Dhanya Lakshmi, Surekha, Triguna, Vikram and Kunti
2 Brown planthopper	ARC 5984, ARC 650, Karivennel, Leb Mue, Nhang, Manoharsali, Oorapandy, Ptb 10, Ptb 18, Ptb 21, Ptb 33, Triveni	Chaitanya, Krishnaveni, Vajram, Vijeta, Cottondora sannaalu, Pratibha, Makom, Ravizham, Manasarovar, Co 42, Chandana, Nagarjuna, Sonasali, Pasmi, Jyothi, Vadra, Neela, Ananga, Daya, Aruna, Kanaka, Remya, Bharatidasan and Karthika

Contd. ...

Table 3 Contd. ...

3. White backed planthopper	Ptb 33	HKR 120, HKR 126
4. Green leafhopper	Ptb 2, W 1263	Vikramaraya, Lalat, Khaira, Nidhi
3. Stem borer	TKM 6	Ratna, Sasyasree, Vikas

However in the last two decades, insect biotypes have caused serious concern as an important biotic factor effecting the host plant resistance and its durability. Biotypes are discrete population or individual distinguished from other population or individuals of its species by some morphological/ physiological trait, usually resulting in adaptation to a particular host, host preference for feeding, oviposition or both. Biotypes reportedly occur in 36 arthropod species belonging to 7 families of six orders. In rice, biotypes of gall midge and brown planthopper have emerged as serious threat to the success of host plant resistance as a tool in rice IPM. Existence of gall midge biotypes in India was suspected as early as in 1969. Studies have revealed that, so far, six biotypes of gall midge (Table 4), have been identified to occur in India, while in case of BPH, single population is known to be distributed across the country (Pasalu and Katti, 2003).

Table 4 Different reaction of rice varieties to Indian gall midge biotypes

Group of differential cultivars		Reaction against gall midge biotypes					
		1	2	3	4	5	6
I.	W 1263/ARC 6605	R	S	R	S	R	R
II.	Phalguna/ARC 5984	R	S	S	S	R	S
III.	CR-MR 1523/ Velluthacheera/ Aganni/Ptb 10/ T-1477	R	R	R	R	S	S
IV.	TN 1	S	S	S	S	S	S
V.	Distribution in states of India	AP, Orissa MP (undivided	Orissa	A.P.	A.P. Mahara-shtra	Kerala	Manipur

S = Susceptible R = Resistant

Recent tests at Warangal showed that all the differentials in the biotype kit were susceptible and hence new sources of resistance have to be explored. Development of gene tags and markers linked to the genes against many of the known gall midge resistance genes will help in marker aided selection for pyramiding two or more genes in the breeding programme (DRR, 2002) which will help to overcome the problem of emerging biotypes. Further, sequential release of varieties with major genes, multiline varieties, development of horizontal resistance and wide hybridization are the foremost requirements to overcome problem of biotype development in rice insect pests.

Cultural practices: Cultural practices are normal agronomic practices which are followed for increasing crop productivity and at the same time useful in pest suppression. For this farmers will not have to spend more. Strategic manipulation of these practices can effectively suppress the multiplication or spread of insect pests. These include:

Early and synchronous planting : Wherever possible, altering the date of planting can often check the development of populations of insect pests like yellow stem borer, gall midge, leaf folder, BPH, WBPH and GLH (Table 5) (Dhaliwal and Arora, 2001). However, this needs community action and often depends upon availability of water in irrigated rice situations. The principle behind it is not to allow coincidence of the vulnerable stage of the crop with the virulent stage (peak period) of the insect pest.

Table 5 Effect of date of transplanting on leaf folder damage in rice varieties

Date of transplanting	Leaf folder damaged leaves (%)						
	IR-8	Jaya	PR 106	PR 4141	Basmati 370	Punjab Basmati-1	Mean
June 15	7.2	6.5	7.9	8.9	7.2	9.2	7.8
July 3	13.6	13.7	15.8	16.8	15.9	17.5	15.6
July 15	21.2	22.3	22.5	22.9	18.6	27.2	22.5

Judicious use of balanced fertilizers: The logic behind it is healthy plants can better withstand the insect pest attack than nutritionally sick plants. Therefore, application of chemical fertilizers at recommended doses for the region or based on the soil test report will help in enhancing the ability of rice plants to compensate for or to tolerate injury due to herbivory by insect pests. Farmers have the normal practice of applying more of N-fertilizers to make the plant lust green and succulent, hence, inviting more pest attack. Therefore, balanced and judicious application of chemical fertilizers at recommended (optimum) doses with nitrogen and potash fertilization in 2-3 splits avoid build up of rice pests like gall midge, leaf folder, BPH and WBPH.

Crop rotation: Avoidance of monocropping of rice or change in cropping pattern with suitable rotation of crops over a piece of land incorporating nonhost crops is important to break continuity in insect pest build up in rice crop.

Field sanitation: Maintaining a clean field free of weeds and destroying the alternate hosts suppresses pest build up. The left over stubbles of the previous rice crop in which larvae of stem borers remain hidden should be destroyed either by harvesting rice close to the ground or cultivating the land immediately after harvest and burn the stubble after collecting them to one place in the field or puddling the stubbles to rot in the field wherever water is available. Sanitation measures help to reduce the incidence of planthoppers, leaf hoppers, leaf folders, hispa, gall midge and stem borer on rice crop.

Presence of diverse vegetation within or near the field may add essential resources for predators or parasitoids and so enable them to find all their requirements near the pest population. Study revealed that *Echinochloa* spp. serve as a niche for both the pests and natural enemies in rice ecosystem (Padmakumari and Pasalu, 2002). Alternatively, weeds may adversely affect the orientation of predators and parasitoids to their prey. The weeds may even directly contribute to pest multiplication by providing preferred surface. Nearly 11 alternative hosts of rice hispa adults were found supporting the pest population in the absence of original host rice (Dutta and Hazarika, 1995). However, Choudhury *et al.*, 2003 reported 10 alternative hosts of rice hispa in Himachal Pradesh (Table 6). Therefore, reducing such weed population helps in reducing pest damage to crops.

Table 6 Alternate hosts of rice hispa in India

Sr No.	Botanical Name	English name	Local name	Family
1.*	*Andropogan gayanus* Kunth	Sadabahar	-	Graminae
2.*	*Chrysopongaon* sp. Trin	Love grass	Moongri	Graminae
3.*	*Carex* sp.	Tussock grass	-	Cyperaceae
4.	*Cyperus rotundus* Linn.	Purplenut Sedge	Dila/Motho	Cyperaceae
5.	*Cynodon dactylon* Pers. (Linn.)	Bermuda grass Dhrubh	Khabbal or Graminae	
6.	*Paspalum distichum* Linn.	Knot grass	Kodri	Graminae
7.	*Saccharum spontaneum* Linn.	Giger grass	Kans	Graminae
8.*	*Sorghum halepens* (Linn.) Pers.	Johnson grass	Baru	Graminae
9.*	*Imperata cylindrical* (Linn.)	Congo grass	-	Graminae
10. *	*Themeda anathero* (Nees) Hock	Red oat grass	-	Graminae

* New record

Plant spacing: Spacing may influence the population and damage by any insect pests by modifying the micro-environment of the crop or affecting health, vigour and strength of crop plants or pattern and duration of crop growth and development. Closer spacing has been reported to increase the incidence of BPH, WBPH, gall midge and leaf folder in rice crop. On the other hand, closer spacing resulted in lower incidence of green leaf hopper, rice hispa and whorl maggot. Provision of alley ways of 30 cm wide after every 2-3 metre is recommended particularly in BPH and WBPH endemic areas.

Trap cropping : Trap crops are plant stands that are grown to attract insects or other organisms, as a result of which the target crop escapes from pest attack. Protection is achieved either by preventing the pest from reaching the crop or by concentrating them in certain parts of the field where they can casually be destroyed. Planting small areas of a crop of other plants on the borders or even in the main crop may serve as a trap crop for same insect pest. Even early or late planting of the same crop in the main crop also act as source of a trap crop (Table 7). Trap cropping of rice for green leafhopper control resulted in 12 % higher economic return than chemical control and 29 % higher return than the untreated control (Hokkanen, 1991).

Table 7 Proposed trap crop systems for rice

Country	Insect pest	Main crop	Trap crop
Philippines	*Nilaparvata lugens* Stal.	Rice	Rice
Philippines	*Nephotettix virescens* (+ Rice tungro virus)	Rice	Rice

Intercropping : It is a process of growing two or more crops simultaneously on the same piece of land with a definite row pattern. It helps in changing the micro climate which reduces the pest population. It is applicable for upland rice situations. The effect of rice based intercrops on yellow stem borer infestation is presented in Table 8. Out of several intercrops, the YSB infestation was recorded lowest (17.3 %) in rice + jute mixed broadcast followed by the same intercrop when sown in lines (17.6 %). But the cost benefit ratio in both the intercrops was low i.e. 1.5 and 1.6, respectively. Therefore, rice + maize (line sown) with little higher YSB infestation (19.0 %) but with highest CBR (2.09) was the best. In another experiment (Table 9) rice + black gram intercrop had the lowest dead heart incidence (10 %) at harvest caused due to stem borer compared to rice + jute intercrop with 26.6 % WEH incidence.

Table 8 Yellow stem borer infestation in rice based intercropping system

Cropping pattern	YSB infestation	Benefit cost ratio
Rice (sole crop)	24.8	1.18
Rice + maize (MB)	18.2	2.02
Rice + jute (MB)	17.3	1.50
Rice + greengram (WB)	20.6	1.73
Rice + early rice (MB)	25.3	1.64
Rice + maize (LS)	19.0	2.09
Rice + jute (LS)	17.6	1.60
Rice + greengram (LS)	20.9	1.99
Rice + early rice (LS)	25.9	1.64
YSB = yellow stem borer.	MB = Mixed broadcast	LS = Line sowing

Table 9 Stem borer infestation in rice based intercropping system

Intercrop	White earhead (%) at harvest
Pure rice	20.0
Rice + jute	26.6
Rice + greengram	21.3
Rice + blackgram	10.0
Rice + maize	16.3

Mixed cropping: In upland situation, where rice is grown under rainfed conditions mixed cropping of rice with green gram (mungbean) or maize without maintaining any definite row proportion has been found to suppress yellow stem borer incidence.

Water management: Simple practices like draining of water from the fields or maintaining appropriate levels of water in the fields is effective to reduce BPH, army worm and case worm infestation.

Other practices : Adoption of cultural practices viz., summer ploughing helps in exposing the egg pods of grass hoppers to its natural enemies. Some of the egg pods will be dried under scorching heat of sun. During or after planting in the main field trimming of bunds help in exposing the egg pods of grasshoppers for their destruction. Another simple practice is to clip off the leaf tips of seedlings before planting in the main field that will help in destruction of egg masses of stem borers laid in the nursery.

These simple cultural practices are cheap and easy for adoption in rainfed rice ecosystem.

***Biological control*:** Use of biocontrol agents to manage crop pests is a key component of IPM. The successful use of several entomophages and entomopathogens in the control of insect pests has projected biological control as a promising alternative to ecologically disruptive chemical control measures. Biological control of rice insect pests can be achieved in the following three ways:

Conservation of native natural enemies: Rice ecosystem is known to possess an equally significant number of natural enemies population as that of the pests that occur within the ecosystem. Due to the limited scope for mass multiplication and release of pest specific biocontrol agents in rice, maximizing the impact of rich and diverse *in situ* natural enemies in rice ecosystem as an essential part of IPM programmes, needs emphasis.

Studies have revealed that native natural enemies can be used profitably in pest management. Therefore, the attention is now being given for the implementation of IPM in rice with specific thrust on the conservation of natural enemies. Several natural enemies identified from rice ecosystems of the country have been documented (Pasalu, 2002).

In case of yellow stem borer, the egg parasitism due to *Tetrastichus*, *Telenomus* and *Trichogramma* is very high in nature and needs to be conserved, while in case of gall midge, the parasitism due to larval parasitoid *Platygaster oryzae* has little impact on the pest in the field. However, if the larval parasitization by *P. oryzae* steadily increases from July onwards and reaches its peak in August (above 50 %), the midge activity declines in spite of conducive weather conditions (Senapati, 1998). The larval and pupal parasitism of leaf folder under natural conditions is also high and effective.

It was found that predatory guilds observed with five different ecosystems in rice included 10 orders and 57 families. It is known that 60% of the pest population is naturally taken care of by natural enemies of the ecosystem. Natural biological pest control system managed insect pests long before the advent of synthetic insecticides. Predators can contribute much more than parasitoids in regulating the population of leaf and planthoppers in rice ecosystem (Ooi and Shepard, 1994). Mathirajan (2001) reported 19 species of spiders in rice ecosystem. Spiders e.g. *Pardosa (Lycosa) pseudoannulata, P. sumatraha, Argiopes, Tetragnatha,, Araenus, Oxyopes* etc and mirid bug, *Cyrtorhinus lividipennis* have been observed to be more common and dominant (Dhandapani and Murugan, 2004). The nymphs and adult spiders can consume 5-15 insect pests (egg, nymphs and adults) a day depending on their size. Green mirid bug consumes 7-10 eggs or 1-5 hoppers/day.

Water striders can predate upon 5-10 prey daily. The ground beetles and rove beetles can consume 3-5 larvae/day. The grubs of the coccinellid beetle can consume 5-10 eggs, nymph, larvae or adults/day. The predatory grasshoppers are known to predate 3-4 egg masses of yellow stem borer/day (Anonymous, 1991). Other general predators like dragon flies, damsel flies, ground beetles, staphylinids and ear wigs have also been found effective in keeping the pest population at lower level. The role of avian predatory (birds) in rice ecosystem in controlling pests is also important.

The destruction of natural enemies due to insecticide misuse and abuse resulted in resurgence of several rice pests including the BPH (Heinrichs and Mochida, 1984), GLH (Kobayashi, 1961) and rice stem borers (Lim *et al*., 1980). Therefore, judicious application of safer insecticides i.e. granular formulations of carbofuran, phorate, cartap hydrochloride, isazophos, fipronil has been found effective and safe to the natural enemies. Even spray formulations of triazophos and acephate have been found to be relatively safe to egg parasitoid of stem borer, predatory mirids and spiders (Pasalu, 2002). It is expected that minimal use of chemical pesticides in paddy ecosystem is essential and augmentation of natural enemies will be accepted by the farmers.

Augmentation: In India, unlike in other crops, the scope of using biocontrol agents through inundative or inoculative releases in rice is restricted to the egg parasitoid, *Trichogramma japonicum* against yellow stem borer and *T. chilonis* against leaf folder. Five to six releases of the egg parasitoid @ 1, 00,000 adult parasites/ha in a crop season is very useful and devoid of any side effects. Careful integration with need-based application of selective insecticides and other components is essential to combat the multiple pest situations in rice.

Inundative release of *Trichogramma* spp. to control stem borers and leaf folders in rice fields is being done across the country by the Central IPM centres of the Directorate of Plant Protection, Quarantine and Storage, Government of India. Establishment of mass multiplication units by State Agricultural Universities is different states has provided the impetus to the use of biocontrol agents in rice. An insecticide resistant strain of *T. chilonis*, endogram with multiple tolerance to endosulfan and monocrotophos is also available.

Microbial pesticides: Use of microbial pesticides viz., Bt (*Bacillus thuringiensis*) formulations with endotoxins is another useful approach. They are specific to insect pests and quite safe to human being, natural enemies of insect pests and other non-target organisms. Evaluation of Bt formulation has revealed that they are effective against leaf folder and moderately effective against stem borers. They can be used in combination with other biopesticides or chemicals for control of rice leaf folder (Rao and Singh, 2003). Some of the fungal pathogens such as *Beuveria bassiana* against rice hispa and its combination with thiamethoxam (12.5 g thiamethoxam + 1000 ml *B. bassiana*/ha) were effective against leaf folder damage and maintained high coccinellid predator population in rice (Rao *et al*., 2003). Similarly, the fungus *Pandora delphacis* has been found effective against BPH (Ambethgar, 1996a). About 60-70 % leaf-folder mortality was obtained by applying spore suspension of *Metarhizium anisopliae* at 1 x 10 x spores/ml and *Erynia radicans*. Ambethgar (1996b) observed 100 % mortality of rice leaf folder by the fungus *Zoophthora radicans*. Cent per cent mortality of leaf folder larvae was recorded with the application of bacteria, *Bacillus cereus*.

An entomopathogenic nematode, *Oscheius* sp. isolated from vertisols was found to infect egg masses of rice YSB (Katti *et al.*, 2003). This nematode @ 40/egg mass prevented hatching upto 16 % of eggs but 11-98 % of larvae of stem borer were infected. There is a need for a suitable mechanism to develop suitable formulation technology, proper marketing and commercialization of these products for their possible use in rice IPM in future.

Mechanical control: The suppression of rice insect population by means of manual devices is referred to as mechanical control. This method utilizes light traps to attract the adults to light source and ultimately to trap them. Removal and destruction of egg masses, caterpillars and adults of insects during surveillance reduce the pest number. Dragging a rope across the rice crop to dislodge the larvae of leaf folder and case worm and adults of rice hispa to kerosenized water is helpful. Setting yellow sticky traps to catch the jumping and flying insects is also done for leaf and planthoppers. In upland situations, hanging fermented snails during milky stage of rice crop attracts the rice gundhi bugs to the trap, thereby preventing damage to rice (Misra, 1999). Providing perching sites to avian predators also aids them to search their prey in the field and control the host by predating them.

Biorational approach

(a) **Pheromones :** Insect sex pheromones are useful for monitoring of insect pest population (Karthikeyan and Purushothaman, 2003). These have been attempted for monitoring of stem borer and leaf folder population as alternative to light traps. Pheromones also showed potential for direct control of stem borer through male annihilation by mass trapping and disrupting mating communication. Mass trapping by installing 20 sleeve traps/ha each with 5 mg pheromone impregnated lures (replaced thrice during crop season) moderately reduce stem borer infestation . A single application of slow release formulation of pheromones @ 40 g a.i./ha within a fortnight after planting with the help of multipoint sources resulted in season long control of stem borer (Pasalu *et al.*, 2005). Grain yields in pheromone treated plots were at par with those achieved with two sprays of conventional insecticides against stem borer (Hall *et al.*, 1994). Sex pheromone components of other stem borers, gall midge, rice hispa and leaf folder have been identified (Krishnaiah, 1995) and a lot of efforts and refinement studies in pheromone utilization technology is needed which will play important role in pest monitoring and surveillance in future. The chemical components of sex pheromones of lepidopterous insect pests of rice is presented in Table 10.

Table 10 Sex pheromone components of important rice pests

Insect pest	Chemicals	Ratio
Scirpophaga incertulas	*Z, 11-16 AL and Z, 9-16AL*	*(3:1)*
Chilo suppressalis	Z, 11-16 AL, Z 13-18 AL and Z 9-16-AL	48:6:5
Sesamia inferens	Z, 11-16AL, Z 11-16 OH and Z-11-16AL	40:10:1
Cnaphalocrocis medinalis	Z 13-18AL and Z, 11-16AL	10:1

(b) **Chitin synthesis inhibitors :** Preliminary studies on buprofezin a chitin synthesis inhibitor has been found to be highly effective against *Nilaparvata lugens*. Rao and

Singh (2003) reported that flufenoxuron @ 100 g a.i./ha recorded the lowest leaf folder damage (7.27 %) compared to 31.36 % leaf folder damage in untreated check.

(c) Botanical pest control : Botanical pesticides are plant derived chemicals that repel approaching insects, deter feeding and oviposition on the plants, disrupt behaviour and physiology of insects in various ways and even prove toxic to different developmental stages of many insects. Among botanicals, neem formulations are novel approach in rice pest management. Neem formulations incapacitate insect pests through repellency, growth retardant effect, feeding deterrency, reproductive inhibition and oviposition deterrence. The repellent and antifeedant effects of neem have been reported against *Nilaparvata lugens*, *Sogatella furcifera, Mythimna separata* etc. with concentrations ranging from 0.001-0.4 % of various neem seed kernel extracts. Reproductive maturation was inhibited in *N. lugens* males by neem products. The active principle of Cinnaberry (*Melia azadirach*) showed a wide range of behavioural, physiological and toxic effects in *N. lugens* and *M. separata*. Plant volatiles extracted as steam distillates from resistant cultivars may act as repellent or may be toxic to BPH (Reddy, 2003).

(d) Avoidance of pest resurgence : In pest endemic areas where insecticide application is unavoidable, every care should be taken to overcome the problem of resurgence, particularly in case of BPH and leaf folder. In case of BPH, WBPH synthetic pyrethroids i.e. deltamethrin, cypermethrin and some organophosphates like quinalphos, chlorpyriphos and methyl parathion have been observed to cause resurgence. Hence, avoiding use of such insecticides in rice ecosystem is the best solution to overcome the problem of resurgence. Similarly, application of granular insecticides such as phorate and carbofuran in early stages of crop growth can lead to resurgence of leaf folder in later stages (Krishnaiah *et al.*, 1999). Since use of these granular insecticides is unavoidable under some circumstances, to nullify their resurgence effect, these could be alternated with cartap or fipronil granules or spray of the latter two chemicals (SP formulations).

Chemical control

Farmers opt for insecticide application as one of the effective and quick methods of reducing insect pest populations. They readily use synthetic chemical pesticides as the first line of defense because; they offer immediate control particularly in times of sudden pest outbreak. However, for better results, it should be ensured that the correct insecticide is chosen, suitable formulation and application techniques are employed and correct dose of the chemical is worked out based on pest biology and crop phenology. Further, understanding the potential environmental hazards of the chemicals used is essential not only to the user and consumer in the context of the rice crop but also the whole farming system.

For management of rice insect pests several insecticides, both granules, spray formulations and dusts have been evaluated and found suitable both in terms of selective and broad spectrum action against different insect pests (Table 11 and 12). Careful selection of proper insecticides their dosages and timing of application will result in optimum benefit to the farming community.

(a) **Seed treatment** - Seed treatment with chemicals like fipronil 5 FS @ 25 g/kg is effective in suppressing stem borer incidence in rice nurseries. Even sprouted rice seeds soaked with imidacloprid 200 SL @ 0.05 % for three hours, then shade dried and broadcast on raised nursery bed protected the rice seedlings from gall midge attack upto one month (Ramesh Babu *et al.*, 2003).

(b) **Seedling root dip :** Seedling root dip technique of treating uprooted seedlings ready for planting with 0.2 % chlorpyriphos solution overnight before planting in the main field provides protection against insect pests in initial stages of crop growth.

(c) **Root zone placement :** Root zone placement or deep placement or band placement of systemic insecticides packed in polymer capsules are placed in rice root zone control insect pests such as stem borers, leafhoppers, plant hoppers and whorl maggots. They are cheaper and more effective when conventionally applied.

(d) **Soil application :** Another effective and economical way is to apply granules of carbofuran 3G, or phorate 10 G or cartap hydrochloride 4 g @ 1.5 kg a.i./ha or fipronil 0.3 G @ 75 g a.i./ha mixed with sand in nursery 5-7 days before uprooting seedlings. This practice will give protection to the crop upto 3 weeks against major rice stem borers and gall midge (DRR, 1993-94).

(e) **Main field application :** In the main field, the crop can be protected either through application of granules like carbofuran, phorate, cartap hydrochloride @ 1 kg a.i./ha or fipronil @ 75 g a.i./ha or need based sprayings with monocrotophos/ chlorpyriphos/ cartap hydrochloride/ triazophos/ profenophos/ fipronil are recommended against stem borer and leaf folder while sprayings with imidacloprid/ thiamethoxam/ ethofenprox/ acephate/ clothianidin/ fenobucarb are effective against plant and leaf hoppers. Against mites, sprays of dicofol/ ethion/ profenophos are recommended. The emphasis in recent years is to find out eco-friendly chemicals belonging to novel groups like neoniotinoids, nereis toxin, phenyl pyrazoles and other derivaties . These are comparatively safer to mammals and effective at 10-15 times lower doses than the traditional compounds belonging to organophosphate/ carbamates. Use of such compounds in IPM on need basis would reduce the pest load in the environment.

Table 11 Spectrum of toxicity of spray formulations against insect pests of rice

Insecticide	Rate g/a.i./ ha	Stem borer	Leaf folder	Hispa	BPH	WBPH	Cut worms	GLH
Quinalphos	500	**	***	**	-	**	-	-
Phosalone	500	***	**	***	**	-	-	***
Monocrotophos	400	***	***	**	***	***	**	***
Chlorpyriphos	500	***	***	**	-	**	**	-
Carbaryl	750	*	**	**	***	***	-	**
Fenitrothion	500	-	**	-	-	-	*	-

Contd. ...

Table 11 Contd. ...

Phosphamidon 40 SL	500	**	***	**	***	***	-	**
Dichlorvos	500	-	**	-	**	-	***	-
Endosulfan	600	***	**	-	-	-	***	-
Ethofenprox	75	*	-	-	***	***	-	***
Cartap hydrochloride	300	***	***	-	-	-	**	**
Fipronil	50	***	***	**	***	***	**	***
Triazophos	400	*	***	**	-	-	**	-
Profenophos	400	**	***	**	-	-	**	-
Imidacloprid	25	-	-	-	***	***	-	***
Thiamethoxam	25	-	-	-	***	***	-	***

* Moderately effective, ** - effective, *** - Highly effective

Table 12 Spectrum of toxicity of granular formulations against insect pests of rice

Insecticide	Rate g/a.i./ ha	Stem borer	Gall midge	Whorl maggot	Leaf folder	Hispa	BPH/ WBPH	GLH
Carbofuran	750	***	**	***	-	**	***	**
Phorate	1250	***	***	-	-	-	**	**
Quinalphos	1000	***	***	-	-	-	-	-
Fenthion	1000	***	**	**	***	-	-	-
Sevidol	1000	***	-	-	-	-	-	-
Cartap hydrochloride	750	***	-	-	***	-	-	***
Isazophos	600	***	***	**	***	-	***	***
Fipronil	75	***	***	**	**	**	***	**

* Moderately effective, ** - effective, *** - Highly effective

Pest surveillance

It is one of the most important components of IPM technology in rice. It involves direct measurement of insect pest or mite occurrence/ damage by them in the field at regular intervals, by sampling rice hills and counting the population of insect pests as well as natural enemies. Usually random sampling of 25 plants in 5 clusters on a diagonal line of the plot, leaving the borders at 7-10 days interval is suitable for ascertaining insect pest levels and natural enemy populations. These form the base for taking decisions to interfere with insecticides into the ecosystem by following tentative economic thresholds as against guidelines (Table 13).

Table 13 Economic thresholds and suggested measures for management of common rice pests

Insect pests	Economic thresholds	Suggested control measures
Stem borer	10 % DH or 1 egg mass or 1 moth/m^2	• Stubble destruction • Clipping of seedling leaf tips before planting • Tolerant varieties (Table 3) • Chemical control as suggested in VI
Gall midge	1 SS/ m^2 in endemic areas or 5 % SS in non endemic areas (at active tillering stage)	• Early planting • Resistant varieties (Table 3) • Chemical control as suggested for stem borer control.
BPH and WBPH	10 insects/hill (vegetative stages 20 insects/hill at later stages	• Resistant varieties (Table 3) • Alleyways formation • Draining the fields • Judicious 'N' fertilizer use • Chemical control as in VI
GLH	2 insects/hill in tungro endemic areas 20-30 insects/hill in other areas	• Resistant varieties • Cehmical control as in VI
Leaf folder	3 damaged leaves/hill in post active tillering stage	• Judicious 'N' fertilizer use • Chemical control as in VI
Gundhi bug	1 nymph or adult per hill	• Removal of alternate hosts • Chemical control like dusting malathion 5 D @ 25 kg/ha or • Spraying DDVP or malathion @ 0.5 kg a.i./ha
Case worm	1-2 cases/ hill	• Water management • Chemical control • Endosulfan/ monocrotophos/ Triazophos @ 1 lit/ha
Hispa	1 adult/leaf or5 % damaged leaf	• Removal of alternate hosts • Chemical control with monocrotophos/ phosalone/ chlorpyriphos @ 1 lit/ha
Cut worms and Army worms	1 larva/ hill	• Spray chlorpyriphos/ triazophos/ endosulfan/ phosalone @ 1 lit/ha and dust the bunds with methyl parathion 2 % D or chlorpyriphos 1 % D.
Whorl maggot	25 % damaged leaf	• Measures suggested for stem borer control

Combination insecticides

Few combination insecticides available in the market have been tested against rice insect pests (Misra and Parida, 2004, Prabhu *et al.*, 2005). These have the advantage of controlling more than one pest with single application as well as the resurgence causing insecticides are taken care of by the combination of other insecticide. It is cost effective and labour saving. But, since two insecticides with different mode of action are mixed, misuse of this product will lead to development of resistance to both groups of insecticides which will be difficult to contain later. Few combination products of promising effect are listed in Table 14.

Table 14 Effective combination insecticides against major rice pests

S. No.	Chemical combinations	Dose (Trade product)	Effective against
1.	Acephate 25 % + Fenvalerate 3 %	2 lit/ha	Stem borer, gall midge, leaf folder, BPH, WBPH, GLH
2.	Cypermethrin 3 % + Quinalphos 20 %	1 lit/ha	Gall midge
3.	Ethion 40 % + Cypermethrin 5 %	1 lit/ha	Gall midge, BPH, WBPH
4.	Profenophos 40 % + Cypermethrin 4 %	1 lit/ha	GLH
5.	Chlorpyriphos 50 % + Cypermethrin 5 %	1 lit/ha	Stem borer, gall midge, leaf folder, GLH
6.	Deltamethrin 1 % + Triazophos 35 %	1 lit/ha	Stem borer, Gall midge, leaf folder, WBPH
7.	Imidacloprid 5 % + Betacyfluthrin 5 %	350 ml/ha	BPH, WBPH
8.	Fipronil 1 % + Fenobucarb 30 % lit/ha	(Tank mix) 1 WBPH, GLH	Stem borer, leaf folder, BPH,
9.	Cyfluthrin 50 g/ha + Imidacloprid 125 ml/ha	Tank mix	Stem borer, leaf folder
10.	Thiamethaxam 25 WG 100 g/ha + Deltamethrin 2.8 EC 250 ml/ha	Tank mix	Stem borer, leaf folder
11.	Imidacloprid 200 SL 125 ml/ha + Deltamethrin 2.8 EC 250 ml/ha	Tank mix	Stem borer, leaf folder
12.	Beta cyfluthrin 25SC 50 g/ha + Thiamethoxam 25 WG 100 g/ha	Tank mix	Stem borer, leaf folder

Biotechnological approaches

Biotechnology has major contribution in production of inherently resistant/tolerant varieties, since it opens up a virtually limitless source of germplasm variability from which insect control genes are selected for introduction into elite crop varieties. Although more than 30,000 accessions of germplasm were screened at IRRI and almost the same number were evaluated in India against yellow stem borer, good sources of resistance were not identified. However, incorporation of novel genes from soil bacterium, *Bacillus thuringiensis* such as Cry IA (b) and Cry IA(c) offers good promise (Sardana and Altosaar, 1996). These genes when deployed with a suitable promoter could check the incidence of leaf folders viz. *Cnaphalocrocis medinalis* and *Marasmia patnalis* (Wunn *et al*., 1996). Another breakthrough is utilization of potato proteinase inhibitor gene for incorporation into rice plants. This was attempted successfully to confer resistance against stripped stem borer, *Chilo suppressalis* and pink stem borer, *Sesamia inferens* (Duan *et al*., 1996). Genetic pyramiding with *B.t.* as well as potato proteinase inhibitor genes may confer long lasting and better resistance against stem borers (Cohen *et al*., 1996). Against BPH, a novel approach utilizing mannose binding snow drop lectin (*Galanthus nivalis*; GNA) an insecticidal gene product has been attempted. GNA has been demonstrated to exert antifeedant effect against BPH (Powell and Gotehouse, 1996).

It is important that refugia of non – *B.t.* rice should be maintained to delay the evolution of stem borer resistance to *B.t.* toxins. Growing of non *B.t.* rice near *B.t.* rice serves as source of susceptible insect to mate with rare resistant insects (if developed). The combined use of refuges with rice plants producing high dose of toxin is known as refuge/ high dose strategy. However, a decision regarding adoption of *B.t.* transgenics of rice to combat rice pests in India is yet to be made by the Government of India.

CONCLUSION

Rice is the major staple food of Asians and is attacked by an array of insect pests at its various growth stages. Farmers usually opt for chemical methods of control as a first line of defense. But, the indiscriminate use of insecticides has resulted in disastors like pest resistance to pesticides, pest resurgence, minor pest assuming major status, residues of pesticides remaining in food, feed and fodder, destruction of natural enemies and environmental pollution. Therefore, the present strategy is aimed at eco-friendly integrated approach to manage rice pests. This approach utilizes resistant varieties, use of cultural, mechanical, biological, microbials, botanicals, biorational as well as chemical methods of pest control in an efficient and cost effective way without harming the ecosystem.

REFERENCES

AICRIP 1980. Progress Report of the all India Coordianted Rice Improvement Project, *kharif* – 1980, ICAR, New Delhi.

Ambethgar, V. 1996a. Natural infection of entomopathogen fungus, *Pandora delphacis* (Hori) Humber in the rice brown planthopper, *Nilaparvata lugens* (Stal.). *J. Ent. Res.*, 20(4): 383-385.

Ambethgar, V. 1996b. First record of entomogenous fungus, *Zoophthora radicans* (Brefeld) Batho on the rice leaf folder, *Cnaphalocrosis medinalis* guenee from India. *Entomon*, 21(3 & 4): 283-284.

Anonymous. 1992. Manual on Integrated Pest Management in Rice. Directorate of Plant Protection, Quarantine and Storage, Faridabad, India, pp.99.

Choudhury, A., Dogra, I., Sharma, P.K. and Kaul, B.K. 2003. Record of some new alternate hosts of rice hispa, *Dicladispa armigera*. Olivier from Himachal Pradesh (India). *J. Ent. Res.*, 26(2): 183-184.

Cohen, M.B., Aguda, R.M., Romena, A.M., Raderick, G.K. and Gould,F.L. 1996. Resistance management strategies for *B.t.* rice: What have we learned so far ? *In*: Rice Genetics III. *Proc. Third Int. Rice Genetics Symp., 16-20 Oct., 1995, Manila (Philippines)*, IARI, pp.749-752.

Cramer, H.H. 1967. Plant protection and world crop protection. Farben Fabriken, Bayer Ag., Westing the insecticidal check method. *Jpn. J. Ecol.*, 12: 1-11.

Dhaliwal, G.S. and Arora, Ramesh. 2001. Integrated Pest Management: Concepts and Approaches, Kalyani Publishers, New Delhi, pp. 109.

Dhandapani, N. and Murugan, M. 2004. Role of predatory fauna in integrated management of rice pests. *IPM Mitr.* 3: 99-122.

DRR. 1986 – 2002. Progress report for 1986-2002. vol.2. Entomology and Pathology. All India Coordinated Rice Improvement Programme (ICAR), Directorate of Rice Research, Rajendra Nagar, Hyderabad, 500 030, A.P., India.

Duan, X, Li, X, Xue, Q., Abo-EI-Saad, M., Xu, D. and Wu, R. 1996. Potato proteinase inhibitor II gene that confers insect resistance in transgenic rice plants. *In* : Rice Genetics III . *Proc. Third Int. Rice Genetics Symp., 16-20 Oct., 1995, Manila* (Philippines), IARI, pp.743-748.

Dutta, B.C. and Hazarika, L.K. 1995. Feeding and oviposition preference of rice hispa, *Disladispa armigera* (Olivier) on some host plants. *J. Agric. Sci. Soc. NE.India*, 8(1): 14-19.

Hall, D.R., Chamberlain, D.J., Cork, A., De Souza, K., Mc. Veigh, L.J., Ahmed, Z., Krishnaiah, K., Brown, N.J., Casagrande, E., Jones, O.T. 1994. The use of pheromones for mating disruption of cotton boll worms and rice stem borer in developing countries. *Proc. Brighton Plant Prot. Conf. – Pests and Diseases*. 9A3: 1231-1238.

Heinrichs, E.A. 1994. Biology and Management of Rice Insects. Wiley Eastern Ltd., New Delhi, pp.779.

Heinrichs, E.A. and Mochida, O. 1984. From secondary to major pest status. The case of insecticide – induced rice brown planthopper *Nilaparvata lugens* resurgence. *Prot Ecol.*, 7: 201-218.

Hokkanen, M.T.H. 1991. Trap cropping in pest management. *Annual Rev. Entomol.*, 36: 119-138.

Katti, G., Padmakumari, A.P. and Jonnalagadda, P.S. 2003. An entomopathogenic nematode infecting rice yellow stem borer, *Scirpophaga incertulas*. *Indian J. Plant Prot.*, 32(2): 80-83.

Kobayashi, T. 1961. The effect of insecticidal application to the rice stem borer on the leaf hoper populations. Special Report 6 . pest Destruction and Forecasting Programs, NAF, Japan. Pp.120-136.

Krishnaiah, K. 1995. Insect pheromones in integrated pest management of cereals. *Proc. Natl. Workshop, Pheromones in IPM*, pp.33-39.

Krishnaiah, K., Reddy, A.P.K., Krishnaiah, N.V. and Pasalu, I.C. 1999. Current problems and future needs in plant protection in rice. *Indian J. Plant Prot.*, 27(1 & 2) : 47-64.

Lim, G.S., Ooi, P.A. C. and Koh, A.K. 1980. Brown planthopper outbreaks and associated yield losses in Malayasia. *Int. Rice Res. Newsl.*, 5: 15-16.

Mathirajan, V.G. 2001. Diversity and Predatory Potential of Spiders in Cotton and Rice Ecosystem Applied with Thiamethoxam. Ph.D. thesis submitted to TNAU, Coimbatore.

Misra, H.P. 1999. Control of gundhi bug in upland rice. *Indian J. Ent.* 61(2): 169-172.

Misra, H.P. and Parida, T.K. 2004. Field screening of combination insecticides against rice stem borer and leaf folder. *Indian J. Plant Prot.*, 32(2): 133-135.

Ooi, P.A. C..and Shepard, B. M. 1994. Predators and parasitoids of rice insect pests. In: *Biology and Management of the rice insects* (Ed., E.A. Heinrichs). Wiley Eastern Limited, New Age International Limited, New Delhi, pp.585-612.

Pasalu, I.C. and Katti, G., 2003. Biotypes of insect pests of rice: an emerging and challenging problem. In: *Proceedings of the National Symposium on Frontier Areas of Entomological Research*, Nov. 5-7, 2003, New Delhi, India, pp. 64-83.

*Padmakumari, A.P. and Pasalu, I.C. 2002. *Echinochloa* – A reservoir of pests and natural enemies of rice. *Indian J. Plant Prot*. 20(1): 79-80

Pasalu, I.C., 2002. Advances in IPM Technology in Rįce. In: *Compendium of short course training on Advances in Implementable Pest Management Technology*. Directorate of Oilseed Research, Hyderabad, AP – 500 030, pp.136-142.

Pasalu, I.C.and Katti, Gururaj. 2004. Recent Advances in rice pest management in India. *Indian Farming*. 54(8): 49-52.

Pasalu, I.C., Varma, N.R.G. and Katti, G. 2005. Insect Sex pheromones in integrated pest management with special reference to rice yellow stem borer. In: *Gleanings in Entomology* (Eds. Ramamurthy *et al.*) Division of Entomology, IARI, New Delhi, pp.20-40.

Pathak, M.D. and Dyck, V.A., 1973. Developing an integrated method of rice insect pest control. *PANS Pest . Artic. News Summ*., 9: 534-544.

Powell, K.S. and Gatehouse, J.A., 1996. Mechanism of mannose- binding snowdrop lectin for use against brown planthoper in rice. *In* Rice Genetics III. *Proc. Third Int. Rice Genetics Symp.,* 16-20 Oct., 1995, Manila (Philippines), IARI, pp.753-758.

Prabhu, S.T., Devaraj, K. and Gundannavar, K.P. 2003. A newer insecticide Koranda 28 EC for the management of insect pests of rice. *Proc*. Natl., Symp. "*Sustainable Pest Mgt*., January 2-4, 2003, BHU, Varanasi, pp.21-22.

Rai, Mangala, 2004. International year of rice – an overview. *Indian Farming*. 54(8): 3-6.

Ramesh Babu, A., Sudhakar, T.R., Ramesh Babu, T. Sreeramulu, M. and Satyanarayana, A. 2003. Efficacy of imidacloprid 200 SL applied in nursery against gall midge in rice. *Indian J. Plant Prot*., 31(2): 109-110.

Rao, N.B. V.C., Singh, V.S. and Subhash Chander. 2003. Studies on the effect of various combination treatments on rice leaf-folder and its coleopteran predators. *In Proc. Natl. Symp. Frontier Areas Ent. Res*., 5-7 Nov., 2003, Divisionof Entomology, IARI, New Delhi, pp.9-10.

Rao, N.B.V.C. and Singh, V.S. 2003. Ecofriendly management of rice leaf folder, *Cnaphalocrocis medinalis* (Guenee). *Indian J. Pl. Prot*., 31(2): 17-19.

Reddy, K.L. 2003. Effect of Plant volatiles extracted as steam distillates from selected rice cultivars against *Nilaparvata lugens* stal. *Insect Env*., 9(3): 132-134.

Sardana, R. and Altosaar, I. 1996. Identification and use of specific promoters and synthetic *Bacillus thuringiensis* toxinogenes in rice biotechnology. *In*: Rice Genetics III. *Proc. Third Int. Rice Genetics Symp*. 16-20 Oct., 1995, Manila (Philippines), IARI, pp.723-729.

Senapati, B. 1998. Rice pests in Orissa and Their Integrated Management. Extension Bull. No.140, Directorate of Extension Education, OUAT, Bhubaneswar-751003, Orissa, pp.1-20.

Wunn, J., Kloti, A., Burkhardt, P.K., Ghosh Biswas, G.C., Bieri, S., Terada,R. and Potrykus, I. 1996.. Evaluation of indica rice variety IR-58 after its transformation with a synthetic Cry IA(b) gene from *Bacillus thuringiensis*. *In*: Rice Genetics III. *Proc. Third, Int. Rice. Genetics, Symp.,* 16-20 October, 1995. Manila (Philippines), IARI, pp.759-762.

Entomology : Novel Approaches, 2007
Eds.: *P.C. Jain and M.C. Bhargava,*(Page 287-318)
New India Publishing Agency, New Delhi
E-mail: newindiapublishingagency@gmail.com
web: www.bookfactoryindia.com

CHAPTER **15**

INTEGRATED PEST MANAGEMENT IN COTTON

S.A. Nimbalkar
Vice-Chancellor, Dr. Panjabrao Deshmukh Krishi Vidyapeeth, Akola, Maharashtra

INTRODUCTION

Cotton is an important commercial crop and plays a key role in the economy of country. Cotton is grown in different states of India viz Punjab, Haryana, Rajasthan, Gujrat, Maharashtra, Madhya Pradesh, Andhra Pradesh, Karnataka and Tamil Nadu. In India the area under cotton crop is 8.76 m ha and the productivity is 436 Kg lint/ha as against the world average of 682 kg lint/ha (Anonymous, 2005). The productivity of cotton is variable in the country in the range of 287 to 720 kg lint/ha (Anonymous, 2005). The major constraints in cotton cultivation in India are as follows

MAJOR CONSTRAINTS IN COTTON CULTIVATION IN INDIA

i. Abnormal weather conditions
ii. Cultivation of cotton under rainfed condition
iii. Cultivation of large number of susceptible and undescript cultivars with varied growth pattern.
iv. Poor soil health
v. Long duration of crop
vi. Long sowing period
vii. Indeterminate nature of cotton plant
viii. Intensive cultivation
ix. Imbalance use of fertilizers
x. Non judicious use of irrigation water
xi. Higher incidence of insect pests due to
    a. Cultivation of large number of cultivator
    b. Staggered sowing
    c. Availability of green and succulent foliage for a long period
    d. Favorable hot and humid climate

e. Nectaries on leaves and fruiting bodies
f. Availability of large number of floral forms (squares, buds, flowers and bolls)
g. Indiscriminate use of insecticides
h. Inadequate spray technology
i. Lack of effective natural biocontrol
j. Poor adoption of non-insecticidal method of pest management
k. Economic status of cotton growers
l. Dependence on traders for inputs particularly insecticides.

WHAT IS IPM

Integrated pest management is the pest management in which all the alternative control measures like cultural, mechanical, physical, biological, pheromones, hormones, sterilants, attractants, resistant varieties etc are utilized in an integrated manner in order to maintain the pest population below the economic injury level. If the pest population increases and reaches to the economic threshold level, measures are used to bring down the pest population. This concept of in integrated system of crop protection is called integrated pest management in which minimum insecticides are used keeping in view the agro-eco system. Actually we do not go for a 100 % kill of insect pests in pest management rather the pest is allowed to remain and feed on crops up to a level which can be tolerated i.e. economic threshold level. This also provides opportunity for parasites and predators to survive feeding upon the insect pests keeping them at a certain equilibrium level. Thus an integrated pest management concept aims at managing the pest through all available tactics with favourable economic, social and ecological consequences.

WHY IPM

The struggle between man and insect began long before the dawn of civilization and still continuing with greater challenges as man wants to protect his health and provide an adequate food supply.

Modern technologies in agriculture have brought out the "Green Revolution" in which high yielding varieties, irrigation, improved agricultural practices, pesticides and fertilizers have all played significant roles. New varieties and hybrids of different crops have been planted throughout the world but while offering higher yields, they are often more susceptible to pest damage. Hence the increased yield obtained through improved seed varieties, fertilizers and farming methods are in danger of being destroyed by pest and diseases No doubt pesticides proved to be effective tools to protect the crops from pest and diseases and paved the way for Green Revolution. Although, pesticides have played a great role in agriculture but indiscriminate and over use of pesticides have led to number of problems including disruption of ecological balance. Many farmers apply pesticides in a routine manner without observing a pest load in the field. Sometimes pesticides are applied when there is actually no pest in the field. Such a practice should be stopped because after some time the problem will be so grave that the pesticide will have no effect on the pest.

The cotton crop in Andhra Pradesh failed during 1987-88 season due to an unprecedented epidemic of American bollworm, *H. armigera* particularly in Guntur and Prakasam districts. There were instances where cotton crop received 30 insecticidal sprayings including 8-10 synthetic pyrethroids at short intervals. Hence there should be judicious use of pesticides in the pest management programme.

Though the developing countries use much less quantity of pesticides in comparison to world's pesticide consumption, they suffer more from pesticide fatalities. In India, only a few percentage of farmers exploit the use of pesticides, while majority use very less or no pesticides. The average pesticide consumption in India is also much less in comparison to other countries. However indiscriminate use of pesticides leads to pesticide hazards like effect on soil, water and air, toxic residues, insecticide resistance in insects, mortality of parasites, predators and pollinators, pest resurgence and out break of secondary pests etc. Hence, to sustain high yields, better pest control is needed in an integrated approach which is called as integrate pest management

LOSSES CAUSED BY INSECT PESTS IN COTTON

Amongst the various constraints in cotton cultivation mentioned above, the losses caused by insect pests are of major importance As many as 1326 insect pests have been recorded on cotton crop in the world (Hargreaves 1948) . In India over 160 species of insect and non insect pests have been reported to damage cotton crop (Agarwal, 1978). Of these 24 phytophagous pests have attained the pest status, out of which nine are key pests, in different zones. Among these cotton jassid [*Amrasca biguttula biguttula* (Ishida)], white fly [*Bemisia tabici* (Gennadius)], pink bollworm [*Pectinophora gossypiella* (Saunders)], spotted bollworm [*Earias vitella* (Fabricius)], (*Earias insulana* Boisudal) and American bollworm [*Helicoverpa armigera* (Hubner)] are serious pests damaging the cotton crop in all the cotton growing zones of the country. Stem weevil (*Pempherulus affinis*), Fabricus, tobacco caterpillar *Spodoptera litura,* Fabricus and thrips (*Scirtothrips dorsalis* Hood) are the key pests in south zone. The polyphagous mites are of prime importance in south and central zones. The aphids (*Aphis gossypii*) and *Thrips tabaci* Lindman also cause moderate to heavy damage in central zone. The various pests cause 50 to 60 % loss in seed cotton. In North zone (Punjab) sucking pests and bollworms cause 11.6 and 49.5 percent loss in seed cotton yield, respectively. Both these pests caused 52.1 percent loss in F 414 hirsutum cotton (Dhawan *et al.*, 1988) . The total loss due to cotton pests (jassids and bollworm complex) in North, Central and South zone is estimated as 52.80, 46.53 and 47.56 percent, respectively (Basu, 1990).

USE OF PESTICIDES ON COTTON CROP

The cotton crop stand first in India where pesticide use is maximum as compared to other crops i.e. 44.5 percent out of total consumption in the country, whereas in world it occupies third position as regards with pesticide use (Table 1)

Table 1 Pattern of pesticide use on major crops

Crops	Area in India (%)	Pesticide use in India		Current pesticide use in world (%)
		1995	2000	
Rice24	17	22.8	13.0	NA
Cotton	05	54	44.5	10.2
Fruits and vegetables	13	8.1	13.0	24.2
Oilseeds	10	2.2	3.5	NA

Source – CICR Technical Bulletin No. 22

NEED FOR ADOPTION OF IPM IN COTTON

It can be seen from the above table that there has been about 10 percent reduction in pesticide consumption in cotton over last 10 years in India. However, only 5 percent area under cotton crop consumes highest amount of pesticides in comparison with other crops. The excessive and indiscriminate use of pesticides on cotton resulted in development of resistance in pests to pesticides, resurgence of secondary and minor pests, adverse effects on natural enemies of the pests and beneficial insects like predator, parasites and pollinators and environmental pollution etc.

In cotton excessive use of synthetic pyrethroids for the control of bollworms enhanced the population of white files (Armes *et al.*, 1994 and Kranthi *et al.*, 1997) reported pyrethroid resistance in the population of *Helicoverpa armigera*. Hence better approach for cotton pest management is needed in an integrated manager for economic and ecofriendly pest management to obtain sustainable cotton production. This can be achieved by adopting integrated pest management approach in cotton pest management

GUIDELINES FOR ADOPTION OF IPM IN COTTON

1. Determination of the economic threshold levels of the pests through regular surveillance by recording periodical observations about the incidence of major pests in cotton field
2. Planning of maintaining the equilibrium level of the pests and their natural enemies in the field
3. When a critical situation arises, suitable measures should be adopted for the management of the pests without or with least disruption of natural balance.

Economic Threshold Level

The judicious use of insecticides is the most important component of pest management to increase the productivity of cotton with least possible disturbance to cotton ecosystem and to enhance the life of chemical insecticides. Extensive work has been done on determination of economic threshold levels against different insect pests of cotton from time to time in different cotton growing states of the country and the same are give below in Table – 2

Table 2 Economic threshold levels of major pests of cotton in India.

Insect pest	ET values		
	North zone	Central Zone	South Zone
1) Jassids	* 2 nymphs per leaf * Appearance of second injury grade on 50 % plants * Curling of 20 % leaves	* 2 nymphs per leaf	* 2 nymphs per leaf
2) Whitefly	• 6 adults per leaf • 20 nymphs per leaf • Appearance of honey dew on 50% plants	* 8 to 10 adults per leaf or 20 nymphs per leaf	* 8 to 10 adults or 20 nymphs per leaf
3) Aphids	• 10 % infested plants • Appearance of honey dew on 50 % plants	* 15 to 20 % affected plants	* 15 to 20 % affected plants
4) Thrips	• 10 thrips per leaf • 20 to 30 % infested plants.	-	• 10 % thrips per leaf • 2 nymphs or adults per leaf
5) Bollworm complex	• 5 % incidence in shed fruiting bodies	* 5 % incidence level in retained and shed fruiting bodies	• 10% incidence level in retained shed fruiting bodies.

Source: Dhawan, 1998

To decide the right insecticide, dosage and time of application monitoring is very essential at macro level. However, now the need is felt that to use the insecticides judiciously, the micro level surveillance is more need based due to wide variation in insect population from field to field in the same village. The field scouting should be done at random in zigzag or diagonal manner. The insecticide should be applied when the action threshold is reached. This helps in reducing the number of sprays and hence cost of plant protection, besides minimizing the adverse effects of insecticides.

VARIOUS MANAGEMENT OPTIONS FOR IPM IN COTTON

Cultural Practices

1. Grazing of cattle immediately after last picking
2. Ratooning or extension of cotton crop should be avoided
3. Deep ploughing
4. Proper crop rotation
5. Time of planting

6. Proper seed rate
7. Purity of seed
8. Use of acid delinted seed
9. Proper plant to plant and row to row spacing
10. Judicious use of fertilizers and irrigation water
11. Destruction of weeds
12. Use of trap crops
13. Intercropping
14. Destruction of sources of carryover

Varietal Resistance

1. Tolerant /Resistant varieties
2. Early maturing varieties
3. Transgenic cotton hybrids

Mechanical and physical methods

Biological control methods

1. Predator
2. Parasites
3. Microbial control (NPV)

Botanicals

Sex pheromones

Legal control

Chemical control

CULTURAL PRACTICES

The pest population can be managed by adopting recommended agronomic practices during the crop season as well as in off season. The adoption of such practices at appropriate time helps to reduce the pest damage. These practices do not cost any extra time or money as they involve only slight modification of routine farm management operations. Some of the useful practices for cotton crop are as given below

Grazing of cattles immediately after last picking

Grazing of cattle particularly sheep and goats in cotton field immediately after last picking so as to eat away all the remaining leaves, fruiting bodies on plants, thereby destroying stages of pests present on these cotton plants. Several workers like Simwat *et al.* (1982), Sidhu and Dhawan (1986) also suggested this practice.

Ratooning or extension of cotton crop should be avoided

The termination of crop as early as economically feasible to reduce the late season pest attack and carryover to next season. Because the incidence of major pests is greater on ratoon or extended cotton crop and moreover it provides shelter and food to pests thereby their life cycles are continued and the incidence of these pests can carry over to next season.

Deep Ploughing

Deep ploughing should be undertaken to expose stages of insect pests in soil to natural enemies like predatory birds & hot sun. *H. armigera* pupate up to 15 cm deep in the soil. These pupae may be killed due to deep ploughing or may be exposed to hot sun and predatory birds.

Proper Crop rotation

In South India, especially in Tamil Nadu, cotton is grown in 2-3 seasons throughout the year which provide a continuous supply of food for the development and multiplication of pests. Hence the cotton crop should be rotated preferably with non-preferred host plants to provide a host free period to curtail the population of insect pests. Basu (1990), Jayaswal and Sundaramurthy (1992) suggested that the cotton crop should be rotated preferably with non-preferred host plants like Ragi or maize to provide host free period to curtail the population of white fly and American bollworms. Sundaramurthy and Chitra (1992) also reported that the crop rotations like cotton rice, cotton groundnut, cotton soybean are less conductive for pest development on cotton

Time of planting

Planting of cotton crop at optimum time is important to avoid condition that synchronies crop-pest association. Careful selection of planting time enables the plants to escape damage during susceptible growth stage and it advance into tolerant stage before the pest attack occurs and reduces the number of generations of pests. Cotton sowing during the shortest span of time with same varietal pattern will reduce the food availability to pest. In North India sowing is delayed due to the non-availability of canal irrigation water and due to staggered sowing pests are able to complete 1-2 generations more in the season. The American bollworm and whitefly are late season pests and by sowing till mid may, crop can escape damage of these pests.

Proper seed rate

Adoption of appropriate seed rate ensures proper crop stand, spacing and crop canopy that helps in adoption of proper spray technology, and checks the unwanted growth of the crop.

Purity of Seed

Non-availability of certified seed to farmers is also a major constraint in increasing cotton productivity

Use of acid delinted seed

Acid delinting of seed mechanically or manually with concentrated sulphuric acid reduces the population of diapausing pink bollworm larvae and incidence of angular leaf spot (Simwat *et al.*, 1991)

Proper plant to plant and row to row spacing

The plant spacing has tremendous impact on host as it influences its health, growth and development as well as microclimate, which affect the build-up of pest species. The close spacing results in bushy growth of crop that affects penetration of light, results in vertical grown of plant, hinders the spraying operation and also results in higher incidence of sucking pests and bollworms.

Jayaswal and Sundaramurthy (1992) reported higher jassid and spotted bollworm incidence at close spacing than at wider spacing. The higher loculi damage due to bollworms was observed in closer plant spacing than in wider plant spacing in Hirsutum cotton varieties by Butter *et al.* (1989). Similarly Sivasubramanian and Uthamasamy (1992) also recorded higher incidence of pink bollworm in closer spacing than in wider spacing.

Judicious use of fertilizer and irrigation water

The dose of fertilizer varies with the cotton cultivars, soil health, rainfed or irrigated condition. In central and south zone large area of cotton is cultivated under rainfed conditions, which is major constraint in increasing the productivity. However when cotton is grown under irrigated condition the excessive use of irrigation water coupled with high nitrogen, close spacing and bushy type of cotton genotype results in unwanted plant growth that favour pest build up and also hamper insecticide sprays. The judicious supply of water is essential during vegetative and flowering phase of cotton crop that depends upon climate and soil types. Late irrigation usually promotes the vegetative grown which is conductive for build up of white fly and American bollworm.

Hence judicious use of fertilizers and irrigation water in cotton crop is necessary as they influence the pest problems in cotton. Moreover the use of increased dose of nitrogenous fertilizers increases the amino nitrogen concentration of plant making it more conductive for quick development and higher fertility of insect pests

Agrawal *et al* (1979) reported that jassid attack to cotton crop was more at higher dose of nitrogen. Butter et al. (1989) also reported higher incidence of bollworms in cotton with 100 kg N/ha as compared to 50 kg N/ha. Similar results were reported by Sharma *et al* (1989) in *arboreum* cotton

Jayaswal and Sundaramurthy (1992) reported enhanced incidence of jassids and bollworms on cotton with the dose of NPK at 160:80:80 or 120:60:60 as compared with 80:40:40 kg NPK/ha

Application of nitrogen @ 200 to 250 kg/ha to cotton crop in Guntur and Prakasan districts of 'Andhra Pradesh was also one of the responsible factors for outbreak of *H. armigera* which resulted in devastation of cotton crop there in 1987-88.

Destruction of weeds

Several weeds are preferred alternate host of American bollworm, spotted bollworm and whitefly. The removal and destruction of these hosts in the vicinity of cotton fields, on sides of road etc, can help to reduce the multiplication and carryover to next season

Jayaswal and Sundaramurthy (1992) also suggested to keep cotton crop weed free and alternate host plants like *Lagesca mollis*, *Tndex procubens*, *Dutura metal* etc should be removed and destroyed periodically

Use of trap crops

The basic strategy in the cultivation of trap crops which is preferred host of pest for feeding, shelter and breeding over a small area to attract and kill the pest. But the area of trap crop should not be more than one tenth of the main crop area

Okra is preferred host plant of cotton jassid, spotted bollworms and American bollworm. The infested okra fruits that have a large population of bollworms should be removed and destroyed periodically (Thimmaiah & Raju, 1991). The jassids can be easily controlled by spraying any systemic insecticides on the trap crop okra

Marigold and *Nicotiana rustica* grown as trap crops are also preferred hosts of *H. armigera.* Planting of castor as trap crop diverted the population of *Spodoptera litura* from cotton (Reddy and Rosaiah, 1987).

Intercropping

Intercropping of cotton with black gram, green gram onion and cowpea is recommended in Southern and Central zone.

Natarajan and Seshadri (1989) reported that intercrop of cowpea should be grown in cotton on which initial heavy infestation of aphids will build up by which natural enemies of aphids, coccinellids develop on this crop which further can take care of aphids on cotton crop.

Venkatesan *et al.* (1987) recommended intercropping of cotton with black gram, green gram, cowpea etc to get higher monetary return and to divert the population of sucking pests and American bollworm from cotton.

Destruction of Sources of carryover

Collection and destruction of cotton debris and utilization of cotton sticks for fuel or other purposes before monsoon reduces the population pink bollworm

VARIETAL RESISTANCE

The resistant varieties are vital tool of Integrated Pest Management. It suppresses the pest population with least disturbance to cotton ecosystem and also reduces the dependence on insecticides.

Tolerant/resistant varieties

Host plant resistance is rated as top priority for IPM. It is highly effective and plays important role in sustaining productivity. More than 3000 different genotypes have been screened against cotton pest in All India Coordinated Research Project and many more were found resistance/tolerant to jassids, whitefly, thrips bollworms and stem weevil (Jayaswal and Sundaramurthy, 1992). During the last 3 decades more than 130 cultivars (70 hirsutum, 26 *arboreum*, 30 hybrids and rest of *barbadense* and *herbaceum* were released under AICCCIP (Venugopal *et al.*, 1996)

The adoption of resistance varieties as components of IPM needs careful screening against all the pests. Cotton crop is attacked by many pest species at a time and development of varieties resistant /tolerant to a number of pests or group of pest cannot be achieved easily. The cultivation of resistant cultivar should be carefully integrated with other components of IPM and some level of resistance/tolerance to key pests is essential for the success of other components of IPM. The varieties that show resistance are to insect pests released for cultivation are given in Table –3 below.

Table 3 Resistant/ tolerant cotton cultivars recommended for cultivation since 1985 in different agro climatic condition.

Gossypium spp	Year	cultivar	Pest species	Zone/state
1. *G. arboreum*	1985	LD-327	Jassids, whitefly, pink bollworm, spotted bollworm, American bollworm	Punjab
2. *G. hirsutum*	1987	Kanchan LK 861 AKH-081	Whitefly & American bollworm Whitefly & American bollworm Pink bollworm & spotted bollworm	South zone South zone Maharashtra
3. G. hirsutum	1988	Abadhita	Pink bollworm, & spotted bollworm	Tamil Nadu
		JK –119	-do-	Karnataka
	1992	LRK-516	Pink bollworm & spotted bollworm	Maharashtra

Source : Jayaswal and Sundaramurthy, 1992; Sundaramurthy and Chitra, 1992; Venugopal *et al.* 1996

Table 3 (a) Resistant/tolerant cotton varieties to insect pests.

Pest	Variety/Hybrids	Area to which recommended	Reference
Jassids (1992)	Bikaneri Nerma	North	Basu & Narayanan
	Digvijay	Central zone	Annoymous (1987)
	Ganganagaragethi Jayalakshimi	Cental & south zone	Basu & Narayanan (1992)
	Krishna (AC122)	A.P. rice fallous	
	Mahalakshmi (1301 DD)	Rainfed area of A.P	Rao *et al.* (1989)

Contd. ...

Table 3 (a) Contd.

...	Mahanandi (NA297)	Rayalaseema/ region	Sundaramuthy & Basu (1990)
	Pryiya	Rayalaseema region of A.P.	Rao *et al.* (1990)
	Savitha	Andhra Pradesh/ Tamalimadu	Basu & Narayan 1992
	Saguna B 1007, Dhy 286 ABH-466, H-777 G. cot-12.G.cot-10. G.Cot RS 875, RST-9 RG-8 Fateh Krisha, Mahanandi (355-E6) Priya (NA 920) Amaravati (AV1661) Mahalakshmi 1301 DD Lam 389, CAHH 468 H4 NHH 44, H6, AHH 468 Savita NHB 12, Eknath (PA-32) Sanjay Sujata, DDH-2 CICRHH 1 (KIRTI) Jayalakshmi (DCH-32) Khandwa 2, F-505	A.P. Godavaritract Central zone Rajasthan A.P. Central and south zone Central zone A.P. MS M.P. Punjab	Rao *et al.* (1989) Rao *et al.* (1989) Mayee *et al.* (2001) Puri *et al.* (1998) Mayee *et al.* (2001) Mayee *et al.* (2001) Puri *et al.* (1998) Mayee *et al.* (2001) Puri *et al.*(1998) Mayee *et al.* (2001) Puri *et al.*(1998) Puri *et al.*(1998) Puri *et al.*(1998) Puri *et al.*(1998) Puri *et al.*(1998) Puri *et al.*(1998) Puri *et al.*(1998) Puri *et al.*(1998) Puri *et al.*(1998) Puri *et al.*(1998)
Thrips	Savitha Sujatha RG-8	AP	Rajasthan Puri *et al.*(1998) Puri *et al.*(1998) Puri *et al.*(1998)
Whitefly	Kanchana (LPS 141) Suguna Kanchana, Supriya LHH-144 Kanchana Amravathi (LK 861), L 389 JKHY-1 RS 875 Vara Lakshmi, LRK 516 LRA 5166	South zone Cental & South zone A.P. Godavari tract Punjab A.P. M.P. Rajastan Central aone M.S.	Jayaswal & Sundemudhy (1992) Basu *et al.* (1992) Rao *et al.* (1989) Mayee *et al.* (2001) Puri *et al.* (1998) Puri *et al.* (1998) Puri *et al.* (1998) Puri *et al.* (1998) Puri *et al.* (1998) Puri *et al.* (1998)

Contd. ...

Table 3 (a) Contd.

...Bollworms	Abhaditha	Central & North zone	Basu,1990
	Mahalakshmi (1301 DD)	Rainfed areas of AP	Basu,1990
	L 604, L 603		Puri *et al.* (1998)
	LH 922, F 414	Punjab	Puri *et al.* (1998)
	Bikaneri Narma	Rajasthan	Puri *et al.* (1998)
	Digvijay	Central zone	Puri *et al.* (1998)
	CAHH 468	-	Puri *et al.* (1998)
	JKHY 1	M.P.	Puri *et al.* (1998)
	AHH 468	Central zone	Puri *et al.* (1998)
	CICRHH 1 (KIRTI)	-	Mayee *et al.* (2001)
	NHH 1 (Godavati)	M.S.	Puri *et al.* (1998)
	Soubhagya (DS 59)	-	Mayee *et al.* (2001)

Source: Booklet on IPM Technology for sustainable cotton production, 2005 published by MAU, Parbhani : pp.24

Early maturing varieties

The releases of early maturing varieties that escape the damage of last generation have reduced the bollworm incidence in initial years in North zone. However, due to the early flowering and availability of floral parts early in the season the population of spotted bollworms has increased later on

Transgenic cotton hybrids

The demand for insecticide free cotton has generated interest in the development of transgenic cotton. The development of genetically engineered or transgenic cotton plants for control of bollworms is a recent bio-technological approach in IPM on cotton. The research on transfer of *Bacillus thuringiensis* gene controlling the production of endotoxin lethal to caterpillars and its expression into plants is being carried out in countries like USA, Australia etc. Monsanto launched first product Bollard in 1996 and now Bt cotton is commercially cultivated in Australia, Mexico, China, South Africa, Argentina and USA

India realizing the economic importance of cotton bollworms and advantages of Bt cotton, Mahyco took initiative and has developed Bt cotton by incorporating cry 1 Ac (Bollgard) gene into their germplasm through a series of back crossing with a donor parent received from Monsanto, USA

The Govt. of India has permitted the three Bt cotton hybrids viz MECH-12 Bt, MECH-162 Bt and MECH –184 Bt for commercial cultivation during 2002 initially for three years. These hybrids were evaluated at Akola Maharashtra under the ETL based protection in rain fed situation during 2001-02 and it was observed that the damage of bollworm complex was less on all these three Bt cotton hybrids as compared to their non Bt. and check hybrids. The yield of seed cotton obtained from all these three Bt cotton hybrids was also significantly more than their non-Bt counterparts. There was also reduction in chemical insecticide sprays for bollworm control in Bt cotton hybrids (Anonymous 2002). However the performance of these Bt cotton hybrids on large scale needs to be assessed but still Bt transgenic cotton hybrids can act as one of the component of IPM in cotton. Hence an IPM module needs to be developed with Bt cotton as one of the component. Because khan (2002) reported that the use of Bt cotton in China and other countries resulted in saving of plant protection expenditure by over 80 percent and enhanced the yield by 15 percent. Similarly Choudhary & Loria (2001) noted that 70 to 100 % decrease in pesticide application for bollworm complex in Bt cotton fields and increase in yield by 40 percent.

Attributes of Bt cotton

Bt cotton has several unique features like

i Bt plant has in built control mechanism for bollworms

ii Protein is expressed in whole plant day and night

iii Neonates cease feeding within hour of ingestion of plant tissues and die within 2 to 3 days Since the pest is killed in its very early stage any potential serious damage to the plant is prevented .

iv Larval populations developing resistance to chemical insecticides are also killed.

v Increase in yield owing effective bollworm control.

vi Safe to bio-control agents, non target organisms, plants and environment

vii The technology is available in seed it self. No special efforts required to grow Bt plants as the seeds are to be sown like any other normal seeds and hence illiterate farmers can easily adopt the technique.

viii Healthier and safer produce can be achieved.

ix Genetic stability.

x Reduction in use of chemical insecticides for bollworm control

Table 4 Reduction in number of insecticidal sprays for Bt cotton in selected countries per season

S.N.	Country	Reduction is number of sprays
1.	China	Up to 14
2.	Indonesia	Up to 8
3.	South Africa	Up to 7
4.	Australia	Up to 7
5.	India	Up to 5
6.	Mexico	Up to 3
7.	Argentina	Up to 3
8.	USA	Up to 2

Source : Compiled by Clive James, 2002, various sources and country case studies.

Constraints of Bt cotton

1. Development of resistance to Bt protein in insects and this could be managed by
   a) High dose of Bt.
   b) Growing refuge crop (non-Bt cotton).

c) Pyramiding of gene i.e. introducing second Bt protein gene into cry-1 Ac cotton.
d) Bt is not a stand alone technology to manage cotton pests. It may be one of the most dependable components in IPM for bollworms management.
e) By rotating Bt with non- Bt .
f) Use of economical thresholds.
g) Establishment of base line susceptibility studies.

2. There is need to control secondary pests i.e. sucking pests by chemical sprays which may kill natural enemies and thus offsets one of the major advantage of transgenic.
3. likely hood of monopoly of seed producers.
4. The cost of producing and deploying transgenic is very high.
5. May reduce effectiveness of transgenic due to insect migration.
6. Possibility of evolution of new insect biotypes.

In future there is a need to develop transgenic cottons for herbicide, disease and sucking pests resistance. The seed of Bt cotton should be made available at lower cost. Straight Bt cotton varieties should be developed. Bt cotton resistant to drought and salinity condition should be developed.

No doubt Bt cotton is valuable pest management tool but Bt cotton is not a panacea for all cotton pests and therefore insect management with other pest control approaches is necessary. However it can be integrated with other pest control techniques and this can be an essential component of integrated pest management is cotton.

MECHANICAL AND PHYSICAL METHODS

The following mechanical and physical methods are useful, effective and economical under integrated pest management in cotton

1. Clipping of spotted bollworm affected terminal shoots along with larvae and its destruction.
2. Collection and destruction of bollworm infested shed squares, flowers, green bolls and leaves with bollworm egg masses can reduce the pest population.
3. Handpicking of big sized larvae of *H. armigera* and their destruction.
4. Removal and destruction of pink bollworm affected rosette flowers along with pink boll worm larvae in it.
5. Installation of 'T' shaped bird perches in cotton field for sitting of predatory birds which will eat away the caterpillars in the field.
6. Erecting of yellow sticky board in cotton field in white whilefly endemic areas. The whiteflies are attracted to the yellow colour on the board and get sticked due to oil smeared on the board.
7. The diapausing larvae of pink bollworm in cotton seed get killed when these cotton seeds are exposed to hot sun thereby reducing the further infestation.

BIOLOGICAL CONTROL METHODS

It involves conservation of natural enemies present in cotton field & use of parasites, predators, microbes to control cotton pests.

Predators

Many general predators like coccinellids, green lace wing, predatory bugs and spiders recorded from cotton crop are summarized below in Table 5. Among all these *Chrysoperla carnea* has been commercially exploited for the management of cotton pests.

Table 5 Predators of major sucking pests of cotton in India

Insect pesto	Predator	Stage of host attacked
1. *Amrasca biguttula biguttula* (Ishida)	*Chrysoperla carnea*	Nymph, Adult
	Chrysoperla cymbela	Nymph, Adult
	Coccinella septempunctata	Nymph, Adult
	Ceilomenes sexmachlatus	Nymph, Adult
	Geocoris SP	Nymph, Adult
	Zelus Sp	Nymph, Adult
	Camponotus Sp	Nymph, Adult
	Spiders	Nymph, Adult
2. *Bemisia tabaci* (Gennadius)	*Chrysoperla Carnea*	Egg, Nymph
	Chrysoperla cymbela	Egg, Nymph
	Geocoris bicolor	Egg, Nymph
	Orius Sp	Egg, Nymph
	Encarsia Sp	Egg, Nymph
	Mallada boninensis	Egg, Nymph
	Geocoris Sp	Egg, Nymph
	Bruonus suturalis	Nymph
	Ceilomenes sexmacilaths	Nymph
	Spiders	Nymph, Adult
3. *Aphis gossypii* (Glover)	*Mallada astur*	Nymph, Adult
	Chilomenes sexmaculalus	Nymph, Adult
	Syrphus confracter	Nymph, Adult
	Syrphus baleatus	Nymph, Adult
	Syrphus searius	Nymph, Adult
	Spiders	Nymph, Adult

Source: Jayaswal and Sundaramurthy, 1992; Sundaramurthy and Chitra, 1992; Singh 1994

The commercial availability coupled with extensive use of insecticides has limited the use of predators for cotton pest management. However the present studies indicate that *Chrysoperla sp* holds a good promise as effective bio-control agent.

Jayaswal and sundaramurthy, 1992 also reported that the predators, *viz.* coccinellids, chrysopids and syrphids feed on soft bodied insects like aphids, whiteflies, thrips, mites, eggs and early instars larvae of bollworms.

Satpute (2003) also reported Chrysopids as important insect predator on soft bodied inset pests like aphids, jassids, white files, mealy bugs & eggs & young larvae of lepidopteran pests.

Parasites

The immense importance of biocontrol agents is recognized in recent years due to the problems associate with indiscriminate use of insecticides. However, still not much field work has done on the utilization of biocontrol agents as component of IPM. In few places, large scale field release of parasitoids has been taken up in systematic manner under IPM. Thirty four different species of egg, larval, egg-larval and pupal parasitoids recorded from bollworms (pink, spotted and American bollworms) which cause parasitization of host to variable extent in the field are given in Table-6 Among these *Trichogramma achaeae, T. chilonis, Bracon greeni, B. hebetor, B. brevicornis, Apanteles angaleti, Chelonus sp* are some important indigenous parasitoids while *T. brasiliensis, Bracon kirkapatricki* and *Chelonus blackburni* are exotic parasitoids. Agrawal and Gupta (1986) reported that mass releases of egg parasitoid *T. chilonis* and *T. achaea* and larval parasitoid *C. blackburni* and *B. kirkpatricki* during initial stage of formation of fruiting bodies reduced the incidences of all the three species of bollworms *T. chilonis* and *C. chloridae* parasitised cent percent eggs and 22.7 to 83.0 percent larvae of *H armigera* as reported by Singh (1990). Verna and Gill (1992) reported that in Punjab the release of *T. chilonis* @ 1.5 lakh/ha at 10 days interval during July – Oct. reduced bollworm damage by 25-45 percent.

Microbial Control

Among various microbial control agents, nuclear polyhedrosis virus (NPV) and bacteria *Bacilllus thuringiensis* (Bt) have been tested extensively for control of cotton pests in different cotton growing areas of the world. Both B.t. and NPV have vast potential as a component of IPM system in cotton in India. However, the research on the utilization of these microbes for pest control of cotton in India is scanty and the results of limited trials conducted have shown inconsistent results. Nuclear polyhedrosis virus has been evaluated alone and in combination with insecticides in North, Central and South zones against American bollworm and most encouraging results are obtained in south zone. However, in North zone, it failed to give desired results. The Bt. have also certain limitations with respect to the efficacy under field conditions, which include inactivation by sunlight, low persistence and slow in action. It is more effective against early instar larvae. The good bio-efficacy in laboratory conditions indicates the need to improve the spray technology to narrow down the gap in bioefficacy under field and laboratory conditions.

The utility of Bt against cotton bollworms has been reported by Rosaiah and Reddy (1993) . Luo et al. (1986) reported effectiveness of Bt against pink bollworm and the damage of pink bollworm was reduced 42.53 % and larval population by 56.1 %. Sarode *et al.* (1994) and Rao & Reddy (1999) reported the effectiveness of HaNPV against *H. armigera* on cotton equivalent to that of recommended chemical insecticices. HaNPV in combination with half dose of recommended insecticide enhances the efficacy and good control of bollworms on cotton. Reddy and Krishnamurthy (1989) recommended the spray of HaNPV @ 250-500 LE/ha at weekly interval during peak oviposition period for the control of *H. armigera* on cotton.

Table 6 Parasitoids of major insect pests of cotton

Parasitoids	Host pest	Stage of host parasitised
Aphelinus sp	*Earias sp*	Egg
Erythmelus empoascae	*Earias sp*	Egg
Gonatocenus Sp	*Earias sp*	Egg
Trichogramma achaeae	Pink bollworm & *Earias Earias sp*	Egg
I. brasiliensis	*Earias sp*	Egg
T. chilonis	*Earias sp*	Egg
T. chilotraeae	Pink bollworm and *Earais* sp	Egg
Telenomus remus	*sylepta derogata*	Egg
Trichogrammatoidae Sp near guamensis	*Earias* sp	Egg
Agathis fabiae	*Earias* sp and pink bollworm	Larva
Apanteles angaleti	Pink bollworm	Larva
Bracon chinensis	Pink bollworm	Larva
Bracon greeni	Pink bollworm & *Earias* sp	Larva
Bracon Kirkpatrick	*Earias* Sp	Larva
Bracon brevicornis	*Earias* sp	Larva
Bracon habator	*Earias* sp	Larva.
Camptolithlipsis gossypiella	Pink bollworm	Larva
Scambus limiepes	Pink bollworm & *H. armigera*	Larva
Rogas aligarhensis	Pink bollworm	Larva
Scambus Liniepes	Pink bollworm & *H armigera*	Larva
Goniozus SP	Pink bollworm & *Earias* sp	Larva
Carcelia illota	*H. armigera*	Larva
Compoletis chloridae	*H. armigera*	Larva
Elasmus johnstoni	Pink bollworm	Larva
E. Philippinensis	*Earias* sp	Larva
Cremastus	*Earias* sp	Larva
Pyemotes ventricosus	Pink bollworm	Larva
Chelomus sp	Pink bollworm & *Earias* sp	Egg – Larval
C. blackburni	Pink bollworm	Egg – Larval
Xanthopimpla punctata	Sylepta derogata	Pupa
Brachymeria Sp	Sylepta derogata	Pupa
B. apantelesi	*Earias* sp	Pupa
B. lasus	*Earias* sp	Pupa
B nephantidis	*Earias* sp	Pupa
B nosatio	*Earias* sp	Pupa

Source: Jayaswal and Sundaramurthy, 1992; Sundaramurthy and Chitra, 1992; Singh, 1994.

Table 7 Some important pathogens on major pests of cotton

Sr. No.	Natural enemy	Insect host (s)	Stage attacked
1	*Nosema Sp*	Spodoptera	Larva
2	*Vairimorpha Sp*	American bollworm	Larva
3	*Bacillus thuringiensis*	Bollworms	Larvae
4	*Beauveria bassiana*	American bollworm	Larvae
5	*Entomophthora aphidis*	Aphids	All stages
6	Nuclear poly hedrosis virus	American bollworm	Larvae
7	*Nomourega rileyi*	Spodoptera	Larvae

Source: Anonymous 1987, 1989; Puri *et al*, 1998

Recently Jagmohan (2001) recorded following parasitoids on *Earias vitella* and *Earias insulana* from Punjab which are given in Table –8

Table 8 Parasites of *E. vitella*

Sr. No.	Name of the parasite	Order	Family	Stage of host attacked
1.	*Agathias* (Microdus) SP	Hymenoptera	Braconidae	Larva
2.	*Actia aegyptia vill*	Hymenoptera	Tachinidae	Larva
3	*Eucepsis* sp.	Diptera	Chalcididae	Larva
4	*Brachymeria* Sp.	Hymenoptera	Chalcididae	Pupa
5	*Brachymeria lasus*	Hymenoptera	Chalcididae	Pupa
6	*Brahymeria tacharidiae* Cam.	Hymenoptera	Chalcididae	Pupa
7	*B. responsator* Walk	Hymenoptera	Chalcididae	Pupa
8	*Centrochaticis* sp.	Hymenoptera	Chalcididae	Pupa
9	*Bracon brevicornis* Wesm	Hymenoptera	Braconidae	Larva
10	*Bracon*-sp (gelechine)	Hymenoptera	Braconidae	Larva
11	*B greeni* Ashm	Hymenoptera	Braconidae	Larva
12	*Rogas aligharensis* Quadri	Hymenoptera	Braconidae	Larva
13	*R. Kanpurensis* Mani	Hymenoptera	Braconidae	Larva
14	*R. Kanpurensis* Mani	Hymenoptera	Braconidae	Larva
15	*Chelonus* sp	Hymenoptera	Braconidae	Egg. Larva
16	*Chelonus rufus* Lyle	Hymenoptera	Braconidae	Egg. Larva
17	*Bracon lefroyi* D. & G.	Hymenoptera	Braconidae	Larva
18	*B. Kitcheneri* D & G.	Hymenoptera	Braconidae	Larva
19	*Centrolchalcis rufescens*	Hymenoptera	Chalcididae	Larva
20	*Chalcis tacharidiae*	Hymenoptera	Chalcididae	Larva

Contd. ...

Table 8 Contd. ...

21	*Elasmus* sp	Hymenoptera	Elasmidae	Larva
22	*Elasmus johnstoni*	Hymenoptera	Elasmidae	Larva
23	*Goryphous* (Melcha) nursei Cam.	Hymenoptera	Inchneumo-nidae	Pupae
24	*Goryphus* sp	Hymenoptera	Inchneumo-nidae	Pupae
25	*Chrysis obliterata*	Hymenoptera	Chrysididae	Pupae
26	*Brachymeria nephantidis*	Hymenoptera	Chrysididae	Pupae
27	*Trichogramma evanescens* west.	Hymenoptera	Trichogra-mmatidae	Egg
28	*Brachymetia apantelesi* Risbec	Hymenoptera	Chalcididae	Pupa
29	*B attevate*	Hymenoptera	Chalcididae	Larva
30	*B. nephantidis* Gahan	Hymenoptera	Chalcididae	Pupa
31	*B. nosatoi* Habu	Hymenoptera	Chalcididae	Pupa
32	*Agathias fibiae* Nixon	Hymenoptera	Braconidae	Larva
33	*Trichogramma achaeae* Nagaraja & Nagarkatti	Hymenoptera	Trichogra-mmatidae	Egg.
34	*T. brasiliensis* Ashmead	Hymenoptera	Trichogra-mmatidae	Egg.

Table 9 Parasites of *P. gossypiella* recorded in India

Sr. No.	Name of the parasite	Order	Family	Stage of host attacked
1	*Agathias* sp	Hymenoptera	Braconidae	Larva
2	*Aleiodes* sp	Hymenoptera	Braconidae	Larva
3	*Aleiodes aligarhensis* Quadri	Hymenoptera	Braconidae	Larva
4	*Bracon breviconis* Wesm	Hymenoptera	Braconidae	Larva
5	*Bracon hebator* Say	Hymenoptera	Braconidae	Larva
6	*Bracon gelechiae* sub sp	Hymenoptera	Braconidae	
7	*Bracon kitcheneri* W	Hymenoptera	Braconidae	Larva
8	Chelonus Sp	Hymenoptera	Braconidae	Egg-Larva
9	*Microbracon efroyi* D & G (*M. gelechidiphagus*)	Hymenoptera	Braconidae	Larva
10	*Elasmus platyedrae* Ferriere	Hymenoptera	Elasmidae	Larva
11	*Rogas aligharensis* Quadri	Hymenoptera	Braconidae	Larva

Contd. ...

Table 9 Contd. ...

12	*Chelonus narayani* Subb Rao	Hymenoptera	Braconidae	Egg & Larva
13	*Bracon* Habrobracon) Gelechine Asmead	Hymenoptera	Braconidae	Larva
14	*Bracon* sp (gelechine)	Hymenoptera	Braconidae	Larva
15	*Apanteies* sp (alter group)	Hymenoptera	Braconidae	Larva
16	*Apanteles* sp	Hymenoptera	Braconidae	Larva
17	*Strobicmyia* sp (Mall)	Diptera	Tachinidae	Larva
18	*Afrovona indica* Mensil	Diptera	Tachinidae	Larva
19	*Apanteles angaleti* Muesebeck	Hymenoptera	Braconidae	Larva
20	*Bracon greeni* Ashmead	Hymenoptera	Braconidae	Larva
21	*Camptothlipsis* sp. Nr Antigastrae Wilk	Hymenoptera	Braconidae	Larva
22	*Chelonus cycloporus* Fraz	Hymenoptera	Braconidae	Larva
23	*Gonizus* sp	Hymenoptera	Braconidae	Larva
24	*Tetarstichus nyemitavus* Rohver	Hymenoptera	Braconidae	Larva
25	*Patalodes gossypiella* Mues	Hymenoptera	Braconidae	Larva
26	*Apanteles pectinophorae* Ferr	Hymenoptera	Braconidae	Larva
27	*Elasmus johnstoni* Ferr	Hymenoptera	Elasmidae	Larva
28	*Trichogramma achaeae* Nagaraja and Nagrakatti	Hymenoptera	Trichogra-mmatidae	Egg
29	*T braziliensis* Ashmed	Hymenoptera	Trichogra-mmatidae	Egg
30	*Tnchogramma* sp	Hymenoptera	Trichogra-mmatidae	Egg
31	*Trichogrammatodea* sp Near guamensis Nagaraja	Hymenoptera	Trichogra-mmatidae	Egg
32	*Chelonus blackbumi* Cameron	Hymenoptera	Braconidae	Egg & Larva
33	*Bracon chinensis* Szepigeti	Hymenoptera	Braconidae	Larva
34	*Rogas* sp	Hymenoptera	Braconidae	Larva
35	*Camptothlipsis* sp	Hymenoptera	Braconidae	Larva
36	*Scambus stratus* Gupta & Tiker	Hymenoptera	Ichneumonidae	Larva
37	*Scambus lineipes* Marlery	Hymenoptera	Ichneumonidae	Larva
38	*Parasierola* sp.	Hymenoptera	Braconidae	Larva

Table 10 Prasites of *H. armigera* recorded in India

Sr. No	Name of the parasite	Stage of host attacked	Order and Family
1	*Odonbtepyris* sp	Larva	HYM Bethylidae
2	*Apanteles* sp	Larva	HYM Braconidae
3	*Apanteles* sp	Larva	HYM Braconidae
4	*Apanteles* sp nr.glomeratus (Linnaeus)	Larva	HYM Braconidae
5	*Apanteles* sp (glomeratus group)	Larva	HYM Braconidae
6	*Apanteles ruficurs* (Haliday)	Larva	HYM Braconidae
7	*Bracon brevicomis* Wesm	Larva	HYM Braconidae
8	*Bracon gelechiae* Ashm	Larva	HYM Braconidae
9	*Bracon greeni* Ashm	Larva	HYM Braconidae
10	*Bracon* sp	Larva	HYM Braconidae
11	*Chelonus heliopae* Gupta	Larva	HYM Braconidae
12	*Chelonus narayani* Subba Rao	Larva	HYM Braconidae
13	*Microplitis* sp	Larva	HYM Braconidae
14	*Brachymeria responsator* Walk.	Larva	HYM Chalcididae
15	*Eribours* (Diotes) *argenteopilosus* Cam.	Larva	HYM Ichneumonidae
16	*Eriborus pilosellus* Cam	Larva	HYM Ichneumonidae
17	*Carcela* sp	Larva	Dip : Tachinidae
18	*Carcelia kockiana* Tns	Larva	Dip : Tachinidae
19	*Carcelia raoi* Mensil	Larva	Dip : Tachinidae
20	*Drino* sp nr unisetosa Bar	Larva	Dip : Tachinidae
21	*Exorista fallax* Mg	Larva	Dip : Tachinidae
22	*Exorista japonica* Tns.	Larva	Dip : Tachinidae
23	*Isomera cinerascens* Rondani	Stage attacked not-known reared from pupa	Dip : Tachinidae
24	*Sisyropa apicata* Baranov	Larva	Dip : Tachinidae
25	*Strobliomyia aegyptia* Villeuve	Larva	Dip : Tachinidae
26	*Thecocarcelia incedens* Rondani	Larva	Dip : Tachinidae
27	*Voria edentata* Baranov	Larva	Dip : Tachinidae
28	*Voria ruralis* Fallen	Larva	Dip : Tachinidae

Contd. ...

Table 10 Contd. ...

29	*Winthemia* sp prob Nr. diversoides Baranov.	Larva	Dip : Tachinidae
30	*Hexamermis* sp	Larva	Nematoda
31	*Campoletis chiorideae*	Larva	Hym:Ichneumonidae
32	*Enicospilus biconatus*	Larva	Hym:Ichneumonidae
33	*Goniopthalmus halli* Meshii	Larval-pupal	Dip : Tachinidae

Numbers of parasites were also recorded from pink bollworm in Maharashtra state by Raodeo and Sarkate, 1979, Naik *et al,* 1976 in Delhi by Narayanan 1960; in Tamil Nadu by Menon and Thangavelu, 1979 in Punjab by Verma *et al.,* 1984; and Jagmohan, 2001. The parasites recorded from pink bollworm in India are given in Table-9.

A large number of parasites have been recorded from *Helicoverpa armigera* in India which are given in Table-10

There are several successful cases of cotton pest management through biological control. About 60 percent of natural control of insects in any crop is due to the biocontrol agents which need to be protected and conserved by avoiding the unnecessary application of chemical pesticides. Bio-control agents fit very well with most of the other components of IPM (Sarkate and Pawar, 2003).

BOTANICALS

Amongst various botanicals, neem which has deterrent, anti-ovipositional, antifeedant, growth disrupting, fecundity and fitness reducing properties on insect can play effective role in the management of pests. Moreover it is ecofriendly, locally available at very low cost and has no adverse effects on natural enemies.

The utility of 5 percent neem seed extract has been reported by Sarode *et al* (1994). The addition of half dose of recommended insecticide with 5 percent neem seed extract found to perform better against major pests of cotton as reported by Sarode *et al.* (1993). Similarly Gupta and Sharma (1996) reported the satisfactory control of bollworm complex as well as increase in cotton yield by using combination of neem, and Bt thereby 84% reduction in the use of synthetic pyrethroids. The spray of NSKE 5 % was also found effective for the control of jassids, aphids, whitefly and *H. armigera* on cotton as reported by Uthamasamy and Gajendran (1992).

More than 63 neem products are in market; however there is no consistency is their concentration and doses. The question of their bio-degradation is also important. Another important point associated with yield of seed cotton treated with neem products are low as compared to that of insecticide treated cotton crop therefore improvement in toxicity of neem products is to be needed.

Method for Preparation of 5 per cent neem seed extract:

i. Take 5 kg of dried neem seeds and clean it thoroughly
ii. A day before schedule spraying, powder it by grinding.
iii. Soak over night in 10 litres of water. stir with wooden plank

iv. Filter through muslin cloth and make its volume to 100 litres by adding fresh water
v. Add 200 g of mild detergent soap and spray the resultant solution to cover upper as well as lower foliar portion of the crop

SEX PHEROMONES

Sex pheromones of bollworms *viz.,* pink, spotted and American bollworm proved extremely useful for monitoring survey and surveillance and assessing the damage of these bollworms in cotton. For this pheromone traps @ 10/ha (for each bollworm) should be installed in cotton field from square initiation or from 45 DAG. The pheromone lures for all these three bollworms are available in the market *viz.,* Helilure for *H arrmigera,* vitlure for *E. vitella* and gossyplure for *P. gossypiilla.* These lures are required to be changed at every 15 to 21 days.

Sex pheromones of *H. armigera* have utilized for studying its behavior and monitoring the population in cotton crop, by Rao (1982)

Patel *et al* (1985) reported that mass trapping of *H. armigera* in cotton in Gujarat reduced pesticide application by 3 to 6 and the average yields obtained from pheromone applied fields were comparatively higher than those where only insecticides were applied

The large deployment of pheromone traps for *H. armigera* in cotton ecosystem has resulted in lowering down the pest incidence due to mass trapping of male moth population (Sarode, 1997)

Bajikar and Sarode (1986) found that the sex pheromone of *E vitella* is useful for monitoring purpose. Gossyplure utility was studied for monitoring the pink bollworm population by demonstrating the different traps by Reed *et. al* (1975) in Andhra Pradesh. Balsubramaniam *et al.* (1979) in Tamil Nadu, Dhawan and Sidhu (1984) in Punjab and Satpute *et al* (1985) in Maharashtra

Besides monitoring, studies were undertaken on seasonal trappings and fixing of economic threshold level for pink bollworm on cotton crop by Taneja and Jayaswal, 1982, Dhawan and Sidhu, 1984[a]

Dhawan and Sidhu (1984)[b] reported that the timing of sprays against pink bollworm done at male moth trap catch level of four to six male month per trap per day resulted in saving of one to two insecticidal applications.

LEGAL CONTROL

It involves the application of regulations for prevention, or restriction of introduction, movement or eradication of pests diseases or weeds. It may be foreign or domestic quarantine. If a pest is limited to a few localities in the country, its spread should be prevented.

CHEMICAL CONTROL

Application of pesticides at ETL

Pesticides will remain as main tool to manage the pest problem in cotton in an emergent situation. However, the insecticides should be applied when the economic threshold level of the pest is reached. This helps in reducing the number of sprays and hence the cost on plant protection, besides minimizing the adverse effects of insecticides.

Resurgence of pests

The over-reliance on insecticides for bollworm control has caused the resurgence sucking pests (jassid and whitefly). Dhawan and Simwat (1996) reported the resurgence of jassid due to excessive use of chlorpyriphos and quinalphos. The severe out break of whitefly as observed in Andhra Pradesh, Tamil Nadu, Gujarat and Punjab is the outcome of the repeated use of synthetic pyrethroids for bollworm control.

Insecticidal Resistance management

The indiscriminate use of insecticides finally resulted in the development of resistance in insects to insecticides particularly in American bollworm and whitefly. Armes *et al.* (1994) and Krathi *et al* (1997) reported pyrethroids resistance in the population of *H. armigera.* The resistance in whitefly in southern states to endosulfan, acephate, dimethoate monocrotophos qninalphos and triazophas has been reported by Devaprasad *et al.* (1993).

Hence judicious use of insecticides based on 'ETL, cultivation of varieties which has some level of resistance to insect pests, need based use of insecticides to reduce the pest load on cotton. Kranthi (2003) proposed some insecticide resistance management strategies for Indian situation to conserve naturally occurring predators and parasitoids. This Insecticide resistance management strategies aim to optimize use of toxic molecules so as to ensure that resistance development is delayed and eco-system is conserved through the use of appropriate molecules at appropriate time.

Spray technology

The adoption of adequate spry technology affects the efficacy of insecticides by providing adequate coverage of target sites and to some extent the amount of pesticide load can be reduced. Most of the farmers use manually operated spray pumps but the coverage is inadequate mainly due to the use of nozzles with high discharge rates which leads to coarse sprays with bigger droplet size hence poor deposition at the desired site. Most of the sprays are done with the help of hired labourers on contract basis. They use bigger orifice nozzle as it helps to operate the pump with less force and also increase the number of sorties/ha. The use of high discharge nozzles coupled with poor maintaince lead to improper coverage and also involve toxic hazards to the operator.

Use of new generation selective chemicals

Although there is a need to reduce dependence on environmentally disruptive chemicals, it seems likely that the further cotton production system will require some insecticides use. The chemical groups mostly in use today are organophosphates, carbamates and synthetic pyrethroids which have broad spectrum activity and are significantly disruptive to most predators (Wilson *et al* 1998) and in some cases have environmental problems. In contrast the new generation pesticides are much selective, less disruptive, more environmentally benign and introduce new mode of action to overcome established resistance problem (Hollway and Forrester, 1998). These new compounds are imidacloprid, spinosans, fipronyl which are likely to be more compatible with IPM. They are also considerably more expensive, which would help to ensure they are used conservatively and so preserve their long term usefulness.

SOME SUCCESSFUL CASES OF IPM IN COTTON

There are several successful cases on cotton pest management through integrated pest management approach in India, some of them are given below.

In India ICAR sponsored village level IPM Operational Research Project (ORP) to demonstrate the efficacy, practicability and economics of IPM in cotton in Punjab during 1975. The project resulted in 73.3 and 12.4 % reduction in number of insecticidal sprays for sucking pests and bollworms with 23.2 per cent higher yields as compared to non-ORP area (Sidhu, 1978). A similar project was initiated at Coimbatore, Tamilnadu in 1980 with a substantial gain in the reduction of pesticide use of 6.3 number of sprays in IPM as compared to 10.7 sprays in Non-IPM areas, with increase in native natural enemies by three fold in IPM area (Simwat, 1994).

Research on IPM of cotton bollworms with the incorporation of exotic parasitoids was explored on large area at MAU, Parbhani under ICDP (ORP) programme for 5 years (1980-85). Results revealed that IPM with the use of biotic agents gave encouraging results i.e. low infestation of bollworms and higher yields as compared to adjacent field of IPM without biotic agents and chemical control as per recommended plant protection schedule (Raodeo *et al.,* 1983) .

Gujarat Agricultural University recommended IPM modules for effective management of sucking pests and bollworms as well as conservation of entomophage diversity in cotton. IPM module-I included picking of infested fruting bodies, random planting of mazie, three releases of *C. carnea* @ 10,000/ha/wk for sucking pests and eight releases of *T. chilonis* @ 1.5 lakh/ha/wk followed by need based application of monocrotophos @ 0.04 and endosulfan 0.07 % after 8th releases of *T. chilonis*, and IPM module II consisting all above components without insecticide use. The ICBR obtained from module-II was 1:15:37 which was much higher than IPM module –1 (1:10.66) (Anonymous 1999)

Sundaramurthy and Chitra (1992) found that IPM measures maintained the yield, reduced cost of pesticides by 50 % avoided environmental pollution by 53 % and conserved bio-control agents safely and also increased their number several fold in field.

IPM was evaluated with use of exotic and indigenous parasitoids undertaken on 15 acres of cotton field during 1992-1993 indicated the establishment of the parasitoids, reduced the bollworm infestation by 14.5 % which was lower than control (38%) and control field (47 %) In IPM only one spray for sucking pests and two sprays for bollworms were required whereas in chemical control two sprays for sucking pests and three sprays for bollworms were required (Anonymous, 1993).

Daware (1996) reported that with the application of 4 sprays of 5 % NSE at 60, 75, 90 and 105 DAG mixed with NPV 450 LE/ha and also release of *T. chilonis* @ 1.5 lakh/ ha 2-3 times at weekly interval starting from 45 to 70 DAG resulted in 8-18 % increased yield of seed cotton.

MAU, Parbhani and NCIPM, New Delhi conducted IPM project with different IPM modules at Cotton Research Station, Nanded at village Barad and Ashta. During 1997-98 and 1998-99. Results indicated that bio intensive module comprising 2 releases of *T. chilonis* @ 1.5 lakh/ha 35-40 and 50-55 days after sowing followed by 3 sprays of neem seed kernel extract 5 % and one spray of HaNPV @ 250 LE/ha gave encouraging results (Anonymous, 1998-99)

Srinivas and Patil (2001) tested three IPM modules during the year 1987-98 at Rachur, Karnataka. The IPM modules comprising bio-intensive IPM, adaptable IPM and recommended package of practices. (RPP) were imposed as per ETL. It was observed that net profit per ha was maximum in adaptable IPM (Rs 30550) which was higher than other two modules. A reduction of 5 insecticidal sprays amounting to 40 % by integrating parasitoids, botanicals and NPV in adaptable module as compared to RPP under irrigated condition.

Rao and Reddy (1999) documented the role of weed control and sanitation, bioagents, botanicals and emphasized non pesticidal approaches like host plant resistance, time of sowing, intercropping, trap crops, fertilizer application, weed control and other methods as important components of IPM in cotton

Demonstration of integrated pest management practices in cotton on one acre farmers field at village Kumaraagul 1996-97 and Kalugasalpuram (1998-99) in Karnataka indicated less bollworm incidence with high seed cotton yield and cost benefit ratio in IPM as compared to farmers methods (Rajaram et al., 2001)

Singh et al., (2001) evaluated IPM module, recommended spray schedule and unsprayed control for pests incidence and yield of seed cotton in Punjab during 1998 and 1999. Less bollworm infestation and higher seed cotton yield was recorded in IPM as compared to spray schedule and untreated control.

During 2001-2002 season, IPM module was tested in village Butki Dist. Dhule of Maharashtra on 63 hectares of unirrigated cotton in tribal area. The IPM module comprising timely sowing of imidacloprid 10gm /kg treated seeds of NHH-44 border rows of maize and cowpeas, at every 10 rows of cotton and a row of sataria sp, installation of pheromone traps @ 5/ha on 45 days, release of *T. chilonis* @ 1.5 lakh/ha on 50 and 60 DAS, NSE 5 % spraying at 70, 80 and 90 DAS, HaNPV 250 LE/ha at 110 DAS. In IPM seed cotton yield was 1950 kg/ha compared to 1025 kg/ha in non IPM plots with the expenditure of Rs. 1800-2000/ha on pest management, farmers got additional income of Rs. 14000-15000/ha (IAUA, 2002).

The IPM module was effectively demonstrated at village level over 100 ha area of cotton at Ashta, Maharashtra during 1998-2001 by joint venture of NCIPM, New Delhi and Cotton Research Station, MAU, Nanded .

This IPM module includes, completing of cotton sowing operation in shortest possible time, utilization of only one or two cultivars, balanced use of fertilizers, regular scouting and monitoring of pest incidence, installation of pheromone traps, augmentations of natural enemies (use of egg parasitoid *T. Chilonis* (@ 1.5 lakh/ha at 10-15 days interval) to reduce reliance on chemicals. These were integrated with a range of cultural methods such as planting border of maize interspersed with cowpea to encourage the actively of natural enemies and to serve as refuge. Use of *sataria* as intercrop and as live perch for predatory birds, use of microbial such as HaNPV and botanicals such as neem seed kennel extract (NSKE 5 %). The comparison of IPM and Non IPM in cotton at Ashta village is given in Table-11

Table 11 Comparison of IPM and Non IPM in cotton at Ashta village.

Particulars	1997	1998		1999		2000		2001	
	Before IPM	IPM	Non IPM	IPM	Non IPM	IPM	Non IPM	IPM	Non IPM
1. Seed cotton yield (kg/ha)	25-125	963	593	1075	806	1002	632	1032	564
2. No of sprays of insecticides	6-12	Nil	-	1	-	Nil	-	1	-
3 No of sprays of HaNPV	Nil	1	-	1	-	1	-	1	-
4. No. of sprays of NSKE	Nil	2-3	-	2	-	1	-	1	-
5. Actual quantity of pesticides used (kg/ha)	9.285	0.021	5.70	0.720	5.780	0.390	7.750	0.247	4.945
6 Cost of plant protection (Rs/ha)	3225	1545	2280	1255	2156	1465	1908	928	1217
7. Cost benefit ratio		1:2.03	1:1.05	1:2.05	1:1.34	1:1.84	1:1.08	1:1.91	1:1.01

Source: Booklet on IPM Technology for sustainable cotton production published by MAU, Parbhani. 2005

Similarly an IPM programme was implemented through NATP-IPM project at village Nawandi Distt. Nanded Maharashtra from 2001-02 to 2003-04. This programme was undertaken on 42 ha area under rainfed situation on cotton on 53 farmers fields. There was increase in yield from 2 to 9 quintals /ha reduction in sprays from 9 to 3, reduction in use of insecticide quantity from 11.170 to 0.932 kg gai/ha there by conservation of natural enemies.

WORK ON COTTON IPM AT DR.PDKV, AKOLA (M.S.)

(1) Three modules were formulated and evaluated for effective and economical integrated management of major pests of rainfed cotton for three years (1996-97 to 1998-99) at cotton Research Unit of Dr. Panjabrao Deshmukh Krishi Vidyapeeth Akola, Maharashtra and it was concluded that for economical and effective integrated management of major pests of hybrid cotton under rainfed condition the module consisting of seed treatment with imidacloprid 70 WS@ 10 gm/kg before sowing, one spray of recommended systemic insecticide for sucking pests based on ETL if required, two releases of *T. chilonis* egg parasitoid @ 1.5 lakh/ha at 45-50 and 55-60 days

after sowing, one spray of HaNPV @ 500 LE/ha, one spray of 5 % neem seed extract and need based one or two sprays of recommended insecticide for bollworm based on E.T. level should be used (Dandale *et al*. 2004). Highest cost benefit ratio was also obtained from IPM module as compared to Bio intensive module & recommended chemical control measures.

(2) An IPM module was developed and recommended by Dr., P.D.K.V. Akola Maharashtra for the management of major pests of cotton during 1998-99. Some refinements were made in this IPM module for more effective and economical management of major pests of rainfed cotton. This refined IPM module so developed was evaluated in comparison with existing recommended IPM module and recommended chemical control measures module for three years (2001-02 to 2003-04) at cotton Research Unit of Dr. PDKV Akola Maharashtra. From these studies it was concluded that the refined IPM module consisting of thiomethoxam 70ws seed treatment @ 4.28 gm/kg seed, one spray of acetamiprid 20SP @ 15 g a.i/ha for sucking pests based on ETL two releases of *T. Chilonis* egg parasitoid @ 1.5 lakh /ha at 45-50 and 55-60 days after germination of crop, for bollworms one spray of neemark 300ppm @ 5ml/L followed by one spray of spinosad 45 SC @ 50 g.a.i. / ha and one spray of Beta-cyfluthrin 2.5 EC @ 0.0025 % should be adopted for effective and economical integrated management of major pests of rainfed cotton (Dandale *et al*., 2004) The cost benefit ratio from refined IPM module was 1:5.96 as against 1:4:83 and 1:4.50 obtained from existing IPM module & recommended chemical central measures module respectively.

CONCLUSION

The indiscriminate use of insecticides on cotton crop has created many problems like resurgence of secondary and minor pests, development of resistance in insect pests to insecticides particularly in *H armigera* and whitefly, adverse effects on natural enemies of the pests and pollinators, pesticide hazards, environmental pollution, etc. The need is to make sincere efforts to promote the judicious need based use of selective insecticides reduce down the reliance on chemical insecticides or to avoid the use of insecticides as for as possible by adopting various components mentioned above under Integrated pest management.

Under integrated pest management in cotton complete cotton sowing operation in shortest possible time use of only one or two cultivars preferably resistant or tolerant ones, balance use of fertilizers and irrigation water, adoption proper spacing, regular scouting and monitoring of pest incidence, use of biotechnological products like Bt NPV, Pheromones, bio-agents like *Trichogramma* and *Chrysoperla*, neem & its formulations transgenic cotton along with other ecofriendly measures like cultural practices mechanical and physical control methods, legal control methods for sustainable cotton production as well increasing the productively and quality of cotton thereby saving the environment from pollution.

References

Agrawal, R.A. 1978 : Cotton insect pests and their control. Richae Harvest 1:22-29

Agrawal, R.A. and Gupta, G.P. 1986: Recent advances in cotton pest management PL. Prot. Bull. 38 (1-4):375-378

Agrawal, R.A. Wankhede, N.P. and Katiyar K.N. 1979: Impact of agronomic practices on the populations of jassid in Cotton. Cott Fibr. Trap 34:375-378.

Anonymous, 1987: Annual report AICRP on Biological control of pest & weeds. Indian Institute of Horticultural Research, Banglore 236pp

Anonymous, 1989: Technology for increasing cotton production in India AICCIP,CICRRegional Station Coimbatore:63.

Anonymous, 1993: Annual report of Entomology Department Submitted to cash crop sub committed MAU Parbhani during the year 1992-93

Anonymous, 1998: Annual report of cotton Research station Nanded submitted to cash crop sub committee MAU Parbhani during the year 1997-98

Anonymous, 1999: Annual report of cotton Research station Nanded submitted to cash crop sub committee MAU, Parbhani during the year 1998-99

Anonymous, 1999[a]: Research accomplishments and recommendations of Research G.A.U. Ahmedabad pp – 23

Anonymous, 2005: AICCIP annual Report of Project Co-ordinator Group meeting of AICCIP at Sri. Ganganagar (7-9 April 2005 Pc-1.2

Anonymous 2002: Annual progress Report of cotton Research (Entomology) submitted to plant protection Research Review Committee Dr.P.D.K.V. Akola held on 24th April 2002

Armes, N.J; Banerjee S.K; Desouza, K.R., Jadhav,. D.R. King A. B. S, Kranthi K.R; Regupathy A; Surlivelu T and Venugopal Rao. N, 1994: Insecticide resistance in *Helicoverpa armigera* in India : Recent development Proc. Brighton crop Prot. Conf Pest and Diseases 4:437-442

Bajikar, M.R. and Sarode, S.V. 1986: Future of pheromones in plant protection. Proc. Indian Natn Sci. Acad. B-52 (1) 129-133

Bhosale, B.B; Rathod K.S. and Pantange N.R. 2005: Booklet on IPM Technology for sustainable cotton production published by MAU, Parbhani.pp 42-44

Balasubramaniam, K, Murugesan, S. and Parmeswaran S 1979 : Synthetic gossyplure to trap cotton bollworm *Pectinophora gossypiella* Saunders J. ISCI special Publication No 2:29:31

Basu A. K. and Narayanan, 1992: Achievements of All India co-ordinated cotton Improvement Project during 1967-92. Silver Jubilee celebrations, CICR Nagpur pp - 176

Basu, A.K. 1990 Action plan for containing *Heliothis* in cotton in Andhra Pradesh in S. Jayaraj. S. Uthemasamy, M. Gopalan and R. J. Rabindra, Eds: Proc. National workshop on *Heliothis* management Feb. 18-19, 1988, Coimbatore pp. 297-300

Basu AK, Narayanan S.S. Kapadia, S.N., and Charyuli, N.R. 1992: Breeding achievements in south zone: In All India coordinated cotton Improvement project, Achievement, silver jubilee (1967-92) Nagpur pp 39-52

Butter, N.S ; Kular, J.S.; Brar, A.S. ; and Singh, T.H. 1989 : Effect of nitrogen, intra row spacing and varieties on the incidence of cotton bollworms under unsprayed condition. J. Cotton Res. Dev. 3 (1) : 27-31

Choudhary and Loria 2001 : Technological development and cotton production in India and china. Current Science, 80 (8) : 925-932

Dandale, H.G, Pradnya Kadam, Sarode S.V., Jane R.N, & Potdukhe N.R 2004 [b] : Development and evaluation of IPM module for effective and economical management of major pests of rainfed cotton PKV Res. J. 28 (1) : 75-80

Dandale . H.G; Thakar, A.Y; and Jane, R.N. 2004[a]. Studies on integrated pest management in rainfed cotton PKV Res. J. 28 (1): 86-90

Daware, D.G. 1996 : Studies of integrated pest management of cotton pests in Marathawada region J. Cott. Res. & Dev. (10): 131-133

Deva Prasad, V., Rambhau, L. and Readdy G.P.V. 1993: An action threshold for *Helicoverapa armigera* based on pheromone trap catch in cotton. Indian J. Pl. Prot. 21 (1) : 17-18

Dhawan, A.K., Sidhu, A.S., and Simwat, G.S. 1998 : Asessment of avoidable loss in cotton (*Gossypium hirsutum* and *G. arboreum*) due to sucking pests and bollworms Indian J. agric. Sci. 58 (4) : 290-92

Dhawan, A.K. and Simwat, G.S. 1996 : Cotton pest status in Punjab and role of insecticides in pest management In : National Seminar of" Century of cotton production in India" (Souvenir), P.E.–2, GAU, Surat, India

Dhawan A.K. and Sidhu A.S. 1984 [a] : Assessment of the capture threshold of pink bollworm moth for timing insecticidal applications on *Gossypium hirsutum* linn. Indian J. agric Sci. 54 (4) 426-433

Dhawan A.K. and Sidhu A.S. 1984 : Timming of sprays against pink bollworm on basis of the moth catch in pheromone. baited traps Agric. Sci. Digest 4 (4) : 203-205

Dhawan A.K. and Sidhu. A. S. 1984[c] : Evaluation of PAU traps for catches of the males of pink bollworm of cotton Indian J. agric Sci. 54 (4) : 318-320

Dhawan, A K ; 1998 : Integrated pest management in cotton status of different management options & future perspectives. Pestology (12) : 21-33

Gupta, G.P. and Kirtisharma 1996. Utilization of biopesticides in managing the cotton pest complex in India. National cotton council 1996 : 1135-1140 Hargreaves, H. 1948: List of reported cotton insects of the world. Commonwealth instt. Ento. London. pp 1-50

Holloway, J and Forester, N.W. 1998 New insecticide chemistry for Australian cotton in sawallo D. (Ed) Proceedings of 9th Australian cotton conference AGGRA, wee waa pp 603 –610

I.A.U.A 2002: IPM on unirrigated cotton in Tribal areas. Mahatma Phule Krishi Vidyapeeth Rahuri pp 7.

Jagmohan, 2001 : Natural enemies of insect pests recorded from Punjab. Biol control contributed paper PAU, Ludhana pp-1

Jayaswal, A.P. and Sundaramurthy, V.T. 1992 : Achievements in insect management in cotton in: All India "Coordinated Cotton improvement Project, Silver Jubilee (1967-1992) Sept 17-19,1992, Nagpur P.P. 117 –151

Khan, M.J. 2002: Bt. Cotton Welcome decision Agriculture Today 5 (4) : 2

Kranthi, K.R.: Armes. N.J.; Rao N.G.V. Sheoraj and Sundarmoorthy, V.T. 1997 Seasonal dynamics of metabolic mechanisma mediating pyrethord resistance in *Helicoverpa armigera* Pestic. Sci. 50:91-98 Lal O.P. 1996 Integrated pest management for sustained crop production APC publication Pvt Ltd, New Delhi 1996 pp 29-44.

Mayee C.D., Rao M.R.K and Yadav M.S. 2001: Cotton march towards new Millenium CICR Nagpur (M.S.) India 19-24 pp

Menon, M.V. and Thangavelu, 1979 : Survey of beneficial arthropods in cotton ecosystem at Coimbatore South Indian Entomon, 4:281-284

Naik, L.M. Parlekar, G.V; Khandge S.V. and Dhunal SV 1976 Further records of predator and parasite of crop pests in central Maharashtra J. Maha Agric Univ. 1 (adi) : 301 –304

Narayanan, E.S. 1960 : Biological control of insect pests and the possibilities of controlling some serious pests of cotton by means of their natural enemies Indian central cotton committee Bombay, pp 1-18

Natarajan and Seshadri, V. 1989 : Abundance of natural enemies of cotton insects under intercropping system J. Biol. Cont. 2: 3-5

Patel, R.C., Yadav. D.N., Joshi, A.D. and Patel, A.K. 1985: Impact of mass trapping of *Helothis armigera* and *Spodoptera litura* from cotton with sex pheomones. Proc. Nat. Seminar Behav. Physiol. Behav. Physiol. Appr, Mgmt. Crop pest TNAU Coimbator 120-122.

Puri, S.N; Sharma O.P; Murthy K.S; and Sheoraj 1998 :Hand book of diagnosis and integrated management of cotton pests, NCIPM, New Delhi, India 100-104 pp

Rajaran V; Ramamurthy R; and Krishnadass D; 2001: Demonstration of integrated pest management in cotton on farmer field J. cotton Res., Dev 15 (2) : 168-170

Rao K.N. 1982 : Pheromonal control of insect pests of agriculture : in India-Problems and prospects Pesticides 16:1-4

Rao M.S. and Reddy, K.D. 1999: Non pesticidal approach in cotton IPM – A review.Agricultural Reviews (Karnal) (1999) 20 (3/4): 203-219

Raodeo, AK and Sarkate M.B 1979: Cotton pests & their natural enemies. Plant Prot. bull (India) 31 (1-4) : 90-98

Raodeo, AK; Sarkate M.B; Kausale, P.P, Bilpate G.G; and Shinde K.S. 1983:Possiblities of biological control of cotton bollworms Cott. Dev. 13 (2) 31-34

Reddy, A.S. and Rosaiah, B. 1987 : Insect pest management In : M Veerabhadra Rao and S. Sithanantham (eds); " Plant protection in field crops", Publ. 4, Plant protection Association of India, Hyderabad, pp. 293-299

Reed, W; Vedmoorty, G; Raghavan M.V. and Rajan M.P. 1975 : Pink bollworm moths: catches in sex attractant traps and nocturnal behavior in South India Cotton Grow Rev. 52:350-359

Reddy, G.P.V. and Krishnamurthy, M.M. 1989 Insect pest management in cotton. Pesticides 23 (10) : 18-19

Rosaih, B. and Reddy, A-S 1993 : Utility of botanicals and biocides as additives to insecticides in *Helicoverpa armigera* (Hub) management of cotton. Paper presented in the "World Neem conference", Feb 24-28 1993 held at Banglore, India.

Sarode, S.V. 1997 : Eco-friendly approach for management of cotton bollworm in integrated pest management in agriculture (eds) Bharad G.M; Bonde, R.S; Nimbalkar, S.A; and Sarode S.V; Dr. PDKV, Akola India pp 157-164

Sarode, S.V., Deotale, R.O; Jumde: Y.,S; and Thakare, H.S. 1994 : Field evaluation of *Heliothis* nuclear polyhedrosis virus (HANPV) for the management of *H. armigera* (Hub) on pigeonpea. Ind. J. Entomol 56 (2) : 176-179

Sarode, S.V. : Thakare, H.S and Kurthatkar T.V. 1993: Neem utility for management of cotton bollworms. Paper presented in the national conference on eco-friendly approaches in the management of pest diseases and industrial effluents held at Kanpur U.P. India PP-42

Satpute US 2003 : Role of biopesticide in agriculture paper presented in state level seminar on pest management for sustainable agriculture Feb 6-7,2003 at MAU Parbhani 68-73 pp

Satpute, U.S; Taley, Y.M. and Bhalerao P.D. 1985: Male catches of pink bollworm with various concentrations of gossyplure. PKV Res. J. 9 (1) : 85-88

Singh R. Daljit Singh and Garg, H.R. 2001: Integrated management of bollworm complex in cotton. Bio-control contributed paper abst, Ludhiana pp 157-158

Simwat, G.S; Sidhu, A.S; and Dhawan, A.K. 1982 : Mortality of diapausing larvae of pink bollworm *Pectinophora gossypiella* (Saund) in cotton sticks during summer in Punjab. J. Res Punjab Agric. Univ. 19 (1) : 35-38

Simwat. G.S, Dhawan, A.K. and Sidhu, A.S. 1991 : Acid delinting of cotton seed and its effect on the survival of diapausing larvae of ping bollworm. J. Insect Sci. 4 (1) : 45-49.

Simwat S.P 1994 Modern concept in insect pest management in cotton in Dhaliwal and Arora (eds.) Trends in Agril insect pest management pp 186

Srinivas S.S. and Patil R.B. 2001: Studies on integrated pest management in cotton on farmers field J. Bio. Cot. 7

Sharma P.K Mukand Singh and Dhawn A.K. 1989 : Management of bollworms in *Gossypium arboreum* L. through nutrient and irrigation in Punjab. J Res. Punjab Agric. Univ. 26 (2) : 204 –205

Sidhu, A.S. and Dhawan, A.K. 1986 : Bollworm damage and carryover in *Hirsutum* cotton in Punjab. J. Res. Punjab Agric. Univ 23:440-450

Singh S.P. 1990 : *Heliothis armigera* (Hubner) prospect of its biological suppression In : S. Jayraj; S. Uthamasamy., M.Gopalan and R.J. Rabindra Eds. Proc. National workshop on *Heliothis* management Feb 18-19 1988, Coimbatore pp 223-229

Singh, S.P. 1994. Fifty years of AICRP on biological control; Project Directorate of Biological Control, Banglore

Sundaramuthy, V.T. and Basu, A.K. 1990 Role of IPM in hybrid cotton production. Paper presented in Training Programme organized on "Hybrid cotton" held at CICR, Coimbatore, Tamil Nadu, India.

Sundaramurthy, V.T. and Chitra, K. 1992 : Integrated pest management in cotton Indian J. Pl Prot 20 (1) : 1-17

Taneja, S.L. and Jayaswal, A.P. 1982 : Capture threshold of pink bollworm moth on hirsutum cotton. Pest Management. 27: 318-324

Thimmaiah, G. and Raju G.T.T. 1991 Possibilities of using Okra (Bhendi) as a trap crop in cotton bollworm management. Curr. ENT. Res. USA 20 (7) 140-142

Uthamasamy. S. and Gajendran G. 1992 : Efficacy of new products in the management of pests of cotton. In National Semianr on Changing Scenario in pest and pest management in India; Jan 31 Feb 1, 1992 Hyderabad P-45

Varma, G.C. and Gill J.S. 1992 : Manipulation of natural enemies in the management of insect pests of sugarcane and cotton in Punjab. In : Proc. Nan. Symp. Recent Advances in Integrated pest management Oct. 12-15, 1992 Ludhiana pp 61-62.

Venkatesan, S, Balsubramanian, G., Sivaprakasan N,; Narayanan A. and Gopalan N. 1987 : Effect of intercropping of pulses and sunflower on the incidence of sucking pests of rainfed cotton. Madras Agric. J. 74 (8-9) : 364-368.

Venugopal, K., Gururajun, K.N. and Kanan, A. 1996 : All India Coordinatedcotton Improvement Project : Impact analysis and further thrusts. In : National Seminar on " Century of cotton production in India" (Souvenir) GAU, Surat, India, pp 66-79

Verma, G.C; Sekhon B.S.; and Sherihnar, M. 1984 : Role of biopesticieds in control of cotton bollworms *pectinophora gossypiella* (Saunders) and *Earias* spp in the Punjab Cott. Dev 18-21

Wilson, L.J; Bauer, L.R; Lally DA 1998: Effect of early season insecticide use on predators and out breaks of spider mites in cotton. Bull Entmol Res. 88 : 477–488

Entomology : Novel Approaches, 2007
Eds.: *P.C. Jain and M.C. Bhargava,*(Page 319-331)
New India Publishing Agency, New Delhi
E-mail: newindiapublishingagency@gmail.com
web: www.bookfactoryindia.com

CHAPTER 16

INTEGRATED PEST MANAGEMENT IN SEED SPICES

P.C. Jain, M.C. Bhargava and R.K. Choudhary
S.K.N. College of Agriculture, Rajasthan Agricultural University, Jobner

INTORDUCTION

Spices are strongly flavored aromatic substances of plant origin which are commonly used for seasoning, cooking food and for preserving other food stuffs. There are about 70 species of spices grown in various parts of the world. Those which are used most commonly throughout the world are grown in India and other parts of South Asia.

The seed spices are annual herbs, whose dried fruits or seeds are mostly used as spices. The seed spices are used for culinary, confectionary, perfumery, cosmetics and pharmaceuticals industry. There are about 20 seed spices grown in India (Table 1) and nine out of them are prominent. The most important among them are belongs to family Apiaceae (coriander, cumin, fennel etc). About 9 lakh ha area is growing in seed spices in the country with annual production of about 5-6 lakh tones (Anonymous, 2003). India is the world's largest producer, consumer and exporter of seed spices. The foreign exchange earned through export of seed spices is more than Rs 200 crores annually. India exports raw spices as well as value added items to nearly 70 countries in the world and is meeting 51% of the global demand of these commodities. There is good potential for increasing export of seed spices, if production as well as quality is increased. The level of productivity in various seed spices is comparatively low as compared to other countries. There are many production constrains attributed to low productivity. The lake of suitable high yielding varieties suiting to different agro-ecological regions and more incidences of diseases and insect pests. Among different insect pests, coriander aphid, *Hydaphis coriandari* (Das) is one of the most destructive and widely distributed pest of coriander (Jain, 1984 and Jain and Yadav, 1989). Recently brown wheat mite was observed attacking coriander and other spices (Jain and Yadav, 1986 and 1989).

Table 1 List of common seed spices

Sr. No.	Botanical name	Common name	Family
1.	*Anethum graveolens* L.	Dill	Apiaceae
2.	*Anethum sowa* Kurz	Indian dill	" " "

Contd. ...

Table 1 Contd. ...

3.	*Apium graveolens* L. var. *dulce* (Miller) Pers	Celery, garden celery	Apiaceae
4.	*Bunicum persicum* (Boiss)	Black caraway	" " "
5.	*Carum bulbocastanum* L.	Black caraway	" " "
6.	*Carum carvi* L.	caraway	" " "
7.	*Coriandrum sativum* L.	Coriander	" " "
8.	*Cuminum cyminum* L.	Cumin	" " "
9.	*Foeniculum vulgare* Miller	Fennel	" " "
10.	*Petroselinum crispum* (Miller)	Parsley	" " "
11.	*Pimpinella anisum* L.	Anise	" " "
12.	*Trachyspermum ammi* (L.) Sperague	Alowain	" " "
13.	*Brassica juncea* (L.) Czernj et Cosson	Indian mustard	Brassicaceae
14.	*Sinapsis alba* L.	White mustard	" " "
15.	*B. nigra* (L.) Koch	Black mustard	" " "
16.	*Nigella sativa* L.	Black cumin	Ranunceelaceae
17.	*Papaver sominiferus*	Poppy seed	Papaveraceae
18.	*Sesamum indicum* L.	Sesame	Pedaliaceae
19.	*Trigonella foenumgraecum* L.	Fenugreek	Fabaceae
20.	*T. corniculata* L.	Kasuri methi	" " "

Anonymous, 2003

Table 2 Status of seed spices production in India

Spices	Major sharing states in the total production in India (%)			
Coriander	Rajasthan	(63)	A.P.	(18)
Cumin	Gujarat	(63)	Rajasthan	(37)
Fennel	Gujarat	(84)	Rajasthan	(37)
Fenugreek	Rajasthan	(80)	Others	(20)
Ajowain	Gujarat	(45)	Rajasthan	(40)
Dill (Sowa)	Rajasthan	(45)	Gujarat	(43)
Celery	Punjab	(85)	Haryana & U.P.	(15)
Kalongi	U.P.	(35)	M.P.	(30)

Anonymous, 2003

INSECT PESTS UNDER FIELD CONDITIONS

The seed spices are prone to attack by insect pests both in field and during storage. Being high value crops, even moderate injury translates in to considerable economic losses, therefore, pest monitoring needs to be done regularly. Wherever available, pest resistant varieties should be used. In areas of regular occurrence of pest, preventive measures should be taken. When injury exceeds the acceptable levels integrated pest management strategies including cultural, mechanical, biological and chemical practices should be applied.

The common pests can broadly classified into categories like subterranean pests (soil dwelling pests damaging roots or shoots such as termite, white grub, cutworm etc.); defoliators (damaging leaves like caterpillars, weevils, beetles, miners); sucking pests (devitalizing plants through sap sucking and transmitting viral or other diseases like aphids, jassids, bugs, white flies, mites) and the borers. Barring the last category spice crops are attacked by all type of pests. Most of the pests of these crops are polyphagous, attacking a variety of vegetation and a very few spices are exclusive to the seed spices. As the some insect attacks more than one seed spice crop, we shall consider individual insect for the purpose of this presentation to avoid repetition under each crop, as the management strategies would be more or less alike for all crops under consideration. The important insect pests of seed spices under field conditions are as follows.

Coriander aphid, *Hyadaphis coriandri* (Das) (Hemiptera: Aphididae)

The coriander aphid, *Hyadaphis coriandri* (Das), was found for the first time in North America on fennel (*Foeniculum vulgare* Mill.) at Orange County residences in Apopka and Orlando. In December 1998, an infestation was also found on coriander (*Coriandrum sativum* L.) and dill (*Anethum graveolens* Linnaeus) at a residence in Tampa, Hillsborough County. There are about thirteen described species of *Hyadaphis* (Remaudière and Remaudière 1997), most of which are obscure Central Asian or European species (Ivanovskaya 1977, Kadyrbekov 1990, Miyazaki 1971, Raychaudhuri *et al.* 1980, Shaposhnikov 1964). There are several species in the genus known to be pests, including the coriander aphid, *Hyadaphis foeniculi* (Passerini), *Hyadaphis passeriniii* (Del Guercio) and *Hyadaphis tataricae* (Aizenberg). Only three *Hyadaphis* species occur in the Western Hemisphere. *Hyadaphis foeniculi*, the honeysuckle aphid, has been in North America for a long time. Its winter host is honeysuckle, and summer hosts are various Umbelliferae including some cultivated crops. There are no Division of Plant Industry specimens or records of this species in Florida. *Hyadaphis tataricae*, the honeysuckle witches broom aphid, was introduced into the Western Hemisphere in the mid 1970s and had colonized northern USA and southern Canada within about ten years. It apparently is restricted to northern states (Voegtlin, 1988).

Coriander aphids are yellow-green in colour, dusted with greyish wax. They have short, dusky, slightly swollen, siphunculi (or cornicles) that are about twice as long as wide. They form dense and often damaging colonies on leaves, heads, and stems of their host plants. In Florida, several other species colonize the common hosts of the coriander aphid. These include various *Aphis* spp., the green peach aphid, *Myzus persicae* (Sulzer) and the rice root aphid, *Rhopalosiphum rufiabdominalis* (Sasaki). Both green peach aphids and the *Aphis* spp. have siphunculi that is much longer than twice their diameter. Rice root aphids live underground.

Life cycle: Overwintering eggs occur on *Lonicera nummulariifolia* Jaub. & Spach. and other *Lonicera* spp. (Remaudière and Halbert 2000). A fundatrix hatches from the egg in the spring. Her offspring are parthenogenetic, winged-female spring migrants. They colonize summer host plants in the family Umbelliferae. During the summer, there are many generations of aphids. All individuals are parthenogenetic females, but they can be either winged or wingless, depending on host plant quality and crowding in the colony. If host plant quality declines, or the colony becomes crowded, winged individuals form that can establish colonies on new plants. In the autumn, parthenogenetic female fall migrants occur in response to cooler temperatures and/or short days. These return to the winter hosts, where they give birth to egg-laying females. Similarly, the summer colonies produce winged males in the fall. The males also return to the winter host plants, where they mate with the egg-laying females to produce the overwintering eggs.

Nature of damage: Species of *Hyadaphis* are associated with Caprifoliaceae (winter hosts) and/or Umbelliferae (summer hosts) include coriander, dill, fennel and parsley, *Petroselinum crispum* (P. Mill.) Nyman ex A.W. Hill. Other reported hosts include celery, *Apium graveolens* Linnaeus var. *dulce* (Miller) DC. and carrots, *Daucus carota* Linnaeus var. *sativus* Hoffm. The coriander aphid is considered a key pest of coriander in India in the spring (Kumar and Sagar, 1994). The nymphs and adults suck the cell sap from the plants and their honey dew secretion provided good medium for the development of shooty mould. Aphid causes both the quantitative and qualitative losses in the seed yield and also deteriorates the green leaves by sucking cell sap and secreting honey dew. Due to their fast multiplication within few days, aphids cover the entire surface of apical shoots and as a result of continuous feeding by such a large population, yellowing, curling, and subsequent drying of leaves takes place, resulting in poor and shriveled seed formation.

Peach leaf curl Aphid, *Myzus persicae (Solzer)* (Hemiptera : Aphididae)

This world wide aphid is a very destructive pest of cumin, and other spices. It is prevalent in plains of India in spring and early summers, the infested plants show curling of the new leaves. This damaged is caused by nymphs and females, which are confined to the growing shoots and leaves, from where they suck the cell sap.

Life cycle: The pest is active from February – March and from June to October. The females produce eggs without fertilization and these hatches inside the body of the mother. Thus, the females give birth to young ones instead of laying eggs. Each viviparous female produces about 50 young aphids in her short life time of about 13 days. Three or four generations are completed on the plants. With the warming up of the season winged males and females are produced. They migrate to other alternative host plants and again start reproducing

Damage: The nymphs and females suck the sap from young leaves and cause them to curl. The severely attacked plants have only small crumpled or twisted leaves and bear very little seeds. The curling of the leaves is also caused by the fungus *Taphrina deformans* Berk and Tull, in which a curled leaf becomes thickened near the midrib and is brittle and lustrous.

***Management*:** Both the aphid species can be managed by following ways

Management Option	Recommendation
Scouting/ Thresholds	No thresholds have been determined. Examine plants weekly, especially after petioles start to form. Intermediate age stalks are most likely to be infested. Check field edges carefully, especially if celery or parsley is being grown next to lettuce.
Cultural Control	Early sowing and selection of less susceptible varieties proved a suitable component for the management of aphid on coriander. In areas of high pest incidence the crop should be sown from the last week of October to the 1st week of November. During the flowering stage a population of 55-70 aphids per 5 plants could reduce the yields by 50%, so crops should be protected just before flowering. Field disking and destruction of crop residues are important for control of all caterpillar, aphid and leafminer pests of leafy vegetables to reduce their migration into nearby crops.
Chemical control	Chemicals with systemic or translaminar penetrating activity are effective for aphid control. Avoid broad-spectrum pesticides, if possible, to preserve natural enemies. As soon as the pest appears on growing points, spray of malathion 50EC @ 0.05 % or dimethoat 30 EC @ 0.03% or oxydemeton-methyl 25EC@ 0.025% can effectively control the pest population. Second spray should be given after 10 days.
Natural Enemies	Many types of natural enemies may control these aphids under low insecticide input situations. However, these aphids reproduce quickly and move into protected areas of the plants, thereby greatly reducing the potential impact of their predators and parasitoids in older stage plants. Pathogens kill many winged adults before they begin producing nymphs. The coccinellid beetles, *Coccinella septempunctata* L. and *Menochilus sexmaculatus* (Fab); *Chrysoperla carnea* (Stephens) and *Ischiodon javana* (Wiedemann) (Diptera: Syrphidae) are the main predators of these aphids.
Resistant varieties	**Coriander**: ND Cor 35, Rajendra Swathi, Haritima, DH-205, JCO-123, JCO-70, ATP-02, UD-28, UD-303, DH-53, Panipat Local, CS-193, UD-684RCr-446, RCr-436 **Fenugreek**: UM-302, HM-291, UM-304, PEB, BDJ-11, BDJ-86-59, PLM 98 Cvs IC 59304, Sel.10, Sel. 38, Sel.95-11, PLM-24, PLM-25, PLM-105, PLM-259 These varieties have been reported to be less susceptible to aphid attack

Indigo Caterpillar, *Spodoptera exigua* (Hubner) (Lepidoptera : Noctuidae)

This polyphagous insect is a serious pest of young plants. The caterpillars feed on leaves, making webs within which they feed gregariously for 2-3 days. The moths have dark spots on fore wings and white on hind wings. They are active at night but remain hidden under various shrubs in the day time. The colour of larva depends on the crop on which they feed.

Life cycle: The pest passes winters as a pupa which is enclosed in webbing, forming a rough pupal cocoon. Some of them emerged as moths in February, but a number of them continue in that state till the beginning of the monsoon. A female lays up to 200 eggs in clusters. The eggs are spherical and resemble poppy seeds in shape and size, having lines, radiating from the center. The egg clusters are covered with buff hairs. The eggs hatch in 1-3 days and the young caterpillar start feeding in groups. They feed voraciously on the epidermis and also make webs. Within 2-3 days, the gregarious larvae spread out and hide under various shelters, from where they come out in the morning and evening to feed on the leaves. The larval stage is completed in 15-20 days and the full larval seek shelter in debris, on the ground. They spin silken webbing and prepare rough cocoons with bits of leaves and other material. The pupal stage lasts 5-7 days and the life cycle is completed in about 30 days. From November onwards, the pest hibernates in the pupal stage. During the active season, a number of generations are completed.

Damage: The damaged crop, on which larvae feeds give a webbed appearance, the older caterpillar, which feed in morning and evening have a voracious appetite and strip off quite large patches on the foliage. However, the young plants suffer the most. The early shown varieties suffer greatest losses then those sown later.

Management

- Natural enemy *Euplectus* sp.
- Collection and destruction of egg masses and the caterpillars, which are feeding with in the web.
- Spray of fenitrothion 50 EC @ 0.05 per cent or endosulfan 35 EC @ 0.035 per cent

Brown wheat mite, *Petrobia latens* (Muller) (Acari: Tetranychidae)

The brown wheat mite, *Petrobia latens* was first reported feeding on coriander (Jain and Yadav 1988). It remains active during winter with peak activity in March. It is a minute, non webbing and swift moving mite and has a tendency to dislodge from the plant when disturbed. The larva, nymphs and adults feed on upper as well as the lower surface of leaves, leaf sheaths and floral parts. Infested leave started withering from top downwards. The plants become chlorotic due to loss of cell sap, resulting in poor seed formation. Heavily infested plants show sickly yellowish or bronzing appearance (Bindra and Kittur 1961 and Anonymous, 1997)

Life cycle: The mite complete its life cycle in eight stages i.e. four active and four quiescent stages. The winter eggs are deposited mainly on the dry leaves. The mite lays two types of eggs "non diapausing winter eggs" and "dipausing summers eggs". The brick red non diapousing eggs are laid singly or in batches. Diapousing summer eggs are white and are laid in clusters on soils clumps. At one time, a female lays on the one type of eggs. Generally no males are found thus reproduction is parthenogenetic. A single female lays 8-13 winter or 2-5 summer eggs per day. The incubation period ranges from 8-13 days at ambient temperature.

Diapousing eggs remain dormant throughout summers and rainy seasons and are coated with a white substance. The newly hatched larvae with three pairs of six segmented leg is colourless and turns brown after feeding. The larval period lasts for 1-2 days followed by a quiescent stage. The active greenish to brown protonymph possessed four pairs of legs. The protonymphal period ranges from 0-3 days followed by a quiescent stage. The deutonymph is more or less similar to protonymph except in size. The deutonymphal period varies from 1-4 days and is followed by a quiescent stage before it develops in to a adult. The adults are very active and start feeding soon after emergence. The fully matured females are metallic brown in colour and measures 0.58 mm in length and 0.46 mm in width. The adult female live for 20 days.

Management

Natural enemies *Lasioseius terrestris* Summer and Schlinger a predatory mite was reported by Menon and Ghai (1968). Sharma *et al.* (1994) and Sharma (2001) observed that *Amblyseius alstoniae* Gupta, as efficient predator in the control of *P. latens* population under field conditions in Jobner area of Rajasthan. Sulphur at high dose of 30-40 lbs/acre effective against the brown wheat mite infestation. Thiometon and dimethoate each at 0.15 % and parathion 0.06% and parathion dust 1.5% @24.7 kg/ha effectively controlled the mite. Monocrotophos @0.15 kg ai/ha and methyl-o-demeton @ 0.10 kg a i /ha are also found effective against the mite.

INSECT PESTS UNDER STORAGE CONDITIONS

In India the domestic consumption of the important seed spices *viz*, cumin, coriander, ajowain, fenugreek etc. is quite high as they are routinely used for culinary and medicinal purposes. Apart from the domestic consumption, spices are being exported to various countries from India. We earn more than Rs. 200 crores from the export of seed spices annually (Baskaran and Rajendran, 2003). The world consumption of spices is growing steadily every year. Expansion of our export of spices to increase or even to retain our share of world market is imperative. This can be achieved only through increased productivity and improved quality. From harvesting to the final product reaching the consumer, several steps are involved for ensuring the satisfactory quality of the final product reaching the end users. Losses of harvested produce may be quantitative or qualitative and these two losses may occur together or separately.

Quantitative losses

Losses in the weight of harvested produce other than that due to moisture reduction are considered as quantitative losses. These may occur due to improper handling and transport resulting in spillage, feeding by insects, rodents or birds and use of poor containers and stores.

Qualitative losses

Deterioration in the quality of produce, apart from causing monetary losses also results in loss of nutrients, seed and very importantly goodwill of the purchasers. The quality of seed spices is adversely affected to many factors *viz*, pests, micro-organisms, foreign matter impurities which get into products from materials used in processing and micro-organisms and dirt introduced into the product through unhygienic practices of the people who handle the produce. As well as loss of quality that results from short-coming in storage practices. Harvested produce get deteriorated between harvest and consumption in various ways in resulting in physical, biological and chemical changes in the final product. These changes are manifested in the form of change in appearance, smell, taste etc. which can be well detected by human sense.

Chemical changes

Due to the high temperature, infestation of moulds, insects or mites, food grains may undergo certain chemical reactions which lead to loss valuable nutrients from the grains and addition of undesirable chemicals. Overexposure to the sun destroys certain vitamins in stored grains. High temperature artificial drying causes loss of thiamine. Insect infestation results in loss of carbohydrates, proteins and vitamins and release of undesirable chemicals like uric acid. It was observed reduction in proteins, fat and ash content in coriander powder and increase in moisture content after 3 and 6 months of infestation by drugstore beetle, *Stegobium paniceum* (L.) and cigarette beetle, *Lasioderma serricorne* Fab. The changes in proximate components were more pronounced after 6 months than after 3 months. The uric acid level exceeded the permissible (PFA) limit of 100mg/100kg following infestation in 3 or 6 months.

Cigarette beetle, *Lasioderma serricorne* Fab. (Coleoptera: Anobiidae)

The cigarette beetle infests under storage conditions a vast variety of commodities including such specialized commodities as spices, saffron, nuts etc. It is distributed practically all over the world. The cigarette beetle is native of Egypt. In fact, a beetle was found in King Tutankhamen's tomb in the 3,500 years since then the beetle has hardly changed.

The adult cigarette beetle is a small, red-brown to yellowish-brown beetle. When viewed from above it appears oval and in profile humpbacked in shape (the head and pronotum are bent forward). It is 3 to 4 mm long. The antennae are serrate (the side edge of each antennal segment is pointed like a saw tooth). The wing covering (elytra) are smooth without longitudinal groove. They are strong fliers and attracted to light at night such as a lamp or TV set. It can be found throughout the year, but seems to be more common in the cool months. The eggs are white oval and too small to be easily seen with the naked eyes. The larvae are white and covered with long hairs give them a fuzzy appearance. When full grown they are about 4mm long. Larvae spin themselves into a cocoon prior to becoming pupae.

Life cycle: The adult beetles live from 2-4 weeks and do not feed. The mother beetle lay up to 100 eggs in all kind of rather hidden safety niches like the space between the folds of the leaves, in nooks and corners, in crevices and depressions, in between grains like coriander etc. The eggs are so small that they often escape notice. These are whitish and opaque and pointed at one end. The incubation period may be as short as 4 days but it can get lengthened to more than two weeks, depending on the temperature and humidity of environment. The larva on hatching out is a grayish white grub about 1 mm in length and its body is covered with a fine hairy coat. The feeding is carried out by the larva results in characteristic tunnel running both longitudinally and across the food. When full grown the larva makes flimsy cell in which it pupates. The fresh pupa is whitish in colour but it turns brownish later on. The pupal period vary from 5 to 31 days.

Nature of damage: As the name would indicate it is a pest of stored tobacco, but also causes considerable damage to other stored products. Both the grubs and beetles attack making holes through the food materials. In serious infestation, numerous small light coloured brown headed grubs or brown beetles can be seen doing damage.

Drugstore beetle, *Stegobium paniceum* (Linnaeus) (Coleoptera: Anobiidae)

The drugstore beetle is a cosmopolitan pest that can infest almost any dry animal or plant material. It is particularly noted as a pest of pharmacies where it infests drugs, and domestic premises where it attacks breakfast foods, biscuits and herbs, amongst its eatholic diet. The beetles are cylindrical 2.25 to 3.6 mm long, and are uniform brown to reddish brown. They have longitudinal rows of fine hairs on the elytra (wing covers). Drugstore beetles are similar in appearance to the cigarette beetle; however, two physical characters can be used to tell the difference between them. The antennae of the cigarette beetle are serrated while the antennae of the drugstore beetles are not and end in a 3 segmented club. The other difference is that the elytra of the drugstore beetle have rows of pits giving them a striated appearance while those of cigarette beetle are smooth.

The larvae of drugstore beetles are white in colour; the later instars are scarab-like. They are similar to cigarette beetle larvae, but have shorter hairs and the marking on the head ends in a straight line across the frons just above the mouthparts.

Life cycle: Females lay up to 75 eggs in the food or substrate. The larval period ranges from 4 to 20 weeks. Larvae tunnel through the substrate and when fully grown build a cocoon and pupate. Pupation takes from 12 to 18 days. Adult females live approximately 13 to 65 days. The entire life cycle is generally less than two months but can be as long as seven months. The duration of the life cycle is highly dependent on the temperature and food source. Development occurs between 60 to 93°F (~15 to 34°C) but is optimal at about 85°F (~30°C) and 60 to 90% relative humidity.

Nature of damage: The drugstore beetle attacks such a wide variety of foods and material that one anonymous quote states that it "eats anything but cast iron." It gets its name from its habit of feeding on prescription drugs. It also feeds on flours, dry mixes, breads, cookies, chocolates and other sweets, and spices. Non-food material includes wool, hair, leather, horn, and museum specimens. It is found in pigeon nests and is known to bore into books, wooden objects and in some cases, tin or aluminum foil and lead sheets. Larval feeding accounts for the greatest amount of damage. Museum and herbarium specimens are vulnerable to attack. Slight damage and contamination can ruin these valuable and priceless items. Drugstore beetles harbor symbiotic yeasts that produce B vitamins. The yeasts are deposited on the eggs as they pass through the oviduct and are consumed by the larvae during egg hatch. These yeasts enable the drugstore beetle to feed and survive on many foods and other items of poor nutritional quality.

MANAGEMENT OF STORAGE INSECT PESTS

Home: Preventing and controlling beetle infestations is relatively simple: Insecticides should be used only as a last resort. Locating the source or sources of infestation is the first and most important step. Heavily infested items should be wrapped in heavy plastic, taken outside and thrown away. All food containers and items should be checked for infestation. Items can be placed in the refrigerator or freezer (16 days at 36°F, 7 days at 25°F or 32°F for four to seven days) to kill all stages. Place the items in a plastic bag to reduce condensation problems during thawing. Heating small quantities of infested material (190°F for one hour, 120°F for 16 to 24 hours) in an oven is also effective. Uninfested items can be cold- or heat-treated to ensure that any undetected stages are killed. To prevent reinfestation, clean up spilled flour, mixes, crumbs, etc. and thoroughly vacuum and clean areas where the contaminated items were stored. Store foods in airtight glass, metal or plastic containers. Clear containers make it easier to check for infestation. Chemical treatment using commercially available insecticides or insect growth regulators is usually not necessary but if used, be sure to follow the label.

Commercial: The cigarette and drugstore beetle are common pest of processed and packaged seed spices. These are also found in flour mills, bakeries, pet food, breakfast cereal manufacturing, and snack food plants, chocolate factories, confectionaries, wholesale distribution centers and sometimes retail stores. Integrated pest management (IPM) programs are often implemented to control infestations at processing, distribution and storage facilities.

Large-scale control

To bring down losses due to insect pests, control measures are adopted at all stages from harvest to storage and during shipment. In the control of stored seed spices insects pests fumigation play a vital role. Seed spices are disinfested mainly with phosphine or methyl bromide as the situation demands. For insect control in stored spices the major chemicals deployed the methyl bromide, phosphine and ethylene oxide.

Methyl bromide is used in all quarantine treatments where time is of primary concern. It is available in cylinders and in canes with or without chloropicrin as the warming gas. Methyl bromide fumigation is carried out under normal atmospheric or at reduced pressures, the exposure period being 4-24 hours. Repeated treatments or application at excessive dosage of methyl bromide must be avoided for spices and spice mixes as it will lead to unacceptable levels of inorganic bromide residues and may cause discoloration of some of the spices and loss of essential oils (Pruthi, 1980; Reeves *et al.*, 1985). Methyl bromide has been reported as carcinogenic and ozone depletor and hence, the fumigant is under severe criticism in recent years. The fumigant is being phased out in stages from 2000 onwards both in the developed and developing countries (Muller, 1998).

Phosphine fumigation is however advantageous for spices as the residue problem is less. However, repeated fumigation under inadequate concentrations may lead to development of insect resistance to phosphine as demonstrated in *L. serricorne* infesting tobacco (Rajendran and Narasimhan, 1994). Insect resistance to phosphine is a major constraint for the continued use of the fumigant (Banks *et al.*, 1994). To over come the insect resistance problem, modified application techniques such as SIROFLO, SIROCIRC and PHYTO EXPLO and newer formulations (eg. Cylinder based phosphine) with reference to grain fumigations have been developed in recent times. However the application of these techniques is yet to be extended for treating stored seed spices.

Ethylene oxide treatment has been used primarily for sterilization or control of microorganisms in spices and during the process insects are also killed. The chemical is relatively less toxic to insect pests. Being an inflammable gas it is always marketed in mixture with Co_2 or Freon. Ethylene oxide is a suspected carcinogen and is known to leave carcinogenic residues such as ethylene bromohydrin/chlorohydrin (Gerhardt and Ladd-Effio, 1983). Therefore many countries have withdrawn the use of ethylene oxide for pest control in food commodities (UNEP, 1994)

Residual insecticides registered for use on these beetles can be applied to cracks, crevices and shelves in storage areas after removal of stored products (check labels for specific use). Insect growth regulators (IGR) also are used as part of an IPM programme. Sticky traps baited with the female sex pheromone, stegobinone, can be used to monitor for drugstore beetles. Mass-trapping which involves setting out large numbers of traps in infested areas with the intent of catching large numbers of beetles to reduce the population has been tried, but effectiveness is limited because only males are attracted to the traps. For mass trapping of drugstore beetles to succeed, a large enough number of males must be caught to ensure that few females will be mated. Pheromone traps are mostly used for detecting and monitoring beetle populations. The traps typically are set out in a grid-like pattern within a facility and monitored at regular intervals for trapped males. If one or more traps in a particular area happen to catch a greater proportion of beetles, then more traps are set out in that area to pinpoint the source of infestation.

Natural Enemies: The beetles are attacked by a variety of generalist natural enemies. Predators include *Tenebriodes* (Tenebrionidae), *Thaneroclerus* (Cleridae) and several carabids. Eggs may be eaten by predatory mites. Parasitoids include wasps in the families Pteromalidae, Eurytomidae and Bethylidae. However, biological control for stored-product pests has not been widely adopted. Part of the problem is that although it reduces the amount of pesticides used for control, the release of beneficial insects to control stored-product pests increases the total amount of insects and insect parts in foodstuffs.

Chemical Ecology: The drugstore beetle sex pheromone, stegobinone (2,3-dihydro-2,3,5-trimethyl-6-(1-methyl-2- oxobutyl)-4H-pyran-4-one) is used in commercially available traps and lures. Stegobinone has also been identified as the sex pheromone of a wood-feeding anobiid, the common furniture beetle, *Anobium punctatum*. Wind-tunnel assays in one study revealed that male furniture beetles were attracted to female drugstore beetle extracts. Thus, stegobinone-baited lures can be marketed for two different pest species. Stegobinone has also been tentatively identified as the sex pheromone of two other wood-destroying anobiids from the mid-Pacific coast of North America, *Hemicoelus gibbicollis* and *Ptilinus basalis.*

References

Anonymous, 1997. Management of brown wheat mite, *Petrobia latens* (Muller) in Rajasthan. *Final report of ICAR adhoc project*, Rajasthan Agricultural University, Bikaner

Anonymus, 2003. Seed Spices News letter. NRC on seed spices, Tabiji, Ajmer, 3 (1): 1-6

Banks, H.G.; Hilton, S.J.; Tarr, C.R. and Thron, B. 1993. Demonstration of Co_2 disinfestation of containerized dried vine fruit in export cartons. *CSIRO Entomol. Division Rept.* No. 54, pp. 9

Baskaran, N.V. and Rajendran, S. 2003. Effect of insect infestation on proximate composition of selected stored spice products. *Journal of food science and technology*, Maysore, 40(2): 239-242

Bindra, O.S. and Kittur, S.U., 1961. Two new records of tetranychids as crop pests in Madhya Pradesh. *Indian J. Ent.*, 23:71

Gerhardt, U. and Ladd-Effio, J.C. 1983. Ethylene oxide residue in spices. *Fleischwirtschaft*, 63: 606-608

Hartsell, P.L.; Vail, P.V.; Tebbets, J.S. and Nelson, H.D. 1991. Mehtyl bromide quarantine treatment for codling moth (Lepidoptera: Tortricidae) in unshelled walnuts. *J. Econ. Entomol.*, 84: 1289-1293

Ivanovskaya, O.I. 1977. Aphids of western Siberia. Volume II. Novosibirsk. (In Russian). 328 p.

Jain, P.C. and Yadav, C.P.S., 1986. Occurrence of brown wheat mite, *Petrobia latens* (Muller) on coriander. *Acarology News Letter*, 2

Jain, P.C. and Yadav, C.P.S., 1988. New record of brown wheat mite, *Petrobia latens* on coriander. *Indian J. Ent.*, 50(3): 396

Jain P.C. and Yadav, C.P.S., 1989. Estimation of losses due to coriander aphid and brown wheat mite in coriander. *Indian J. appl. Ent.*, 3: 5-10

Jain, P.C., 1984. Incidence of pests and their control on coriander (*Coriandrum sativum* L.). *Ph.D. Thesis, Sukharia University, Udaipur*

Kadyrbekov, R.X. 1990. Aphid fauna of trees in natural ecosystems of southeastern Kazakhstan. *Kazakhstan Academy of Sciences, Institute of Zoology, Almaty*. 38 p.

Kumar, N. and Sagar, P. 1994. Seasonal history and host range of the coriander aphid, *Hyadaphis coriandri* (Das). *Journal of Research, Punjab Agricultural University* 31: 283-284.

Menon, R.M.G. and Ghai, S., 1968. Further records of distribution of *Petrobia latens* (Muller) (Acarina: Tetranychidae), a pest of wheat in India together with the description of a new species of predatory mite on the same. *Indian J. Ent.*, 30:77-79

Miyazaki, M. 1971. A revision of the tribe Macrosiphini of Japan (Homoptera: Aphididae, Aphidinae). *Insecta Matsumurana* 34: 1-247.

Muller, D.K. 1998. *Fumigants and pheromones Newsletter*, 50:2-3

Purthi, J.S. 1980. Spices and condiments – Chemistry, Microbiology, and Technology. Academic Press Inc., New York, pp. 450

Rajendran, S. and Narasimhan, K.S. 1994. Phosphine resistance in *Lasioderma serricorne* and overcoming control failure during fumigation of stored tobacco. *Intern. J. Pest Managt.*, 40: 207-210

Raychaudhuri, D.N.; Ghosh, A.K.; Basu, R.C.; Ghosh, M.R.; Chatterjee, M.; Chakrabarti, S. and Pal, P.K. 1980. *Aphids of North-east India and Bhutan.* The Zoological Society, Calcutta. 521 p.

Reeves, R.G.; McDonell, C.A. and Ford, J.H. 1985. Organic and inorganic bromide residues in spices fumigated with methyl bromide. *J. Agric. and Food Chem.*, 33: 780-783

Remaudière, G. and Halbert, S.E. 2000. *Hyadaphis coriandri* (Das), espèce hètèroecique distincte de H. tataricae (Aizenberg) [Hemiptera, Aphididae]. *Ann. Soc. Entomol. Fr.* 36: 157-163.

Remaudière, G. and Remaudière, M. 1997. Catalogue des Aphididae du Monde. *INRA. Versailles.* 478 p.

Shaposhnikov, G CH. 1964. Suborder Aphidinea, pp. 616-799. *In* G. Ya. Bei-Bienko [ed.], Keys to the insects of the European USSR Vol. I Apterygota, Palaeoptera, Hemimetabola. *Israel Program for Scientific Translations, Jerusalem.*

Sharma, A., 2001. *Amblyseius alstoniac* Gupta, a mite predator of *Petrobia latens* (Muller). *National Conference: Plant Protection New Horizons in the Millennium*, February 23-25, 2001, pp 57 (Abstract)

Sharma, A., Pareek, R.K. and Nanda, D., 1994. *Amblyseius alstoniac* Gupta, a potent predator of *Petrobia latens* (Muller). *Indian J. appl. Ent.*, 8: 71-72

UNEP (United Nations Environment Programme) 1994. Report of the Methyl bromide technical options committee (1995. Assessment Montreal protocol on substances that deplete the ozone layer) November, 1994 pp.304

Voegtlin, D.J. 1988. *Hyadaphis tataricae* (Homoptera: Aphididae): 10 years after its introduction into North America. *Proceedings of the Entomological Society of Washington* 90: 256-257

Entomology : Novel Approaches, 2007
Eds.: *P.C. Jain and M.C. Bhargava,*(Page 333-351)
New India Publishing Agency, New Delhi
E-mail: newindiapublishingagency@gmail.com
web: www.bookfactoryindia.com

CHAPTER 17

RECENT ADVANCES IN MANAGEMENT OF INSECT PESTS OF FRUIT CROPS

D.R. Sharma and R.C. Batra
Department of Horticulture, Punjab Agricultural University, Ludhiana 141 004

INTRODUCTION

In India, climatic conditions are very diverse and fruit crops of temperate, sub-tropical and tropical regions are grown. Frequently, the fruit crops of one regional environmental requirement are grown in another region. This type of crop husbandry has some typical pest problems as the insect and mite pests have easy access to any of these regions depending upon the fruit crop. The application of management approaches in isolation of insect pests on fruit crops demand a unified strategy to suppress the different pest populations under the concept of 'crop management'. Thus, to sustain the crop management for the better fruit quality, a judicious integration of safe chemical/biopesticide molecules must be used with other non-chemical approaches namely host plant resistance, bioagents, cultural and mechanical practices etc.

Pest suppression strategy

Direct/curative	Indirect/preventive
Mechanical	**Cultural**
• Collecting	• Sanitation
• Trapping	• Vigorous growth
• Spiking	• Plant density
• Pruning	• Tillage
• Banding	• Irrigation
Biological	**Mechanical**
• Parasitoids	• Collection & destruction
• Predators	• Banding
• Pathogens	• Fencing
Chemtherapeutics	**Chemical**
• Dusting	• Repellant
• Spraying	• Antifeedant
• Fumigating	• Protective chemicals
• Using poison baits	**Regulatory**
• Killing	• Quarantine
• Repelling	• Inspection
	• Certification
	Host plant resistance
	Reproductive suppression

The scope of success of non-chemical approaches. Prior to take up any control method for any insect pest of fruit crops, some important points like the identification of the most vulnerable stage and seasonal activity of the pest in relation to host plants, maintaining a low threshold level, a principle of no eradication, precise knowledge of orchard ecosystem, identifying principal factors, to determine economic threshold, regular assessment of pest and natural enemies populations, to develop sequence of compatibility of available control methods and modification of management programme according to the condition and available inputs. Tactics commonly adopted for the control of fruit pests are classified into curative/ direct and preventive/indirect methods. These methods include cultural, mechanical and physical; chemotherapeautic, regulatory, biological, plant resistance and reproductive, either to curb pest outbreaks or create conditions unfavourable for outbreaks of the pest are described with some pertinent examples.

NON-CHEMICAL APPROACHES

Various non-chemical approaches are being utilized in the management of fruit insect and mite pests. These methods are helpful in providing a way to reduce the pest population through reduction in carryover of the pest as well as the conditions suitable for the conservation of natural enemies.

Cultural control

Crop husbandry of fruit crops is radically different to that of field crops and under many situations the pest problems are of some permanent nature. Several cultural practices if performed regularly in orchard ecosystem would be helpful in either elimination or atleast reducing the carryover of pest population.

Clean Cultivation: The practices which create conditions unfavourable for pest development, are very important in regulating the pest populations. Clean culture, which eliminates the weeds and other secondary hosts, is useful in avoiding the carryover of some pests. It is useful in either reducing the opportunity of pest to lay eggs or feeding in the off-season. It has been found promising in controlling banana leaf and scaring beetle; scales and mealy bugs on citrus, pineapple and mango; thrips and shot hole borer on grapes and apple; fruit sucking moth on citrus; pomegranate butterfly [*Deudorix isocrates* (Fabricius)]; San Jose scale [*Quadraspidiotus pernisiosis* (Comstock)] on apple, peach leaf curl aphid (*Myzus persicae* Sulzer) on peach and hoppers and mealy bug on mango.

Tillage: It has been proved useful in controlling the dormant stages of pests either by burying them deep in the soil so that they are unable to emerge or brought to the soil surface for exposing them to inclement weather conditions and/or natural enemies. Many insects like mango mealy bug [*Drosicha mangiferae* (Green)], gall makers, leaf-cutting weevil, litchi nut borer (*Blastobasis* sp. and *Gatesclarkeana* sp.), fruit fly of guava, pear and peach [*Bactrocera dorsalis* (Hendel)] and *ber* (*Carpomyia vesuviana* Costa) are found to be controlled by plouging the soil. Mango mealy bug (*D. mangiferae)* can effectively be controlled by ploughing the soil during May-June as the gorged female lay eggs in the soil. The eggs are exposed to either intense heat or natural enemies. Similarly, the pre-pupal and pupal populations of *B. dorsalis* occurred in the soil and shallow ploughing with cultivator immediately after harvest is effective in exposing and killing the pupating larvae/pupae which are mostly present at 4-6 cm depth. Ploughing also destroys the hibernating mango stone weevil [*Sternochaetus mangiferae* (Fabricius)] while raking the soil during October-November and again in March can manage the pest in mango orchards.

Plant density: Plant densities change the micro-climate habitat e.g. dense banana planting provides congenial conditions for banana aphid (*Pentalonia nigronervosa* Coquillett) population as closer plantation develops canopy favourable for multiplication of aphid in shady humid climate. The infestation of mango hopper [*Amritodus atkinsoni* (Lethierry)], scales, mealy bugs, citrus whitefly [*Dialeurodes citri* (Ashmead)] and blackfly (*Aleurocanthus woglumi* Ashby) is facilitated in close plantations.

Good agricultural practices: Good cultural or agronomical practices namely irrigation, fertilization and pruning have been found to stimulate the rapid plant growth and maturity. Vigorous growth thus tolerates the pest attack of aphids, mealy bugs, leaf folders, borers etc. due to the fact that out growth withstands the insect injury. The neglected and poorly maintained plants are severely attacked. Encouraging the plants to put intermittent flushes by regular irrigation and split dosages of nitrogenous fertilizers enhances the attack of mango hoppers and citrus leaf miner (*Phyllocnistis citrella* Stainton). Heavy irrigation of orchards during October, however, kills the eggs of *D. mangiferae* and pupae of mango midges.

***Orchard sanitation*:** Orchard sanitation i.e. keeping plantations clear of any plant debris will be helpful in reducing the pest incidence or found effective in managing many insect pests e.g. cuttings of banana pseudostems after reaping remain in the field, which otherwise provide suitable site for banana rhizome and pseudostem weevil to lay eggs. The regular removal of infested twigs and fruits are also effective against mealy bugs, fruit flies and litchi nut borer.

Pruning: This horticultural practice is used to maintain maximum yield of high quality fruit as well as to control tree size and shape. In fruit trees, which are perennial in nature, proper pruning of undesirable plant parts often removes some active and overwintering stages of insect pests like mealy bugs and scales on several fruit crops; citrus leaf miner, mites, some borers on citrus and shoot gall psylla and leaf webber on mango. This may also facilitate better spray coverage and light penetration in the canopy of tree. Pruning or removing of some of the mango branches enhances the light perception and is effective against mango hoppers. Pruning of infested leaves, twigs and branches and destroying the same, reduces the incidence of mango scale (*Aspidiotus destructor* Signoret).

Mosaic planting: This practice has been found to have direct impact on the survival of pest, parasitoids and predators. For example, planting of blackberry in grapevine orchards is useful for the control of grapevine leaf hopper.

Removing of sick plants: Fruit trees heavily infested by some insect pests should be uprooted and destroyed. The practice is useful against shot hole borers, trunk borer, bark eating caterpillar on various fruit crops and papaya plants infected with virus which is being transmitted either by aphid (*Aphis gossypii* Glover) or whitefly [*Bemisia tabaci* (Gennadius)]to healthy plants.

Mechanical and physical control

In fruit crops the mechanical and physical methods are largely practiced against a number of insect pests. It is easy, quick and effective but labour intensive.

Removal or destruction: These practices generally involve the removal and destruction of pests mostly by using manual labour. At small scale, it is very effective and profitable especially to homestead fruit growers. Collection and destruction of insect stages is a predominating and effective method in checking the outbreak of many pests e.g. hand picking of lemon butterfly caterpillar (*Papilio demoleus* Linnaeus), brushing of scales of mango, dropped guava, *ber*, pear, peach etc., fruits infested by fruit fly maggots (*Bactrocera sp.* and *C.vesuviana)*, dropped mango leaves mined by grubs of leaf cutting weevil, mummies of litchi for litchi nut borer, rolled leaves by leaf roller and fruit borers on many fruit crops, mealy bugs and scales on various fruits, gall makers on fig and, bud and fruit borer leaves on sapota. Removal and destruction of infested plant parts is also effective in regulating the pest populations of citrus leaf miner (*P. citrella)*, mango stem borer [*Batocera rufomaculata* (de Geer)] and guava (*Microcolona technographa* Meyrick) and mango (*Chlumetia transversa* Walker) shoot borers. Regular removal and burying of fallen fruits at least at 60 cm depth has been proved to be very effective against fruit fly, *B. dorsalis* in guava, peach and pear orchards. The collection and destruction of infested and fallen leaves (litchi mite, mango midges and sapota leaf webber) and fruits (mango stone weevil) reduces the source of pest carryover.

Bagging: Bagging the fruits of pomegranate and citrus with butter paper or cloth immediately after fruit formation prevents the damage of butterfly (*D. isocrate)*and fruit sucking moth (*Eudocima* sp.), respectively.

Banding: Slippery alkathene sheet or sticky Ostico band, encircling the trunks of trees are used to prevent certain stages of insects from climbing to the foliage from the ground. This method is widely used to control insects that spend a part of their lives in the soil and move to the tree tops by creeping up the trunks. By using slippery alkathene sheet, swarming population of newly hatched nymphs of mango mealy bug (*D. mangiferae)* is barred to climb the tree due to slippery nature of sheet. The crawlers die due to starvation as the suitable site of feeding i.e. tender apical parts for feeding are not available. The slippery band consists of 15-20 cm wide alkathene sheet and is applied at the basal end of the stem of the mango tree. The top or upper part of the sheet is secured with 1-3 nails (2 cm) while lower end should be covered with compact soil so as to prevent the crawlers to climb. Occasional wiping of the band is desirable during rains. Fixing of sticky bands on the main trunk prevents the movement of ants and thus the dispersal of scales and mealy bugs in sapota orchards is reduced.

Spiking: The adults of palm rhinoceros beetle, [*Oryctes rhinoceros* (Linnaeus)] which feed within bored hole just below the crown, and cause the death of tree could be spiked out with the help of iron hook. It is very economic when practiced after detecting the pest in the hole. The beetle can be detected either by fresh chewed fibrous material coming out from the bored hole or by audible gnawing sound of the beetles. This method not only controls the pest population but also reduces its carryover.

Shaking and beating: It can be practiced to control the sluggish adults of *ber* beetle, which cause heavy damage to the foliage during night hours. The adults can be dislodged by shaking or beating the trees with a stick and are collected on a cloth spread below the tree in the early morning and are destroyed.

Traps: Various trapping devices, viz. light, bait, colour traps etc. are being used to attract the insects. Yellow sticky or pan water trap is used in daytime to catch aphids on various fruit crops. Traps baited with some attractive material namely fermented sugar, molasses etc. and poisoned with chemicals have also been found effective in capturing fruit fly adults. Likewise, freshly chopped pseudostems of banana when spread in the banana orchard during night, are found an effective means of catching the adult weevils.

Plant resistance

Despite the great achievement in plant resistance in field crops, this aspect has gained little attention in fruit crops. Baring the historical examples of plant resistance in apple variety Winter Majetin against wooly apple aphid *(Eriosoma lanigerum* Hausmann*)* and wild American grapes *(Viteus viniferata)* against *Phylloxera vitifolia* Fitch, much has not been achieved in fruit crops and these two examples are still legendary. An overview of host plant resistance of fruit crops in India (Sharma *et al*, 2004) revealed that information on sources of resistance based on the insect damage is available in citrus, mango, apple, guava, peach, plum, banana, grapes, date palm, *ber*, sapota, etc. but little efforts have been made to incorporate the resistant/tolerant (Table 1) germplasm in breeding programme. The bases and mechanisms of host plant resistance in some fruit crops are discussed.

Table 1 Resistant/tolerant stock of different fruit crops against different insect and mite pests in India

Fruit crop	Insect/mite pest	Resistant/tolerant stock
Citrus	Whitefly	Redblush, Foster, Marsh seedless, Fallglo, Nova mandarin, Star ruby
	Psylla	Cleopatra, Rubidoux, Orange Michal, Coorg-citron, Deshndo, Nagpur mandarin, Gal-gal, Kagzi lime
	Leaf-miner	*Jatti Khatti*, Citrumelo, Carrizo, Troyer, Orage-Michal, Coorg citron, Deshndo, Eureka lemon, Savage, Sacaton, Rangatra L.J.
	Lemon caterpillar	Citron, Redblush, Marsh seedless
	Aphids	Nagpur mandarin, Ikeda Unshiu, Long sport-valencia, Campbel valencia, Washington Naval
	Scales	Washington Naval, Hazara, Jenru tenga
	Mites	Chakotra, Blood red, Kagzi lime, Malta, Mosambi, Kinnow, Galgal, Sweet lime, Etrong citron, Coorg, Wilking, Sylhet, Valencia champman

Contd. ...

Table 1 Contd. ...

Datepalm	Stone beetle	Late ripening Medjool, Thoory
	Greater date moth	Barhee
Sapota	Chikoo moth	Cricket ball, PKM-I, Guthi, Oval, Kirthabarthi
	Bud borer	PKM-I, Kirthabarthi, Cricket ball, Guthi, Oval
Banana	Rhizome weevil	Locatan, Cavendish, Cavendish (R), Poovan, Chakkia, Pedohapack, AAB clone
	Rhinoceros beetle	Chowghat dwarf orange x West coast tall
	Thrips	ABBB clone
	Lacewing bug	Jurmony, Malakali, Padathi, Kali, Krishan vazahai, Agneswar, Malbhog, Chenichampa, Bhimkal, ABBB clone
Ber	Fruit fly	Seb, Jogia, Tikadi, Illaichi, Nazuk
	Fruit borer	Illaichi, Jogia, Dandan
	Leaf roller	ZG-2, Golar, Sanaur-1, Nazuk, Ladoo, Kala Gola
	Mites	Chomuocal
Guava	Fruit fly	Nasik, China Surkha, Behat Coconut, Pear shaped, Red flesh, Smooth Green hybrid
	Tea mosquito bug Shoot borer	Bangalore Round, Bapatala, AC 10, Seedless Red flush
Pomegranate	Butterfly	Sindh, Swarupkathi
	Castor capsule borer	Guinea, Red flesh
Mango	Hoppers	Pulhora, Kala Hapus, Keshar Basti, Annanas
	Gall insect	Annanas, Anain, Delicious, Gulabkhas, KO7, KO11, Maharaja of Mysore, Salem, Banglora, Vellakachi
	Fruitfly	Toranjo, Monteiro, Manjurad
	Leaf gall midge	Deshi Malgoba, Alam Baneshan
	Mites	Mohandas, Malgoa, Pulliadi
Apple	Wooly apple aphid	MM_9, M_{21}, M_{997}, M_{793}, MM_{111}, MM_{115}, M_9, M_{16}, MM_{101}, MM_{106}, Crab apple, Yellow Delicious, Isreal type 2, 4, 5
	San Jose scale	Benoni, Green sweet, Pecks Pleasant, Summer Golden Pippin, Pippin, Cox's orange, Mcintosh, Cortlant, Tompkins

Citrus: The oviposition by the citrus leaf miner was the highest on Rangpur lime and slightly lower on *Jatti Khatti* and *Karna Khatta*. Cleopatra and Troyer were moderately preferred, while the lowest number of eggs was recorded on *Murraya exotica* followed by *Jasminum sambac*, but no mining was observed on these two hosts. The mean duration of egg stage did not vary much on different hosts. The larval duration was highest in Troyer, indicating prolongation, which was possibly due to antibiosis. The minimum larval survival was also recorded in Troyer. There was not much difference in the pre-pupal stage. The pupal period was prolonged on Troyer compared to other hosts. It was also found that the resistance in Troyer Citrange to *P.citrella* was owing to prolonged life stages compared to in *Jatti Khatti* and minimum percent larval survival, could be due to antibiosis (Fig.1). All the commercial citrus rootstocks in the nursery from the seedling to the budding stage were infested with leaf miner. Development and fecundity during spring and autumn of citrus leaf miner revealed that there was shorter duration of egg stage on Kagzi lime, *C.aurantifolia* (3.41 and 3.18 days) than that on Addinimma lime (3.98 and 3.69 days). The longer duration of larval and pupal stages was observed on Addinimma lime and Zumukhiya lime compared to Kagzi lime. Life span of *P.citrella* on Kagzi lime and Mexican lime was shorter than that on Addinimma lime and Zumukhiya lime. The fecundity (38.75 eggs/female) and maximum larval survival were higher on Kagzi lime than on Addinimma lime. The growth index was the highest on Kagzi lime.

The total phenolic contents estimated from the healthy leaves had a positive correlation with the percent leaf injury by *P. citrella*. Eureka lemon, Troyer, *P. trifoliata* var. Rubidoux and *S. buxifilia* var. Box Orange having 0.185, 0.205, 0.225 and 0.155g/100g of total phenols registered 17.5, 1.5, 0.5 and 13.6 percent of leaf injury, respectively. The contents of orthodihydroxy phenolic compounds had no relationship with the extent of injury recorded in the field. Under laboratory conditions, cultivars Cleopatra mandarin Morocco and Trifoliate orange were the least preferred compared to other varieties. The presence of higher phenolic compounds in the six Rough lemon strains viz., *Jatti Khatti*, Miri, Volkamar, South Africa I and South Africa II seem to contribute towards the tolerance to *P.citrella*. The total phenolic content of the healthy leaves also showed a positive correlation with the percent incidence of citrus leaf miner in different species/cultivars viz., Box Orange, Campbell, Valencia, Carrizo, Italian mandarin, *Jatti Khatti*, Jullundhari Khatti, Kandanarum, Kara, *Karna Khatta,* Kinnow mandarin, Mosambi, Heaven sweet orange, Pineapple, Redblush, Rubidoux, Scaton, Savage and Willow leaf.

The percent successful development of citrus caterpillar, *P. demoleus* on different citrus cultivars revealed that Cirton had minimum development (40%) followed by Khasi Papeda and Pink fleshed, while it was more on Khasi mandarin, Mandarin King, Theipi, Calamondin and Citrange. Larval population counts showed that Rough lemon was highly susceptible to *P. demoleus* compared to Pectinifera and Cleopatra rootstocks. Sweet lime and Sour lemon were found comparatively more susceptible to mites, *Eutetranychus orientalis* Klien due to high fecundity, longevity, feeding and oviposition.

Ber: The preference of egg laying and subsequent larval development on a cultivar also depicts its level of tolerance. The reaction of cultivars were presumed to vary according to the antibiosis and the percent fruit infestation, number of eggs or egg punctures per fruit, number of larvae/fruit and percent larval hatching in fruits of fruitfly, *C.vesuviana*. The fruitfly preferred egg laying on Dandan, Gola, Nazuk and Rashmi. The preference varied from 6.7 to 7.3 percent on these cultivars. The percent hatching on different cultivars also varied from 6.1 to 83.2. In cultivars Tikadi, Illaichi and Umran, only a small percentage of eggs developed into larvae and, therefore, were less susceptible to the fruitfly. *C. vesuviana* infestation was positively correlated with fruit weight, pulp stone ratio, total soluble solids and total sugars, while it was negatively correlated to acidity, Vitamin C and total phenols. The varieties high in pulp, TSS, total sugars, low acidity, vitamin C and total phenols were highly susceptible to the pest.

Guava: The effect of guava cultivars viz., AC 10, Chittidar, Lucknow 46 and Lucknow 49 on larval survival, adult occlusion and size of *Bactrocera correcta* Bezzi were investigated and significant variations in larval and pupal periods were observed between cultivars, but there was no discernible

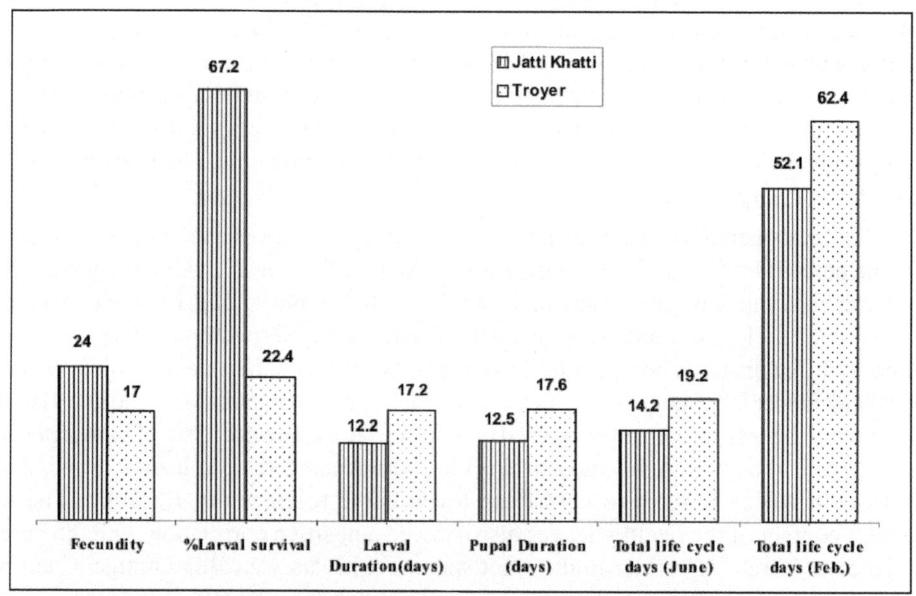

Fig.1 Biological parametres of citrus leaf miner on two commercial rootstocks of citrus (Batra and Sandhu, 1983)

difference in pupation rates (Fig. 2). Adult emergence was higher on Chittidar (90.6%) and minimum on Lucknow 46 (59%). Chittidar yielded the largest adult (>2.17mm head width). The role played by shape of the guava fruit in the ovipositional behaviour of *B. correcta* revealed that Lucknow 49 (round big) was the most resistant cultivars with 0.07 punctures/fruit, 0.71 maggots/fruit and 4.8 per cent fruit damage. Smooth skinned *cv.* Chittidar was most susceptible with 1.28 punctures/fruit, 2.18 maggots/fruit and 49.38 per cent fruit damage. Fruits of Lucknow 46 (bell shaped) and AC 10 (oblong) were intermediate in reaction.

Fruits of Red Flesh and Seedless guava with rough and wrinkled skin had very low fruit fly, *B. dorsalis* infestation whereas Pink Flesh (Gulabi) and Strawberry cultivars with rough and gritty skin, respectively, exhibited low infestation. Baramasi and Guinea varieties with intermediate mellow skin had moderate infestation while smooth skinned Allahabad Sufeda, Apple colour and Lucknow 49 were severally infested. Fruit morphological characters like skin surface and flesh colour significantly influenced the preference of fruitfly. Fruits with red flesh and rough skin had less damage compared to those with white flesh and smooth skin. However, fruit shape did not show any effect on the incidence. EC 147037, EC 147039, Kamsari, Spear Acid and Superior Sour Lucidum were the least affected (< 10% damage) while Bangalore Local, Seedless Smooth Green and White Flesh were susceptible with 41.0, 40.66, 37.67 and 35.33 per cent fruit damage respectively,

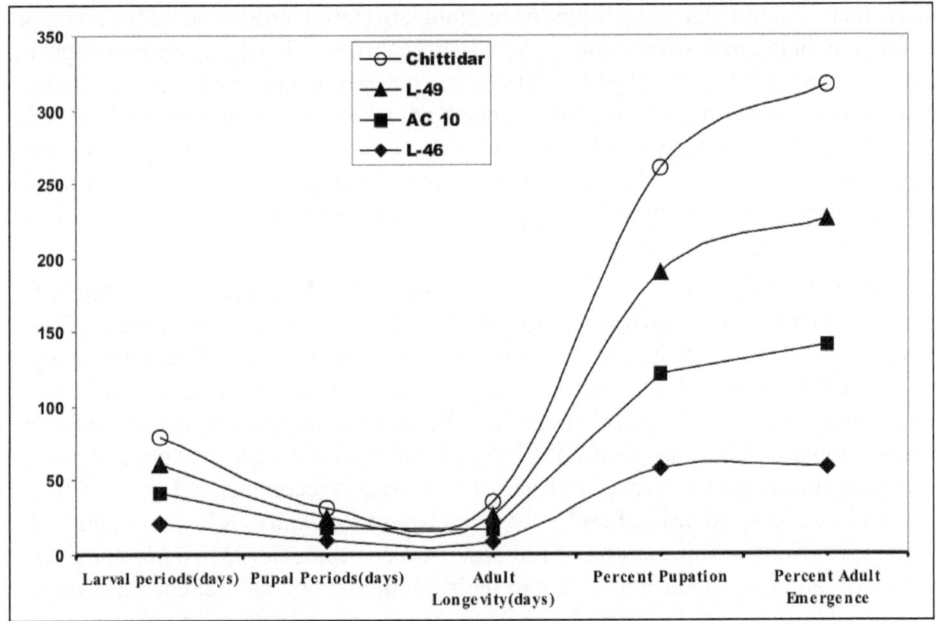

Fig. 2 Development of guava fruit fly Bactrocera correcta in different cultivars (Jalaluddin and Sadakathullu, 1999)

High levels of vitamin C, total soluble solids and total phenols in the fruits of different guava cultivars were found to contribute the resistance to *B. correcta*. The percent infestation of *B. dorsalis* was positively correlated to total sugars, TSS and total proteins but was negatively correlated to total phenols, orthodihydroxy phenols and flavonols. Acidity and vitamin C in fruits showed a very weak but negative impact on fruit fly incidence.

Mango: Chemical analysis of inflorescence showed that the varieties such as Baneshan Bangalora, Chinnarasam and Khander resistant to leaf hoppers had less nitrogen, more phosphorus, potassium, calcium, total phenol and orthodihydroxy phenols and less of reducing sugars. Magnesium and non-reducing sugar content were not related to resistance. A moderately resistant cultivar Gaddemar also had similar composition as that of a resistant cultivar but had more reducing sugars. Moreover, another moderately resistant cultivar Rumani contained more nitrogen than that in resistant one but had less than that in a susceptible cultivar. Susceptible cultivar such as Himayudin had more of nitrogen but less of potassium.

Biological control

In fruit crops, several predator and parasitoids of different insect pests have been identified and appeared to be ideal for the control of pests, but actually there are few instances where the biocontrol agents are being used to reduce the pest population below the economic injury level. The bioagents are generally used indirectly, except few instances where they are used directly for the immediate control of pest species e.g. predatory coccinellid, *Cryptolaemus montrouzieri* Mulsant has been used against mealy bugs and scales on citrus, grapes, pomegranate, sapota, *ber,* custard apple, guava etc. Among different bioagents, parasitoids have received increasingly more attention than predatory insects. Hitherto many entomophagous spiders, mites, fungi, bacterial, protozoan, virus, nematodes and other vertebrates have been found to play a significant role in balancing the pest population of the fruit crops. The biological pest suppression is not only an important tool for sustainable agriculture but also eco-friendly and safe for mankind. A complex of natural enemies affecting all life stages of insect pests would be desirable to promote population regulation below damaging level rather than reliance on a single parasitoid species to function on its own. A number of natural enemies have been reported on fruit crops in India but only the successful examples are discussed in this text.

Citrus: Cottony cushion scale, *Icerya purchasi* Maskell appeared in epidemic form in the Nilgris Hills in 1927 on many crops and the introduction of ladybird beetle, *Rodolia cardinalis* Mulsant brought its population under control and thus established the biological control as valid method of pest control in India. *I. purchasi* is appearing on citrus, mango and other fruit crops and *R. cardinalis* would be helpful in wiping out its attack. Californian red scale, *Aonidiella aurantii* (Maskell) is found parasitised by *Aphytis* spp. and predated by *Chilocorus nigrita* (Fabricius). Release of *C. nigrita* is recommended @10-15/tree for the control of citrus red scale. Field releases of *Aphytis melinus* DeBach @ 2000/tree is also suggested for the suppression of red scale. Green scale, *Coccus viridis* (Green) is a serious pest of citrus, sapota and coffee in India. Of the several natural enemies *Aneristus ceroplastic* Howard and *Encyrtus lecanorium* Mayr. cause up to 43 percent parasitism in the field. Two coccinellid predators, *C. nigrita* and *C. circumdatus* (Schonh) also play a significant role in bringing down the population of *C. viridis* on acid lime. *Verticillium lecanii* (Zimmemlan) @ 3 x 10^8 spores/ml is found effective against green scale in humid areas.

Mealy bugs, namely *Planococcus citri* (Risso), *Planococcus lilacinus* (Cockrelle), *Nipaecoccus viridis* (Maskell), *Maconellicoccus hirsutus* (Green) and *Ferrisia virgata* (Cockrelle) are known to occur on citrus in India. Of the several natural enemies attacking *P. citri* in citrus orchards, the encyrtid, *Coccidoxenoides peregrinus* (Timberlake) played an important role in suppressing its population. An exotic parasitoid, *Leptomastix dactylopii* Howard a solitary endoparasitoid, obtained from West Indies for the control of *P. citri* in 1984 found attacking nymphal stages and young adult mealy bugs. The parasitoid was released in Karnataka and Tamil Nadu on sweet orange, acid lime and lemon and within four months, a complete control of *P. citri* was obtained. It was also able to migrate to nearby orchards. The parasitoid has now permanently established in several orchards and the release @ 5000 parasitoids/ha is recommended. The Australian ladybird beetle, *C. montrouzieri* @ 20 adults/tree is recommended for the control of *N. viridis* and *M. hirsutus.* Similarly, the releases of *Tetracnemoidea indica* (Ayyar) or *C. montrouzieri* are suggested for the sup-pression of the oriental mealy bug, *P.lilacinus,* if the local natural enemies are not found in adequate numbers.

The parasitoids namely, *Trichogramma chilonis* Ishii, *Telenomus incommodus* Nixon and *Distatrix papilionis* (Viereck) cause a cumulative parasitism of 88 percent on *Papilio demoleus* Linnaeus and *Papilio polites* Linnaeus but *D. papilionis* is a common and gregari-ous larval parasitoid. Release of *T. chilonis* @ 500 adults/tree is recommended for the suppression of *Papilio* spp. Among the various natural enemies, the eulopid *Tamarixia (=Tetrastichus) radiata* (Waterson) is the most common parasitoid of citrus psylla, *Diaphorina citri* Kuwayana causing about 95 percent parasitism in Punjab. In Maharashtra also *T.radiata* causes 30-90 percent parasitism on citrus psylla. *T. radiata* and the coccinellid, *Serangium parcesetosum* Sicard keep the pest under check at some locations in Maharashtra.

Encarsia divergence Silvestri and *Encarsia merceti* Silvestri are dominant in Assam and cause about 90 percent parasitism of *A.woglumi*. Other parasitoids namely, *Amitus hesperdium* Silvestri and *Encarsia clypealis* Silvestri also cause 80-90 percent parasitism in Karnataka and Maharashtra. Releases of *Encarsia opulenta* Silvestri has considerably brought down the blackfly population from 68 to 35 percent at Nagpur. The use of *Mallada boninensis* (Okomato) @ 25-30 larvae/tree also keeps the *A.woglumi* under check. Among the various parasitoids, *Tetrastichus phyllocnistoides* (Narayanan) and *Ageniaspis* sp are important and cause up to 80 percent parasitism of leaf miner, *P.citrella* in Punjab. Release of *M. boninensis* @20-50/plant depending upon the crop canopy, age and the pest density is recommended for the control of leaf miner.

Mango: Several natural enemies are found to attack mango hoppers in India. Among them, the reduviid, *Irantha armiper* Stal. preys voraciously on *A. atkinsoni*. In nature, *Verticillium lecaniiI@* 1 x 10^8 spores/ml is observed to wipe out mango hoppers around Lucknow. Among the different natural enemies, *Thompsonisca desantiellus* Shafee is commonly found on *Aspidiotus destructor* Signoret. *Aulacaspis tubercularis* (Newstead) is found predated by *C. nigrita, C. circumdatus* and *Scymnus* sp. An aphelinid, *Pteropteryx koebeli* (Howard) is found to cause about 90 percent parasitism in nature on *A. tubercularis*. *C. nigrita* is recommended for the suppression of both *A. tubercularis* and *A. destructor.*

Grapevine: The pink mealy bug, *M. hirsutus* is a major pest of grapevine. Among the local natural enemies, the encyrtid parasitoid, *Anagyrus dactylopii* (Howard) and the coccinellid predator, *Scymnus coccivora* (Ramakrishana Ayyar) are of considerable importance. *A. dactylopii* causes up to 70 percent parasitism in nature while the coccinellid predator, *C. montrouzieri* feeds about 1000 eggs or 300-500 mealy bug nymphs. Adult beetles (10 days old) at the rate of 5000/ha are recommended for release in the evening hours to control the grape mealy bug and is giving good control in Maharashtra, Andhra Pradesh, Kamataka and Tamil Nadu. *N. viridis* and *P. citri* also appear in serious form on grapes in certain locations and the release of *C. montrouzieri* helps in controlling these two mealy bugs. Both *Spodoptera litura* (Fabricius) and *Helicoverpa armigera* (Hubner) cause severe damage to the leaves and bunches of grapevine and application of nuclear polyhedrosis virus controls these two pests on grapevine.

Guava: Green shield scale, *Chloropulvinaria psidii* (Maskell) occurs on the fruits, leaves, shoots and trunk. They excrete honeydew resulting in development of sooty mould on the leaves and thereby hindering the photosynthetic activity. Several natural enemies have been found to attack scale, however, *C. montrouzieri* is known to feed about 4000 eggs during its development and the release @ 20 adult beetles/tree is found very effective against C. *psidii.* Wax scale, *Drepanococcus* (=*Ceroplastodes) chiton* (Green) is also very serious in some orchards around Bangalore. Parasitoids, *Anicetus ceylonensis* Howard and *Cephaleta brunniventris* Motschulsky are highly effective in suppressing *D. chiton* population. Local predator, *C. nigrita* observed in large numbers, also helps in suppressing its population and in the subsequent years, the scale population does not appear at all. Among different mealy bug, *F. virgata* is very serious in many guava orchards in South India. It occurs on shoots and fruits. Fruits infested with the mealy bugs are covered with sooty mould and loose the market value. In guava orchards, several local parasitoids and predators in nature very well regulate the mealy bugs.

Among the different natural enemies, the parasitoids, *Aenasius advena* Compere, *Blepyrus insularis* (Cam.) and the predators, *S. coccivora, M. boninensis*, *Brumus suturalis* (Fabricius) and *Spalgis epius* (Westwood) play a significant role in suppressing the *F. virgata* population. *A. advena* causes up to 50 percent parasitism in nature while *B. insularis* also causes about 32 percent parasitism. The parasitoid can be bred easily on 5-10 days old mealy bugs in large numbers. Release of *C. montrouzieri* @ 20-30/tree gives excellent control of *F. virgata* on guava.

In guava orchards, parasitoids, *C. peregrinus, Allotropa citri* Muesebeck and the predators, *S. epius* and *C. montrouzieri* regulate *P. citri* in nature. The release of exotic parasitoid, *L. dactylopii* also found effective and in many orchards, *P. citri* disappeared subse-quently due to its activity. The predators, *S. epius* and *C. montrouzieri* are commonly associated with the mealy bugs, *P. lilacinus* and *M. hirsutus.* Release of *C. montrouzieri* helps in managing these two mealy bug species on guava.

Sapota: The green scale, *Coccus viridis* and the green shield scale, *Chloropulvinaria psidii* are found to infest leaves and tender shoots of young sapota plants. The green scale is suppressed by *C. nigrita* and C. *montrouzieri* while the green shield scale is controlled by *C. montrouzieri.* Mealy bugs, *P. lilacinus* and *P. citri* are parasitized by two encyrtids, *L. dactylopii* and *C. peregrinus* and are able to suppress the populations of *P. citri* while *C. montrouzieri* is highly effective in checking the populations of *P. lilacinus.*

Pomegranate: Ash whitefly, *Siphoninus sphillyreae* (Haliday) is serious on the leaves and the entire plant becomes sick due to the development of black sooty mould on the honeydew excreted by the whiteflies. The aphelinid, *Encarsia azimi* Hayat cause as high as 90 per cent parasitism in nature. Mealy bugs, namely *P. lilacinus, P. citri, N. viridis* and *M. hirsutus* are checked by *S. epius* and *C. montrouzieri* efficiently. The parasitoid, *L.dactylopii* is recommended for the suppression of *P. citri* while *C. montrouzieri* for all other mealy bug species in pomegranate orchards.

Anar butterfly, *Deudorix epijarbas* Moore is a serious pest in north India especially Himachal Pradesh and Jammu & Kashmir. The eggs of *D. epijarbas* are found parasitised up to 62 percent and the larvae up to 15 percent in nature by various parasitoids. Release of *Trichogramma* spp. also reduces the pomegranate fruit borer infestation. *Deudorix isocrates* (Fabricius) another butterfly is parasitized by the egg parasitoids namely, *Telenomus* sp., *Ooencyrtus papilionis* Ashmead and *Trichogramma chilotraeae* Nagaraja and

Nagarkatti up to 60 percent in south India. Release of *T. chilonis* @ 2.5 lakhs/ha four times at 10 days interval reduces the fruit borer infestation.

Ber: Wax scale, *D. chiton* is a serious pest on the shoots and leaves. Two parasitoids, namely *Cephaleta brunniventris* Motschulsky and *Anicetus ceylonensis* (Howard) cause heavy parasitism on the scales and thus disappeared subse-quently in the orchards. The local parasitoids and the predator *S. epius* play a significant role in reducing the pest population of mealy bugs, *N.viridis, M. hirsutus, P. lilacinus* and *P. citri*. Release of *C. montrouzieri* is also found to be effective against *N. viridis* and *P. citri. Leptomastix dactylopii* is being used against *P. citri* in *ber* orchards.

Custard Apple: The predators, *S. epius,* C. *montrouzieri* and *S. coccivora* are commonly found feeding on the mealy bugs, *N.viridis, M. hirsutus, P. lilacinus, P.pacificus* and *P. citri*. Release of *C. montrouzieri* helps in clearing the mealy bugs quickly on the fruits. *Aphytis* sp is known to cause up to 30 percent parasitism on scale insects namely, *Aonidiella orientalis* (Newstead) and *Parasaissetia nigra* (Nietner) while *Scutellista cyanea* Motschulsky causes about 42 percent parasitism on *P. nigra.* Release of *C.nigrita* also helps in bringing down the scale population.

Apple: A large number of predators and parasites are recorded on San Jose scale, *Q. pernisiosis*. Among them the parasitoids *Encarsia* (= *Prospaltella*) *perniciosi* Tower, *Aphytis* sp. *Proclia* group, *Aphytis diaspidus* Howard, *Aphytis mystilaspidus* (Leb), *Chiloneurus* sp. and predators *Chilocorus bijugus* Mulsant, *C. circumdatus*, and *C. septumpuncta* are causing significant control of scale. The Russian, Chinese and Japanese strains of *E. pemiciosi* introduced in 1959 in Himachal Pradesh caused 8-89 per cent parasitism. Predators, *C. montrouzieri*, *Chilocorus kuwanae* Silvestri and *Cybacephalus gibbulus* Errick were also released for scale control. *Aphytis proclia* group Azotus, *Perniciosus, E. perniciosi* and *Teleterebratus perversus* and predators *C. bijugus. C. septempunctata, P. flexibilis* and *Sticholotis marginalis* are effective natural enemies of San Jose scale in Jammu and Kashmir, while in Himachal Pradesh *A. proclia. C. bijugus. C. septempunctata* and *P. flexibilis* are reported attacking the pest. In Kashmir, the release of *E. perniciosi* caused upto 9.1 per cent parasitism after its establishment in apple orchards. The release of predator, *C. bijugus* and *P.flexibilis* in Himachal Pradesh had 10-39.7 per cent reduction of scale population after 3 years of release.

In 1937, attempts were made to introduce and establish exotic parasitoid, *Aphelinus mali* Haldmann from U K in Kullu valley. It established successfully on woolly apple aphid, *E.lanigerum* within a short time and provided up to 98 per cent control. Afterwards, the parasitoid was also introduced against *E.lanigerum* at Mashobra and Kotgarh in Shimla hills, Chaubattia (Uttaranchal). Coimbatore (Tamil Nadu) and Shilong (Meghalaya) and provided significant control of aphid at the initial stages but later on failed to establish due to varying climatic conditions. Various predators have also been reported to exercise fairly good control of woolly apple aphid but none is released at commercial scale. However, the combined effect of *A. mali* and predators, *Chrysopa scelestes* and *Syrphus confrater* Weid on woolly apple aphid population caused 23.8-34.8 per cent reduction. *Aphelinus mali* is well established in valley where insecticides are used at a minimal level. It can withstand lower winter temperatures than its host. Parasitization reaches high levels in the late summers, when other factors have slowed down the aphid increase. The best would be to minimize insecticidal sprayers to conserve *A. mali.* Weekly releases of the egg parasitoids namely *Trichogramma embryophagum and T. cacoeciae pallidum* @ 2000 adults/tree are recommended.

REPRODUCTIVE OR AUTOCIDAL CONTROL

The reduction of insect population by using substances that cause sterility by altering sexual behaviour or otherwise disrupt the normal reproduction in insects is found effective in many insect pests. Male sterility has successfully been used on oriental fruitfly, melonfruit fly and codling moth. It involves the artificial rearing, sterilization with direct application of γ-radiation or by using some chemosterilants like TEPA, METAPA and apholate and releasing otherwise healthy male insects in numbers that will compete with natural males. The females thus mated with such sterile males, produce unfertilized and nonviable eggs and ultimately the suppression of insect population. No work on autocidal control on fruit crops is done in India. Using of sex attractants or pheromone is also found effective in monitoring the pest populations of codling moth and Mediterranean fruitfly.

REGULATORY CONTROL

It is a lawful regulation of eradication or to prevent or to control infestation or to reduce damage by insects including some other problems arising out of pest control programme. Initially plants and plant products of all kinds could be moved freely both between and within the countries of the world. As a result, a number of insect and mite pests introduced in areas where they were not present e.g. San Jose scale band woolly apple aphid. They reproduce and establish at alarming rate and become as major pests. To check this spread, regulations can be either through quarantine (foreign and/or domestic) and embargoes to prevent invasions of new pests or inspection and certification to ensure that fruit plants and products including other plant materials are not infested with noxious insects. By enforcing such laws, tremendous benefits can be obtained in the elimination of infestations. The successful campaigns have been conducted against major fruit pests like Oriental fruitfly, citrus blackfly, mango stone weevil, apple leaf roller, cottony cushion scale and San Jose scale.

CHEMICAL CONTROL

Chemical control of insect and mite pests of fruit crops is normally done by the use of insecticides, repellents, attractants and antifeedants. There was a phenomenal increase in fruit production after the use of synthetic organic chemicals mostly organochlorine, organophosphates and carbamates. Off late, synthetic pyrethroids and other new chemicals are available in the market. Several insecticides have been adequately tested in the orchards but many of them are either too expensive or too dangerous to animals, man and beneficial organisms. Despite of these limitations, consumers demand a top quality fruit, free from blemishes or contamination of any kind of insect damage. The orchardists thus resort to use more and more chemical agents to provide immediate control of pests. The indiscriminate use of these insecticides against a wide range of fruit pests, are causing serious problems like health hazards, environmental pollution, rapid development of resistance and resurgence of target or non-target insect species. Furthermore, toxicity of pesticides as such and their residues in/on fruits pose a serious threat particularly in fruit crops, wherein the crop products are normally consumed without allowing reasonable time for degradation of the toxic pesticides applied near the harvest. Keeping in view, the deleterious effects of pesticide residues, an ideal orchard pesticide is never possible. However, the application made at economic threshold levels (ETL) or otherwise providing a long pre-harvest interval would produce table fruits free from the insecticide residues. The information in literature showed that little efforts are made on determining the ETL on fruit crops (Table 2) and there is a need to develop the strategies to avoid the insecticides residues on fruits. Moreover, to obtain high efficacy of insecticides on fruit trees the following factors must be taken into consideration:

- Right stage, site and activity period of the pest.
- Proper selection of insecticide, dosage and application equipment.
- Coverage of the tree canopy.

It has been seen that due to heavy foliage of fruit trees the spray fluid does not reach at the site of attack of the pest e.g. the attack is underside the leaves or outer/inner canopy of the tree or terminal portion of shoots or top of the trees. Depending upon these factors the Entomologists have evaluated a large number of insecticides and their dosages against different insect and mite pests of fruit crop (which are not discussed in this text), only some practical aspects of insecticidal usage are elaborated.

Besides dusting or foliar spraying as normally done on other field crops, some other methods of application of insecticides in fruit crops are applied due to the fact that the fruit trees are perennial in nature. Thus insecticidal application is also made to roots, trunk, bark, stems, branches, twigs, leaves, flowers, fruits and seeds. The different methods of application found effective in many cases are discussed as under:

Dipping: Certain pests are carried over through nursery plant from one place to another. Prior to planting, these plants may be treated by dips or washes with some safer insecticides, which may not be harmful to the tender nursery plants. Aqueous solutions are often used in planting stocks of banana. Before shipment in and abroad, nursery stocks or harvested fruits are also treated by dipping to disinfect the stock of any pest.

Tree injection/root feeding: Injection of insecticides into sap stream by removing a band of bark around the tree trunk at a proper place and applied through root system, kill insects infesting trunks. It also protects the tree against subsequent infestation by sap or boring insects and infection by rotting fungi. Aqueous solution of monocrotophos is applied through roots for grubs of red palm weevil or solution of methyl parathion is injected for the control of mango stem borer, *B. rufomaculata*. Insertion of cotton soaked in carbon disulphide, chloroform, formalin or petrol and sealing the holes with mud is effective against bark eating caterpillar, *Indarbela* spp.

Baiting: Pests like fruitflies, *Bactrocera* spp. and fruit sucking moths, *Eudocima* spp. are attracted to and feed upon various baits. By addition of desired amount of suitable toxic chemicals like malathion to the attractive mixtures (pheromones or other chemicals), the insects can be killed. For this purpose dry (saw dust, wheat bran etc.) and wet ingredient along with sweet juices (mostly molasses or gur or sugar) are conveniently blended with poisonous chemicals and spread over the target places. This practice is being done in guava, peach and citrus orchards.

Painting: In the protection of fruit trees, one of the most efficacious barriers against insect attacks is a coat of paint containing low percentage of active ingredient of insecticide together with filler often used over the surface of susceptible region. The substances like oil solution of creosote, lindane, chlorpyriphos, , pentachlorophenol, coal tar, lime sulfur etc. having repellency and residual toxic effects against bark eating insects that deposit their eggs and/or feed on the bark are being used. Swabbing the trunk after removing the all loose bark with IIHR swab mixture (carbaryl 6gm + copper oxychloride 10 gm + neem oil 1 ml + kerosene oil 1 ml + sticker 1 ml per litre of water) after pruning of grapevine mealy bug *M.hirsutus* and grapevine shot hole borer, *Xyleborus crassinsculus* (Motschulsky).

5.5 Soil application: Soil inhabiting pests like white grubs, white ants, root eaters or other dormant stages of insects may be killed or the soil may be made unsuitable for insect species by mixing insecticides directly in the top layer of soil or applied through irrigation water. Soil application of insecticides like chlorpyriphos or lindane may provide the fruit grower a powerful weapon against some soil arthropod like termites. Broadcasting of chlorpyriphos or lindane in sand followed by irrigation or in standing water or application of these insecticides to basal part of the tree has been found very effective against termites.

5.6. *Antifeedants:* Certain insecticides or botanicals that act as indirect control agents against insect pest are popularly known as antifeedants or deterrants. These substances when sprayed to protect the crop by creating a barrier through preventing the insect to feed or to lay eggs. Extract from neem seed kernels act as an effective antifeedant against citrus leaf miner, *P.citrella.*

Table 2 Economic threshold levels of different insect and mite pests

Insect pests	Economic threshold levels
Aphis spiraecola	5-10% infested shoots
Aphis gossypii	25% infested shoots
Toxoptera aurantii	25% infested shoots
Dialeurodes citri	5-10 nymphs/leaf on mandarin
Aleurocanthus spp. *Parabemisia myricae*	First colonies occurrence First colonies occurrence
Planococcus citri	5-10% infested fruits
Saissetia oleae (soft scale)	3-5 nymphs /leaf
Aonidiella aurantii (armoured scale)	3-5 nymphs /leaf
Prays citri (flower moth)	50% infested flowers.
Ceratitis capitata	20 adults/trap/week (Clementine)
Eriophyes sheldoni	50-70% infested buds
Aculops pelekassi	2-3% infested fruits.
Panonychus citri	3 specimens/leaf or 50% infested leaves
Tetranychus urticae	2% infested fruits or 10% infested leaves.
Odoiporus longicollis	5% infested plants
Cosmopolites sordidus	4 weevils/trap

CONCLUSION

Most of the fruit crops provide a relatively stable micro-ecosystem for the effective pest management programme. Tactics commonly adopted for the control of fruit pests are classified as curative/direct and preventive/indirect methods. These methods largely include non-chemical (cultural, mechanical and physical, biological and plant resistance) or chemical approaches or in few cases regulatory enforcement or reproductive suppression of pest population. An overview of the literature showed that fruit crops are more amenable for most of the non-chemical approaches. Barring the success of biological control in few instances (e.g. mealy bugs, scales and aphids) and screening the sources of plant resistance to insects, the knowledge of mechanisms and bases of resistance, biological control of many other insects and determination of injury levels of different insect pests is still inadequate and concerted efforts must be laid on these important management components.

REFERENCES

Anonymous, 1987. Annual Report of citrus blackfly scheme, Nagpur University, 27 p

Anonymous, 1999. *Crop production Techniques of Horticultural Crops*. Dept. of Horticulture and Plantation Crops, Chennai and Tamil Nadu Agricultural University, Coimbatore, 260p.

Anonymous, 2001. *Integrated Pest Management. Package for Citrus (IPM Package No: 28)*. Directorate of Plant Protection, Quarantine and Storage, Government of India

Anonymous, 2001. *Integrated Pest Management. Package for Grapes (IPM Package No: 32)*. Directorate of Plant Protection, Quarantine and Storage, Government of India

Anonymous, 2001. *Integrated Pest Management. Package for Litchi (IPM Package No: 37)*. Directorate of Plant Protection, Quarantine and Storage, Government of India

Anonymous, 2001. *Integrated Pest Management. Package for Mango (IPM Package No: 34)*. Directorate of Plant Protection, Quarantine and Storage, Government of India

Anonymous, 2001. *Integrated Pest Management. Package for Pomegranate (IPM Package No: 31)*. Directorate of Plant Protection, Quarantine and Storage, Government of India

Anonymous, 2001. *Integrated Pest Management. Package for Sapota (IPM Package No: 30)*. Directorate of Plant Protection, Quarantine and Storage, Government of India

Ayyar, T. V. R. 1963. *Hand book of Economic Entomology for South India*. 2nd Edition Govt. Press, Madras, 516p.

Babu, T.R. 1986. Population density of grape mealybug, *Maconellicoccus hirsutus* (Green) and its control by a predator *Cryptolaemus montrouzieri* Muls. Ph.D. Thesis, APAU, Hyderabad.

Batra, R.C. and Sandhu, G.S. 1983. Mechanism of resistance in troyer citrange to citrus leaf-miner. *J. Res. Punjab agric. Univ.* 20: 558-559.

Batra, R.C. and Sharma D.R. 1993. Current status of insect resistance in fruit crops. Advances in Host Plant Resistance to Insects

Batra, R.C. and Sharma, D.R. 2001. Insect pest management in fruit crops. *National Conference on Plant Protection – New Horizon in the Millenium* (eds. S.C.Bhardwaj, R.C.Saxena and R. Swaminathan held at Rajasthan College of Agriculture. Udaipur. Feb.23-25, (2001). pp140-142

Batra, R.C., Sharma, D.R. and Singh, S. 2005. Current status of integrated pest management on grapes in India. Lead paper presented at *National Seminar on Viticulture in Non-Traditional Marginal and Degraded Soils* held on 25-26 May, 2005, PAU, Ludhiana. pp 149-157 (eds. Y.R. Chanana, J.S. Bal, J.K. Sharma and M.I.S. Gill)

Bottrell, D.R.1979. *Integrated Pest Management*. Counc. Environ. Qual. Washington D.C.GPO:120.

Butani D.K. 1979. *Insects and fruits*. Periodical Export Book Agency. New Delhi, 415p.

Chattopadhyay, T.K., 1994. *A textbook on Pomology, Vol. 1*, Kalyani Publishers, Ludhiana, pp. 316.

David, B.V. 2001. *Elements of Economic Entomology* Popular Book Depot, Madras, 562p.

Deol, G.S. and Sharma, D.R.2005. Advances in integrated pest management of fruit crops. In: *Advances in Integrated Pest Management of Horticultural, Spices and Plantation Crops* (eds. B.S.Chillar, V.K.Kalra, S.S.Sharma and Ram Singh. Centre of Advanced Studies, Deptt. of Entomology CCS HAU, Hisar pp 1-8

Dhaliwal, G.S. and Arora, R. 2001. *Integrated pest management concepts and approaches*. Kalyani Publishers, Ludhiana pp. 427.

Dout, R.L. Nakata, J and Skinner, F.E. 1969. *Integrated pest control in grapes*. Calif. Agric.23 (4): 4-11 and 16-17. Hameed, S.F. and Singh S.P. 1998. Hand Book of Pest Management. Kalyani Publishers, New Delhi.

Garg, D.O., 1978. *Insect Pests of Citrus Fruits*. Agri. Horticultural Publishing House, Nagpur, pp. 176.

Hameed, S.F. and Singh, S.P. 1998. *Hand book of Pest Manageament.* Kalyani Publishers, Ludhiana, pp. 169.

Hill, D.S. 1985. *Agricultural Insect Pests of Tropics and their control* – Cambridge University Press, London, 746p.

Huffaker, C.B. 1972."Ecological management of pest systems (In *Challenging Biological problems, Directors Toward their Solution* (ed Behnke, John, A.). Oxford University Press, New York: 313-42.

Jalaluddin, S. M. and Sadakathulla, S. 1999. Development and survival of *Bactrocera correcta* Bezzi Diptera: Tephritidae on selected guava cultivars. *Pest Manage. Hort. Ecosyst.* 5 (1): 24-27.

Johansen, C.1978."Principles of insect control" (In *Fundamentals of applied Entomology* ed. Pfadt, R.E), Macmilan Publishing Co. Inc. New York: 189-208.

Kakar, D.K. and Sharma, J.P. 1988. Feasibility to *Trichogramma* spp. for the biological suppression of fruit borers of pomegranate. Paper presented in the *National symposium on tree protection.* Dr. Y.S. Parmar Univ. Hort.& Forestry, Solan, March 10-11, 1988.

Krishnamoorthy, A. 1988. Host range, development and sex ratio of *Leptomastix dactylopii* on different stages of citrus mealybug, *Planococcus citri. J. Biological Control,* 2:8-11

Krishnamoorthy, A. and Singh, S.P. 1987. Biological control of citrus mealybug with an introduced parasite, *Leptomastix dactylopii* in India. *Entomophaga,* 32:143-148.

Krishnamoorthy, A. and Singh, S.P. 1988. Observational studies on the occurrence of parasitoids of *Papilio* spp in citrus. *Indian J.Plant Prot.,* 16:79-80.

Mani, M. 1993. Studies on mealybugs and their natural enemies in *ber* orchards. *J.Biological Control,* 7:75-80.

Mani, M. 1994a. Effectiveness of the exotic parasitoid *Leptomastix dactylopii* in the control of *Planococcus citri* in guava orchards. *J.Entomol. Res.,* 18:351-355

Mani, M. 1994b. Recovery of indigenous *Coccidoxenoides peregrinus* and the exotic *Leptomastix dactylopii* in lemon and acid lime orchards. *Biocontrol Sci. Techn.* 4:49-52.

Mani, M. 1995. Studies on the natural enemies of the wax scale *Drepanococcus chiton* (Green) (Homoptera, Coccidae) on *ber* and guava in India. *Entomon,* 20:55-58

Mani, M. and Krishnamoorthy, A, 1998 Biological control studies on the mango green shield scale, *Chloropulvinaria polygonata* (Ckll.) (Homoptera: Coccidae) in India. *Entomon,* 23:105-110.

Mani, M. and Krishnamoorthy, A. 1989. Occurrence of mealybugs and their natural enemies on custard apple around Bangalore, S. India, *J. Biological Control,* 3:77.

Mani, M. and Krishnamoorthy, A. 1994. Natural enemies of *Siphoninus phillyreae* (Homop., Aleyrodidae) and *Aphis punicae* (Homop., Aphidae) on pomogranate. *Entomon,* 20: 31-34.

Mani, M. and Krishnamoorthy, A. 1996. First record of *Ooencyrtus palilionis* on pomegranate butterfly, *Deudorix isocrates* Fabr. (Lycaenidae, Lepidoptera). *Entomon,* 21:275-276.

Mani, M. and Krishnamoorthy, A. 1997 Biological suppression of the soft green scale *Coccus viridis* (Green) and the green shield scale, *Chloropulvinaria psidii (*Maskell) on sapota. *Pest Manag Hort. Ecosys.,* 3:114-116.

Mani, M. and Krishnamoorthy, A. 1999. Biological suppression of spherical mealybug *Nipaecoccus viridis* Newstead (Hemiptera: Pseudococcidae) on acid lime in India. Paper presented in *International Symposium on Citriculture.* November 23-27, 1999 at Nagpur (Maharashtra).

Nair, M.R. G.K. 1988. *Insects and Mites of Crops in India* ICAR, New Delhi, 408p.

Nayar. K.K., T.T. Ananthakrishnan and B.V. David. 1976. *General and applied entomology.* Tata McGraw Publishing Hill Co. Ltd. New Delhi. 589p.

Sexena. A. B. 1996. *Harmful insects.* Anmol Publishing Pvt. Ltd. New Delhi. 334p.

Shamra, D.R., Batra, R.C., Gill, H.S. and Arora, P.K. 2004. Advances in fruit crop resistance to insect pests, pp 439-491. In *Host Plant Resistance to Insects: Concepts and Application* (Eds. G.S. Dhaliwal and Ram Singh), Panima Publishing Corporation, New Delhi.

Sharma, D.R. and Batra, R.C. 2005. Management of insect and mite pests of litchi. In: Upadhaya, R.K., Naqvi, S.A.M. and Dubey, O.P.(eds *IPM system of Agriculture. Fruits* Vol. 10. Aditya Books Private Limited (In Press)

Sharma, D.R., Batra, R.C. and Dhaliwal, G.S. 2005. Management of insect pests of guava. In: Upadhaya, R.K., Naqvi, S.A.M. and Dubey, O.P.(eds *IPM system of Agriculture. Fruits Vol.* 10. Aditya Books Private Limited (In Press)

Sharma, D.R., Singh, S. and Arora, P.K. 2005. Insect Pests . *40 Years of Research on fruit crops* (eds. Y.R. Chanana, J.S. Bal, V.K. Mehan, J.K. Sharma and M.I.S. Gill), PAU, Ludhiana. pp 81-93

Sharma, D.R.2005. Advances in integrated pest management of citrus. In: *Advances in Integrated Pest Management of Horticultural, Spices and Plantation Crops* (eds. B.S.Chillar, V.K.Kalra, S.S.Sharma and Ram Singh. Centre of Advanced Studies, Deptt. of Entomology CCS HAU, Hisar pp 26-34

Sharma, D.R.2005. Advances in integrated pest management of litchi, papaya and pomegranate. In: *Advances in Integrated Pest Management of Horticultural, Spices and Plantation Crops* (eds. B.S.Chillar, V.K.Kalra, S.S.Sharma and Ram Singh. Centre of Advanced Studies, Deptt. of Entomology CCS HAU, Hisar pp 18-25

Sharma, D.R.2005. Advances in integrated pest management of mango, sapota and grapevines. In: *Advances in Integrated Pest Management of Horticultural, Spices and Plantation Crops* (eds. B.S.Chillar, V.K.Kalra, S.S.Sharma and Ram Singh. Centre of Advanced Studies, Deptt. of Entomology CCS HAU, Hisar pp 9-17

Shivankar, V.J., Rao, C.N. and Singh, S. 2002. *Mannual on Citrus Insect Management (Mannual No. 1).* NRCC Nagpur, Maharashtra, pp. 76.

Singh S.P. 1993. Integrated Pest Management in Horticultural Crops. *Indian Horticulture,* April-June, pp.25-28.

Singh, S.P. 1993. Biological control of Pests. In *Advances in Horticulture III* (K.L.Chadha and O.P. Pareek, Eds.), Malhotra Publishing House, New Delhi, pp 1591-1616.

Tandon, P.L. and Srivastava, R.P. 1980. New records of parasites and predators of important insect pests of mango. *Entomon,* 5:243-244.

Wadhi S.R.1972. Investigations to develop quarantine infestation procedure for mango stone weevil, *Sternochaetus mangiferae* (Fab.) and the oriental fruit fly. *Dacus dorsalis* Hendel. *Entomologists Newsletter,* 2(3): 14.

Entomology : Novel Approaches, 2007
Eds.: *P.C. Jain and M.C. Bhargava*,(Page 353-375)
New India Publishing Agency, New Delhi
E-mail: newindiapublishingagency@gmail.com
web: www.bookfactoryindia.com

CHAPTER 18

INTEGRATED PEST MANAGEMENT IN VEGETABLE CROPS

T.M.Shivalingaswamy and S.Satpathy
Indian Institute of Vegetable Research, Varanasi-221005.

INTRODUCTION

India ranks second in the production of vegetables after China. The existing area under vegetable cultivation in India is around 4.5 million ha. There is a need of around 5-6 million tons of food to feed 1.3 billion India's population by 2020 AD. The major constraint in vegetable production, include the extensive crop losses due to increased pest infestations directly or due to viral diseases vectored by insects. The extent of crop losses in vegetables varies with the plant type, location, damage potential of the pest involved and cropping season. Vegetables are more prone to insect pests and diseases mainly due to their succulence as compared to other crops. The crop losses to the tune of 40 per cent have been observed in vegetable crops. The increased demand and limited productivity of local cultivars, led to introduction of high yielding varieties and hybrids. As a result, dramatic changes in pest scenario occurred like minor pests assuming major status (eg., Serpentine leaf miner, *Liriomyza trifolii* a serious problem in tomato hybrids). Chemical pesticides are the sole panacea to mange the pest by vegetable farmers even today. Indiscriminate use of pesticides has led to severe ecological consequences like destruction of natural enemy fauna, effect on non-target organisms, pesticide residues in vegetables and resistance of the pest to pesticides. The modern pest management is aimed at educating the farmers on various issues of pest management like identification of the pests, natural enemies, judicious use of pesticides with alternate methods of pest control like cultural, mechanical, biological and behavioral (pheromones) management.

PESTS OF BRINJAL

Shoot and fruit borer: *Leucinodes orbonalis* Guenee (Lepidoptera: Pyralidae)
Major host plants: Brinjal and potato

Young caterpillars are creamy white, but full-grown are light pinkish and measure about 18-23 mm in length. The moth is medium sized with white wings having triangular brown and red markings on the fore wings. The moth measures bout 20-22 mm across the wings.

Life cycle. A female moth lays about 8-120 creamy white eggs, singly or in batches of 2-4 on the underside of leaves, on green stems, flower buds or the calyces of fruits during its life span of 2-5 days. The incubation period is 3-6 days. The young caterpillars bore into the tender shoots near the growing points, flower buds or the fruits. There are five larval instars and mature in 9-28 days. A single caterpillar may destroy as many as 4-6 fruits. The larva comes out of the damaged shoot or fruit and pupates in a gray tough cocoon in soil or plant debris. The pupal period is 6-17 days. During the active season, the total Life cycle is completed in 20-43 days. There are five overlapping generations in a year.

Damage: The larvae bore into the petiole and midribs of large leaves and tender shoots and cause wilting. In the later stage, they also bore into the flower buds and fruits. The flower buds shed due to attack. The fruit is damaged considerably and the entry hole is plugged with excreta. Attacked fruits become unfit for consumption and marketing. Infestation up to 70 percent has been recorded.

Natural enemies: *Trathala flavoorbitalis* and *Goryphus nursei,* are some of the larval parasitoids of this pest.

Management

- Use of tolerant varieties like Arka Kesav, Arka Kusmakar, Pusa Purple Cluster, Pusa Purple Round, Punjab Barsati, Kalyanpur-2, Punjab Chamkila and Gote-2
- Intercropping with coriander /fennel as a single line, double line or border crop
- Removal and destruction of infested shoots and fruits regularly.
- Installation of sex pheromone traps @100/ha (10 x10m) and lure replacement after 45 days
- 4-5 Releases of *Trichogramma chilonis* @ 50,000 parasitized eggs/ha coinciding with infestation
- Application of NSKE 4 % or Bt formulation @500g/ha
- Need based application of Cartap hydrochloride @ 2g/litre of water

Stem borer: *Euzophera perticella* Rag. (Lepidoptera: Pyralidae)

Occasionally this pest attains serious status in some areas. Infestation is noticed usually in the late stage of the crop.

Life cycle: Grown up caterpillars are creamy white with a few bristly hairs and measure about 20-22 mm in length. The moth is small having pale straw yellow fore wings and whitish hind wings. The insect is active from March to October and hibernates as larva in the stem of old plant from November to the beginning of March. The moths emerge sometime in second half of March and soon after mating, the females start laying eggs. The eggs are cream in colour and are laid singly or in batches on tender leaves, petiole, and branches. The incubation period is 3-10 days. The young larvae feed for a few minutes on exposed parts of plants and then bore into the stem making longitudinal tunnels. They pupate inside the feeding galleries after making silken cocoons or in cracks and crevices in the soil. Moths emerge in 6-8 days. The insect completes its life cycle in 35- 75 days.

Damage: The caterpillars cause the damage and feed exclusively in the main stem. It enters the main stem and makes tunnel, which result either in stunting of growth or wilting of whole plant.

Natural enemies: *Pristomerus testaceus* and *P. euzopherae* are parasitic on the larvae of this pest.

Management

- Uproot and destroy the infested plants and destroy them
- If the pest is of regular occurrence, apply neem cake (500 kg/ha) or carbofuran@ 1kg/ha at the time of transplanting around the base of the plant.

***Epilachna* beetles: *Epilachna vigintioctopunctata*(F.)** (Coleoptera: Coccinellidae)

Major host plants: Brinjal, tomato, beans and occasionally on cucurbits

The grubs are almost 6 mm long, yellowish in colour and have six rows of long branched spines. Beetles measure about 8-9 mm in length and 5.5 mm in width and are deep red and usually have 7-14 black spots on each fore wing whose tip is somewhat pointed.

Life cycle. The activity of beetle starts during March-April. A single female can lay up to 400 eggs in her life time. The eggs are yellow, cigar-shaped, laid mostly on the underside of the leaves, in batches of 5-40 each. The incubation lasts for 3-5 days; a larval period of 7-18 days and pupal period of 5-12 days have been recorded.

Damage. Both beetles and the grubs cause the damage. They feed on the upper surface of leaves. The leaves are eaten between the veins, sometimes being completely stripped to the mid-rib. The leaves present a lace-like appearance; turn brown, dries up and falloff. Thus, the plant is completely skeletonized.

Natural enemies: *Pediobius foveolatus* and *Uga menoni* are the two important parasitoids on the grubs.

Management

- Collection and destruction of various stages (eggs, grubs and pupae are visible on the plant surface) of the pest.
- Spray with 200 ml of malathion 50 EC or or 1 kg/ha carbaryl 50 WP or dusting with malathion 5% @25 kg/ha is effective.

Minor pests of brinjal

Midge: *Asphondylia* sp. (Diptera: Cecidomyiidae)- The pest attacks the flower buds or young fruits resulting in drying of flowers and malformation of the fruits.

Leaf roller: *Eublemma olivacea* (Lepidoptera:Noctuidae)- Leaves are rolled longitudinally with silk by the larvae.

Aphids: *Aphis gossypii* (Homoptera:Aphididae)- Infests the younger foliage and occasionally serious.

Leafhopper: *Amrasca biguttula biguttula* (Homoptera: Cicadellidae)- Both nymphs and adults suck the sap from the leaves. Severely infested leaves curl upward along the margins (cupping).

Mites: *Tetranychus urticae* (Acari: Tetranychidae)- Small red mites found feeding on the lower side of the leaves. Under severe conditions, they form webbing on the leaf surface and ballooning (congregation) at the leaf tips, which helps in dispersion by wind.

PESTS OF TOMATO

Fruit borer: *Helicoverpa armigera* (Hub.) (Lepidoptera: Noctuidae)

Major host plants: okra, onion, brinjal, potato, chillies, cowpea, beans and several cereals, pulses and wild hosts.

The caterpillars are greenish to variable colours with dark broken gray lines along the sides of the body. They measure about 35-45 cm in length and the moth is brown with a V-shaped marking and dull black border on the hind wings.

Life cycle. Female moth lays eggs on the leaves, flowers and developing fruits. Eggs are yellowish and spherical. A female lays an average of 750 eggs over a week's time. The younger larvae feed initially on the foliage for a while and later bore into the green fruits. The larva while feeding, thrusts its head into the fruit leaving the rest of its body outside. The fully matured larva pupates in a earthen cell. *H.armigera* has a facultative pupal diapause which is induced by short day lengths (11-14 hours/day) and low temperatures (15-23°C) experienced as a larva. The total developmental period varies from 25-40 days depending upon the temperature and the host.

Damage: Early instars feed on leaves and fruiting bodies. Later they bore into fruits. The bored fruits invite secondary infection by other organisms lead to rotting.

Natural enemies: *Bracon brevicornis*, *Campoletis chlorideae*, *Trichogramma chilonis*, *Trichogrammatoidea* sp. are some of the important parasitoids recorded from India.

Management

- Use of tolerant varieties like Punjab Kesri, Punjab Chuhara, Pant Bahar, Azad, Pusa Hybrid-4
- Deep summer ploughing to expose the larvae and pupae to sunlight and predation by birds
- Planting of marigold (40 days old) as trap crop with every 16 rows of tomato (25days old) attracts the pest. Collection and destruction of larvae from marigold flowers
- Use of pheromone traps (5/ha) for early pest detection
- *Trichogramma brasiliense* @ 2,50,000-parasitised eggs/ha (Inundative release) during peak flowering stage
- Spray HaNPV @ 250LE with jaggery (10g/lit), soap powder (5g/lit) and a tinopal (1 ml/lit) during evening hours or need based application of endosulfan @ 700g ai/ha

Defoliator: ***Spodoptera litura*** (Lepidoptera: Noctuidae)

Major host plants: Brinjal, cabbage, cauliflower, *Capsicum*, cucurbits, okra, *Phaseolus*, potato, sweet potato and species of *Vigna* and other wild hosts.

Life cycle: Commonly known as the tobacco caterpillar, *S. litura*, is one of the most important insect pests of agricultural crops in the Asian tropics. The fully-grown larva is stout, cylindrical, pale-greenish brown with dark markings, which measures about 35-40 mm in length. The moth is stout, dark with wavy white markings on the forewings and white hind wings. The pest is active throughout the year. Fecundity varies from 2000-2600 eggs, and oviposition days vary from 6 to 8 days. The female moth after mating at night, lays about 300 eggs in clusters. These clusters are covered with brown hairs. The incubation period is about 3-5 days. After hatching, the larvae feed gregariously together and later they spread out and feed individually. The larvae go through six instars. The last instar larvae enter the soil for pupation. The pupal period and the longevity of moths are 7-15 days and 7-10 days, respectively. The total life cycle covers 32-60 days. There are 8 generations in a year.

Damage: The caterpillars feed on leaves in the nursery and in the main field and cause serious damage. In crops like tomato and chillies, they bore into fruits and spoil them. Larvae are very active at night. Heavy infestation in the form of outbreaks can severely defoliate the crop.

Natural enemies: *Bracon brevicornis*, *Campoletis chlorideae* and *Eriborus argenteopilosus* are the larval parasitoids, *Trichogramma chilonis* an egg parasitod, *Bacillus thuringiensis* (bacteria), *Beauveria bassiana, Metarhizium anisopliae, Nomuraea rileyi* (fungi) and *Nucleopolyhedrosis* virus are some of the entomopathogens recorded.

Management:

- Clean cultivation to expose *Spodoptera* pupae to natural enemies.
- Pheromone traps to predict *Spodoptera* egg laying
- Hand picking and mechanical destruction of caterpillars during early stage of attack helps in reducing the infestation.
- Application of neem kernel extract during the early stages of crop growth
- Spray SlNPV @ 250LE with jaggery (10g/lit), soap powder (5g/lit) and a tinopal (1 ml/lit) during evening hours.
- Spray 400 ml malathion 50 EC or 370ml endosulfan 35 EC in 250 litres of water per acre and repeat the spraying at 10 days interval, if necessary.

Serpentine leaf miner: *Liriomyza trifolii* (Diptera: Agromyzidae)

Major host plants: Okra, onion, cucurbits, beans, potato, brinjal, crucifers and other crops

Life cycle. *Liriomyza trifolii* is one of the polyphagous agromyzids causing serious damage on several crops including ornamentals. Eggs are inserted just below the leaf surface. Eggs hatch in 2-5 days according to temperature. Many eggs may be laid on a single leaf. Neonate larva is transparent later turn to a yellow-orange in later instars. After hatching, the larva start mining into the leaves. The typical mine resembles a serpent shape, hence the name. Pupation occurs in the soil beneath the plant. Pupal development will vary according to season. Adult emergence takes place 7-15 days after pupation. Life cycle completes in 12-15 days.

Damage: Damage is due to mining into leaves and petiole by the larvae. The photosynthetic ability of the plants is often greatly reduced as the chlorophyll containing cells are destroyed. Severely infested leaves may dry and fall. *L. trifolii* is also known to be a vector of many plant viruses.

Management:

- Judicious nitrogen application reduces pest build up in endemic areas
- Infested leaves should be removed and destroyed
- Application of NSKE 4% with a sticker found to be effective.
- Use of imidacloprid during early stages of crop growth before flowering.

Whiteflies: *Bemisia tabaci* (Homoptera:Aleyrodidae)

Major host plants: Okra, crucifers, cucurbits, solanaceous crops, pulses

Life cycle. Eggs are laid in circular groups, on the undersides of leaves by a pedicel inserted into a fine slit made by the female. Eggs are whitish in colour when laid later turn brown. Each female lays up to 160 eggs. Hatching occurs after 5-9 days at 30°C depending on host species, temperature and humidity. On hatching, the first instar larva (only mobile stage) or crawler moves to a suitable feeding site on the lower leaf surface where it moults and becomes stationary throughout the remaining stages. The first three larval stages last 2-4 days each (depending on temperature). The fourth larval stage is called as puparium. Pupation lasts for about 6 days. The adult emerges through a 'T'shaped rupture in the puparium. A female may live up to 60 days while the male lives shorter (9 to 17 days). There are 11 to 15 generations in a year.

Damage: Apart from the direct damage to the crops by way of sucking the sap, *B. tabaci* is an important vector of a wide array of plant virus diseases. Eg. Tomato yellow leaf curl virus (TYLCV), Bean Golden Mosaic (BGMV), African cassava mosaic virus (ACMV) etc.

Infected plants exhibit any one or a combination of the symptoms like vein yellowing, inter-vein yellowing, leaf yellowing, yellow blotching of leaves, yellow mosaic of leaves, leaf curling, vein thickening, leaf enations, leaf cupping, plant stunting.

Natural enemies:

Encarsia brevivena, *Eretmocerus corni*, *Eretmocerus mundus* are some of the parasitoids, *Chrysoperla carnea*, *Mallada boninensis*, *Coccinella septempunctata* are the predators and *Beauveria bassiana* and *Paecilomyces farinosus* are the pathogens recorded from *B. tabaci.*

Management:

- Many weed plants harbor whiteflies, Removal of weed hosts found to reduce both the incidence of whiteflies and associated viral diseases.
- Seed treatment with imidacloprid @2.5 g per kg of seed provides protection for 25-30 days.
- Protection of seedling nursery by nylon net (200 mesh) covering for 25-30 days followed by need based application of systemic insecticides after transplanting.

PESTS OF CHILLI

Thrips: *Scirtothrips dorsalis* Hood (Thysanoptera: Thripidae)

Major host plants: This is a polyphagous pest. Besides chillies, it feeds on tomato, cotton, castor, sunflower, mango and citrus.

Life cycle. The pest is active throughout the year except the monsoon season. The nymphs are like adults in shape and colour but are wingless and smaller in size. The adults are slender, yellowish brown, having apically pointed wings and measure about 1 mm in length. The female having long narrow wings with the fore margin fringed with long hairs. A female lay about 45 to 50 eggs over a period of 20-30 days inside the tissues of the leaves and shoots. Entire life cycle completes in 15-20 days. The pest has several overlapping generations in a year.

Damage: The nymphs and adults lacerate and suck the cell sap from leaves and tender parts of the plants and cause the leaves to shrivel. In case of severe infestation, there is a malformation of leaves, buds and fruits. The attacked shoots hardly develop and the leaves falloff. These thrips are also responsible for transmitting the *leaf curl disease* in chillies.

Natural enemies: *Franklinothrips megalops*, *Mymarothrips garuda, Scolothrips indicus* (Predatory thrips), *Orius maxidentex*, *Geocoris ochropterus* (Predatory bugs) are some of the effective predators.

Management:

- Use of tolerant varieties like Kashi Anmol
- Protecting the nursery from thrips and whiteflies using nylon nets reduces the leaf curl disease
- Seed treatment with imidacloprid @ 3 g/kg of seeds
- Foliar spray with monocrotophos 0.05% at 15 days interval if necessary.

Broad mite: ***Polyphagotarsonemus latus*** (Acari: Tarsonemidae)

Major host plants: Polyphagous pest with a wide array of host plants belonging to solanaceae and cucurbitaceae.

Life cycle: Tiny yellow coloured mites are found in large numbers in the ventral side of leaves. They suck the cell sap and cause deformity of leaves creating confusion with leaf curl. On chilli, the developmental period from egg to adult averages 3-4 days. Adult longevity is 10-15 days, respectively. Each female lays 25 eggs. Eggs are laid on the underside of leaves, tender stems, flowers and fruits. Attack occurs during a short period of time. Discoloration of tissues due to mite feeding, fruits become deformed or fail to develop. Severely infected fruits fall. When leaves are attacked, the leaf tissues disintegrate and the epidermal layer of the infested leaves thickens. Downward curling of leaves like inverted boat and petiole elongation in older leaves are characteristic symptom of mite damage.

Damage: Edges of damaged young leaves usually curl. The foliage often becomes rigid and appears bronzed or scorched. Yield is significantly reduced.

Natural enemies: *Amblyseius* sp., an efficient predatory mite that regulate the pest under field conditions.

Management

- Release of native predatory mites if available
- Spraying with dicofol @ 500 g ai/ha should be done at 10 days interval at the initiation of symptoms.
- Spraying with 0.3% wettable sulphur. Sulphur spraying should be done alternatively with dicofol.

PESTS OF CRUCIFERS (CABBAGE, CAULIFLOWER ETC.)

Diamondback moth: *Plutella xylostella* (L.) (Lepidoptera: Plutellidae)

P. xylostella is considered as the main insect pest of crucifers, particularly cabbage, broccoli and cauliflower.

Life- cycle: The moths are brown or gray, with distinct white spots on the fore wings. They measure about 8-12mm in length. When the wings are folded on its body at rest, three white diamond shaped median dorsal patches are seen and hence the name "diamondback moth". The female lays minute yellowish-white eggs singly or in batches of 2-40 on the underside of leaves. A single female lay about 20-300 eggs. The eggs hatch in 3-10 days. The freshly hatched larvae move to the under surface of the leaf, mine into the leaf tissue. The three later instars feed on the underside of the leaf making holes. The full grown larvae are pale yellow-green with fine black hairs all over the body and measure about 8 mm in length. The larval development is completed in 9-17 days. Pupation takes place inside a barrel shaped, loose silken cocoon constructed by the larva. Normally, the cocoon is attached to the leaf surface. The pupal period is 4-5 days. The moths may live up to 20 days. The total life cycle is completed in 2-3 weeks. There are 8 to 12 generations in a year.

Damage: The newly hatched larvae cause serious damage to crops like cabbage, cauliflower by scraping the leaf tissues. As they grow older, bite holes in them and render the vegetables unfit for human consumption. A severe infestation results in the formation of under-sized flowers in cauliflower and in cabbage, head formation does not take place.

Natural enemies: The larvae are parasitized by *Cotesia plutellae, Diadegma semiclausum, Oomyzus sokolowskii* and the pupae are parasitized by *Brachymeria excarinata.*

Management

- Removal and destruction of the crop residues after harvest
- Ploughing the field after harvest helps in exposing the insects to predatory birds
- Intercropping of Cole crops with tomato or carrot reduces the infestation
- Growing of paired mustard rows in both sides of 16 cabbage rows, first row is sown 15 days prior to cabbage planting while second is sown 25 days after planting helps in trapping the DBM adults in mustard. Spray mustard crop with 0.1% dichlorvos to control the pest
- Neem seed kernel extract 5% or Bt @ 500g ai/ha alternatively also proved to be effective during primordial stage.
- Avoid pesticide use during peak activity of the parasitoid (*Cotesia plutellae)*

Leaf Webber: *Crocidolomia pavonana* (F.) (Lepidoptera: Pyralidae)

Major host plants: Cabbage, cauliflower, mustard

Life cycle: *C. pavonana* is of great economic importance in Central and Southern India. This pest is active with the availability of cruciferous crops in the field. The larva is green with red head and longitudinal red stripes on the body. It is about 2 cm in length. The moths are small with light-brownish forewings. The moth lays eggs in masses of 40-100 each, on the lower surface of the leaves. The eggs hatch in 5-15 days. Young larvae feed gregariously on the leaf tissues. As they grow older, they start webbing the leaves and feeding on them. The larval stage is completed in 24 to 27 days in summer and in about 50 days in winter. The pupation takes place in the soil in an earthen cocoon. In Central and Northern India, *C. pavonana* undergoes diapause in summer (38-45°C) from March-May. The adult emerges in 14-40 days. The total life cycle is completed in 43-82 days. There may be more than one generation in a year.

Damage: The larvae web the leaves with silken strands and feed on the lower surface of the leaves completely skeletonize them. In cauliflower, larvae nibble the growing tip of seedlings and bore into the curd resulting in discoloration of curd. Even a single mature larva per plant is capable of causing economic loss to cabbage at pre- and post-heading stages.

Natural enemies: *Microbracon mellus, Apanteles crocidolomiae, Palexorista solennis* are the parasitoids on the larvae of this pest.

Management:

- Removal and destruction of webbed leaves with larvae inside.
- Use of safer insecticides for conservation of these beneficial insects
- Use of Indian mustard as trap crop for DBM management also attracts the leaf webber moths for egg laying and reduce the infestation in the main crop
- Foliar spray of neem seed kernal extract (4%) along with sticker or *Bacillus thuringiensis* formulations @ 500 g ai/ha
- Need based application of cypermethrin 10 EC (0.005%) or malathion 50 EC (0.1%) is effective. For increased efficacy, spray the insecticides prior to web formation.

Cabbage caterpillar: *Pieris brassicae* (L.) (Lepidoptera: Pieridae)

Major host plants: A specific pest on cruciferous crops

Life cycle: A sporadic pest and causes serious damage in northern and eastern parts of India. Newly hatched larvae are pale yellow and turn green yellow later on. Mature larva measures 40-50 mm in length. Adults are medium sized butterflies pale white with black markings. Wings in female have two black circular dots on the dorsal side of each fore wing whereas in male the black spots are on the underside of each fore wing. After mating, the butterflies lay about 100-150 yellowish eggs, which are conical in shape in clusters on the upper or lower side of the leaf. The eggs hatch in 11-17 days in winter and 3-7 days in summer months. They feed gregariously in early stages. There are five instars and are full-grown in 2-3 weeks in March-April and in 4-6 weeks during November-February.

Damage: Caterpillars cause the damage. Freshly hatched caterpillars scrape the leaf surface. The older larvae eat up the leaves from margins, leaving intact the main vein. Sometimes all the leaves of entire plant are eaten away.

Natural enemies: *Cotesia glomeratus, Phryxe vulgaris* and *Pteromalus puparum* are the hymenopteran parasitoids on the larvae of this pest in the field.

Management

- Hand picking and mechanical destruction of caterpillars during early stage of attack helps in reducing the infestation.
- Foliar spray of *B. thuringiensis* formulation @ 500 g ai/ha along with sticker control the larvae effectively.
- Need based application of cypermethrin (0.005%) or malathion (0.05%) provides good control of this pest.

Cabbage head borer: *Hellula undalis* F. (Lepidoptera: Pyralidae)

Major Host Plants: A major pest of cruciferous crops in tropical and subtropical regions.

Life cycle: The mature larva is creamy-yellow with a pinkish tinge and has seven purplish brown longitudinal stripes. It measures 12-15 mm long. The adult moth is slender, pale yellowish brown, having gray wavy lines on the fore wings and hind wings pale dusky. The female lays pinkish oval eggs either singly or in groups on the under surface of the leaves. The eggs hatch after 2-3 days and caterpillars feed inside the cabbage and mature in 7-15 days after passing through five larval instars. The full-grown caterpillar forms a cocoon among the leaves or even inside the larval tunnel. The adult emerges in about six days. Total Life cycle is completed in 15 to 25 days.

Damage: Younger plants are most susceptible to the attack of this pest. Particularly in seedling stage .The caterpillar in initial first two instars mine the leaves. Later on, they feed on the leaf surface and bore into the heads of the cauliflower and cabbage. The plants lose their vigour resulting in deformed heads which are unfit for consumption and marketing.

Natural enemies: *Bracon gelechiae* and *B. hebetor* are the larval parasitoids,

Management

- Monitoring of infestation in seedling and early growth stage
- Collection and mechanical destruction of caterpillars in early stage of attack helps to check the infestation.
- Spraying of synthetic pyrethroids like fenvalerate 20 EC, cypermethrin 10 EC or deltamethrin 28 EC @ 0.5 ml/lit at 10 days interval up to 50-60 days after transplanting controls this pest effectively.

Semilooper: *Thysanoplusia orichalcea* (F.) (Lepidoptera: Noctuidae)

Major host plants: *T. orichalcea* is a polyphagous pest and defoliates a wide range of crop plants such as cabbage, turnip, lettuce, cotton, chicory, groundnut, sun hemp, tobacco, sunflower, safflower, tomato, potato, chickpea and soybean.

Life cycle. The caterpillars are pale-greenish, plumpy and move with a characteristic half-loop movement hence the name *'semilooper'*. Females lay eggs on leaves of host plants and the larvae feed individually making holes of varying sizes. Pupation takes place in debris on the ground. The total generation time is 24-35 days. The moth has mottled brown forewings with pale straw-coloured hindwings. The forewing has a conspicuous, roughly triangular, golden patch, which makes the species readily recognizable. The adults are active during evening and are seen visiting in large numbers on flowers.

Damage. The younger larvae inflict damage by biting holes into the leaves. The later stage larvae completely defoliate the plant during severe infestation.

Natuaral enemies: *Voria ruralis* is a larval parasitoid.

Management: *Refer to cabbage caterpillar, Pieris brassicae*

Flea beetles: *Phyllotreta cruciferae* Goeze, (Coleoptera :Chrysomelidae)

Major Host Plants: *P. cruciferae* feeds primarily on crucifers

Life cycle: The adults are beetles vary in colour from shiny black to black, have a very stout femora, with which they jump like a flea. The female lays 50 to 80 creamy white eggs, in the soil around the host plant. The eggs hatch in 5-7 days and larva is dirty white in colour with pale-white head. It measures about 5 mm in length. There are three larval instars. Total larval period lasts for 9 to 15 days and later it pupates in the soil in an earthen cell near the infested plants. The pre pupal and pupal periods lasts for 2-4 days and 8-14 days respectively. Adults of *P. cruciferae* are known to over winter in plant debris on the soil surface or in cracks in the soil. The pest completes 7-8 generations in a year.

Damage: The larvae live in the soil and feed upon the roots of the host plants. The adults feed on the leaves of young plants leading to a characteristic 'shot hole' appearance. Severe infestation results in loss of plant vigour, wilting and death. The stem and flowers may also be attacked. The old eaten away leaves dry up and the young leaves rendered unfit for consumption.

Management

- Deep ploughing during summer helps to expose and to kill the over wintered population of the pest.
- In endemic areas, off-season cultivation of cabbage and cauliflower particularly late crop should be avoided.
- In young seedlings and early planted crop endosulfan (4%) or malathion (5%) dust should be applied @ 10-15 kg/ha.
- Foliar spray of dichlorvos (0.1%) or endosulfan (0.05%) may be sprayed if necessary.

Aphids: *Myzus persicae, Brevicoryne brassicae and Lipaphis erysimi* **(Homoptera:Aphididae)**

Aphids appear in November and remain active till April. Cloudy and moist weather favours rapid multiplication of this pest. Both nymphs and adults are similar in appearance. Nymphs are 1.5 to 2.0 mm long and adults are 2.0 to 3.0 mm. A single female produces 5 to 10 young ones per day during February to May and 2 to 5 per day during July to January. Their activity goes down in extreme hot conditions of May-June. The life cycle is completed in 11 to 18 days during April to March. Under favourable conditions, 11 to 20 generations are completed in one year. During October to March, pre-reproduction, reproduction and post reproduction period last for 8 to 18, 13 to 30 and 5 to 18 days, respectively whereas during, April to September these stages occupy only 6 to 8, 4 to 9 and

4 to 5 days, respectively. A female produces on an average 7, 13 and 26 young ones during October to December, January to March and April to September, respectively.

Damage: Direct feeding on young growth leads to wilting of plants. Early attack may lead to stunted growth. Symptoms of viruses transmitted by *B. brassicae* include mosaic, chlorotic and necrotic lesions on leaves. *M. persicae* is the most important aphid virus vector. It has been shown to transmit well over 100 plant virus diseases

Natural enemies: *Chrysoperla carnea* (Chrysopidae*), Coccinella septempunctata* (Coccinellidae*), Episyrphus balteatus*(Syrphidae*)* are some of the important predators in pest suppression in nature.

Management

- Cut and destroy the infested shoot mechanically as soon as aphid infestation appears.
- Foliar spray with methyl demeton (0.025%) or endosulfan (0.05%) when 5% plants are infested. Repeat foliar spraying fortnightly 5% infestation persists.

Cut worms: *Spodoptera litura* (F.) (Lepidoptera: Noctuidae)

See under pests of tomato

Minor pests of crucifers

Agrotis ipsilon (cutworms of seedlings)

Spilarctia obliqua (Bihar hairy caterpillar) ,*Trichoplusia ni* (semi-looper), *Athalia lugens proxima* (mustard saw fly) - Defoliators

Bagrada cruciferarum (Painted bugs), *Thrips tabaci(*Thrips)- Sap feeders

PESTS OF OKRA

Shoot and fruit borer: *Earias vittella* (F.) (Lepidoptera: Noctuidae)

A major pest on Malvaceous crops like okra and cotton.

Life cycle: Eggs are spherical, about 0.5 mm in diameter, light bluish green in colour, with ridges. The larva is grayish brown with dark brown to black spots on the body. Full-grown caterpillars are 18-24 mm long, stout, spindle shaped. Pupa is 13-16 mm long and chocolate brown in colour. Adults are medium sized moths 13-15 mm long, forewings are pale white with a broad wedge shaped horizontal green patch in the middle, hind wings are silver creamy white in colour. Wing expansion is 30-35 mm. A female lays on an average 400 eggs, incubation, larval and pupal period lasts for 3-9, 9-20 and 8-12 days respectively. Single life cycle takes 22-25 days extending up to 74 days during winter.

Damage: Caterpillars bore into tender shoots and fruits results in wilting of the shoot and affected fruits become deformed in shape and remain stunted in growth. The damaged fruits become unfit for consumption and marketing.

Natural enemies: *Trichogramma chilonis* (Egg parasitoid*),Apanteles angaleti*, *Bracon lefroyi* and *Goryphus nursei* are some of the larval paraitoids.

Management

- Use of tolerant varieties like Parkins Long Green, AE57, PMS- 8, Karnal Special
- Deep ploughing during summer to expose the puape to sunlight and predation by birds
- Avoid overuse of nitrogenous fertilizers
- Soil application with carbofuran @ 1 kg ai/ha gives protection against shoot damage.
- Application of fenvalerate 50 g or deltamethrin 10 g or cypermethrin @ 30 g ai/ha from 40 days after sowing at 20 days interval gives effective fruit borer control.

Leafhoppers: *Amrasca biguttula biguttula* Ishida (Hemiptera: Cicadellidae)

Major host plants: Okra, brinjal, tomato, beans and cowpea

Life cycle: Eggs are pear shaped, elongated and yellowish white in colour. Nymphs are whitish, pale green wingless and move laterally in a peculiar fashion. Adults are wedge shaped, 2-3 mm long. Pale green in colour with black dot on the apical region of the fore wing. A female lays 15-30 eggs. Incubation and nymphal periods lasts for 4-10 and 7 to 21 days respectively. Longivity of the adults varies between 5 and 8 weeks.

Damage: Both nymphs and adults suck the cell sap from the ventral surface of the leaves and while feeding inject toxic saliva into plant tissues. Affected leaves turn yellowish and curl. The pest appears with the onset of cloudy weather and their population is adversely affected after heavy monsoon showers. Severe infestation leads to cupping of leaves and reduced photosynthesis and stunted growth of the plants.

Management

- In endemic area seed should be treated with imidacloprid @ 3 g/kg which will protect the crop for 40-50 days after sowing.
- Soil application of carbofuran @ 1 kg ai/ha at the time of sowing.
- During the vegetative stage, folir spray of monocrotophos @ 500 g ai/ha at 15 days interval is found to be effective.

PESTS OF ONION

Thrips: ***Thrips tabaci*** L. (Thysanoptera: Thripidae)

These are cosmopolitan, polyphagous pests, often abundant in warm, dry regions.

Major host plants: Onion, garlic, cabbage, cucurbits, tobacco and cotton.

Life cycle: The nymphs are slender, yellowish-brown and look similar to adults but are wingless and slightly smaller in size. The adults are yellowish-brown and measure about 1 mm in length. The males are wingless while the females have long, narrow fringed wings. Female live for 12-30 days and lay 50-60 kidney shaped eggs singly inside leaf tissue with its sharp ovipositor. The incubation period is 4-10 days and the nymphs feed on the plant sap by lacerating the leaf tissues. Usually they congregate at the leaf sheath or in the flowers. Nymphs mature in 4-5 days after passing through four instars and pupate in the soil. There is a pre-pupal stage, which lasts for 1-2 days, and pupal period is 2-4 days. Several generations are completed in a year.

Damage: Both nymphs and adults cause the damage. As a result of its attack, the leaves show a spotted appearance initially and later develop into pale-white patches (Silvery shoot) due to drainage of sap. Subsequently the leaf tips fade and gradually dry up. The infested plants do not form bulbs and there is no seed setting in seed crop.

Natural enemies: *Orius albidipennis* and *Scymnus roepkei* are some of the predators and *Aspergillus ochraceus* , *Beauveria bassiana* and *Metarhizium anisopliae* are the entomopathogenic fungi recorded on thrips.

Management

- Seed treatment with imidacloprid
- Spray malathion 50 EC @ 250 ml /acre as soon as the pest appears.

Onion maggot*:* *Delia antiqua* (Meigen) (Diptera: Anthomyiidae)

This is a specific pest on onion.

Life cycle: The flies are slender grayish and large winged and about 6 mm in length. The females lay white and elongate eggs near the base of the plant. The incubation period is 2-8 days. Maggots are small, white and measure about 8 mm in length. The maggots after hatching crawl up and enter the leaf sheath and reach the bulb and feed inside for about 15-20 days and then crawl out of the bulb and pupate in the soil. The adult flies emerge and start the new generation. The last generation of the pest shortly before the harvest leads to rotting of produce in storage.

Damage: The attacked plants become flabby and yellowish. The maggots bore into the bulbs leaving only outer sheath. The infestation varies from 8-10%. The infestation is not completely destructive to the bulbs but it leads to subsequent rotting in the storage.

Management

- Application of phorate 10 G to the soil followed by light irrigation found to be effective.

PESTS OF LEGUMINACEOUS VEGETABLES

Whiteflies: *Bemisia tabaci* Genn. (Homoptera: Aleyrodidae)

Refer under pests of tomato

Bean Aphids: *Aphis craccivora* Koch. (Homoptera: Aphididae)

These aphids are distributed worldwide particularly in the tropics. Leguminaceous crops are the major host plants of this pest.

Life cycle: Young colonies of *A.craccivora* concentrate on growing points of plants and are often tended by ants (mutualism). This symbiotic association with ants helps in dispersal of the aphids from plant to plant. Parthenogenetic reproduction is common and noticed throughout the year. The aphid is ovo-viviparous, with females retaining eggs inside their bodies and giving birth to nymphs. Males are winged and females are both winged and wingless. The total life cycle varies from 10-12 days depending on the weather conditions.

Damage: Aphids feed on young leaves, shoots, flowers and immature pods by sucking the sap. Symptoms of damage include stunting of the plants and seed shriveling in developing pods. Yield losses are severe when aphid colonies concentrate on the growing tips of plants. Indirect damage caused by aphids is due to the production of honeydew, which hampers plant respiration and photosynthetic efficiency. *A.craccivora* are known to transmit several plant viruses (Bean common mosaic virus, Bean yellow mosaic virus, cucumber mosaic virus etc.) leading to complete crop losses.

Natural enemies: *Trioxys indicus* is an important parasitoid, *Cheilomenes sexmaculata, C. septempuncta* (lady bird beetles), *Episyrphus balteatus, Ischiodon scutellaris,* (syrphid larvae) are the predators and *Fusarium pallidoroseum* is a fungal pathogen.

Management:

- Cut and destroy the infested shoot mechanically as soon as aphid infestation appears.
- Foliar spray with methyl demeton (0.025%) or endosulfan (0.05%)(safer to natural enemies) when 5% plants are infested. Repeat foliar spraying fortnightly 5% infestation persists.

Red Spider mites: *Tetranychus urticae* (Acari: Tetranychidae)

Host plants: The pest is polyphagous and is known to feed on more than 150 species of plants including cucurbits, brinjal and okra.

Life cycle: The red spider mite is also called the two-spotted spider mite. The mite is active from March to October and passes the winter as gravid female. As the season warms up in March, it spins webs on the undersurface of leaves of various host plants and lays 60-80 eggs. The eggs are spherical and hatch in 2-6 days. The emerging larvae are light brown and have three pairs of leg. Larvae feed on lower surface underneath of the webs and changes into nymph within 3-4 days, which have four pairs of legs. The nymphs grow to maturity within 4-9 days and the adults live for 9-11 days. Red spider mite completes their life cycle within 9-19 days during active period. It is believed that in Punjab red spider mite complete 32 generation in a year.

Damage: The damage is caused both by nymphs and adults. A large number of webs are formed on the leaves under which the pest sucks the cell sap. The affected leaves dry up and finally wither away giving an unhealthy appearance. Webbing interferes with the plant growth and reduction in yield.

Management

Destruction of severely attacked plant part mechanically during initial stage of infestation reduces mite population.

Spraying with dicofol @ 500 g ai/ha and wettable sulphur 80 WP @ 0.3 % alternatively at 10 days interval at the initiation of symptoms.

In endemic areas the crop should not be sown adjacent to dusty roads and the field should be kept wet regularly.

Pea stem fly: *Ophiomyia phaseoli* (Tryon.) (Diptera: Agromyzidae)

Major host plants: Peas, Cowpea, Beans

Life cycle. The larvae are leaf and stem miners. The metallic flies are active in summer and mate 2-6 days after emergence. The female inserts her ovipositor into the leaf and sometimes in the stem and petiole and makes an elliptical cavity under the epidermis. A female lays 14-64 elongate, oval and white eggs. The incubation period is 2-4 days. The maggot mines through the leaf towards the petiole and later on reaches the stem. They pass through three instars. The larval development is completed in 6- 7 days in March-April and in 9-12 days in November-December. The larva pupates within its gallery. The pupal period lasts 5-9 days and 18-19days in March- April and November-December respectively. The longevity of adults varies from 8-12 days. There are 8-9 generations in a year.

Damage. The maggots bore into the stem causing drying of affected shoots. The adults also cause damage by puncturing the leaves and feed on the liquid that oozes. Owing to the feeding, the injured portion turns yellow. Seedlings are more affected than grown up plants.

Management

- Removal and destruction of all the affected branches during the initial stage of attack.
- Seed dressing with 2 kg of 50% phorate or 8 kg of oxy-disulfoton per 100 kg seed provides effective protection. oxydisulfoton 1 kg or phorate 2kg applied as pre-sowing treatment is also effective.
- First application of 300 ml of methyl oxydemeton 25 EC in 250 litres of water per acre should be given just after germination on the crop and repeat twice at an interval of two weeks each.

Leaf miner: *Chromatomyia horticola* Gour. (Diptera: Agromyzidae)

Major host plants: Leguminous crops, cucurbits, crucifers, tomato, lettuce,

The larvae are minute and slender. The adults are two winged flies having grayish black mesonotum and yellowish frons.

Life cycle. The pest is generally active from December to May. The rest of the period passes in the soil as a pupa. The adults emerge at the beginning of December and after mating start-laying eggs singly in leaf tissue. The incubation period is 2-3 days. After hatching the larvae feed between the lower and upper epidermis by making zig-zag tunnels. They are full-fed in about 5 days and pupate within the galleries. Pupal period lasts in 6 days. The insect completes one generation in about 13-14 days and pass through several generations during the period of its activity.

Damage: The larvae are the only destructive form. They make prominent whitish tunnels in the leaves. The tunnels made by the larvae interfere with photosynthesis, and also normal development of the plants. Mining causes wilting in seedlings.

Natural enemies: *Chrysocharis horticola*, *Closterocerus agromyzae*, *Diglyphus isaea* and *Tetrastichus* sp. are the important parasitoids of this leaf miner.

Management

- Spray 125 ml of phosphamidon 85 WSC or 400 ml of dimethoate 30 EC per acre at seedling stage and repeat spray at 15-days interval.

Pea pod borer: *Etiella zinckenella* Treits. (Lepidoptera: Pyralidae)

Major host plants: Leguminous crops like peas, beans and pulses

Life cycle: *E. zinckenella* is a cosmopolitan pest of worldwide distribution. The larvae are pinkish in colour. The moths are brownish with a white line on the anterior margin and a yellow transverse band of the forewing. Hind wings are pale. The moths are active during night and after mating the females lay eggs on different plant parts including the immature pods. The eggs are whitish and elliptical laid singly. The incubation and larval and pupal periods last for 5-6 and 10-25 and days respectively. The fully matured larvae drop down and pupate in the soil, the pupal period last for 10-15days.

Damage: The larvae feed on the floral parts and seeds of the younger pods resulting in economic loss. The damaged pod has a large emergence hole made by the pupating larva.

Natural enemies: *Bracon pectoralis* and *Phanerotoma planifrons* are the parasitoids

Management

- Spray the crop at flower initiation with 500 g carbaryl 150 WP per acre.

Pod borer: *Maruca vitrata* F. (Lepidoptera: Pyralidae)

Major host plants: A serious pest of leguminous crops.

Life cycle: The eggs of are laid individually or in small batches on flowers or flower buds. The eggs hatch in 2- 3 days. Neonate larvae may be found in-group on flowers, later they disperse moving from one flower to another so that each larva damages number of flowers. Young larvae may feed on any part of the flowers or foliage, but later-instar larvae are more common in the pods. The larval stage lasts for 12-15 days. There is a brief pre-pupal for 1-2 days and the pupal stage takes about 6-7 days. Moth in medium sized with brownish-black wings. Fore wings have a conspicuous transverse white elongate marking.

Damage: Larvae make round holes on in the petals of the flowers. The pods are bored and covered with frass.

Natural enemies: *Cedria paradoxa* and *Phanerotoma hendecasiella* are the important parasitoids recorded from this pest.

Management

- Spray the crop at flower initiation with 500 g carbaryl 150 WP per acre.

PESTS OF CUCURBITS

Melon fruit fly: *Bactrocera cucurbitae* (Coq.) (Diptera : Tephritidae)

B. cucurbitae is a serious pest of cucurbit crops particularly bitter gourd and musk melon.

Life cycle. The adult flies are reddish-brown with lemon yellow markings on the thorax and have dark patch on the outer margins of their wings. The female fly makes a cavity with the help of her sharp ovipositor and thrusts the white elongate eggs(cigar shaped) singly or in groups of 4 to 10 into the flowers and tender fruits. After laying the eggs, the female secretes a gummy secretion on the ovipopsition puncture and makes the entrance waterproof. During the life span of 15-60 days, a female on an average lays 58-95 eggs. The incubation period is 1-9 days. The freshly hatched maggots bore into the pulp, forming galleries. Depending on weather, the larval period extends from 3 days in summer to 21 days in winter. The full-grown maggots come out of the fruits and drop to the ground and pupate in the soil at a depth of 1.5 to 15 cm. The pupal period is 5 to 9 days in summer and may extend to 30 days in cold weather. The total Life cycle occupies 12- 34 days depending on weather conditions. There are several generations in a year.

Damage: The larvae tunnel in the fruit, contaminating them with frass and providing entry points for fungi and bacteria, which cause the fruit to rot. Owing to feeding on pulp, there is a pre-mature dropping of fruits and make them unfit for consumption. The young infested fruits can be destroyed in a few days, but older fruits show less obvious symptoms but on splitting open, a mass of maggots in pulp is found. The infestation varies from 50-100 per cent in different cucurbitaceous crops.

Management

- All the infested fruits should be collected and destroyed.
- Earthing up of the soil around the vine helps in exposing the pupae for desiccation and predation by birds.
- Ploughing the infested field after the crop is harvested can help in killing the pupae.
- Apply the bait spray containing 20 ml Malathion 50 EC + 10 g protein hydrolysate + 500g of molasses/jaggery in 20 litres of water per acre. When the attack is serious, it should be repeated at weekly intervals.

Red pumpkin beetle: *Aulacophora foveicollis* (Lucas) (Coleoptera : Chrysomelidae)

Host plants: pumpkin, cucumber, ash guard, bitter gourd and melons

Life cycle: The grubs feed on the roots and sometimes on the stems of the host plants. They are creamy white, with a slightly darker oval shield at the back. When full- grown, they measure about 12mm in length. The beetles are oblong and 5.8 mm long. Their dorsal body surface is brilliant orange red and the ventral surface is black. The female beetle lay about 300 oval, yellow eggs singly or in batches of 8-9 in moist soil, near the base of the plant during their life span of 60-85 days. The incubation period lasts 6-15 days. The larval and pupal periods are 13-25 and 7-17 days, respectively. The pupation takes place in earthen chambers in the soil at a depth of about 20-25 cm. On emergence, the beetles began to feed on leaves and flowers. The Life cycle is completed in 26- 3 7 days. There are five generations from March to October.

Damage: The grubs live in soil feeding on roots and underground stems of vines and sometimes on fruits in contact with the soil. The beetles feed on aerial parts of the plants like cotyledons, flowers and foliage by biting holes into them. The damage by beetles is detrimental to the cotyledon stage of the seedlings.

Management

- Deep ploughing after harvest to kill the grubs and pupae in the soil.
- Apply 2.75 kg of carbofuran 3G per acre 3-4 cm deep in the soil near the base of the plant after germination and irrigate or spray 75-150 g carbaryl 50 WP in 50-100 litres of water per acre.

Leaf eating caterpillar: *Diaphania indica* (Saunders) (Lepidoptera: Pyralidae)

It is an irregular and sporadic pest of cucurbits becoming serious currently on cucumber, gherkins and bitter gourd.

Life cycle: The adults are medium sized moths; forewings are white and transparent with big brown patches on the margin. Females have a tuft of orange coloured hair at the caudal end. The caterpillars are elongated bright green and bear a pair of narrow longitudinal white streaks along the mid-dorsal line. The female lays up to 360 eggs, singly or in groups on the lower surface of leaves. The egg, larval and pupal periods last for 3-6, 9-14 and 5-13 days respectively. The pupation takes place in coccons formed in the leaf folds.

Damage: Initially after hatching, caterpillars feed on the leaves by binding them together. Some time the flowers are also infested. The vines are cut from one side and the terminal parts become yellowish and dry up. The fruits show clear holes made by the larvae. The bored fruits become unfit for consumption.

Natural enemies: The gregarious endoparasitoid, *Apanteles taragamae*, *A.machaeralis, Elasmus brevicornis,Trathala flavo-orbitalis* are the larval parasitoids and the entomopathogenic fungus, *Paecilomyces farinosus* cause excellent natural mortality in the field.

Management

- The pest should be carefully monitored in the field. In the early stage infestation the caterpillars and infested leaves along with the larvae inside can be handpicked and destroyed.
- Spraying with *B. thuringiensis* formulations @ 500 g ai/ha.
- Foliar spray of cypermethrin (0.01%) or endosulfan (0.05%) or carbaryl (0.1%) once or twice based on pest pressure.

References

Ahuja D.B. and Noor A; 1991. Effect of different host plants on the development of *Spodoptera litura* (Fab.). *Journal of Insect Science*. 4: 2, 176-177.

Alam, S.N., Rashid, M.A., Rouf, F.M.A., Jhala, R.C., Patel, J.R., Satpathy, S., Shivalingaswamy, T.M., Rai, S., Wahundeniya, I., Cork, A., Ammaranan, C. and Talekar, N.S. (2003). Development of an integrated pest management strategy for eggplant fruit and shoot borer in South Asia. Shanhua, Taiwan: AVRDC – the World Vegetable Center. Tech. Bull. No. 28, AVRDC Publication No. 03-548. 56 pp.

Attygalle A.B., Schwarz J. and Gunawardena N.E. (1988). Sex pheromone of brinjal shoot and pod borer *Leuconodes orbonalis* Guenee (Lepidoptera: Pyralidae: Pyraustinae). *Zeitschrift fur Naturforschung*, C (Biosciences). 43(9-10): 790-792.

Awasthi AK (2000). Preliminary screening of brinjal genotypes to *Leucinodes orbonalis* (Guen.). *Insect Environment*. 6(1): 33-34.

Bajaj KL, Singh D and Kaur G (1989). Biochemical basis of relative field resistance of eggplant (*Solanum melongena*) to the shoot and fruit borer (*Leucinodes orbonalis* Guen.). *Veg. Sci.* 16(2): 145-149.

Bhaskar H and Viraktamath CA, 2002. Diversity and abundance of aphidophagous coccinellids in cabbage field. *Insect Environment*., 8: 1, 31.

Brar JS, Bhalla JS and Singh H (1992). Chemical control of *Leucinodes orbonalis* Guen. in brinjal. *J. Insect Sci.* 5(2): 225-226.

Brar KS, Dhillon GS, Singh M and Dhillon TS, 1995, A note on the resistance in pea genotypes to leaf miner, *Chromatomyia horticola* (Goureau). *Haryana Journal of Horticultural Science* 24: 2, 156-158.

Chakraborti S (2001). A biorational approach for the management of brinjal fruit and shoot borer, *Leucinodes orbonalis* Guen. *J. Entomol. Res.* 25(1): 73-76.

Cork A, Alam SN, Das A, Das CS, Ghosh GC, Farman DI, Hall DR, Maslen NR, Vedham K, Phythian SJ, Rouf FMA and Srinivasan K (2001). Female sex pheromone of brinjal fruit and shoot borer, *Leucinodes orbonalis* blend optimization. *J. Chem. Ecol.* 27(9): 1867-1877.

Dahiya B, Chauhan R and Goel SC, 1992. Chemical control of pod borer complex in pea Pisum sativum L. Bioecology and control of insect pests: Proceedings of the National Symposium on Growth, Development & Control Technology of Insect Pests. 170-174.

Dash PC, 1994. Studies on the host range of American serpentine leafminer in Orissa. *Orissa Journal of Agricultural Research* 7: 92-94.

David PMM, 1991. Resurgence of yellow mite *Polyphagotarsonemus latus* (Acarina: Tarsonemidae) on chilli following application of insecticides. *Madras Agricultural Journal* 78: 1-4, 88-91.

Deepesh-Sharma and Usha-Sharma, 1997. Status of *Liriomyza trifolii* (Burgess) and its host plants in Jabalpur district of Madhya Pradesh. *Crop Research* (Hisar). 14: 2, 351-355.

Dhamdhere SV, Koshta VK and Rawat RR (1990). Effect of food plants on the biology of *Epilachna vigintioctopunctata* Fabr. (Coleoptera: Coccinellidae). *J. Entomol. Res.* 14(2): 142-145.

Gajendra Singh and Mukherjee A (1987). On the oligophagous nature of *Henosepilachna dodecastigma* (Wiedemann) and *Henosepilachna vigintioctopunctata* (Fabricius) (Coleoptera: Coccinellidae). *Indian J. Entomol 49*(1): 118-126.

Gill CK, Dhooria MS and Chadha ML (1984). Relative susceptibility of some varieties of brinjal to *Euzophera perticella* Ragonot. *Indian J. Entomol*. 46(2): 239-240.

Gopal S and Senguttuvan T, 1997. Integrated management of tomato fruit borer with insecticides, neem products and virus. *Madras Agricultural Journal*. 84: 2, 82-84.

Gopalakrishnan C, 1989. Susceptibility of cabbage diamondback moth *Plutella xylostella* L. to the entomofungal pathogen *Verticillium lecanii* (Zimmerm.) Viegas. *Current Science*. 58: 22, 1256-1257.

Gupta M and Pawar AD, 1989. Monitoring *Spodoptera litura* (F.) in vegetable crops through synthetic sex pheromone traps. *Indian Journal of Plant Protection*. 17: 1, 117-118.

Gupta SC, Singh PP and Kumar A (1999). Effect of intercropping with spices on the incidence of *Leucinodes orbonalis* Guen. in brinjal and its economics. Shashpa 6(1): 45-48.

Jagannatha R and Viraktamath CA, 1997. Host preference by the female serpentine leafminer, *Liriomyza trifolii* (Diptera: Agromyzidae). *Insect Environment*. 2: 4, 137.

Justin CGL, Rabindra RJ and Jayaraj S, 1990. *Bacillus thuringiensis* Berliner and some insecticides against the diamond-back moth, *Plutella xylostella* (L.) on cauliflower. *Journal of Biological Control*. 4: 1, 40-43.

Karmakar K, 1997. Effect of micronutrients on the biology of *Polyphagotarsonemus latus* (Banks) (Acari: Tarsonemidae). *Environment and Ecology*. 15: 3, 699-701.

Kashyap RK, Banerjee MK, Kalloo G and Verma AN, 1990. Survival and development of fruit borer, *Heliothis armigera* (Hubner). (Lepidoptera: Noctuidae) on *Lycopersicon* spp. *Insect Science and its Application*. 11: 6, 877-881.

Katole SR, 1992. Effect of various larval food plants on the development & reproduction of *Heliothis armigera* Hb. *Comparative Physiology and Ecology*. 17: 3, 97-101.

Krishnaiah K, 1980. Methodology for assessing crop losses due to pests of vegetables. In: Govindu HC , Veeresh GK ,Walker PT and Jenkyn JF(Eds.), Assessment of crop losses due to pests and diseases. pp 259-267.

Krishnamoorthy A and Mani M (1998). New record of parasitoid *Diadegma apostata* (G.) on brinjal shoot and fruit borer. *Insect Environ.* 4(3): 87.

Krishnaprasad-NK; Rai-PS, 1978, Effect of different host plants on the duration of life stages of pumpkin leaf caterpillar, *Margaronia indica* Saunders (Pyralidae: Lepidoptera). *Current Research.* 7: 4, 68.

Kumar NKK, 1995. Yield loss in chilli and sweet pepper due to *Scirtothrips dorsalis* Hood (Thysanoptera: Thripidae). *Pest Management in Horticultural Ecosystems.* 1: 2, 61-69.

Kumar NKK, Aradya M, Deshpande AA, Anand N and Ramachandar PR, 1996. Initial screening of chilli and sweet pepper germplasm for resistance to chilli thrips, *Scirtothrips dorsalis* Hood. *Euphytica.* 89: 3, 319-324.

Kumar SP and Babu PCS (1998). Ovipositional deterrent and ovicidal effects of Neem Azal against brinjal shoot and fruit borer and leaf beetle. *Shashpa* 5(2): 193-196.

Kumbhar TT, Kokate AS and Dumbre AD, 1991. Studies on the varietal resistance in okra (*Abelmoschus esculentus*) (L.) (Moench) to shoot and fruit borer (*Earias* spp.) .*Maharashtra Journal of Horticulture.* 5: 2, 78-82.

Lal OP, 1991. Studies on field resistance of different cultivars and crosses of cabbage against cabbage caterpillar, Pieris brassicae L. (Lepidoptera: Pieridae). *Zeitschrift-fur Pflanzenkrankheiten und Pflanzenschutz.* 98: 3, 293-300.

Lal OP, Gill HS, Sharma SR and Singh R, 1991. Field resistance in different cultivars of cauliflower against head borer, *Hellula undalis* Fabricus (Lepidoptera: Pyralidae). *Journal of Entomological Research.*, 15: 4, 277-281.

Mallapur CP, 2000. Screening of chilli genotypes against thrips and mites. Insect-Environment. 5: 4, 154-155.

Mallik SN, Kumar M, Sinha AN and Karn BP (1989). *Trathala flavo-orbitalis* Cameron (Ichneumonidae) - a parasite of *Leucinodes orbonalis* Guen. from Bihar. *Curr. Sci.* 58(19): 1098-1099.

Mehta PK, Chandel YS and Sharma DC (1993). Occurrence of parasites of brinjal hadda, *Henosepilachna vigintioctopunctata* F. in Himachal Pradesh. *Indian J. Entomol.* 55(1): 103.

Mohanty JN, Prakash A and Rao J (2000). Management of brinjal shoot and fruit borer, *Leucinodes orbonalis* Guen. by release of *Bracon brevicornis* Wesm. in the field. *J. Appl. Zool. Res.* 11(2/3): 96-97.

Moorthy PNK and Srinivasan K, 1989. Petiole-mining thresholds and potassium chloride sprays for management of bean fly (Diptera: Agromyzidae) in snap beans and cowpea. *Journal of Economic Entomology.* 82: 1, 246-250.

Moorthy PNK and Srinivasan K, 1993. Plant ash sprays for the management of bean fly, *Ophiomyia phaseoli* (Tryon) in beans. *Botanical pesticides in integrated pest management.* 385-390.

Pareek BL and Bhargava MC, 2003, Estimation of avoidable losses in vegetable crops caused by borers under semi-arid conditions of Rajasthan. *Insect Environment.* 9: 2, 59-60.

Patel JR Borad PK, Ratanpara HC, Shah-BR, 1991. Need based control of *Helicoverpa (Heliothis) armigera* Hubner on tomato with endosulfan. *Indian Journal of Plant Protection.* 19: 1, 79-80.

Patnaik HP and Mohapatra-S, 1997. Effect of fertilizers on the incidence of thrips and leaf-curl in chilli under protected and un-protected conditions. *Mysore Journal of Agricultural Sciences.* 31: 2, 159-163.

Patnaik HP, 1997. A note on field evaluation of insecticides and *Bacillus thuringiensis* Berliner to control *Spodoptera litura* Fab. in tomato. *Orissa Journal of Horticulture.* 25: 1, 79-80.

Patnaik HP, 2000. Host preference of serpentine leaf miner, *Liriomyza trifolii* under field conditions. *Insect Environment*. 6: 1, 31.

Peter C, Balasubramanian R and Prasad VG, 1987. Studies on the biology of the cabbage borer *Hellula undalis* Fabricius (Lepidoptera: Pyralidae) in Karnataka. *Mysore Journal of Agricultural Sciences*., 21: 3, 309-312.

Rai PS and Chandra BKN, 1976. Bionomics of the cabbage leaf-webbing caterpillar, *Crocidolomia binotalis* Zeller. *Indian Journal of Entomology*. 38: 3, 233-235.

Rajagopal D and Trivedi TP (1989). Status, bioecology and management of epilachna beetle, *Epilachna vigintioctopunctata* (Fab.) (Coleoptera: Coccinellidae) on potato in India: a review. *Trop. Pest Managt.* 35(4): 410-413

Rajavel DS and Gopalan M, 1991. Control of cabbage diamondback moth, *Plutella xylostella* (L.) with synthetic pyrethroids. *Madras Agricultural Journal*. 78: 1-4, 98-99.

Ram S, Patnaik NC, Mohapatra AKB, Sahoo S and Samal KC, 1996. Effects of fertility levels and chilli varieties on the yield of green chillies and incidence of chilli thrips *Scirtothrips dorsalis* Hood in Eastern Ghat Highland Zone of Orissa. *Environment and Ecology*. 14: 3, 642-645.

Ramnath S, Chitra K, Uthamasamy, S, 1992. Behavioural response of *Helicoverpa armigera* (Hub.) to certain host plants. *Journal of Insect Science*. 5: 2, 147-149.

Rao-SV, Azam KM and Ahmed K, 1989. Mating disruption through behavioural manipulation of *Spodoptera litura* Fabricius. *Current Science*. 58: 16, 933-935.

Ravi KC, Puttaswamy; Viraktamath-CA; Mallik-B; Ambika-T, 1998, Influence of host plants on the development of *Diaphania indica* (Saunders) (Lepidoptera: Pyralidae). In: Reddy-PP Kumar-NKK and Verghese-A(Eds.) Advances in IPM for horticultural crops. Proceedings of the First National Symposium on Pest Management in Horticultural Crops: environmental implications and thrusts, Bangalore, India, 15-17 October 1997. 135-136.

Rawat US and Pawar AD, 1993. Biocontrol of tomato fruit borer, Heliothis armigera (Hubner) in Himachal Pradesh, India. *Plant Protection Bulletin Faridabad*. 45: 2-3, 34.

Ray D and Banerjee TC, 1993. Bionomics and feeding capabilities of *Spodoptera litura* (Fabr.) (Lepidoptera: Noctuidae). *Phytophaga*. 5: 1, 51-62.

Rizvi SMA and Sanyal PK (1977). *Euzophera perticella* Rag. (Lepidoptera: Pyralidae) as a root borer of brinjal. *Indian J. Pl. Prot*. 5(2): 213.

Samuthiravelu P and David BV, 1991, Bioefficacy of neem oil and endosulfan against the fruit borer *Earias vittella* (Fab.) (Noctuidae: Lepidoptera) on okra. *Madras Agricultural Journal*. 78: 1-4, 77-78.

Sharma ML, Sharma TR, Sharma RC and Nandkar DN, 1990. Field evaluation of tomato-cultivars to fruit borer (*Heliothis armigera*) Hubner in Satpur Plateau region of Madhya Pradesh. *Orissa Journal of Horticulture*.18: 1-2, 22-25.

Singh D, Narang DD and Chahal BS, 1984, Control of red pumpkin beetle *Raphidopalpa foveicollis* (Lucas) by foliar and soil application of insecticides. Journal-of-Research, *Punjab Agricultural University*. 21: 4, 525-532.

Singh L and Singh H, 1990. Biology of head borer, *Hellula undalis* Fab. on cauliflower. *Journal of Insect Science*. 3: 1, 72-76.

Singh OP and Parihar SBS, 1988. Effect of different hosts on the development of *Heliothis armigera* Hub. *Bulletin of Entomology*. 29: 2, 168-172.

Singh W, Kotwal DR and Singh R, 1991. Evaluation of some contact insecticides for the control of jassid, *Amrasca biguttula biguttula* Ishida on okra. *Indian Journal of Plant Protection*. 19: 2, 182-184.

Sivaprakasam N, 1996. Ovipositional preference of *Helicoverpa armigera* to tomato cultivars. *Madras Agricultural Journal*. 83: 5, 306-307.

Sivaprakasam N, 1997. Parasitoids of fruit-borer (*Helicoverpa armigera*) of tomato (*Lycopersicon esculentum*) in Tamil Nadu. *Indian Journal of Agricultural Sciences*. 67: 4, 177.

Sood A and Singh G, 1997. The intrinsic rate of *Myzus persicae* (Sulzer) (Homoptera: Aphididae) on cabbage Brassica oleracea var. capitata. *Annals of Entomology*., 15: 2, 21-26.

Sood A and Singh G, 1999. Biology of *Myzus persicae* (Sulzer) on winter and summer cabbage in Pantnagar. *Annals of Agricultural Research*., 20: 2, 243-244.

Srinivasan K & Murthy PNK 1991. Mustard plants trap major cabbage pests. *Indian Farming*. 40: 11, 11-12.

Srinivasan K, Viraktamath CA, Gupta M and Tewari GC, 1995. Geographical distribution, host range and parasitoids of serpentine leaf miner, *Liriomyza trifolii* (Burgess) in South India. *Pest Management in Horticultural Ecosystems*. 1: 2, 93-100.

Thakur JC, Khattra AS and Brar KS, 1992, Comparative resistance to fruit fly in bitter gourd. *Haryana Journal of Horticultural Sciences*. 21: 3-4, 285-288.

Thakur JC, Khattra AS and Brar KS, 1996, Correlation studies between economic traits, fruit fly infestation and yield in bittergourd. *Punjab Vegetable Grower*. 31: 37-40.

Thakur NSA and Deka TC, 1997. Ecological studies on the cabbage butterfly, *Pieris brassicae* (Linnaeus) in north eastern India. *Pest Management in Horticultural Ecosystems*. 3: 1, 13-15.

Uthamasamy S, 1988. Seasonal incidence and damage potential of leafhopper (*Amrasca biguttula biguttula*) on okra. *Indian Journal of Ecology*. 15: 1, 117-118.

Varadharajan S and Veeravel R, 1995. Population dynamics of chilli thrips, *Scirtothrips dorsalis hood* in Annamalainagar. *Indian Journal of Ecology*. 22: 1, 27-30.

Venkatesh HM, Muniyappa V, Ravi KS and Krishnaprasad PR, 1998. Management of chilli leaf curl complex. In: Advances in IPM for horticultural crops: Proceedings of the First National Symposium on Pest Management in Horticultural Crops: environmental implications and thrusts, Bangalore, India, 15-17 October 1997. 111-117.

Vyas SH and Patel JR, 1991. Intensity and damage of *Earias vittella* Fab. on various cultivars of Bhendi. *Gujarat Agricultural University Research Journal*. 17: 1, 140-141.

Entomology : Novel Approaches, 2007
Eds.: *P.C. Jain and M.C. Bhargava,*(Page 377-398)
New India Publishing Agency, New Delhi
E-mail: newindiapublishingagency@gmail.com
web: www.bookfactoryindia.com

CHAPTER 19

INTEGRATED PEST MANAGEMENT IN POTATO

R.S. Chandel[1], K.R. Dhiman[2], V.K. Chandla[3] and Vinod Kumar[4]
[1] Sr. Scientist, [2] Principal Scientist and Head, [3] Principal Scientist, [4] Scientist
Central Potato Research Station, Kufri, Shimla – 171 012.

INTRODUCTION

Potato is a crop of major significance in human nutrition, ranking fourth in world production, after wheat, maize and rice. Potatoes are grown on over 44 million acres in more than 125 countries, with annual production of about 250 million tons. Potatoes are grown across India, with production concentrated in several areas of the North Indian plains and West Bengal. Indian potato production is significant and accounts for 8.5 per cent of the world total (Dahiya *et al.,* 1999).

The potato has hundreds of recognized pests including many insects, mites, nematodes and rodents. Some of these were transported to new locations alongwith seed tubers. Others were already present in locations where the potato was introduced and then proliferated on this new host plant. Because a potato crop is vegetatively propagated from tubers, which easily carry some pathogens and pests, many pest problems have followed potatoes to areas wherever they are grown. The potato tuber moth was introduced accidentally into India from Italy in 1906, probably through seed potatoes (Lefroy, 1907), and by the middle of century had become well established throughout India (Kumar and Nirula, 1967). Of all the major food crops, potato is heaviest user of chemical pesticides, which often total up to 20 per cent of its cost of production (Anonymous, 1991).

Recognizing that sustainable production can not be achieved as long as pest control depend on chemical pesticides, the Central Potato Research Institute has focused its activities on the development of IPM strategies to help potato producers in India to reduce the use of pesticides. A combination of management practices including biological, manipulation of crop and pest environments, careful selection of chemical and its time of application can significantly reduce the pesticide load. These control components must be integrated in farmer's fields. This chapter provides up date information on advances made in the development of IPM technologies and options relevant to potato pest management.

MAJOR INSECT PESTS

Potato is vulnerable to attack of pests, both in fields as well as in stores. A great diversity of pests attacking potato exists in India due to varying agro climatic conditions. These pests damage the potato crop by feeding on leaves, thus reducing photosynthetic efficiency, by attacking the stems thus weakening the plant, inhibiting growth of potato tubers and by feeding on tubers. Accordingly, potato pests are grouped into soil pests, foliage feeders, sap feeders and storage pests, besides nematode pests are dealt separately. In seed production, the pests of greatest concern are usually the aphid vectors of potato viruses especially *Myzus persicae* (Sulzer). In ware production, the key pests may be insects which attack tubers, such as tuber moth, white grubs and cut worms. In some situations, foliage feeders such as noctuids and coccinellids are important.

SOIL PESTS

Cut worms: Cut worms (*Agrotis* spp), are polyphagous insects of cosmopolitan distribution. In India, *Agrotis ipsilon* (Hufn.), *A. segetum* (Schiff.), *A. flammatra* Schiff., *A interacta*Wlk., and *A. spinifera* Hb. occur on potato (Dass 2000). But, *A. segetum* and *A. ipsilon* are more serious. The former is common in the hills and the latter in the plains (Chandla, 1985). These are generally destructive at seedling stage and several crops like potato, cotton, rice, tobacco, cereals and vegetables have been found infested with these pests (Trivedi and Rajagopal, 1999). The attack is more pronounced during dry period when potato vines are quite tender. Tuber damage does not occur in rainy season crop, but larvae inflict considerable tuber damage in spring crop. In plains, it is active from October onwards and with onset of summer, it migrates to hilly regions. In India, cut worms are more serious in northern region than in south (Butani and Jotwani, 1984). The larvae feed voraciously and cut potato plants in early stage of the crop growth. The plants are eaten off just above, or at short distance below the soil surface. In certain cases entire row of plants is cut. Tuber damage is manifested in the form of deep holes (Fig.1). The peak activity is found during May-June in Shimla hills, August in peninsular India and March-April in Patna and Jalandhar (Singh, 1987). Moths are nocturnal and appear soon after dusk, mate and lay eggs on ventral surface of leaves or moist soil (Misra *et al*., 2003). In general moths have brownish forewings and whitish hind wings and moths of *A. flammatra* are much bigger in size as compare to other species.

Surface cut worm, *Agrotis spinifera* Hb.: This species occurs in Punjab, Bihar, Andhra Pradesh and Karnataka. Moths appear in August and peak population is found in September which gradually declines during October – November. Larvae and pupae are found during February and April, respectively (Trivedi and Rajagopal, 1999). The female lays 431 – 901 eggs in 4 – 10 days. The mated males and females survive for 4 – 18 and 6 – 12 days, respectively. Larvae feed on leaves, stem and tubers.

Greasy cut worm, *Agrotis ipsilon* (Hufn.): This is generally a cool climate pest active from October onwards in plains and migrates to hills in summer. In India, it is more serious in north than in the south (Butani and Jotwani, 1984). Moths are medium sized, stout, dark greenish brown with reddish tinge and have grayish – brown wavy lines and spots on forewings; hind wings are hyaline having dark terminal fringe. The female starts laying eggs

4–6 days after emergence and lays 300–450 eggs in 4 – 11 days. Eggs are laid during night either singly or in batches of 30–50 eggs. Egg hatches in 3–5 days. The caterpillars are light brown with reddish tinge which turns greenish thereafter. Full fed larvae are 40–50mm long. The larvae inhabit in the upper layer of soil and fourth instar is voracious feeder (Trivedi and Rajagopal, 1999). Larval period is 22 – 30 days. Occasionally the caterpillars may also nibble tubers. Pupation takes place in soil and lasts for 12 – 15 days. Life cycle is completed in 39 – 53 days (Singh, 1987).

Black cut worm, *Agrotis segetum* (Schiff.): The moths are pale whitish brown with forewings ochreous – brown and having double waved sub basal ante and post medial lines and marginal series of specks; hind wings are iridescent white with dark marginal lines; wing expanse is 40 – 48mm (Fig.2). Eggs are laid in clusters of 18 – 40 on leaves and stems of the plants during summer. Some of the cutworms feed until August then pupate and emerge to produce a second generation of moths. Most, however, over-winter as caterpillars in the soil and pupate in the following April – May; the moths emerge in May and June. The eggs hatch in 4 – 7 days. The eggs are dome shaped and creamy white in color. Full grown caterpillars are 35mm in length. The larvae are full fed in 22 – 30 days. The pre-pupal and pupal periods range between 2 – 3 and 12 – 20 days. The pupation takes place in soil or between the folds of dried potato leaves. Male moths live for 2 – 4 days and females live for 5 – 8 days. Mean fecundity is 161 eggs / female. Two generations are completed during potato crop season (Misra *et al.,* 1995).

Gram cut worm, *Agrotis flammatra* (Schiff.): This pest is distributed in Punjab and sub Himalayan region. The pest is active from October – April in the plains and migrates to hills in summer. Moths of *A. flammatra* are much bigger in size, average wing expanse is 56mm. Its forewings have characteristic markings and smoky patches, two – thirds of the proximal areas being pale. On each wing, there is a semi circular spot below the pale area and a grayish brown kidney shaped spot towards the apical area. The caterpillars are dark grey or dull green measuring 40 – 50mm. In October, the moths lay eggs on the underside of leaves, on shoots, stems or in the soil. A female lays up to 980 eggs in her life time. The eggs hatch in 4 – 7 days during summer and 10 – 14 days during winter. Larvae are full fed in 4 – 7 weeks. The pupal stage lasts 12 – 15 days. The moths live for 7 – 13 days and maximum eggs are laid between 5 – 9 days after emergence. The life cycle in completed in 7 – 11 weeks and there are generally two generations in a year.

***Agrotis interacta* Wlk:** The moths are more or less similar in appearance and size as *A. segetum*. It is exclusively subterranean and feed generally on roots and tubers by chewing inside cavities.

Management of cut worms: Cut worms are subject to attack by other predators. Thus, fork the soil during period of attack so that the larvae are exposed and readily fed by their avian predators. Efficient control can be achieved by properly applied sprays when the young caterpillars are still on the haulms and therefore vulnerable. Once below ground, cut worms are unlikely to be significantly affected by the insecticides applied to the soil. Older caterpillars are generally less susceptible to insecticides than young caterpillars. Treatments made routinely are likely to be applied at the wrong time or when the risk of damage is less. Young caterpillars can be killed by rain or irrigation and an insecticidal spray would be unnecessary in these conditions (Hancock, 1990).

Treatments should be applied when the soil is dry and the weather is warm. Good control of cutworms depend on a thorough foliage coverage with a high volume application, using at least 1000L / ha of water. Chlorpyriphos @ 2.5 L / ha, cypermethrin 0.25 L / ha and triazophos @ 2.5 L / ha can be used for the control of cut worms on potatoes. They should not be applied to crops suffering from drought because the plants can then be damaged. Severe losses caused in spring crop can be reduced significantly by using metasystox @ 0.03 per cent. The ETL is one larva / 10 plants (Trivedi and Rajagopal, 1999). In many instances insecticides may be applied to soil at 2 per cent plant damage (Chandel and Chandla, 2003). In nature, cutworms are parasitized by *Pericepsia carbonaria, Macrocentrus collaris, Bracon kitcheneri* Will. and *Dlieanta ruficauda,* whereas, *Broscus punctatus* Klug and *Liogryllus bimaculatus* are important predators of this pest. (Butani and Jotwani, 1984; Misra *et al.,* 2003).

White grubs: White grubs are most destructive and troublesome soil insects, threatening potato production in hilly states. These white grubs are present in the soil at a depth of 5 – 20 cms during the crop season. Large holes are made in the tubers which ultimately may be entirely transversed by wide deep mines (Fig.3). They live concealed, feed on the roots or decaying vegetative matter and suddenly increase their population in places having enough food and least soil disturbance. This is a polyphagous pest both in grub and adult stage and inflicts heavy damage on various fruit / forest trees, their nurseries, vegetables, lawns and field crops (Chandel *et al.,* 1995).

Beetles usually remain unnoticed throughout the year and their appearance in large numbers occurs just after break in monsoon and attack persists for a month or two. In India, 20 species of white grubs have been reported on potato. Out of these, *Brahmina coriacea* (Hope) and *Holotrichia longipennis* (Blanchard) are most destructive in north western hills (Misra and Chandel 2003). *Holotrichia serrata* (Fab) damages potato in Karnataka (Regupathy *et al.,* 1997). All the requirements of the life cycle except mating and feeding are met with underground (Chandel and Kashyap, 1997).

***Brahmina coriacea* (Hope):** In the spring after the apple and other fruits have put fourth leaves, the adults become active, fly about during night and feed on the foliage of apple and leaves of some other plants (Chandel and Kashyap, 1997). They leave the soil just at dusk and remain on the leaves during night, mating and feeding. At the first streaks of dawn, they return very promptly to the soil where the females lay their pearly white eggs. The eggs are generally laid in grassland or patches of grassy weeds in cultivated fields and hatch in about 2 weeks (Chandel *et al.,* 1995).

The grubs after hatching move into the soil. There are three distinct larval instars. The newly formed grubs in three successive instars are on an overage 5.6, 12.6 and 22.9 long and full fed grubs are about 30mm in length. Development of 1st and 2nd instar grubs requires about 14.4 and 20 days in mid hills (Chandel *et al.*, 1995). In higher hills, 1st and 2nd instars have been of 20.1 and 29.6 days, respectively (Chandla *et al.*, 1988). The third instar grubs are most active and have immense capacity to cause damage. These grubs feed very well up to October and are therefore responsible for heavy damage. Most of the third instar grubs attain their full size (Fig.4) in about 50 days before the on set of winter (Misra and Chandel, 2003). During later part of winter season, full fed grubs stop feeding, construct earthen cell, remain inside these cells till spring and then change into pupae.

Formation of pupa takes place in March – April. Ecdysis from pupa to the adult occurs in earthen cells in 20 days. Adults remain confined in these cells for sometime. With the onset of rains, the earthen cell is softened and the beetles emerge out of these cells. Adults are well known black beetles of apple (Fig.5). Maximum adult emergence takes place in mid-June (Chandel *et al.,* 2003).

Holotrichia longipennis* (Blanch)*: The beetle emergence starts during first fortnight of May. The beetles (Fig.6) have been reported to feed upon the foliage of a wide variety of fruit / forest trees, but, *Rubus ellipticus*, apple, walnut, chestnut and plum are its preferred hosts (Shah and Shah, 1990). Eggs are laid singly inside earthen cells and the incubation period is 11 – 18 days. The young grubs soon begin to feed on organic manure available in soil and grow rapidly. The first instar grub stage lasts for 38 – 44 days. Two moults of grubs occur during rainy season and the second instar can be seen damaging tubers by the end of July. The full grown grubs measure 38.12mm in length. When they cease to feed, move down 15 – 20 cm deeper in soil and construct earthen cells for over wintering by mid November. They can go up to 2.5 meters deep for over wintering. Total larval period is 243 – 282 days (Shah and Shah, 1990). The diapausing grubs start their upward movement by the end of March and pupate at a soil depth of 20 – 30 cm. Pupal period is of 22 – 28 days. In the first week of May, almost all the pupae get transformed into adults (Mishra and Singh, 1993). The beetles live for 28 – 56 days.

***Holotrichia serrata* (Fab):** The grubs cause extensive damage to vegetables, pulses, oilseed, cereals, millets, tobacco, sugarcane and sorghum (Yadava and Sharma, 1995). The adults are attracted to neem, palas, babul and guava. A female lays up to 40 eggs in her life time. Fresh eggs are spherical and measure on an average 2mm. Eggs hatch in 12-15 days and one or two days before hatching the egg swells up to 4.0mm in diameter. The first instar larva comes out of the earthen cell if there is sufficient moisture, under drought it remains inside the cell until favorable condition occur. First, second and third instar grubs are 10.8, 20.0, and 47.0mm long and average duration of each instar is 22.5, 35 and 124.5 days. The grubs become full grown in October, stop feeding, burrow deeper in the soil and construct an earthen cell for pupation. Pupa remains for 15.5 days. Adults are 22.4 long and 14.0mm in body width. The colour is dull brown with light brown abdomen (Yadava and Sharma, 1995).

Management of white grubs: As a rule, these beetles will not deposit eggs in fields of clovers unless there is a considerable admixture of grasses or weeds in such fields. Thus grow clovers or alfalfa if possible to prevent the beetles from laying eggs. One of the best ways to clean grubs out of field is to graze the land with pigs. When pigs are allowed to run on heavily infested land they will usually root out and eat the grubs nearly freeing the land of grubs (Chandel and Kashyap, 1997).

Ploughing infested fields especially when most the grubs are pupating kills many of the pupae and newly formed adults (Mişra and Chandel, 2003). The beetles can be collected at night and killed in kerosinized water. The collection should be done on community basis. The FYM used should be well rotten. The preferred hosts which attract these beetles should be planted to trap and kill them. Soon after the attack is noticed, methyl parathion (0.05%) or carbaryl (0.1%) or monocrotophos (0.05%) can be sprayed (Chandel *et al.*, 1994; Chandla *et al.,* 1988; Anonymous, 2000). In potato, their damage can be minimized by application of phorate at the time of earthing up. Chlorpyriphos is also equi-effective. To obtain best results, insecticide application should be properly timed. Hence, apply insecticides soon after the emergence of beetles coinciding with their egg laying or egg hatching (Misra and Chandel, 2003).

Termites and Ants: Several species of termites such as *Microtermes obesi* (Holmgren), *Odontotermes obesus* (Rambur) and *Eromotermes* spp. have been reported damaging potato crop especially in sandy soil or in fields having poor aeration and drainage (Chandel and Chandla, 2003). The worker caste of termites is responsible for the damage by making deep holes. They are about 2.5cm long, dirty white in color with brown head. They are wingless and can be easily recognized by their vertically carried head with small broad jaws. These are sporadically severe. These termites feed on roots and tubers. The tubers become hollow and are often filled with soil. As a result the leaves of such plants start yellowing and wilting and ultimately dry up (Butani and Jotwani, 1984).

Red ants, *Dorylus orientalis* Westwood and *D. labiatus* Shuckard have termite like habit of attacking plants underground. Red ants are reported as a pest of potato, cauliflower, groundnut, sugarcane and coconut seedlings in north eastern states, Bihar and Uttar Pradesh (Roonwal, 1976). The pest appears during December and remains active till April, causing more than 10 per cent damage in irrigated crop. High temperature and dry weather favours population build up (Ram Kishore *et al.,* 1993). The pest damages stem and tubers by making holes. Severely damaged plants show wilting during bright sunshine and finally plants dry up (Trivedi and Rajagopal, 1999).

Management: To avoid damage by the termites and ants, raising potatoes in sandy soils or areas infested with these pests (especially termite) should be avoided. Avoid use of partially decomposed FYM. The damage can also be reduced by irrigation. Charge irrigation water with crude oil @ 12 L / ha (Pandey, 2002). Soil treatment / spray with chlorpyriphos 20 EC @ 2.5 L / ha has been found effective (Raj and Parihar, 1993).

Minor soil pests: The minor soil pests recorded on potato crop are wire worms, *Agonichis* sp; grey ash weevil, *Myllocerus subfasciatus*, lesser bulb fly, *Eumerus* spp; tobacco caterpillar, S*podoptera litura* and mole cricket, *Gryllotalpa africana*

STORAGE PESTS

Potato tuber moth : Potato tuber moth, *Phthorimaea operculella* (Zeller) is a serious pest of stored potato tubers and does considerable damage to potato plants and developing tubers in field. The damaging stage of the pest is larva which feeds on potato foliage and attack tubers in the field before and shortly after harvesting. The foliage feeding larvae make transparent galleries in infested leaves. The tuber mining larvae bore mostly near the eyes of tubers filling the tunnels with excrement. Fungus grows in burrow and discolours it so that course of work may be easily followed. Later the skin of potato partially dries and sinks so that the scars become very prominent; this is called sub-epidermal injury as shown in figure 7 (Chandel and Chandla, 2003).

The infestation of PTM starts in the field on leaves and acts as an initial source of infestation. Moths emerge from over-wintering larvae in early spring and lay eggs, chiefly on underside of leaves or upon exposed tubers. The eggs hatch in 4 – 7 days. Newly hatched larva is white with dark brown head. Full grown larvae are about 12mm in length and 3mm in thickness. It is pink in colour with dark colouration on both sides (Fig. 8). The larvae are full fed in 15 – 20 days. They pupate in a grayish, silken, dirt covered cocoons, entangled in dead leaves or among trash etc. on ground and emerge as adults in 6 – 8 days. The adults are very small narrow – winged, night active moths, grayish brown in colour and mottled with darker brown (Fig. 9). The forewings are light black in colour with fringed apical margin. A characteristic feature of the moth is that its hind wings are densely fringed. Female moths are larger (8.2 mm) than male moths (7.63mm). More over newly emerged females has a dark black patch of scales in the middle of forewings. The abdominal tip is conical in females and broad in males (Chauhan and Verma, 1991). An entire generation may develop in 22 – 25 days in warm weather and 5 – 6 in a year in Kangra valley of Himachal Pradesh (Chandel *et al.,* 2001). The damage is most severe in years of low rainfall and high temperatures. In Assam, as many as 11 generations, in the northern plains 8-9 generations and in plateau region 10-13 generation are reported to occur in a year (Gubbaiah and Thontadarya, 1977; Isahaque, 1978). In Bangladesh, 18 generations are completed in a year (Kabir, 1994).

Management of potato tuber moth: To control this pest, healthy seed tubers should be planted at 10cm depth and potato fields should be kept well-cultivated and deeply hilled during their growth. Irrigation may be done at regular intervals to avoid cracking of the soil. Irrigating the field at 60 – 80 per cent field capacity has been observed to reduce tuber infestation (Temerak, 1983). It is observed that tuber infestation is more in furrow irrigation than with sprinkler irrigation (Shelton and Wyman, 1979). Infestations in growing potatoes may be controlled by spraying the foliage after mid-March when worms begin to infest leaves with monocrotophos (0.05%) or synthetic pyrethroids (Chandel *et. al.,* 2000). The infested wilting vines should be cut and removed from the field a few days before digging and never piled over dug potatoes. Harvesting should be done before three – fourth of the foliage dries up (Shaheen, 1983). The tubers should not be left exposed to egg-laying moth during late-afternoon or over-night.

In storage, it is necessary to remove damaged tubers before storing. Storage of healthy tubers in cold stores is the best way to control tuber moth. However, in traditional rustic stores, tuber infestation can be saved by covering them with 2 cm thick layer of dry leaves of *Lantana* or *Eucalyptus* (Anonymous, 2000). The application of insecticides to tubers destined for human consumption is highly undesirable and banned. Granulosis virus (GV) is extremely effective in reducing PTM damage under storage condition (Raman, 1988). The traditional country stores should be improved by blocking the air holes. Openings should have good air circulation to keep inside temperature cool, but wire mesh screens are to be used to limit access by PTM (Chandel *et al.,* 2005). Seed potatoes not to be used for food may be treated at the time of storage with malathion dust 5% @ 125g/100 kg tubers. A through cleanup of potato storage sheds helps prevent re-infestation (Chandel *et al.,* 2000).

Several natural enemies have been reported by various workers. *Copidosoma koehleri,* an exotic egg-larval parasitoid is effective in Maharashtra (Dalaya and Patil, 1973). *Chelonus blackburnii* is another exotic parasitoid causing moderate parasitization in Bangalore (Khandge *et al.,* 1979). Other important parasitoids are *Bracon hebetor, Orgilus jennieae* and *Apanteles subandinus* (Trivedi and Rajagopal, 1999).

LEAF EATING INSECTS

Hadda beetles: *Epilachna* beetles and its grubs form important pests of potato. The two types of *Epilachna* beetles commonly found all over India are the 12 spotted (*Epilachna ocellata* Redt. Fig.10) and 28 spotted beetles (*Epilachna vigntioctopunctato* Fab.). The former is generally found in higher hills and later is, however, restricted to mid hills or plains (Misra *et al.*, 2003). This has also been recorded on brinjal, tomato, tobacco, pumpkin and bitter gourds. The damage is caused by the adult and grubs feeding on leaf tissues and skeletonizing the leaves totally (Chandla, 1985). The adults are active flies and move about from plant to plant. The grubs (Fig.11) are persistent on the leaves and occur in large numbers. The grubs eat out somewhat regular areas, leaving slender parallel strips and uneaten portion between them, giving the plants a characteristic lace like skeletonized appearance. When abundant, the plants are shredded and dried out so that they die within a month after the attack begins, often before crop is matured.

The bioecology of *Epilachna* beetles was reviewed by Rajagopal and Trivedi (1989). The adults are yellowish brown dorsally and pale ventrally. The body length ranges from 5.6mm and breadth varies from 4 - 4.5mm. Eggs are laid in clusters on the lower surface of leaves. The yellowish, cigar shaped eggs are laid erect on leaf surface close to one another. A female lays 500-750 eggs. Egg period lasts 3-4 days. The larval period lasts from 8-10 days on potato. Pupation taken places on leaves. Pupal period lasts 3-6 days and longevity of adults has been found to be 3 – 4 weeks. But in case of *E. ocellata*, the life span is longer. (Trivedi and Rajagopal, 1999). The life cycle occupies 21 to 36 days. Adults over winter under grass and weeds.

Control: Where the beetles are abundant, it is suggested to collect and destroy the eggs and grubs mechanically. Good control may be obtained by thoroughly spraying the foliage with proper contact insecticide such as malathion, dichlorvos, endosulfan, chlorpyriphos @ 0.05% or carbaryl @ 0.1% (Dass, 2000; Pandey, 2002). The application should be made as soon as the beetles or their eggs are found on the plants.

The natural enemies may play an important role in reducing the pest population in the fields. Eggs are parasitized by a eulophid, *Tetrastichus oularum*. Important larval parasitoids are *Pleurotropis foveolatus, P. epilachnae* and *Eupelmus urozonus*. A predatory bug, *Rhinocoris fuscipes* feed on all the stages of *Epilachna* beetles.

Leaf eating caterpillars: Several leaf eating caterpillars such as semiloopers, *Plusia orichalcea* (F.); tobacco caterpillar, *Spodoptera litura* (Hb); gram pod borer, *Helicoverpa armigera* (Hb.); Bihar hairy caterpillar, *Spilosoma obliqua* Walk.; and brinjal shoot and fruit borer, *Leucinodes orbonalis* Guen. have been reported to feed on potato foliage from different regions of the country. Of these, *H. armigera* and *P. orichalcea* are quite important. In plains, caterpillars of *H. armigera* migrate from chickpea to potato in the spring season and feeds on potato foliage. In hills, the moths appear in March on ornamental plants and females lay eggs singly on the lower side of leaves. On hatching, the caterpillars feed on potato foliage. *P. orichalcea* caterpillars cause severe damage to foliage in the summer potato crop in Meghalaya and to spring crop in Punjab and Himachal Pradesh. It is also a polyphagous pest having a wide host range including pea, cabbage, cauliflower, radish, turnip, carrot, celery, indigo, linseed, etc.

Minor defoliating pests: There are many leaf eating insects of localized importance reported from different parts of the country. The flea beetle, *Psylloides plana* Maulik is prevalent in north-western hills; weevil, *Myllocerus subfasciatus* Guerin in Nilgiris, blister beetle, *Epicauta hirticornis* Hagg in Shimla hills, army worm, *Mythimna separata* (Walker) in north western plains. *Chalaenosoma metallium* Fab., a green leaf beetle and *Monolepta orientalis*, the flee beetle feed on potato in south India. Another weevil reported from Orissa is *Atmetonychus peregrinus* Olveier (Nair, 1995). All these damage the crop by feeding on its foliage and tender shoots.

Management of leaf eating insects: Hand picking of larvae and their destruction is the cheapest and bęst method of control. Efforts should be made to collect early gregarious stages. The fields should be properly ploughed to expose and kill the pupae in the soil. In case of severe infestation, spray endosulfan, monocrotophos, malathion and carbaryl @ 0.05% and repeat at an interval of two weeks, if needed (Pandey, 2002).

SAP SUCKING INSECTS

Aphids: The primary concern with aphids is usually their role as virus vectors in potato seed production. Five aphid species are found on potato in India; the green peach aphid (*Myzus persicae* Sulzer), cotton aphid (*Aphis gossypii* Glover), bean aphid (*Aphis fabae* Linn.*),* potato root aphid (*Rhopalosiphum rufiabdominalis* Sasaki), tuber aphid (*Rhopalosiphoninus latysiphon* Davidson). The cosmopolitan aphid, *M. persicae* is sufficiently important a pest on potatoes and other crops (Verma and Chandla, 1999).

Green peach aphid, *Myzus persicae* Sulzer: In *M. persicae* (Fig 12) only eggs are produced by sexual reproduction whereas all subsequent reproduction is viviparous and parthenogenetic (Harris, 1992). The aphids have both winged and wingless forms. Wingless forms are predominant on potato during most part of the year. *M. persicae* over winter as eggs on a very restricted number of primary host species, often woody plants (Peach etc.). In spring, wingless aphids called stem mothers hatch from eggs, feed on the primary host, mature and produce young ones asexually. Offspring's of stem mothers are generally all wingless. Aphids moult four times mean number of offspring's per wingless aphid ranges from 60-75 (Chandla and Verma, 2000). Optimum temperature for reproduction is around 70°F. There may be several generations on the primary host, but eventually winged adults (spring migrants) develop which fly away to colonize "secondary host plants" (Radcliffe *et al.,* 1993). Winged spring migrants are not produced until at least the second generation on the primary host, and the peak production occurs in the third generation. Spring migrants leave the primary host in search of suitable secondary hosts, usually herbaceous plants, one of which is the potato.

Once an acceptable host is found, the spring migrants settle and reproduce. Their offspring's are all wingless, but a small proportion of each succeeding generation is winged. As the quality of host plant declines, more winged summer migrants are produced, which then fly to other secondary host plants. As the day length shortens, fall migrants are produced, both males and females. They return to the primary host plant, on which the females give birth asexually to wingless females, which then mate with the male fall migrants and lay fertilized over wintering eggs. The eggs are deposited near buds of the primary host. In mild climates, winged migrants also develop from aphid populations that survive the winter as adults. Sometimes, the life cycle is incomplete, in which asexual breeding takes place throughout the year and over winter takes place on stored, sprouting potatoes, on hardy herbaceous plants or on glass house crops (Fig.13).

Transmission of viruses by aphids : Both PLRV and PVY are transmitted by several aphid species, but the most important vector throughout India is the *M. persicae*. It is the most efficient vector of PLRV. Both winged and wingless *M. persicae* transmit viruses efficiently. Aphids transmit viruses in two basic ways referred to as persistent and non-persistent transmission (Fig.14). PLRV is the only potato virus transmitted by aphids in persistent manner (Harris, 1992). An aphid can acquire this virus only while feeding from the phloem cells in the conducting vessels of the host plant. After ingestion of PLRV particles, virus passes through the aphids' gut wall and into the blood stream. From there, the virus enters many insect tissues. When it enters the salivary glands, the aphids can transmit it to the other plants while feeding. Aphids can acquire PLRV after feeding for only a few minutes. But, it takes at least 12 hours for transmission efficiency to develop. Once the PLRV is acquired, an individual aphid generally retains the virus throughout its life. Non-persistent viruses, such as PVY, are concentrated in epidermal tissues and any aphid that is sap sampling is a potential vector, regardless of whether it belongs to a species that colonize the crop.

Management of aphids: Young plants are particularly susceptible to infection with a virus and seed plots must therefore be kept free of aphids, in particular early in the season. It is best to cultivate seed potatoes in areas where only a few aphids occur. The selection and rouging of plants infected with a virus must also be done under aphid free conditions. An action threshold of 20 aphids/100 compound leaves has been strictly used in seed production (Chandla *et al.,* 2004). The most vulnerable period in the aphid life cycle is passed on Peach. The application of chemical defoliant to peach trees in fall, to deny foliage to fall migrants and pruning of trees to remove most over wintering eggs is useful. When peach trees are pruned, the pruned twigs must be destroyed, or else the eggs can still hatch (Radcliffe *et al.,* 1993). Several soil applied systemic insecticides give good early season control of aphids. Late season aphid pressure is often more severe in potatoes treated with a soil systemic insecticide at planting. Late season out breaks in aphid population may be a consequence of early season control measures that prevent the establishment of natural enemies of the aphids (Hancock, 1990). They are effective in limiting the spread of PLRV but are much less useful in limiting the spread of non-persistent viruses. Therefore, foliar insecticides may be applied following the use of a soil systemic.

Many insecticides repel aphids but are not lethal to them and this tends to promote plant to plant movement and increased flight activity. Pyrethriod insecticides are generally not effective against aphids and their use may be avoided even for control of other insects, in potato. Even continuous presence of residues sufficient to kill aphids may not reduce the transmission of non-persistent viruses. In fact, the insecticide may, agitate aphids and cause them to increase their movements and sap sampling, thus resulting in increased spread of these viruses. Both foliar sprays of dimethoate or metasystox @ 0.03 per cent and phorate 10G @ 20 kg/ha applied in-furrows at planting are widely recommended in seed potato for aphid control (Chandla *et al.,* 2004). Imidacloprid and thiamethoxam also provide effective protection for about two weeks.

Several important parasitoids of aphids are *Aphidium colemani, A. matricariae* and *Aphelinus abdominalis.* Among predatory coccinellids *Adalia quadrispilota, Adonia variegate, Ballia euchasis, Coelophora sexareats, Illeis cincta, Menochilus sexmaculatus, coccinella septumpunctata* and *Harmonia octamculata* are effective.

Leaf hoppers : These are common pests of potato which are distributed in India and other tropical and subtropical countries. Unlike aphids, leaf hoppers are important mainly because of the direct feeding damage they cause. In India, *Amrasca biguttula biguttala* and *Empoasca devastans* are the major species of leaf hoppers (Trivedi and Rajagopal, 1999). They have a broad host range and are found to infest cotton, groundnut, rice, okra, brinjal and tomato. No *Empoasca* spp. is known to transmit any potato pathogens (Raddcliffe *et al.,* 1993). Prolonged feeding by the potato leaf hopper causes a condition known as "hopper burn" manifested in the form of brown triangular lesion at the tip of the leaf. Both adults and nymphs are injurious. Damage results from disruption of phloem (Trivedi and Rajagopal, 1999). Toxins in the saliva of potato leaf hopper induce swelling of cells, which eventually crushes the phloem. There is depletion of plant reserves due to increase in plant respiration subjected to hopper attack

On potato, they usually complete 2-4 generations in a year. The population density is dependent upon the date of arrival on the crop and the temperature. Over wintering takes place in adult stage, often at distance away from the potato crop. In summer, potato leaf hopper adults live for 30-40 days, and a single female can lay up to 200 eggs @ 2-3 eggs per day. During summer, eggs hatch in about 10 days. The nymphs are yellowish green when young but closely resemble the adult leaf hopper when older. They require about 12 days developing into adults. New adults begin laying eggs when they are 6 days old (Radcliffee *et al.,* 1993).

Control : Early maturing cultivars are generally more susceptible, but these cultivars bulk more rapidly and their yield may actually be affected less. It is more important to control leaf hoppers under stress because potato is more susceptible to leaf hopper injury under these conditions. In seed crop, soil systemics applied in furrow at planting or side dressed at plant emergence give 6-8 weeks of control. On fresh market potatoes, standard practice is to apply foliar sprays. Dimethoate @ 0.03 per cent and methyl demeton @ 0.025 per cent, applied on the appearance of the pest, and phorate 10 G @ 20 kg/ha applied at planting are approved for the control of leaf hoppers (Pandey, 2002).

Thrips : Thrips are the vectors of tospo viruses causing stem necrosis in potato. Seven species of thrips are associated with potato. Of these, *Thrips palmi* Kamy, *Scirtothrips dorsalis* Hood, *Coleothrips collaris* Priesner and *Haplothrips* sp. are important. Both adults and nymphs scrap the epidermal tissues of leaves usually near the tips and rasp the oozing sap (Butani and Jotwani, 1984). The surface of leaves becomes whitened and somewhat flecked in appearance. The tips of leaves wither, curl up, and die. The under side of leaves will be found spotted with small brownish-black specks. They rasp and puncture the surface of the leaf with their stabber like mouth parts and swallow the sap, together with bits of leaf tissue. Under conditions of high incidence, the whole field gives a "dry blight" appearance where most of the infected plants have dry leaves hanging on blighted stems. The plants wilt and die (Khurana *et al.,* 2001).

Parthenogenesis is common. The males are wingless and very scarce, while females regularly reproduce without mating. The eggs are laid on tissues of plants, or in some species inserted into slits made by a sharp ovipositor. The high temperature (30-35°C) and dry weather during September-October are highly favourable for thrips activity and consequently, higher disease incidence (Khurana *et al.,* 2001).

Control : In potato, there is a strong positive correlation between early planting and thrips activity which declines with delayed planting to end of October or early November. Therefore, early planting (September/October) must be avoided. Certain varieties are resistant to injury by this pest. In plains Kufri Sutlej, Kufri Badshah and Kufri Jawahar are comparatively resistant(Singh *et al.,* 1997) and Kufri Chandramukhi and Kufri Bahar are susceptible. After the plants start growing, they should be watched carefully and if the thrips appear upon plants, the crop should be sprayed with imidacloprid @ 0.002 per cent to reduce the pest and disease incidence (Singh *et al.,* 2000). Imidacloprid can also be used for side dressing. Generally, chemical application is confined during first and second week of crop emergence, because this is the only period where thrips activity is maximum and viruliferous thrips land on the germinating crop.

White flies, *Bemisia tabaci*: It is a highly polyphagous and serious pest of cotton, tobacco, okra and various other vegetables and weed plants (Puri *et al.,* 1995). The population of *B. tabaci* is highly diverse and many biotypes have been identified. Infestation is more on early potato crop planted in September. Maximum population on potato occurs in November and there is sharp decline in population by December.

Both nymphs and adults suck the sap usually from ventral surface of leaves and devitalize the plants (Butani and Jotwani, 1984). In addition, these insects also act as a vector mainly for potato Gemini viruses in plains. The virus is transmitted in a persistent manner (Puri *et al.,* 1995). The affected plants remain stunted and their leaves show distinct upward or downward curling. Leaves of affected plants show dark green veins as compared to normal translucent veins of healthy plants.

The female white fly attaches eggs to the underside of leaves by a short stalk. On hatching the nymphs are flat and nearly transparent. They settle upon the leaf, near the point where they hatch and suck sap from the leaves for about two weeks. The scale like nymphs grows through four instars (Fig.15) and form pupae on the leaves. In 2-3 days, adult whiteflies emerge from the pupae (Sharma and Rishi, 2004). The adult white flies remain beneath the leaf surfaces (Fig.16) and disperse in clouds when disturbed. Both males and females fly actively, living for about 3-4 weeks. The life cycle is completed in about three weeks and many overlapping generations are completed in a year.

Management of white flies: White flies are very difficult to control with conventional insecticides. Under conditions of severe white fly attack, none of the control measures are effective to prevent virus spread. The white fly population can be managed to a certain level by adopting a package of practices specifically targeted for white fly control. The cultivation of varieties susceptible to leaf curl virus should be avoided. Varieties like Kufri Bahar has shown high tolerance (Lakra, 2003), while cultivars like Kufri Sutlej and Kufri

Anand are highly susceptible. Alternate hosts of virus and the vector, white fly should be discouraged to avoid possible build up of virus and vectors. Excessive use of nitrogenous fertilizers should be avoided (Puri *et al.,* 1995). The virus infected plants of potato should be uprooted immediately and burnt. Seed treatment with imidacloprid @ 0.004 per cent and foliar application @ 0.002 per cent at emergence with repeated application after 15 days has been found quite effective(Malik *et al.,* 2005). Weeds harbouring white fly population near potato fields should be eradicated, immediately.

Minor sap feeders: Some of the minor sap feeders are green potato bug, *Nezara viridula*; green house whitefly, *Trialeurodes vaporarium* (Westwood), and bulb mite, *Rhizoglyphus echinopus* F & R..

NON-INSECT PESTS

Mite, *Polyphagotarsonemus latus*: This "broad mite" cause *tambera* in potato and *murda* in chilies. The damage is severe to the crop in the Deccan Plateau as well as in the early planted crop in the plains of India (Pandey, 2002). It is a poly phagous pest, and its collateral hosts are chilies, citrus, tomato, beans, pepper and cotton (Trivedi and Rajagopal, 1999). Both nymphs and adult (Fig.17) damage the crop. The margins of fresh leaves are cupped and distorted with corky area between main veins on underside of the leaves. There are characteristic copper colour deposits on the lower side of leaves. Under severe mite attack, the infested leaves dry up resulting into ultimate death of plant that can be easily spotted in the infested fields due to their bronze colour (Fig.18).

The peak activity of the mites occurs in August when sun shine is bright (Trivedi and Rajagopal, 1999). Eggs are laid singly on lower surface of leaves and are 0.7mm long, and hatch in 2 – 3 days (Raj, 2000). Eggs are oval in shape, but flattened on lower side, translucent with white tubercles on dorsal side. Nymphs are minute, white, pear shaped with a distinct white stripe along dorsal mid rib (Raj, 2000). They are found more on apical leaves and the nymphal period is of 2 – 3 days (Trivedi and Rajagopal, 1999). The pupae are stuck to underside of leaves and pupal stage lasts for 24 hours. Adults are yellow and their life span is about 10 days. In females hind legs are thinner than other 3 pairs and terminate in two setae. The males are generally smaller than females having hind legs modified to form a pair of claspers. The life cycle is completed in 5-8 days.

Control: The mite infestation in potato crop can be avoided by delayed planting towards mid October in plains. This broad mite in potato can be controlled by spray of dicofol 18.5EC @ 0.360 kg a.i./ ha, endosulfan 35EC @ 1.0kg a.i./ ha, Carbaryl 50WP @ 2kg a.i./ ha and quinalphos 25EC @ 0.50kg a.i./ ha. Application of mancozeb (80WP) @ 2.0kg/ha or micronized wettable sulphur (80%) @ 2.5kg/ha is also effective. Application of quinalphos is useful if there is problem of potato tuber moth and mite. To control mite with fungal leaf spots, mancozeb is beneficial (Raj, 2000). The treatment may be repeated after every 7 – 10 days depending upon the severity of infestation.

Nematodes: Nematode damage to potato ranges from a very subtle restriction in plant growth to severe stunting of plants or obvious disfiguration of tubers (Fig.19). The nematodes are present almost in all potato production areas of India. Some species are limited in geographic distribution, where as others are found in all areas. Of the many types of nematodes that can infest potatoes, the two most significant are root knot nematodes (*Meloidogyne* spp) and cyst nematodes (*Globodera* spp).

Root knot Nematodes: three species of root knot nematodes viz. *Meliodogyne incognita, M. javanica* and *M. hapla* are distributed in hills and plains of Himachal Pradesh, Jammu & Kashmir, Tamil Nadu and other states. All these species are polyphagous attacking vegetables, fruit crops, grasses and cereals. Invasion and feeding by these nematodes cause swelling of cells at the feeding site and an increase in cell multiplication around the site. This results in a visible root swelling or gall (Fig.20).

Root knot nematodes are sedentary endoparasites and they hatch from eggs as second stage juveniles. The eggs are held together by a sticky substance produced by the female. The egg masses adhere to the posterior end of the female and may be embedded in root tissue, exposed on the surface of root or detached and free in the soil. Hatched second stage juveniles migrate through the soil in search of infection sites on host roots. After penetrating a root, each juvenile establishes a feeding site composed of 2 – 6 cells. The bodies of juveniles mature through the third and fourth juveniles stages before reaching adult stage. Most root knot nematodes are composed largely of females. Adult females are swollen and round or oblong. Each female produces one egg mass, which may contain from 200 – 1000 eggs. Males, when they are produced, are worm like and do not feed. ne – five generations may be produced per year. These nematodes reproduce inside roots and tubers. Eggs and second stage juveniles persisting through the winter in soil root debris and tubers act as inoculums in subsequent crops.

Management of root knot nematodes: The classical method is crop rotation to keep nematode population below damaging levels. By cultivating non-host crops or leaving the land fallow, nematode densities can be reduced by 80 – 90% annually. The greatest reductions are effectuated on moist and bare land, when the larvae hatch at temperatures above 8°C. As a consequence damage to sensitive crops can be limited by sowing or planting late enough after tillage to exhaust the initial population by starvation. Inter cropping with marigold (*Tagetes* sp.) reduces nematodes in potato (Pandey, 2002). Aldicarb @ 1 – 4 kg/ ha or carbofuran 3G @ 60 – 100 kg/ha in two splits i.e. half at planting and the remaining half at earthing up is recommended for the control of this nematode.

Golden Nematodes: The golden (Potato cyst) nematode (*Globodera rostochinensis*) and white potato cyst nematode (*G. pallida*) damage potato only in Nilgiris hills of Tamil Nadu and Kodaikanal area (Jones, 1961; Stone, 1972). The host range is restricted to plants closely related to potato (tomato, brinjal and some weeds). The infested plants show yellow foliage, wilting appearance and die prematurely. The most diagnostic feature of infection by the golden nematode is round pin head-sized females (cysts) on potato roots around the time of flowering. The females are white when alive but turn yellow as the cysts mature (Fig.21). The mature cysts remain white in case of *G. pallida*.

The golden nematode is a sedentary endoparasite with a similar life cycle to that of root knot nematodes. The infection does not cause root to form galls. Rather, the body of the female breaks through the root surface and it grows and matures. Eggs are retained inside the female's body, which then becomes a cyst, serving as a protective covering for the eggs. Eventually, the cyst separates from the root and is free in the soil. Each cyst contains about 200 – 600 eggs. The root exudates stimulate the eggs to hatch. In the absence of hosts, only small proportion of eggs hatch every year.

Management of cyst nematodes: In case the presence of cysts is obvious, disinfect the infested soil and cultivate a variety resistant to pathotype (s) concerned. The crop rotation with wheat, maize, beans, cabbage and cauliflower reduces nematode population. The resistant varieties Kufri Swarana and Kufri Thenmalai are recommended and widely used in Nilgiri hills. Domestic quarantine is strictly followed so that infested potato seeds do not get disseminated to other areas. To contain the spread of potato cyst nematodes in Nilgiris, potato grown in infested fields should not be used as seed. The volunteer potato plants should not be allowed to grow after harvest. Application of carbofuran 3G @ 2kg a.i./ ha at the time of planting brings down nematode populations to economically manageable levels (Prasad, 2000).

References

Anonymous, 1991. *Annual Report.* International Potato Centre (CIP), Lima, Peru.

Anonymous, 2000. *Package of Practices For Rabi Crops.* Directorate of Extension Education, HPKV Palampur (Himachal Pradesh).

Butani, D.K. and Jotwani, M.G. 1984. Insects in Vegetables. Colour Publications, Mumbai: 356p

Chandel, R.S. and Chandla, V.K. 2003. Managing tuber damaging pests of potato. *Indian Hort.*, 48 (2): 15-17.

Chandel, R.S.; Chandla, V.K. and Sharma, A. 2003. Population dynamics of potato white grubs in Shimla hills. *J. Indian Potato Assoc.*, 30 (1-2): 151-152.

Chandel, R.S.; Chandla, V.K. and Singh, B.P. 2005. Potato tuber moth – *Phthorimaea operculella* (Zeller). Tech. Bull. No.65, CPRI, Shimla.

Chandel, R.S.; Chander, R, Gupta, P.R. and Verma, T.D. 1994. Relative efficacy of some insecticides against apple defoliating beetle *Brahmina coriacea* (Hope) on apple. *Indian J. Plant Prot.*, 22 (1): 45-49.

Chandel, R.S., Gupta, P.R. and Chander, R. 1995. Behaviour and biology of the defoliating beetle, *Brahmina coriacea* (Hope) (Coleoptera: Scarabaeidae) in Himachal Pradesh. *J. Soil. Biol. Ecol.,* 15 (1): 82 - 89.

Chandel, R.S. and Kashyap N.P. 1997. About white grubs and their management. *Farmer and Parliament,* XXXVII (10): 29 – 30.

Chandel, R.S., Kumar, Rajnish and Kashyap N.P. 2000. Saving potato from tuber moth. *Indian Farming,* 50 (2): 9 & 28.

Chandel, R.S., Kumar, Rajnish and Kashyap, N.P. 2001. Bioecology of potato tuber moth, *Phthorimaea operculella* Zeller in mid hills of Himachal Pradesh. *J. Ent. Res.*, 25 (3): 195 – 203.

Chandel, R.S., Kumar, Rajnish and Mehta, P.K. 2001. Monitoring of incidence of potato tuber moth, *Phthorimaea operculella* Zeller in mid hills of Himachal Pradesh. *Pest mgnt. Econ. Zoo.* 9: 71-77

Chandla, V.K. 1985. Potato pests and their management. *Indian Farming*, 34 (12): 31 – 32.

Chandla, V.K., Misra, S.S., Bist, S.S., Bhalla, O.P. and Thakur, J.R. 1988. White grub *Brahmina coriacea* (Hope) infesting potato in Shimla Hills. *Seeds and Farms,* 14: 12 – 13.

Chandla, V.K.; Khurana, S.M. Paul and Garg, I.D. 2004. Aphids, their importance, monitoring and management in seed potato crop. *Tech. Bull. No.* 61, CPRI, Shimla : 12 p.

Chandla, V.K. and Verma, K.D. 2000. Potato aphids. In : *Diseases and Pests of Potato – a manual.* Ed. Khurana, S.M. Paul. CPRI, Shimla : 48-52 pp.

Chauha, U. and Verma, L.R. 1991. Biology of potato tuber moth, *Phthormiaea operculella* (Zeller) with reference to pupal eye pigmentation and adult sexual dimorphism. *Entomon*, 16: 63-67.

Dahiya, P.S.; Pandey, N.K.; Srinivas, K. and Tondon, M.L. 1999. World potato statistics. *Tech. Bull. No.*52, *CPRI, Shimla.*

Dalaya, V.P. and Patil, S.P. 1973. Laboratory rearing and field releases of *Copidosom koehleri* an exotic parasite for the control of *Gnorimoschema operculella* (Zell.). *J. Mahatama Phule Agril. Univ*., 4: 99-107.

Dass, P.C. 2000. Potato in India. Kalyani Publishers, New Delhi : 193p.

Gubbaiah and Thontadarya, T.S. 1977. Bionomics of the potato tuberworm, *Gnorimoschema operculella* Zeller (Lepidoptera; Gelechiidae) in Karnataka. *Mysore J. Agric. Sci.,* 11: 380-386.

Hancock, M.1990. Pests of the potato crop. In: *Crop Protection Hand Book – Potatoes.* Ed. John S Gunn. Lavenham Press Ltd., Lavenham, Suffolk, Great Britain: 79-89 pp.

Harris, P.M. 1992. *The potato crop*. Second Edition. Chapman & Hall, London : 909 p.

Isahaque, N. Md. M. 1978. Observations on the bionomics of *Phthorimaea operculella* Zeller (Lepidoptera; Gelechiidae), a pest of potato in Assam. *Pesticides*, 12: 57-60.

Jones, P.G.W. 1961. Potato root eelworm *Heterodera rostochinesis* Wolf in India sp. 14. *Curr. Sci*., 30(5): 187.

Kabir, A. 1994. Laboratory studies on the oviposition and generation production of potato tuber moth, *Phthorimaea operculella* Zeller (Lepidoptera: Gelechiidae). *Bangladesh J. Zoo*., 22 : 25-28.

Khandge, S.V., Parlekar, G.V. and Naik, N.M. 1979. Inundative release of *Copidosoma koehleri* Bl. (Hymenoptera: Eneyrtidae) for control of potato tuber worm *Phthorimaea operculella* Zeller. *J. Maharashtra Agril. Univ.,* 4: 165-169.

Khurana, S.M. Paul, Bhale, Usha and Garg, I.D. 2001. Stem Necrosis disease of potato. *Tech. Bull.* No.54, CPRI, Shimla.

Kumar, R. and Nirula, K.K. 1967. Control of potato tuber moth in the field. *Indian J. Agric. Sci*., 37: 553-554.

Lakra, B.S. 2003. Effect of date of planting on white fly population, leaf curl incidence and yield of potato cultivars. *J. Indian Potato Assoc*., 30: 115-116.

Lefroy, H.M. 1907. The potato tuber moth. *Indian Agric. J.,* 2: 294-295.

Malik, Kamlesh; Chandel, R.S.; Singh, B.P. and Chandla, V.K. 2005. Studies on potato apical leaf curl virus disease and its white fly vector *Bemisia tabaci*(Gennadius). In : *Proc. Ann. Meet of Indian Soc. of Plant Pathologists & Centenary Sym. on Plant Path.,* 7-8 April 2005. CPRI, Shimla : 17.

Mishra, P.N. and Singh, M.P. 1993. Field biology of white grubs, *Holotrichia longipennis* on potatoes in UP hills. *J. Indian Potato Assoc*., 20 (3): 249-251.

Misra, S.S. 1995. Potato pests and their management. Tech. Bull. No.45, CPRI, Shimla.

Misra, S.S. and Chandel, R.S. 2003. Potato white grubs in India. Tech. Bull. No. 60, CPRI, Shimla.

Misra, S.S., Chandla, V.K. and Singh, A.K. 1995. Potato pests and their management. Tech. Bull. No.45, CPRI, Shimla.

Misra, S.S., Singh, D.B., Chandla, V.K. and Raj, B.T. 2003. Potato pests and their management. *In: The Potato – Production and Utilization in subtropics.* Eds. S.M. Paul Khurana, J.S. Minhas and S.K. Pandey. Mehta Publishers, New Delhi : 252-269.

Nair, M.R. G.K. 1995. Insect and mites of crops in India. Indian Council of Agricultural Research, New Delhi: 404p.

Pandey, R.P. 2002. The Potato. Kalyani Publishers, New Delhi: 222p.

Prasad, K. S. Krishna. 2000. Root-knot-nematodes. *In: Diseases and pest of potato– manual.* Ed: SM Paul Khurana. CPRI, Shimla: 41-44pp.

Puri, S.N.; Baranwal, V.K. and Surender, Kumar. 1995. Cotton leaf curl disease and its management. NCIPM Extension Folder I. NCIPM, New Delhi.

Radcliffe, E.B.; Radsdale, D.W. and Flanders, K.L. 1993. Management of aphids and leaf hoppers. In : *Potato health management*. Ed. Rowe, C. Randall. American Phytopathological Society : 117-126.

Raj, B.T. 2000. Mites. *In: Diseases and pests of potato.* Ed. SM. Paul Khurana. CPRI, Shimla: 64-66pp.

Raj, B.T. and Parihar, S.B.S. 1993. Red ant damage on potato in western Uttar Pradesh. *J. Indian Potato Assoc.,* 20: 61 – 62.

Rajagopal, D. and Trivedi, T.P. 1989. Status, biology and management of *Epilachna* beetle, *Epilachna vigintioctopunctata* (Fab.) (Coleoptera: Coccinellidae) on potato in India. *Trop. Pest Mgnt.*, 35: 410-413.

Ram Kishore, Misra, S.S. and Singh, Lallan. 1993. Incidence of red ant, *Dorylus orientalis* on different potato genotypes. *J. Indian Potato Assoc.,* 20: 62

Raman, K.V. 1988. Integrated insect pest management for potato in developing countries. *CIP Circular*, 16(1): 11p.

Regupathy, A., Palanisamy, S., Chandramohan, N. and Gunathilagaraj, K. 1997. *A guide on crop pests*. M/S Sooriya Desk Top Publishers, Coimbatore : 290 p.

Roonwal, M.L. 1976. Plant pest status of root eating ants, *Dorylus orientalis* with notes on taxonomy, distribution and habits (Insects: Hymenoptera). *J. Bombay Nat. Hist. Soc.,* 72:305 – 313.

Shah, N.K. and Shah, Lata, 1990. Bionomics of *Holotrichia longipennis* (Coleptera: Melolonthinae) in western Himalayas. *Indian J. Forestry,* 13: 234 – 237.

Shaheen, A.H. 1983. Some ecological and biological studies on the potato tubermoth, *Phthorimaea operculella* Zell. in Egypt. *Bull. de la Soc. Entomol. Egypt,* 62: 135 – 45.

Sharma, P. and Rishi, Narayan. 2004. Population build up of the cotton whitefly, *Bemisia tabaci* in relation to weather factors at Hissar, Haryana. *Pest Mgnt. Eco. Zoo.*, 12 : 33-38.

Shelton, A.M. and Wyman, J.A. 1979. Potato tuberworm damage to potatoes under different irrigation and cultural practices. *J. Econ. Ent.*, 72: 261 – 64

Singh, R.B.; Khurana, S.M. Paul; Pandey, S.K. and Srivastava, K.K. 1997. Screening germplasm for potato stem necrosis resistance. *Indian Phytopath.*, 51 : 222-224.

Singh, R.B.; Khurana, S.M. Paul; Pandey, S.K. and Srivastava, K.K. 2000. Tuber treatment with imidacloprid is effective for control of potato stem necrosis disease. *Indian Phytopath.*, 53 : 142-145.

Singh, S.P. 1987. Studies on some aspects of biology - ecology of potato cut worms in India. *J. Soil. Biol. Ecol.,* 7: 135 – 143.

Stone, A.R. 1972. *Hetrodera pallida* N. sp. (Nematode: Heteroderidae) a second species of potato cyst nematode. *Nematologica,* 18: 591-606.

Temerak, S.A. 1983. Field management of potato tuber worm *Phthorimaea operculella* Zeller in testing potato varities through manipulation of different level of irrigation. *Bull. de la Soc. Entomol. Egypte,* 62: 153 – 60.

Trivedi, T.P. and Rajagopal, D. 1999. Integrated Pest Management in Potato. *In: IPM System in Agriculture, Cash Crops, Vol. 6.* Eds. Rajeev K. Upadhyay, K.G. Mukerji and O.P. Dubey.. Aditya Books Pvt. Ltd., New Delhi: 299 – 313.

Verma, K.D. and Chandla, V.K. 1999. Potato aphids and their management. Tech. Bull. No.26 (Revised) CPRI, Shimla.

Yadava, C.P.S. and Sharma, K. 1995. Indian white grubs and their management. Technical bulletin No.2. Project coordinating centre, AICRP on white grubs, Jaipur.

❑❑❑

Fig.1: Cutworm rolled up on a tuber and tubers showing cutworm damage

Fig 2: Moth of *Agrotis segetum*

Fig 3: Tubers showing white grub damage

Fig 4: White grubs, *Brahmina coriacea*

Fig 5: Adults of *Brahmina coriacea*

Fig.6: Adults of *Holotrichia longipennis*

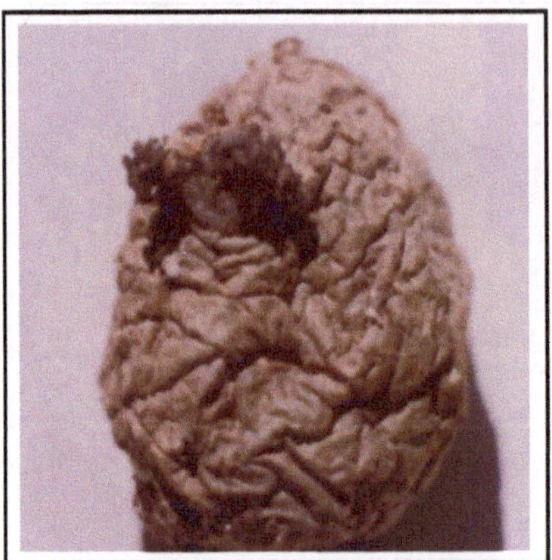

Fig 7: Tuber showing PTM damage

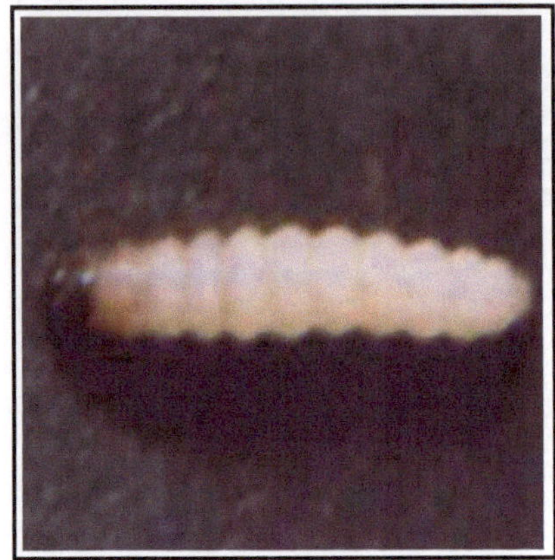

Fig 8: Larva of *Phthorimaea operculella*

Fig 9: Moth of *Phthorimaea operculella*

Fig.10: Adult hadda beetle, *E. ocellata*

Fig.11: Hadda beetle grubs feeding on a potato leaf

Fig.12: Wingless aphids of *M. persicae* on under side of a leaf

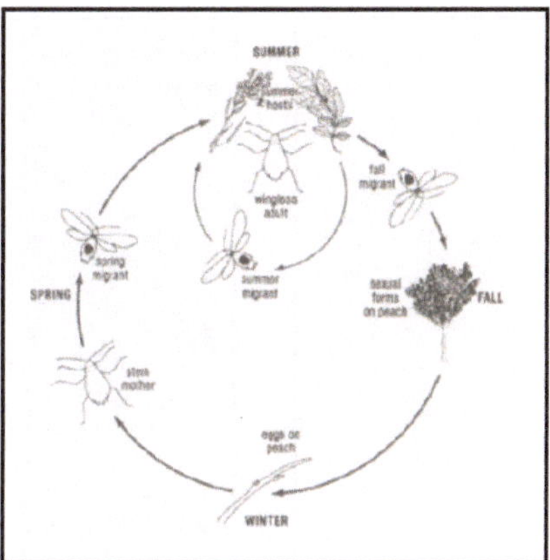

Fig.13: Life cycle of the *M. persicae* in high hills

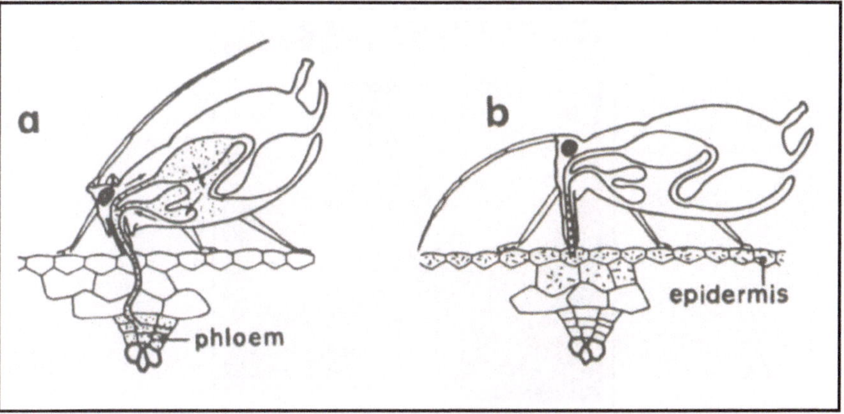

Fig.14: Acquisition of persistent (a) and non-persistent (b) viruses in aphids

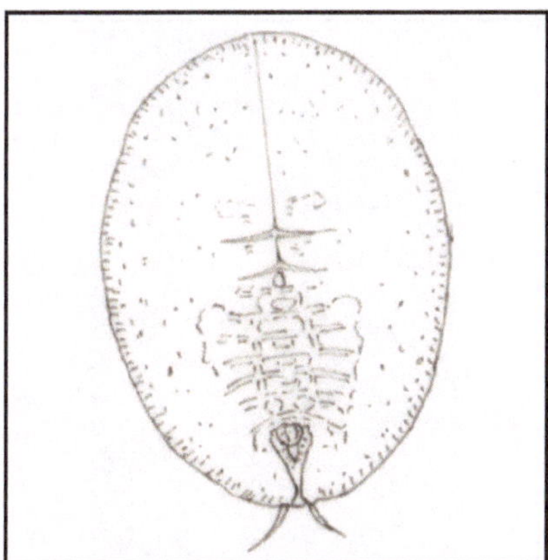

Fig.15: Dorsum of 4th-instar nymph of *B. tabaci*

Fig.16: Adults of *Bemisia tabaci*

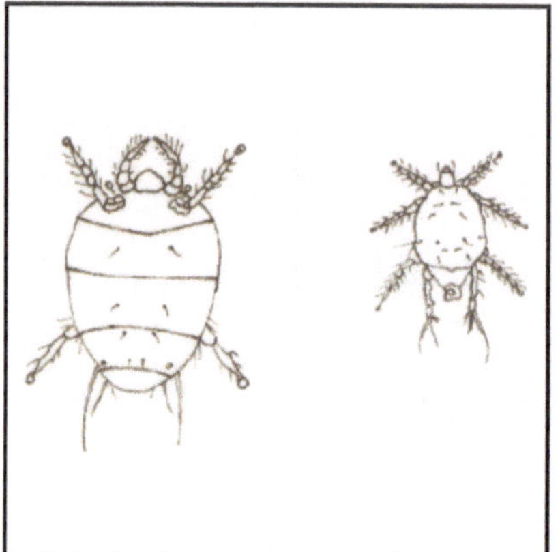

Fig.17: Adult and nymph of broad mite, *Polyphagotarsonemus latus*

Fig.18: Mite damage on potato

Fig.19 : Tubers affected by root knot nematodes

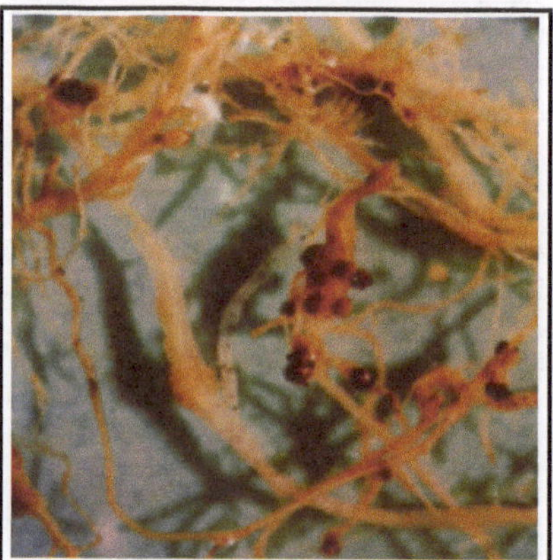

Fig.20: Roots affected by root knot nematodes

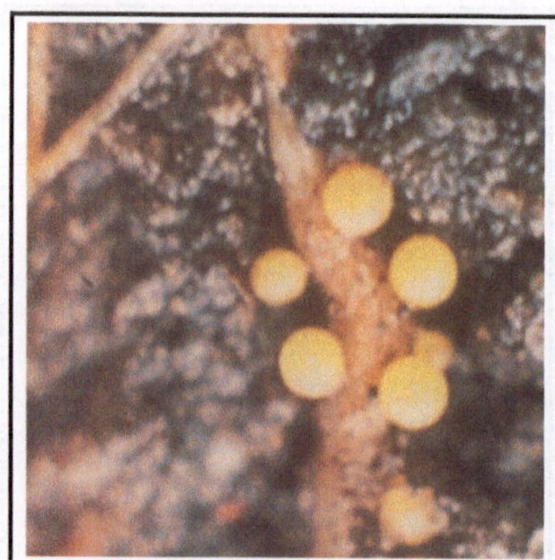

Fig.21: Roots with cysts of *G. rostochiensis*

Entomology : Novel Approaches, 2007
Eds.: *P.C. Jain and M.C. Bhargava,*(Page 399-423)
New India Publishing Agency, New Delhi
E-mail: newindiapublishingagency@gmail.com
web: www.bookfactoryindia.com

CHAPTER 20

INTEGRATED INSECT PEST MANAGEMENT IN RAPESEED AND MUSTARD

G.C. Sachan and J.P. Purwar
Department of Entomology, College of Agriculture, GBPUA&T.
Pantnagar-263 145, Uttaranchal

INTRODUCTION

Rapeseed-mustard is attacked by more than 43 insect species. Out of which, mustard aphid, *Lipaphis erysimi* (Kaltenbach); mustard sawfly, *Athalia proxima* (Klug); painted bug, *Bagrada hilaris* (Burnmerister); pea leaf miner, *Chromatomyia horticola* (Goureau) and Bihar hairy caterpillar, *Spilarctia obliqua* (Walker) are of major importance. Other pests are of sporadic occurrence. Status of important insect pests attacking rapseed mustard in India have been given in detail by Bakhetia and Arora (1993) and Singh 2005. The occurrence of various insect pests on rapeseed-mustard is stage-specific. This has been a gift provided by nature to help avoid inter-competition among them for their fullest exploitation of the host. The insect pests attacking rapeseed- mustard as given in Table 1 and at various stages of the crop in Table 2.

Table 1 Important Insect Pests of rapeseed-mustard

Pest status	Common name	Scientific name
Key pest	Mustard aphid	*Lipaphis erysimi*(Kalt)
Major Pests	Saw fly	*Athalia lugens proxima* (Klug)
	Painted Bug	*Bagrada hilaris* (Burm)
	Pea leaf miner	*Chromatomyia horticola* (Goueau)
	Bihar hairy caterpillar	*Spilarctia obliqua* (Walker)
Minor pests	Cabbage butterfly	*Pieris brassicae* (Linn.)
	Flea beetle	*Phylloteta crucifeae*(goeze)
	Green peach aphid	*Myzus persicae*(Sulzer)
	Diamond back moth	*Plutella xylostella* (Linn)

Contd. ...

Table 1 Contd. ...

New pests	The larger moth	*Crocidolomia binotalis* (Zeller)
	Cabbage top borer	*Hellula undalis* (Fab.)
	Cabbage aphid	*Brevicoryne brassicae* (Linn.)
	White fly	*Bemisia tabaci* (Gen.)
	Cabbage semilooper	*Thysanoplusia orichalcea* (Fab.)
	Turnip moth	*Agrotis segetum* (Dennis & Schiff)
	Gram pod borer	*Helicoverpa armigera* (Hub.)
	Termites	*Odontotermus obesus* (Rambur)

Table 2 Plant stages and important insect pests

Stages	Insects
0= Pre emergence	Termites and other soil insects
1= Seedling	Saw fly, painted bug, flea beetle, cut worm
2= Rosette	Aphids, diamond back moth
3= Bud	Aphids, bud fly
4= Flowering/podding	Aphids, pea leaf miner, cabbage butterfly, pod borer, larger moth
5= Ripening	Painted bug, pea leaf miner, larger moth

Out of nine oleiferous brassicae, *Brassica juncea* and *B. campestris* var. toria and sarson are cultivated under large scale while *B. napus* and *B. carinata* are of recent introduction and now being accepted by the farmers on a limited scale.

CHANGING PEST STATUS

Number of insect pests associated with rapeseed mustard is increasing. The total number of insects reported in 1976 (Rai, 1976) were 24 which increased to 40 in 1991 (Bakhetia and Arora, 1993) with the addition of *Myllocerus setailifer* Debrochers (AICORPO, 1993) in the list, the total number rose to 41. But now this number is 43. Thus there is substantial increase in pest number on these crops which is more pronounced between 1984 to 1993 (Table 3).

Changing in cropping system, cropping pattern and increased cropping intensity have led to increase in insect number (Bakhetia and Arora, 1993). Intercropping of mustard with potato and other cruciferous vegetables has led to increase in the population of *Myzus persicae* Sulzer and *Plutella xylostella* (Linn) in north whereas intercropping of safflower with cruciferous vegetables increased the population of *Peregia capensis, Athalia proxima* Klug and *Mylabris pustulata* (Thunb) in the south. Pests like *Crocidolomia binotalis* (Zeller), *P. xylostella* and *Helicoverpa armigera* (Hubner) are gaining importance particularly under rainfed conditions. Introduction of late maturing varities of *B. juncea* and long duration crops like *B. napus* and *B. carinata* accentuated the attack of mustard aphid. With the introduction of *B. napus* in Haryana the incidence of *P. xylostella, Helulla undalis* (Fab.) increased considerably (Singh and Singh, 1983). Earlier the incidence of these pests was restricted in central and central-eastern parts of the country (Rout and Pant, 1968).

Some new pests like *Agrotis segetum* Dennis & Schift, *C. binotalis* and *Dasyineura hisarensis* Sharma and Singh were recorded from Haryana (Singh and Singh, 1983; Singh *et al*., 1990) and latter two insects from Uttar Pradesh (Anonymous, 1991). In addition to these a painted bug, *Eurydema pulchrum* Westwood was also noticed at Pantnagar, Uttaranchal (Sachan, 1990). A polyphagous pest, *H. armigera* has been reported to damage mustard pods in Maharastra (Jindal and Upadhyaya, 1989), and terminal flowers and pods of toria at Ludhiana (Bakhetia, 1991a), Pantnagar and Dehradun of Uttaranchal where 10 per cent of the pods were damaged and probably the population on toria were the migrants from the adjoing soybean field (Anonymous, 1991) and Haryana (Singh *et al.,* 1994) where this insect first appeared in 1984 and then in 1987 and 1991 on cotton and then migrated to toria crop.

Table 3 Changing insect pests scenario on rapeseed mustard

Year	Insect species number	Per cent increase over base year
1968	24	-
1981	28	16.6
1984	31	29.2
1987	38	58.3
1991	40	66.7
1993	41	70.8

(Bakhetia and Arora, 1993) and AICRPO (1993)

EXTENT OF YIELD LOSSESS

Lossess caused by insect pest complex on rapeseed mustard have been extensively reviewed by Bakhetia, (1983). Bakhetia and Singh (1992) and Bakhetia and Arora (1993). The damage caused by aphids varied from 9 to 95 per cent depending upon various biotic and abiotic factors at different locations (Bakhetia and Arora, 1993). Painted bug (Joshi *et al*., 1989), saw fly (AICORPO, 1976) and leaf miner (Yadav and Bhola, 1977) cause 70, 32 and 16 per cent lossess, respectively. Based on researches done under AICORPO from 1990 to 1993 the lossess in yield are depicted in the Table 4.

Table 4 Mean yield loss due to insect pest complex (1990-93)

Species	Per Cent Loss In Yield				
	Agroclimatic Zones				
	I	II	III	V	Mean
Yellow sarson	44.1	53.0	35.5	63.2	48.95
Brown sarson	40.1	36.3	24.6	-	33.67
Rai	30.8	24.8	22.4	60.3	34.57
Gobhi sarson	29.3	40.8	26.9	-	32.33
Mean loss	36.07	38.72	27.35	61.75	37.38

I= Kangra II = Ludhiana, Bathinda, Hisar, Navgaon

III = Pantnagar, Kanpur, Faizabad, Morena V= Berhampur

Losses are much higher in zone V followed by zone II, I and III. On crop wise yield lossess are higher on yellow sarson (48.95%). Recently Singh (1992) pointed out the lossess in oil content, grain yield and oil yield in yellow sarson and rai by mustard aphids (Table 5) which are more severe in former than latter.

Table 5 Per cent average loss in oil content, seed yield and oil yield of yellow sarson and rai due to mustard aphid

Variety	Oil content		Grain yield		Oil yield	
Yellow sarson						
YST-151	5.6	(6.3)	29.8	(29.5)	39.6	(40.4)
PYS-6	7.1		29.2		41.2	
Rai						
Varuna	5.2		26.9		37.1	
Krishna	2.8	(3.5)	23.1	(25.6)	29.2	(33.9)
Kranti	3.1		23.4		33.7	
PR-35	3.2		29.0		35.7	

Data in parentheses are mean values Singh (1992)

MUSTARD APHIDS

COMPONENTS OF APHID MANAGEMENT

Surveillance

Regular surveillance of aphid must be carried out throughout the year in order to assess the future population. Yellow water trap and yellow sticky trap, found to be most effective and should be used for this purpose. Yellow sticky trap should be fixed in the field at a height of 1.5m and 6-8 traps should be maintained throughout the year.

Regular survey of this should also be conducted particularly during off season as this aphid has been found thriving on *Chenopodium album, Tropaeolum*, beans, potato, peas, beet, spinach, celery, onion, cucumber and also on stray plants of crucifers grown in humid conditions.

Planting time

It is now established that early sown crop either escapes aphid attack or has low degree of attack. Toria escapes aphid attack if sown in mid September (Bhattachrjee,1961 and Sachan, 1990). For remaining rapeseed mustard crops the optimum sowing time is between middle and the end of October depending upon the regions (Tripathi and Singh, 1964; Ghosh and Ghosh, 1981; Kalra *et al.*, 1983; Rawat and Singh, 1983; Sachan *et al.*, 1983a; Phadke and Prasad, 1987; Bakhetia, 1986; Rohilla *et al.*, 1988; Joshi *et al.*, 1989; Rohilla, 1990 and Bhadouria *et al.*, 1992). Sowing of mustard crop can be advanced by 10-15 days for escaping from aphid without any appreciable yield loss (Singh *et al.*, 1984).

Manual removal of infected twigs

It is advisable to nip the aphid problem in the bud. Mustard aphid after alighting on the crop multiplies in colony and thereafter dispersing over the entire crop. The infested twigs and inflorescence should be plucked and destroyed to check dispersal. This should be carried out at 15 days interval (Singh *et al.,* 1993). But this practice is labour consuming and ineffective for long duration crops like *B. napus* and *B. alba / B. carinata* or other late maturing mustard crops where aphid comes late and almost on all the plants.

Crop Geometry

Crop geometry and intercropping have no influence on aphid infestation (Kalra *et al.,* 1983 and Bhattachrjee, 1961). However, Bindra and Deole (1962) observed that intercrops of sarson and raya when 7 m apart in wheat crop had more aphid than when rows were 3-4 m spaced. Allelopathic effect of taramira was seen by Dilawari and Dhaliwal (1992) when taramira was grown in between raya crop and aphid infestation reduced on latter crop. These observations need further research support as taramira is being heavily attacked by mustard aphid in Punjab.

As recommended practice the sowing of rapeseed mustard crop is preferred in rows of 30 or 45 cm with plant to plant spacing of 15 or 10 cm as per local need. Under broad cast condition aphid thrives well due to development of congenial microclimate for aphid multiplication where the crop is dense.

Water Management

Water management is one of the important factors responsible for plant growth and higher yield. Rapeseed mustard grown under rainfed conditions suffers badly due to poor absorption of food material and water from the soil and desapping by the aphids, result plant to succumb very early, even at lower aphid population. Decreasing fecundity has been observed with increased water stress to plant (Sidhu and Kaur, 1976 and Arora and Sidhu, 1982). This could be due to increase in the concentration of amino acids (Chadda and Arora, 1982) which had adverse effect and secondly the plant sap gets so concentrated with low cell turgidity that aphid feeding efficiency is badly hampered (Bakhetia and Brar, 1988). Irrigation once has no affect on aphid (Singh, 1982) but more irrigations favour aphid multiplication (Kalra *et al.,* 1983) without appreciable loss in yield. Hence, timely irrigation must be provided to the crop as per need.

Fertilization

Application of NPK to the plant have crucial role in the growth of the plant. In general application of nitrogenous fertilizer increases the plant growth and makes the plant susceptible to attack by insect pests and diseases. On the other hand K imparts tolerance to the plant. The effect of nitrogen on the population buildup of mustard aphid is contradictory. On one hand Ramachandran (1952), Singh and Sidhu (1958), Rawat *et al.,*(1968), Tandon (1973), Sidhu and Kaur (1977) and Kalra *et al.,* (1983) reported that higher doses of nitrogen applied increased aphid population and yield loss also but on the better hand Kundu and Pant (1967). Bakhetia and Sharma (1978), Sachan *et al.,* (1983a) and Rawat and Singh (1983) noted no increase in aphid population. Even the various combination of NPK had no increasing effects on the aphid number. Mathur *et al.,* (1991) indicated that appilication of nitrogen (80 Kg/ha) significantly increased aphid population irrespective of the P application but K decreased the aphid number even when applied with nitrogen and P. Higher doses of P and K reduced the incidence on fodder mustard (Ram and Gupta, 1992). Application of thiourea @ 0.1% does not affect the aphid population (Sachan and Gautam, 2001).

Varietal Resistance

As early as 1967 one variety of mustard, T 6342 was notified as resistant to mustard aphid. Later on RLM 198 was also included. Since the large number of germplasm have been screened and several tolerant/ resistant varieties have been indicated by the entomologists but they failed to prove themselves resistant. Clearly speaking we do not have resistant source of high degree which can be exploited in the development of aphid resistant material. No doubt we have some early material in *juncea* which escape aphid attack but succumb when planted late. Long duration crops are badly damaged.

In general, the cultivars of *B. campestris* var toria, yellow sarson and brown sarson are highly susceptible to aphid as compared to *B. juncea*. So, where aphid is a serious problem *B. juncea* should be grown.

Bakhetia and Singh (1992), Bakhetia and Arora (1993) and Bakhetia and Singh (1993) have excellently reviewed the aphid resistant lines/strains of cruciferous oilseed and gave a definite status to each line. These strains can be used. Sachan (2005) discussed the different screening techniques for resistance against major insect pests of *Brassicas*. Several morphological and biochemical characters in plants associated with *L. erysimi* resistance have been reviewed by Bakhetia (1991b) and Bakhetia and Singh (1993) and summary of these valuable reviews is presented in Table 6.

Natural Enemies

Among the predators more than 7 coccinellids, 14 syrphids, 4 chrysopelids and one chamaemylid have been recorded. Parasitoids number is more than 8 and 7 fungi (*Beauveria bassiana, Metarhizium anisopliae, Verticillium lecanii and Nomurea reliy*) are found attacking mustard aphid. Coccinellids, *Coccinella septempunctata* L., *C. repanda* Thub., *Brumoides suturalis* F., *Menochilus sexmaculatus* (F.) are active predators feeding on aphid (Atwal and Sethi, 1963 and Nath and Sen, 1976) whereas *C. transversalis* is not effective (Roy, 1975). Population dynamics of C. *septempunctata, Hippodamia variegate* (Coeze) and *M. sexmaculatus* have been studied by Kalra (1988). Chrysopids, *Chrysoperla* sp., *Chrysoperla carnea* (St), *Anisochrysa boniensis* Ckomata are effective aphid predators (Roy, 1975 and Kalra, 1988). Among syrphids, *Syrphus rebesii* L. and *Episyrphus balteatus* (Degea) are important (Sundby, 1966 and Kalra, 1988). Syrphids population is influenced by aphid number and particularly on late sown crop harbouring more aphid (Kumar *et al.,* 1988 and 1989). Chamaemylid, *:Leucopis* sp. has been found very effective in controlling aphid (Kalra, 1988).

Parasitoid, *Diaeretiella rapae* (Mintosh) is very effective in controlling aphid and parasitization goes as high as 93% (Stray and Ghosh, 1978; Dhiman and Kumar, 1985; Chandra and Kushwaha, 1987). Parasitization is more in mid January to February end (Veda and Shaw, 1988) and parasitoids are killed at 29°C (Tripathi and Tripathi, 1991).

Fungi, *Cephalosporium aphidicola, Entomophthora aphidis, Beauveria bassiana, Metarhizium anisopliae, Verticillium lecanii, Nomurea reliy* and *Conidiobolus thromboides* are found attacking mustard aphid. These fungi are more prevelant at 27°C and 81 % RH. *Verticillium lecanii* is found effective against mustard aphid in field as well as in laboratory (Purwar and Sachan, 2004a).

By the time natural enemies arrive and become active the aphid population started declining due to maturity of host, increase in temperature, decrease in humidity and intraspecific and interspecific competition and hence, they remain ineffective in controlling the aphid.

Table 6 Mechanism of resistance in R & M plants against mustard aphid

Character	Reaction with aphid		No reaction
	Positive	Negative	
Morphological	Bright Yellow Flower	Apeteolous white flower	Trichome density
	Compact flower	More branches	PH of sap
	Yellowish leaves	Wider angle of orientation	Moisture
	Greenish leaves Lighter leaves	Loosely packed Flower buds	
	Higher carotein	Purple leaf	
	Soft Twig	Stiffer twig	
		Chlorophyll	
		Darker leaves	
Histological		Non waxy plants	
		Deeper Phloem cells	
		Compact parenchymatus cells	
Biochemical	More Protein	Phenols, Glucosinolates	N, Cu, Mn, Zn
	More sugar	Ascorbic acid,Flavanoids	
	More oil	Total sugar, reducing sugars	
	More N,P,Na	K,Ca,Mg,Zn,S	

Based on Bakhetia (1991b) and Bakhetia and Singh (1993)

Chemical control

Judicious use of insecticides for the control of mustard aphid is advocated. For need base application to get maximum benefit Economic threshold level and economic injury levels have been worked out extensively by each Experiment Station working on mustard aphid management. These values differ from region to region because of different agroclimatic conditions, cost of production, cost of produce etc. Recently Bakhetia (1991a), Bakhetia and Singh (1992), Bakhetia and Singh (1993), Bakhetia and Arora (1993) and Singh (2005) have reviewed the work done in various states (Table 7). Interestingly these values are changing every year even in the same research station. Thus, these values which are location specific must be used as a guideline for insecticidal applications. Bakhetia and Sachan (1997) have advocated the adoption of economic threshold level values, wherever these are not available/ computed, for insecticidal application. These values are 50-60 aphids/ plant, 0.5-1 cm colony length/twig or 40-50% infested plants.

Several insecticides are effective against mustard aphid. This combination with various fungi is encouraging. When endosulfan was mixed with *Beauveria bassiana,* the toxicity of mixture increased tremendously giving cent percent kill of mustard aphid. Other insecticides except lufeneuron and oxydemeton methyl the increase in toxicity was 2.73, 1.72, 14.74 folds in imidacloprid, diflubenzuron and dimethoate with this fungus. But when these chemicals were mixed with *Metarhizium anisopliae* except in dimethoate the toxicity increased in endosulfan (24.86x), imidacloprid (5.17x), diflubenzuron (18.8x) and oxydemeton methyl (17.9x) but it decreased in lufeneuron (6.8x) (Purwar and Sachan, 2004b)

Application of granular insecticides, if applied at the time of sowing, are ineffective in managing the aphid and must be applied when incidence is seen (Bakhetia, 1984). Aphid control by granular insecticides is costly, needs irrigation and effective when aphid population is very high. On the other hand, it is very safe to natural enemies of aphids and pollinators.

The commonly used insecticides like 0.025% methyl-o-demeton, 0.025% phosphamidon, 0.03% dimethoate, 0.025% quinalphos, 0.035% endosulfan and 0.02% chlorpyriphos are compatible with fungicide, Dithane -45 (mancozeb). Insecticides with fungicides can be sprayed together for the control of aphid and fungi like alternaria, white rust, powdery mildew etc. (Tripathi *et al.,* 1983; Tripathi *et al.,* 1985 and Mathur *et al,* 1991).

Table 7 Economic threshold of *L. erysimi* under different agro ecological situations in India

State/ crop/ Variety	Economic threshold	References
Haryana		
B. campestris	9-19 aphids/ central shoot or 20% plant infestation	Singh, 1982
B. napus (GSL-1)	4 aphids/ central shoot or 10% plant infestation	Rohilla, 1990
B. juncea (RL 1359)	9 aphids/central shoot or 20% plant infestation	Rohilla, 1990
Uttar Pradesh		
Mustard	9-13 aphids/15 cm top terminal	Singh and Mishra, 1986
Punjab		
B. juncea (RLM 198) Ghorbandi,	50-60 aphids/10 cm shoot or 40-50 plant infestation or 0.5 -1.0 cm of apids colony on central shoot	Bakhetia 1983; Singh *et al,* 1983; Bakhetia and 1986

Contd. ...

Table 7 Contd. ...

Radish *Raphanus sativus*	50 aphids /10 cm terminal portion of central shoot	
Himanchal Pradesh		
Brown sarson	4 mm shoot infestation	Suri, 1986
B. campestris	< 20% infestation	Lee, 1988
B. oleracea	< 30% infestation	Singh (2005)

Botanicals as insecticides

Very less number of plant extract have been tested for their effectivity against mustard aphid. Pandey *et al.,* (1987) indicated that 1 and 1.5% extract of *Azadirachta indica, Lantana camera* and *Ipomea cornea* are highly effective in killing mustard aphid under laboratory condition. These extracts also reduced the fecundity in treated insects. Narang and Atwal (1986) reported that *Eruca sativa* extract also adversely affect the biology of this insect.

Effect of insecticides on the quality and quantity of seed

Several insecticides are recommended for the control of insect pests on rapeseed mustard and some of the insecticides alter the quality and quantity of the seed produced from the treated crop.

Oil Content

Several insecticides change the nutritive value of the crop and the seed produced. Dimethoate increased the oil content by 5.42% and with other insecticides it ranged from 0.3 to 4.36% (Arora *et al.,* 1969). Increase in percent oil content in the varuna seed was observed by 10.47, 10.28, 10.2, 10.2, 9.97, 9.94, 8.74, 8.34 and 7.92 in methyl – o- demton, endosulfan, dimethoate, malthion, monocrotophos, phosphamidon, dichlorvos, quinalphos and chlopyriphos treatment respectively over control when they were used to protect the plants from aphid (Singh, 1992). Under aphid free condition at LC_{50} level methyl-o-demeton, cypermethrin and decamethrin this increase was 1.27, 1.07 and 1.07 times respectively (Tripathi, 1985).

Protein Content

The treatment of dimethoate, thiometon, BHC, menazon, phosphamidon, diazinon, methyl demeton, parathion and malathion adveraely affected the protein content in the seed. Lowest protein content was observed in 0.1% malathion treated plots (Arora *et al.*, 1969). Insecticidal treatments also affected the protein content in leaf and inflorescence of mustard which was more at higher concentration (Sachan and Tripathi, 1987).

Germination and Vigour

Peshwani *et al.,* (1968) observed that 96 per cent of seeds were viable in the treated plot and 76 per cent in the untreated plot. Singh (1993) reported that good aphid managed crop's seed has better germination than unmanaged crop seeds. He further reported that there was no effect on the seed vigour of the seeds obtained from protected and unprotected crops.

Yield Parameters

There is increase in number of branches (Singh and Sidhu, 1959, Suri *et al.,* 1988 and Singh, 1992), number of pods per plant (Singh and Sidhu, 1959; Singh *et al.,* 1983; Suri *et al.* 1988 and Singh, 1992,) and number of seeds per pod (Singh and Sidhu, 1959; Singh *et al.*, 1983 and Singh, 1992). Kalra *et al.,* (1983) did not find increase in the number of pods per plant or seeds per pod. Singh *et al.,* (1984) reported that sprays and lower doses of granules had no effect on the number of seeds per pod but significant effect at higher dose of granules was observed. Increase in seed weight due to insecticidal control was observed by Singh and Sidhu (1959), Kalra *et al.,* (1983) and Singh (1992).

More plant height was observed in treated plots (Suri *et al.*, 1988 and Singh, 1992) than in untreated ones. This could be due to aphid free conditions which allowed the plant to grow to its fullest potential.

Insecticide resistance

For the first time resistance to endosulfan and tolerance to malathion and dimethoate were reported from Punjab by Udeaan and Narang, (1986). Based on biochemical studies Phokela and Mehrotra (1987) indicated that mustard aphid has developed resistance to organophosphates and other insecticides. Carboxyesterase kinetics studies made indicated that this insect has developed resistance to synthetic pyretheroids like cypermethrin, decamethrin and fenvalerate (Phokela *et al.,* 1981; Mehrotra and Phokela, 1985). Development of resistance to this group of chemicals is of cross resistance phenomenon. The development of resistance in this insect is very serious problem and efforts have to be made to overcome it. Mehrotra (1991) has suggested some remedies to manage resistance and they are, judicious use of pesticides, use of synergists and alternation of chemicals with unrelated mode of action.

Resurgence

Farmers are using low dose of insecticides than recommended due to cost factor, improper coverage due to ineffective machinery and defective application methods and certain abiotic factors which reduce the lethality of toxicants in short period are responsible for aphid resungence like situation. Tripathi and Sachan (1990) reported thet sublethal doses of insecticides like decamethrin, fenvelerate and phosphamidon at LC_{10} and LC_{20} levels enhanced fecundity and reduced nymphal duration . In another study synthetic pyretheroids, cypermethrin and deltamethrin stimulated aphid reproduction and increased fecundity but not with fenvalerate. In pyrethroid the increase in fecundity was due to increase in the concentration of amino acids viz., cystein, histidine, isoleucine/leucine, methionine and phenyl alanine in the treated plants. In endosulfan direct application over aphid increased fecundity (Singh, 1992, quoted by Bakhetia and Arora, 1993).

Insecticidal residues

Residue analysis of various insecticides used for the control of mustard aphid on rapeseed mustard is very limited. The available information is restricted to foliar and soil applied insecticides. Because of lipophilic nature practically no information is available on the residues of chlorinated hydrocarbon except endosulfan which is highly biodegradable and leaves no residue after a desirable period. Metasystox sprayed at 0.03 and 0.05% at flowering stage did not leave its residue after 15 days and when two sprayings were given of each concentration the residue became non detectable on pods on 20th day of second spray. In both the cases no residue be detected in seed at the time of harvesting (Dikshit *et al.,* 1986). Similarly the application of quinalphos at 0.15 and 0.25 Kg a.i./ha at flowering stage, 45 days after sowing, left no detectable residue after 25 days, even when second spray was given no residue could be observed on 35th day of application on pods and also in seeds at harvest (Dikshit *et al.,* 1986). He further emphasized that malathion treated crop can be consumed within 24 hours provided it is thoroughly washed and boiled. Kavadia and Lal (1968a) cautioned that the mustard crop treated with malathion, endosulfan, methamidophos, oxydemeton methyl and fenitrothion is harmful to utilize early as residues were found above the tolerance level within 3-7 days of application. They in the year 1968b reported that parathion residue persisted upto 28 days of application. Endosulfan residue reached below tolerance level of 1 ppm in 25 days and no detectable residue was observed in seed at harvest on mustard (Handa *et* al., 1980a and Udeaan *et* al., 1991) and on Indian Colza (Dikshit *et al.,* 1982). Handa *et al*., (1980a) furthure pointed out that carbofuran residue was maximum at 30th day of application but degraded rapidly, reached below tolerance limit of 0.5 ppm in 60 days in leaves and pods and no detectable residue in seeds at harvest. 2.75 Kg a.i./ha carbaryl and 1.1 Kg a.i./ha fenitrothion applied at flowering and again after 20 days left no detectable residue in the seeds of *B. napus* (Handa *et al*., 1980b). Disulfoton and phorate had harmful residue on leaves after 50 days on application in soil at the time of sowing (Jain *et al.,* 1974). Soil application of phorate, disulfoton, aldicarb, carbofuran and spraying of lindane20 EC, chlorpyriphos 20 EC, methyl-o-demeton 25 EC, phosphamidon 100 EC, phosalone 35 EC, fenitrothion 50 EC, carbaryl 50 WP and dusting of BHC+ parathion are safe at harvest time provided they are applied at recommended doses (Mukherjee, 1981).

Effect on insect pollinators

Granular systemic insecticides, phorate and carbofuran when applied in soil followed by light irrigation did not disrupt foraging activity of bees. Similarly methyl-o- demeton, dimethoate, endosulfan and phosphamidon at 0.03% did not deter bees visiting mustard crop (Rajput, 1988). Kalra and Gupta (1983) and Singh *et al.,* (1987) also indicated that 0.025% methyl-o-demeton, 0.03% phosphamidon and 0.03% endosulfan are safer to pollinators.

Safety to aphid predators

Coccinellids predators are capable of exerting considerable check on aphid population builup. A considerable loss in population may occur when insecticides is applied to control aphid on crop. The predators suffer in two ways (I) by coming in direct contact with insecticides and (II) by feeding on aphids poisoned with insecticides. Now the main emphasis is given to select out insecticides having lower toxicity to predators but show higher toxicity to aphid. The safer insecticides to predators is based on three sets of conditions namely contact toxicity under laboratory condition, when fed on poisoned aphids and when insecticides sprayed on the crop where both aphid and predator get poisoned are given in table 8.

Table 8 Safe insecticides to aphid predators

Mode of Testing	Insecticides
Contact Toxicity	Gamma BHC
	Endosulfan
	Trichlorfon
	Menazon
	Methyl-o-demeton
	Dimethoate
	Carbaryl
	Pirimicarb
Fed on treated	Endosulfan
	Methyl-O-demeton
On sprayed crop	0.05% phosphamidon
	0.05% Methyl-O-demeton
	0.1%˙ endosulfan
	0.04% endosulfan
	0.025% methyl–o-demeton
	205g/m in soil phorate
	Dimetoate and methyl-o-demton (sublethal doses)

Growth regulators in aphid management

Various insect hormones particularly juvenile hormone which is responsible for maintaining juvenile characters in younger stage and reproduction in adult stage and ecdysone, a moulting hormone are of great significane. Alteration in their titre at an inappropriate time either by incorporation of these or by non synthesis in the body lead to several abnormalities, change in behaviour and ultimately death of the insect. Several juvenile hormone analogues like altozar (Hydroprene) (Bindra and Singh, 1976); several terpenoid lactenes, guanolide (Arora *et al.,* 1982) and JHA-R 20458 (Singh and Sidhu, 1990); JHA-ZR_512(R) (Sidhu and Arora, 1990) and JHR-R 20458 (Singh, 1991) have been tested against mustard aphid. These JHA have shown toxicity, varied deformities, drastic reduction in longevity and fecundity and some repellent effects on this aphid by these workers. Recently Rup and Sohal (1989) have tested anti- juvenile hormone (AJH) compound, Precocene II and reported both toxic and anti juvenilizing effects on mustard aphid.

No reports are available on the effects of ecdysone analogue and anti-moulting hormone compound.

Pandey *et al.,* (1987) indicated that 1% extract of neem caused 80% mortality and reduced fecundity in *L. erysimi.* Preliminary studies are also available on the effects of

JHA on *P. xylostella* (Singh *et al.,* 1983 and Bhargava *et al.,* 1988) and *P. brassicae* (Dhaliwal *et al.*, 1983), particularly on the growth and development of these insects.

These limited informations indicated that more detailed studies are needed so that this method of insect control can be utilized satisfactorily and can be made a part of IPM programme.

Use of Biotechnology

One of the novel strategies to increase the host plant resistance against insect pests is genetic engineering method which enables the desired gene (s) to be transferred from any source into plant. The transfer of gene do not cause any undesirable effect as undesirable traits are not transferred. Three types of technologies are generally employed for the plant improvement particularly for increasing host resistance. They are the utilization of somaclonal variation, protoplast fusion and transformation. The transformation technology is being widely used in the production of transgenic plant. This is achieved either by direct transfer technique where in DNA is directly introduced either by uptake of DNA into chemically or electrically treated protoplast or microinjection or vector mediated gene transfer through the use of *Agrobacterium tumefaciens* bacterium. The first gene transfer in plant from plant source was that in tobacco plant and the gene was cowpea trypsin inhibitor gene (CPTI). This protease inhibitor acts like soybean and cowpea trypsin inhibition effect in the digestive proteases in insects, thus less attack of the insect on the transgenic plant. Another approach is the introduction of bacterial gene of *Bacillus thuringiensis* (*Bt*) which code the production of *Bt* toxin in the plant. *Bt* endotoxic gene introduced for insect control through genetic engineering in tomato for *Manduca sexta* and *H. virescence*, Cry. IA (b)gene, delta endotoxic gene in maize for *Ostrinia nubilalis*,Bt gene in rice for *Chilo* suppressalis and *Cnaphalocrosis medinalis* and in cotton for *Pectinophora* suppression are some of the successful examples of the use of biotechnology in increasing the host resistance. This type of work has also been started in crucifers. Recently Metz *et al.,* (1994) have synthesized a *Bt* ICP gene (Insecticidal crystal protein). These plants showed resistance to *P. xylostella* larvae. Similar efforts can be made to exploit the possibilities of intercropping the desired gene (s) to increase the resistance in cruciferous oilseed crops to key pests particularly mustard aphid, sawfly and painted bug. For this some resistant source must be made available or toxin producing organisms searched and then transferred in to these plants to obtain deleterious effects on the insects feeding on them. In future, these genetic engineered plants may be of limited significance as insect may develop resistance to these plants having *Bt* insecticidal crystal protein gene. Under field condition *P. xylostella* has developed resistance to *Bt* sprays (Shelton *et al.,* 1993) and when transgenic plants containing *Bt* ICP gene were offered to *P. xylostella* larvae, the larval survival was more than 90% (Metz *et al.,* 1994). Now there is ample fear that in the near future insects may develop resistance to plant produced insecticides. So in future our efforts would be to manage this type of resistance also.

MUSTARD SAWFLY

Athalia proxima, a serious pest of rapeseed mustard, is distributed throughout India. It attacks the crop only at the seedling stage and causing 4.93 to 34.2% yield loss in toria

(Verma, 1987) and 10 to 25% in different oilseed brssicas (Sachan, 1990) at Pantnagar, Uttaranchal; 14.68% to 17.62% at Navgaon, Rajsthan (AICORPO, 1987-1988) and on country basis losses due to this pest have been estimated as 32% (AICORPO, 1976). But under severe attack yield loss is cent per cent and farmer has to resow the crop and making the crop more prone to aphids.

Management

Resistant Varieties

Several toria and rai germplasm have been screened at Pantnagar both under field as well as laboratory conditions and some prominent materials have been identified but they need more testing at various locations (Sachan *et al.,* 1975). Among the various oilseed brassicas toria, yellow sarson, brown sarson and rai are highly preferred over *B. napus* and *B.carinata* for feeding and egg laying (Sachan and Sumati, 1985 and Sachan, 1990).

Cultural practices

Since the adult fly prefers the seedlings for egg laying and larvae for feeding asynchrony of the crop stage and adult emergence and egg laying will result in escape of the crop from damage and hence early sowing is advocated (Bakhetia and Sachan, 1997). Timely irrigation is also useful in reducing the larval population by drowning and killing them (Singh, 2005). Late sown mustard crop supplied with high dose of nitrogen suffers more (Kalra, 1985).

Chemical Control

Sachan *et al.,* (1983b) reported that carbaryl, quinalphos, sevithion each @ 0.75 Kg a.i./ha, monocrotophos and ripcord at 0.5 Kg a.i./ha and sevisulf at 0.75 Kg a.i./ha were better than MIPC (0.75 Kg a.i./ha) and gave good control of sawfly grubs. BHC dust @ 0.75 Kg/ha was ineffective. In another trial conducted during 1987-1988 all the insecticides tested viz., 10% BHC dust @ 20 Kg/ha, 0.02% chlorpyriphos, 0.03% dichlorvos, 0.035% endosulfan, 0.04% monocrotophos, 0.05% phenthoate and 0.025% quinalphos gave excellent control with their respective cost benefit ratios of 48.89, 17.5, 26.74, 17.1, 11.67, 13.28 and 15.18 (Sachan, 1990). Soil application of phorate, carbofuran, disulfoton and aldicarb each at 1 Kg a.i./ha at the time of sowing gave good control of this insect. (Annonymous,1977; Sarup *et al.,* 1971 and Pareek and Gupta, 1977). Carbofuran, endosulfan and quinalphos each at 0.5, 1.0 and 1.5g a.i./ Kg seed have also been used as seed treatment (Sachan and Sumati, 1985).

Botanical feeding deterrent / repellent / insecticides

Various plant parts extracts have shown to possess feeding deterrent properties and they have tested under laboratory as well as field conditions. The antifeeding properties for this insect were found in *Acorus calamus* (L), *Aloe barbadensis, Crinum defixum* Kerl-Gawl, *Derris indica, Euphorbia royleana* (Boist), *Lantana camera* (L) and *Momardica charentia* (L.) (Banerjee *et al.,* 1982; Pandey *et al.,* 1977; Sudhakar *et al.,*1978 ans Kumar *et al.,* 1979). The repellent property was maximum in *A. calamus* (0.5 and 1%) followed by undiluted extracts of *E. royleana, L. camera, C. bulbispermum* and A. vera (Pandey *et al.,* 1977). They further indicated the higher insecticidal property in *A. calamus* under laboratory condition. Under field condition ether extract of *A. calamus* rhizome gave highest mortality (Pandey *et al.,* 1977), 0.2% bitter gourd seed oil emulsion gave cent per cent kill to grubs in field as well as in laboratory. Lower doses of 0.5 and 1% also gave significant control (Kumar *et al.,* 1979).

Banerjee *et al.,* (1982) reported that 0.5% alcoholic extract of *D. indica* seeds, 0.5 to 1.0% petroleum ether extract of *A. calamus* and 2.5% alcoholic extract of *Allium sativum* gave 100% protection to the crop because of their antifeeding properties. Feeding deterrent effect was more when glucosiniolates were extracted from the seeds of *E. sativa*, *B. napus* and *Trapeolum majus* and isothiocyanates were given with leaf discs sharp reduction in feeding was observed (Ahuja and Sehgal, 1982).

Natural enemies

A bacterium, *Serratia marcescens* Bizio caused heavy mortality in larvae (Bogawat *et al.,* 1966 and Rangarajan *et al.,* 1968). The larvae turned pink. 0.05 and 0.2% spray of *Bacillus thuringiensis* could protect the mustard crop for 5 days only (Saxena, 1979).

Exacrodus populans Morley and *Perilissus cingulator* Morley Ichneumonid parasites are parasitizing the grubs of saw fly (Dutt, 1914). Pentatomid bug *Canthecona furcellata* Distant is predacious on grubs (David and Basheer, 1961).

PAINTED BUG

Bagrada cruciferarum Kirk (= *B. hilaris* Burm.) is an occasional pest of cruciferous crops (Singh and Malik, 1992). Both nymphs as well as adults suck the sap of the tender parts particularly seedlings in the early stage. Attacked leaves develop white blotch then wilt and die. This necessitating is resowing of the crop. Attack in podding stage and on harvested material piled on threshing floor results in shrivelling of grain and reduction in oil content. This insect caused yield loss varying from 27 to 70% in Rajasthan (Joshi *et al.*, 1989) and 30% in Haryana (Singh *et al.*, 1980). In addition to quantitative losses it also caused reduction in oil content 2.75%, protein content 3.5%, reducing sugar 1.17% and total sugar by 2.9% (Singh *et al.,* 1980). There was an out break of this insect in 1991 and 1992 in south western zone of Punjab, Haryana and western zone of UP, tarai region of Uttaranchal on toria and rai forcing the farmer to resow their crops. It is a polyphagous pest feeding on paddy, pearlmillet, green gram, black gram, coffee, sugarcane, sunflower, maize and sorghum in addition to crucifers (Rakshpal, 1949; Narayanan, 1954; Anderson, 1917; Pillai, 1922 and Gupta and Gupta, 1970) and hence it is found through out the year.

Cultural control

Clean cultivation and timely sowing of crop is very important to protect it from bug attack at the seedling stage. Irrigation helps in survival of the crop as it reduces the chances of wilting of the crop due to desapping (Sekhon *et al.,* 1993). Singh *et al.,* (1993) advocated irrigation at 40-50 days after sowing for effective control of this bug. Quick threshing of the harvested material and disposal of the crop refuge help in checking the attack (Bakhetia and Sachan, 1997).

Rawat *et al.* (1968) and Makhmoor and Shinde (1986) reported that the bug infestation increased with increase in nitrogen level applied and thus balanced fertilization is required.

Resistant varieties

Under field condition *E. sativa* is tolerant to this bug in addition, Bathinda Raya strain No. 846, 1131 and 1167 are moderately resistant (Sekhon *et al.*, 1993). They further pointed out that among *B. napus, B. carinata, Sinapis alba, B. campestris* and *B. juncea. S. alba* was highly susceptible and *B. juncea* the least. Among the 8 brassicas and one *E. sativa* the soaked seeds of *E. sativa* were highly suitable for the growth and development whereas *B. nigra* seeds were highly unsuitable. When seedlings were provided as food, BSH1 seedlings were highly suitable and of *B. nigra* and RW 32-3 the least. The seedlings of yellow sarson (YST-151) and of *B. alba* were intermediate (Tripathi, 1992).

Chemical control

The work on the chemical control of this bug is restricted at the seedling stage of the crop and practically no information is available to control this pest at podding or on harvested material. Saxena *et al.*, (1993) reported that 2% methyl parathion dust @ 25 Kg/ha 14 days after sowing was very effective and economic (C:B= 11.2) followed by 0.07% endosulfan spray (C:B= 10.6). Application of dimethoate @ 0.05 Kg a.i./ha gave good results (Gupta and Sharma, 1971). Sarup *et al.*, (1971) indicated that fenitrothion was best followed by mevinphos whereas methyl parathion and carbaryl proved least effective. Among the granular insecticides Nakat and Peshwani (1972) reported that phorate 10G @ 1.15 g/m row in furrow, seed treatment with carbofuran, 4g/100g seed and furrow application of disulfoton @ 1.95 g/m row were more toxic than lindane, 1.95g/m row. The first, second and third treatments were 1.83, 1.42 and 1.21 times as toxic as fourth treatment. Application of phorate protected the crop upto 54 days.

Several plant products have been found effective against this insect. Both under laboratory and field conditions rhizome extract of *A. gramineus,* dried leaf extract of *Solanum xanthocarpum* each at 0.5, 1 and 2% concentrations, *Melia azadarach* ether extract and seed and leaf extract of *Argemone maxicana* each at 2% were highly effective in killing the bug (Pandey *et al.*, 1980, 81 and 82). Cent per cent mortality was observed with acetone extract of *A. calamus* rhizome and *Gynandropsis gynandra* (Verma and Pandey, 1981).

Harvested material piled on the threshing floor can be protected by safe insecticide application.

Painted bug is parasitized by *Alophora* species (Tachinidae: Diptera)(Rakshpal, 1949).

MUSTARD APHID

- Timely sowing, as per recommendation, late sown crop suffers more.
- Growing of short duration crops
- Use of tolerant crops. Instead of sarson grow mustard or gobhi sarson in areas where crops are heavily attacked by aphid.
- Balanced fertilization, over dosing of nitrogen shoud be avoided.
- Proper water management, rainfed crops suffer more.
- Plucking of infested twigs in the beginning to check dispersal.
- Chemical control

- Need base application as per economic threshold values *i.e.* 40-50% infestation, 50-60 aphids/plant or 0.5-1.0 cm twig covered with aphids.
- Use safer insecticides to bioagents and pollinators
- Should not leave residue
- Mixing of fungicide with insecticide.
- Adopt proper time of application
- Do not apply granular systemic insecticide at the time of sowing but apply when aphid population reaches to threshold level and then irrigate.

MUSTARD SAW FLY

- Field sanitation
- Timely irrigation and proper fertilization
- Monitoring of grub attack almost daily to adopt control measures immediately
- Grow gobhi sarson/karan rai which are resistant to this insect.
- Apply dust formulation early in the mornig. 2% malathion dust is effective.

PAINTED BUG

- Clean cultivation because it is polyphagous
- Chemical control : 2% malathion dust or other insecticides.
- Irrigation at seedling stage when attack is seen.
- Balance dose of nitrogen should be applied.
- Quick threshing of harvested material and its proper disposal.
- When bullocks are used for threshing purpose the dung should be buried deep in the soil to avoid congregation.

Future Research Needed

- Study on crop and insect life tables to establish asynchrony in crop stages and pest arrival.
- Use of biopesticides, insect growth regulator and use of natural enemies.
- Exploitation of conventional cultural practices.
- Mass screening of available germplasm for resistance to aphid.
- Dietics studies on aphid.
- Monitoring and management of insecticide resistance.
- Aphid biotype.
- Frontline demonstration to make farmers aware of the new technologies.

CONCLUSION

With the increase in area and intensification of rapeseed mustard the pest problem is increasing at very fast rate. The increase in number of potential pests has been much more during the last 8 years. Insect like *Helicoverpa* is becoming of major concern. The key pest, mustard aphid can be managed by monitoring of aphid during off season, timely planting, balance fertilization, appropriate water management, proper spacing and growing of early material. Need base application of insecticides with fungicide which are non hazardous, leaves no residue and safer to natural enemies and pollinators should be done. A number of germplasms have been screened for aphid resistance but without success and unfortunately no good source has yet been identified. Moreover mustard aphid is also developing resistance to insecticides very fast and resurgenece like situations are also observed. Such conditions are due to misuse of insecticides on rapeseed and mustard crops. Efforts are needed in this direction. Some new method in aphid control like the use of Insect Growth Regulators is at preliminary stage and the work on the synthesis of brassicas transgenic plant has been just started. Bioagents are ineffective in managing the aphid. Insecticides of plant origin are being tried and more information is to be generated. Management of sawfly and painted bug is simple provided control measures are adopted timely.

Management strategies for major insect pests of rapeseed mustard as describe by Bakhetia and Sachan (1997) and Singh (2005).

References

Ahuja, D.B. and Sehgal, V.K. 1982. Effect of seed extracts containing mustard oil glucosides on feeding responses of mustard sawfly grub, *Athalia proxima* Klug. *Indian J. Ent.,* 44: 34-40.

AICORPO, 1976. *Seminar cum Workshop*. All India coordinated Research Project on Oilseeds (AICORPO). 1974-75, April 5-9, 1975, Nagpur, Maharashtra, ICAR, New Dehi.

AICORPO, 1988. Annual progress Report.1987-88, ICAR. Directorate of Oilseed Research, Hyderabad, pp 309-369.

AICORPO, 1993. Annual progress Report.1992-93, ICAR. Directorate of Oilseed Research, Hyderabad, pp 242.

Anderson, T.J. 1917. Notes on insects injurious to coffee. Dept. Agric. Bri. East Afr., *Nairobi Agric. Bull.,* 2: 20.

Annonymous, 1977. Research on rapeseed and mustard. Experiment Station, GBPUA&T., Pantnagar. pp 61.

Annonymous, 1991. Annual Research Progress Report (Rapeseed – Mustard). 190-91. Directorate of Experiment Station, GBPUA&T., Pantnagar, pp 68-70.

Arora, R. and Sidhu, H.S. 1982. Effect of plant water deficit on the fecundity of mustard aphid, *Lipaphis erysimi*(Kalt). *Indian J. Ecol.*, 9: 346-347.

Arora, S.K.; Saini, M.L.; Sandhu, R.S. and Singh, M. 1969. Effect of insecticides on yield and quality of Indian mustard, *Brassica juncea* (L) Czern & Coss. *Indian J.Agric. Sci.*, 39: 444-447.

Arora, R.; Singh,G.; Kailey, J.S and Kalsi, P.S. 1982. Biological activity of terpenoides lactenes as juvenile hormone analogues against the mustard aphid, *Lipaphis erysimi*. *Phytoparasitica,* 10 (1): 57-60.

Atwal, A.S. and Sethi, S.L. 1963. Predation by *Coccinella seetempunctata* L. on the cabbage aphid, *Lipaphis erysimi* (Kalt) in India. *J. Animal Ecol.,* 32(3): 481-488.

Bakhetia, D.R.C.1983. Losses in rapeseed mustard due to *Lipaphis erysimi* (Kalt.) in India. A literature study. Proceedings of VI International Rapeseed Conference held at Paris from 17-19 May, 1983, pp 1142-1147.

Bakhetia, D.R.C. 1984. Chemical control of *Lipaphis erysimi* (Kalt) on rapeseed and mustard crops. *J. Res. Punjab Agric. Univ.,* 21(1): 63-75.

Bakhetia, D.R.C. 1986. Pest management in rapeseed and mustard. *Pesticides*, 20: 32-39.

Bakhetia, D.R.C. 1991a. Attack of *Heliothis* on toria crop. Report submitted to Director of Research, P.A.U. Ludhiana, Punjab, India, p 1.

Bakhetia, D.R.C. 1991b. Insect Pests on Oilseed in Indian Agriculture (Eds. V.L. Chopra and Shyam Prakash). Pub Har Anand, New Delhi, pp 211-239.

Bakhetia, D.R.C. and Arora, R. 1993. Changing scenario of insect pests and pest management in rapeseed-mustard. In Pests and Pest Management in India- The Changing Scenario. (Ed. H.C. Sharma and M.V. Rao). Lead Paper of the National Seminar on Changing Scenario in Pest and Pest Management in India, 31 Jan-1 feb., 1992. Plant Protection Association of India, Hyderabad, pp 56-69.

Bakhetia, D.R.C. and Brar, K.S. 1988. Effect of water stress in Ethiopian mustard (*Brassica carinata*) and Indian mustard (*B. juncea* sub sp. juncea) on infestation by *Lipaphis erysimi. Indian J. Agric. Sci.,* 58(8): 67-70.

Bakhetia, D.R.C. and Ghorbandi, A.W..1986. Economic threshold of *Lipaphis erysimi* (Kalt) on Indian mustard. *INowl. Aph. Soc.,* India. 5 (2): 5-7.

Bakhetia, D.R.C. and Sachan, G,C. 1997. Identification and Management of Strategies for Insect Pests of Rapeseed and Mustard. Directorate of Extenesion Dept. Agril. Min. of Agri., Govt of India, New Delhi. 43p.

Bakhetia, D.R.C and Sharma, A.K. 1978. Preliminary observations on the aphids infestation of *Eruca sativa* Mill. *Indian J. Ent.,* 41(3); 288-289.

Bakhetia, D.R.C. and Singh, J. 1992. Technological advances in Pest Management in rapeseed and mustard. In Advances in Oilseeds Research (Ed. D. Kumar and M. Rai.), 1: 119-149. Scientific Publishers, Jodhpur.

Bakhetia, D.R.C. and Singh, P. 1993. Advances in host plant resistance to insect pests in oilseed crops. In G.S. Dhaliwal and V.K. Dilwari. Eds., Advances in Host Plant Resistance to Insects, Kalyani Publishers, New Delhi.

Banerjee, R.; Mishra, G.; Nigam, S.K.; Prasad, N.; Pandey, R.S. and Mathur, Y.K. 1982. Indigenous plant as antifeedants. *Indian J. Ent.,* 44(1): 71-76.

Bath, D.S. and Singh, D.1989. Studies on the economic threshold of mustard aphid, *Lipaphis erysimi* on mustard and radish. In: Aphids ed. B.K. Beaura. The Zoological Society of Orissa, utakl University, Bhubneshwar, India p 43-51.

Bhadouria, N. S.; Bahadur, J.; Dhamdhere, S.V. and Jakhmola, S.S.1992. Effect of different sowing dates of mustard crop on infestation by the mustard aphid, *Lipaphis erysimi* (Kalt). *J. Insect Sci.,*5 (1): 37-39.

Bhargava, M.C.; Srivastava, R.P. and Mathur, N.M.1988. Effect of two juvenile hormone analogues on reproductive potentiality of *Plutella maculipennis* Custie (Lepidoptera: Plutellidae). In proceedings of National Symposium on Integrated Pest Control, Progress and Prospective (Eds. N. Mohandas and G. Koshy). Association of for Advancement of Entomology, Trivendrum, Kerla, India, pp 415-417.

Bhattachrjee, N.S. 1961. Control of mustard aphids by cultural practices. *Indian Oilseeds J.,* 5: 133-138.

Bindra, O.S. and Deole, J.Y. 1962. Investigations on the cultural control of mustard aphid, *Lipaphis erysimi* (Kalt) (Hemiptera: Aphididae). Proceedings Second All India Congress of Zoology, Varanasi, UP, India pt. 2 (Issued Nov. 1966).

Bindra, O.S. and Singh, D. 1976. Effect of altozar, a juvenile hormone analogue on the mustard aphid, *Lipaphis erysimi* (Kalt). *Botyukagaku,* 41: 65-66.

Bogawat, J. K.; Rangarajan, M.; and Chakravarti, B.P. 1966. Bacterial mortality of the grubs of mustard sawfly, *Athalia proxima* Klug (Hymenoptera : Tenthredinidae). *Indian J. Ent.,* 28: 417-421.

Chadda, I.C. and Arora, R. 1982. Influence of water stress in the host plant in the mustard aphid, *Lipaphis erysimi* (Kalt).*Entomon.*, 7: 75-78.

Chandra, S. and Kushwaha, K.S. 1987. Impact of environmental resistance on aphid complex of cruciferous crops under the agroclimatic conditions of Udaipur II. Biotic Components. *Indian J. Ent.*, 49(1): 86-113.

David, B.V and Basheer, M. 1961. Mass occurrence of the predatory stink bug, *Cantheconida* (*Canthecona*) *furcellata* (Wolff.) on *Amsacta albistriga*(Walk) in south India. *J. Bombay Nat. Hist. Soc.*, 58: 817-819.

Dhaliwal, Z.S.; Singh, D. and Deol, B.S. 1983. Evaluation of juvenile hormone analogue, RO- 20458 against larvae of *Pieris brassicae* Linn. *Indian J. Ent.*, 45: 514-516.

Dhiman, S.C. and Kumar, V. 1985. Biological control of mustard aphid, *Lipaphis erysimi* (Kalt) by suing parasitoid, *Aphidius* (*Diaeretiella*) *rapae* (Curtis). Influence of parasitism on aphids. In: Proc. National Symposium Pesticide Residue and Enviromental Pollution, India, October, 2-4, 1985. Muzaffernagar, pp 219-222.

Dikshit, A.K.; Handa, S.K. and Verma, S. 1986. Residues of methamidophos and effect of wasing and cooking in cauliflower, cabbage and Indian Colza. . *Indian J. Agric. Sci.*, 56: 661-666.

Dikshit, A.K.; Handa, S.K.; Awasthi, M.D. and Verma, S. 1982. Residue of endosulfan on Indian Colza. *Indian J. Agric. Sci.*, 52(7): 463-465.

Dilawari, V.K. and Dhaliwal, G.S. 1992. Allelochemical interactions between taramira and *Lipaphis erysimi* (Kalt). Paper presented at Nat. Symposium on Allelopathy and Agroecosystem (Agriculture and Forestry), February, 12-14, 1992. Haryana Agriculture University, Hisar, India.

Dutt, G.R.1914. A annotated list of Ichneumonidae in the Pusa Collection. Mem. Dept. Agic., India, 8(2): 13-18.

Ghosh, A. K. and Ghosh, M. R. 1981. Effect of time of sowing and insecticidal treatment on the pests of Indian mustard (*B. juncea* L) and on seed yield. *Entomon,* 6 (4): 357-362.

Gupta, B. M. and Sharma, J.C. 1971. Efficacy of certain insecticidal control schedule against pests of cabbage. *Madras Agric. J.*, 58: 889-891.

Gupta, J. C. and Gupta, D. S. 1970. Notes on some new hosts of painted bug, *Bagrada cruciferarum* Kirk, (Pentatomidae, Heteroptera). *Indian J. Agric Sci.,*40 (7): 645-646.

Handa, S.K.; Dikshit, A.K.; Awasthi, M.D.; Singh, R.N.; Singh, S.; Singh, N.P. and Singh, K.M.1980a. Residue following application of carbofuran and endosulfan to mustard crops in relation to irrigation. *Indian J. Ent.*, 42(4): 766-769.

Handa, S.K.; Dikshit, A.K.; Awasthi, M.D. and Verma, S. 1980b. Residues of carbaryl in /on Indian Colza crop. *Indain J. Agric. Sci.,*50 (11): 973-975.

Jain, H.K.; Pandey, S.Y.; Agnihotri, N. P.; Dewan, R.S.; Saxena, A.N. and Peshwani, K.M. 1974. Dissipation of phorate and disulfoton in rapeseed crop (*Brassica campestris*). *Indian J. Pl. Protec.*, 1(2): 37-43.

Jindal, G.S. and Upadhyaya, V.R. 1989. New host plants of *Heliothis armigera* in India. *Tropical Pest Management*, 35: 213.

Joshi, M.L.; Ahuja, D.B. and Mathur, B.N. 1989. New host plant of *Heliothis armigrea* in India.. *Tropical Pest Management.* 35: 213.

Kalra, V.K. 1985. Influence of cultural practices on the attack of mustard sawfly, *Athalia proxima* Klug. *Cruciferae Newsletter,* 10: 122-123.

Kalra, V.K.1988. Population dynamics of various predators associated with mustard aphid, *Lipaphis erysimi* (Kalt). *J. Biol. Cont.*, 2(2): 77-79.

Kalra,V.K and Gupta, D.S.1983. Integrated Control of mustard aphid, *Lipaphis erysimi* (Kalt). In: Proc. 10th Int. Con. Pl. Prot., Nov. 20-25, 1983 Brighton, England.1198.

Kalra, V.K.; Gupta, D.S. and Yadav, T.P. 1983. Effect of cultural practices and aphid infestation on seed yield and its component traits in *Brassica juncea* (L) Czern & Coss. *Haryana Agric. Uni. J. Res.,* 13(1): 115-120.

Kavadia, V.S. and Lal, R. 1968a. Residues of malathion on Indian rapeseed, (*Brassica campestris* L. var. *Dichotoma* Watt.). *Indian J. Agric. Sci.,* 38(4): 708-713.

Kavadia, V.S. and Lal, R. 1968b. Estimation and dissipation of parathion residues from the crop of Indian rapeseed. *Indian J. Ent.,* 30(2): 113-118.

Kumar, A., Kapoor, V.C. and Mahal, M.S. 1988. Population buildup and dispersion of immature stages of aphidophagous syrphids (Syrphidae: Diptera) on raya, *Brassica juncea* Coss. *J. Insect. Sci.,* 1(1): 39-48.

Kumar, A. Kapoor, V.C. and Mahal, M.S. 1989. Pupal population of aphidophagous syrphids and their hyperparasitism on *Brassica juncea* Coss. *J. Insect. Sci.,* 2(2): 114-118.

Kumar, A., Tewari, G.C. and Pandey, N.D. 1979. Antifeeding and insecticidal properties of bitter gourd, *Momordica charantia* Linn. against *Athalia proxima* Klug. *Indian J. Ent.,* 41(2): 103-106.

Kundu, G.G. and Pant, N.C. 1967. Studies on *Lipaphis erysimi* (Kalt.) with special reference to insect plant relationship. II. Effect of various levels of N, P and K on fecundity. *Indian J. Ent.,* 29: 285-289.

Lee, H.S. 1988. Action thresholds for turnip aphid and green peach aphid on cabbage and leaf mustard. *J. Agric Res. China.*37: 91-99.

Makhmoor, H.D. and Shinde, C.D. 1986. Preliminary study on the effect of different doses of N, P and K and their combinations in mustard on the incidence of *Athalia proxima* Klug., *Crocidolomia binotalis* Zell., and *Phytomyza atricornis. Res. Dev. Rept.,* 3: 66-67.

Mathur, Y.K., Singh, S.V., Singh, R.S., Agrawal, N. and Katiyar, R.R. 1991. Aphid management in mustard, *Brassica juncea* Czer and Coss. In Proceedings of National Seminar on Strategies for Making India Self-reliant in Vegetable Oils. (Eds. V. Ranga Rao and M.V.R. Prasad) Sept. 5-9, 1988. Indian Society of Oilseeds Research, Directorate of Oilseeds Research and ICAR, pp. 158-169.

Mehrotra, K.N. 1991. Current status of pesticide resistance in insect pests in India. *J. Insect Sci.,* 4(1): 1-14.

Mehrotra, K.N. and Phokela, A. 1985. An anionic site in the *Lipaphis erysimi* Kalt. Carboxyesterase (EC.3.1.1.1). *Indian J. Expt. Biol.,* 23: 48-50.

Metz, T., Dixit, J., Goldsmith, J., Rousch, R. and Earle, E. 1994. Production of transgenic *Brassica oleracea* expressing *Bacillus thuringiensis* insecticidal crystal protein genes. *Cruciferae Newsl.,* 16 :63-64.

Mukherjee, S.K. 1981. Residues of insecticides in oilseeds and pulses. *Pesticides information,* 7(3): 26-32.

Nakat,S.S. and Peshwani, K.M.1972. Relative efficacy of some granular systemic insecticides against some important pests of mustard. *Indian J. Ent.,* 34: 118-122.

Narang, D.D. and Atwal, A.S. 1986. Effect of leaf extracts containing glucosinolates on the biology of mustard aphid, *Lipaphis erysimi* (Kalt.). *Indian J. Ecol.,* 13(2): 307-312.

Narayanan, E.S. 1954. The painted bug, *Bagrada cruciferarum* (Kirk.). *Indian Farming,* 3: 8-9.

Nath, D.K. and Sen, B. 1976. Some observations on aphidophagous coccinellid beetles in mustard cultivation. *Sci. Cult.,* 42: 288-290.

Pandey, N.D., Singh, L., Singh, Y.P. and Tripathi, R.A. 1987. Effect of certain plant extracts against *Lipaphis erysimi* Kalt. under laboratory conditions. *Indian J. Ent.,* 49: 238-242.

Pandey, N.D., Singh, M. and Tewari, G.C. 1977. Antifeeding, repellent and insecticidal properties of some indigenous materials against mustard asawfly, *Athalia proxima* Klug. *Indian J. Ent.*, 39: 60-64.

Pandey, U.K., Pandey, M. and Chauhan, S.P.S. 1981. Insecticidal properties of some plant material extracts against painted bug, *Bagrada cruciferarum* Kirk. *Indian J. Ent.*, 43(4): 404-407.

Pandey, U.K., Verma, G.S. and Pandey, M. 1980. Efficacy of some plant origin insecticides on *Bagrada cruciferarum* Kirk. *Indian J. Ent.*, 42(4): 775-776.

Pandey, U.K., Verma, G.S., Lekha, C. and Singh, A.K. 1982. Note on the use of some insecticides against *Bagrada cruciferarum* Kirk. *Indian J. Agric. Sci.*, 52(3): 205-206.

Pareek, B.L. and Gupta, H.C. 1977. Evaluation of some granular insecticides against *Athalia proxima* Klug. Infesting mustard. *Indian J. Ent.*, 39(4): 392-394.

Peshwani, K.M.; Kumar, K. and Lal, R. 1968. Successful mustard cultivation in Delhi villages by aphid control. *Indian Farming,* 18: 17-18.

Phadke, K.G. and Prasad, S.K. 1987. Effect of sowing date on aphid incidence and yield in some varieties of rapeseed and mustard. *J. Aphidology,* 1: 233-28.

Phokela, A. and Mehrotra, K.N. 1987. Inhibition of aphid caboxyesterase (EC.3.1.1.1) by carbamate and organophosphate inhibition. *Proc. Indian Natn. Sci. Acad.*, B, 53: 35-38.

Phokela, A., Saxena, H.P. and Mehrotra, K.N. 1981. Kinetic studies on the carboxyesterases (EC.3.1.1.1) from the aphids, *Aphis craccivora* Koch. and *Lipaphis erysimi* Kalt. *J. Ent. Res.*, 5: 16-24.

Pillai, N.K. 1922. Insect Pests, Rep. Dep. Agric. Fish. Travancore, 1921-22 pp. 19-20.

Purwar, J.P. and Sachan, G.C. 2004a. Bioefficacy of entomopathogenic fungi against mustard aphid, *Lipaphis erysimi* (Kalt) on *Brassica campestris. J. Aphidology,* 18: 5-10.

Purwar, J.P. and Sachan, G.C.. 2004b. Synergistic effect of entomogenous fungi with some insecticides for management of mustard aphid, *Lipaphis erysimi* (Kalt). *J. Aphidology,* 18 : 11-14.

Rai, B.K. 1976. *Pests of Oilseed Crops in India and their Control.* Indian Council of Agricultural Research, New Delhi, p. 121.

Rajput, K.S. 1988 Investigations on the thermoregulation and foraging activities of honey bees during winter. M.Sc. Thesis, G.B.P.U.A.&T., Pantnagar.

Rakshpal, R. 1949. Note on the biology of *Bagrada cruciferarum* (Kirtk.). *Indian J. Ent.,* 11: 11-16.

Ram, S. and Gupta, M.P. 1992. Role of fertilizer (NPK) in insect pest management of mustard for fodder production. *Indian J. Agric. Res.*, 26(12): 955-956.

Ramachandran, S. 1952. Manuring in relation to insect pests. *Madras Agric. J.*, 39(3): 91-95.

Rangarajan, M., Srivastava, R.P. and Chakravarti, B.P. 1968. A note on the different insect hosts of *Serratia marcescens* Bizio. *Labdev, J. Sci. Tech.* 6(3): 159-160.

Rawat, R.R. and Singh, O.P. 1983. Effect of different dates of sowing and combinations of fertilizers on the incidence of mustard aphid, *Lipaphis erysimi* (Kalt.) and the grain yield of mustard, In B.K. Behura, Ed. *The Aphids*, The zoological Society of Orissa, Utkal University; Bhubeneshwar, India, pp. 295-302.

Rawat, R.R., Mishra, U.S., Thakare, A.V. and Dhamdhere, S.V. 1968. Preliminary study on the effect of different doses of nitrogen on the incidence of major pests of mustard. *Madras agric. J.,* 55(8): 363-366.

Rohilla, H.R. 1990. Management of mustard aphid: Research on rapeseed mustard. In: I.Ohlson and P.R. Kumar, Eds., Proc. Indo-Sweedish *Sysm. Univ. Agric. Sci.,* Upsals, Sweden, pp. 92-98.

Rohilla, H.R., Kumar, P.R. and Singh, H. 1988. Insect pest management strategies of rapeseed mustard group of crops. In Nat. Workshop on Pest and Disease Surveillance for IPM. Tamil Nadu Agriculture University, Coimbatore, pp. 91.

Rout, G. and Pant, S.C. 1968. Insecticide trails to control mustard aphid, *Lipaphis erysimi* Kalt. and Larger moth, *Crocidolomis binotalis* Zell. affecting mustard crop. *Indian J. Agric. Sci.,* 38: 828-831.

Roy, P. 1975. Life history and feeding behaviour of *Coccinella transversalis* Fab. (*Coccinellidae: Coleoptera*), a predator of mustard aphid, *Lipaphis erysimi* (Kalt.). *Indian Biologists,* 8(1-2): 59-61.

Rup, R.J. and Sohal. 1989. Morphogenetic effect of Precocene II on *Lipaphis erysimi* (Homoptera: Aphididae). *Acta Entomologica Bohemslovakis*, 86: 172-178.

Sachan, G.C. 2005. Screening Techniques for Insect resistance in *Brassicas.* In: Winter School Advances in Rapeseed-Mustard Research Tehchnology for sustainable production of oilseed. December 15, 2004-Jan, 04, 2005. National Research Centre on Rapeseed-mustard, Bharatpur.

Sachan, G.C. and Gautam, V. 2001. Reactions of *Lipaphis erysimi* (Kaltenbach) (Homoptera:Aphididae) to some agrochemicals. *J. Aphidology.* 15: 159-162.

Sachan, G.C. and Sumati. 1985. Incidence of *Athalia proxima* Klug on species and varieties of *Brassica. Cruciferae Newsl.,* 10:128-129.

Sachan, G.C. and Tripathi, N.L.M. 1987. Effect of sublethal concentrations of insecticides on protein content of mustard leaf and inflorescence *Cruciferae Newsl.,* 12: 80.

Sachan, G.C., Pathak, P.K. and Chibber, R.C. 1983a. Effect of nitrogen levels and date of sowing on the incidence of *Lipaphis erysimi* and yield of mustard. In *Annual Report Res.,* Directorate of Expt. Sta., G.B. Pant Univ. Agric. Tech., Pantnagar. Pp. 78.

Sachan, G.C., Pathak, P.K. and Chhibber, R.C. 1983b. Chemcial control of *Athalia proxima* in toria crop. In *Annual Reprot Res. Directorate of Expt. St.* G.B.P.U.A.&T., Pantnagar. pp.79.

Sachan, G.C., Singh, C.P., Tripathi, N.L.M. and Sachan, S.K. 1975. Entomology. In Oilseeds research at Pantnagar. *Expt. Stn. Tech. Bull.* 111/ Directorate of Experiment Station, G.B. Pant Univ. Agric. Tech. Pantnagar. Pp.84-89.

Sachan, G.C.1990. Problems of insect pests in *Brassica* and research work at Pantnagar. In: proceedings of the IDRC three meeting held at Pantnagar and Hyderabad, India, 4-17 January, 1989. pp 56-65.

Sarup, P.; Sircar, P.; Sharma, D.N., Singh, D.S.; Amarpuri, S., Dewan, R.S., and Lal, R. 1971. Effect of formulations on the toxicity of pesticidal granules to some important pests of mustard. *Indian J. Ent.,* 33(1): 82-89.

Saxena, R.C. 1979. Efficiency of *Bacillus thringiensis* Berliner for the control of *Athalia proxima* Klug. *Indian J. Pl. Protec.*,6(1): 33-35.

Saxena, A.K., Dhamdhere, S.V.; Bhadouria, S.S. and Dixit, S.C. 1993. Painted bug, *Bagrada hilaris* Burm., a new threat to toria cultivation. Natn. Semin. Oilseeds Res. And Dev. In India: Status and Strategies, Aug. 2-5, 1993. Directorate of Oilseeds Research, Hyderabad. pp.200-201.

Sekhon, B.S., Arora, P.K., Dhillon, S.S., Singh, G. and Karmuka, K.S. 1993. Studies on resistance in rapeseed mustard varieties to painted bug, *Bagrada hilaris* Burm. Natn. Semin. Oilseeds Res. And Dev. In India: Status and strategies, Aug. 2-5, 1993. Directorate of Oilseeds Research, Hyderabad. pp. 175-176.

Shelton, A.M., Robertson, J.L., Tang, J.D., Perz, C., Eigenbrode, S.D., Preisler, H.K., Wilsey, W.T. and Cooley, R.J. 1993. *J. Eco. Ent.,* 86: 697-705.

Sidhu, H.S. and Arora, R. 1990. Toxic and demorphogenetic effect of juvenile hormone analogues against alate mustard aphid, *Lipaphis erysimi* (Kalt.). *J. Res. Punjab Agric. Univ.,* 27: 65-71.

Sidhu, H.S. and Kaur, P. 1977. Influence of nitrogen application to the host plant on the fecundity of mustard aphid, *Lipaphis erysimi* (Kalt.). *J. Res. Punjab Agric. Univ.,* 14: 445-448.

Sidhu, H.S. and Kaur, P. 1976. The influence of water stress in the host plant on the reproduction of the mustard aphid, *Lipaphis erysimi* (Kalt.). *Indian J. Ecol.,* 3: 163-167.

Singh, B., Singh, R. and Mahal, M.S. 1983. Assessment of losses in yield of *Brassica juncea* by *Lipaphis erysimi* (Kalt.) II Economics of aphid control. *Indian J. Ecol.*, 10: 279-284.

Singh, C.P. 1992. Assessment of yield losses due to *Lipaphis erysimi* (Kaltenbach) and its chemical control. Ph.D. Thesis, G.B.P.U.A.&T., Pantnagar. P. 130.

Singh, D. 1993. Studies on the role of biological control agents in the population suppression of mustard aphid, *Lipaphis erysimi* (Kalt.) on mustard. Ph.D. Thesis, CCS H.A.U. Hisar. pp. 1-69.

Singh, G. 1991. Repellent effect of juvenile hormone analogue against mustard aphid, *Lipaphis erysimi* (Kalt.). *Pestology,* 15(1): 31-32.

Singh, G.S. and Sidhu, H.S. 1990. Toxic effect of juvenile hormone analogues on mustard aphid, *Lipaphis erysimi* (Kalt.). *J. Insect Sci.,* 3: 18-32.

Singh, H. 1982. Studies on insect pest complex in *Brassica campestris* L. Var. brown sarson. Ph.D. Thesis, Haryana Agriculture University, Hisar.

Singh, H. and Malik, V.S. 1992. Biology of painted bug, *Bagrada cruciferarum* Burm. *Indian J. Agric. Sci.,* (In press).

Singh, H. and Singh, Z. 1983. New Records of insect pests of rapeseed. *Indian J. Agric. Sci.,* 53(9): 870.

Singh, H., Gupta, D.S., Yadava, T.P. and Dhawan, K. 1980. Post harvest losses caused by painted bug, *Bagrada cruciferarum* Kirk. to mustard. *Haryana Agric. Univ. J. Res.,* 10(3): 407-409.

Singh, H., Rohilla, H.R. and Singh, H. 1994. Some records and observations on insect pests of rapeseed-mustard crops. *Cruciferae Newsl.,* 10: 37-38.

Singh, H., Rohilla, H.R., Kalra, V.K. and Yadava, T.P. 1984. Response of *Brassica* varieties sown on different dates to the attack of mustard aphid, *Lipaphis erysimi* (Kalt.). *J. Oilseeds Res.,* 1: 49-56.

Singh, H., Singh, H. and Verma, A.N. 1990. Incidence of budfly, *Dasineura* sp. on *Brassica* crops in Haryana state. *Indian J. Pl. Prot.,* 18: 135.

Singh, H., Singh, H., Rohilla, H.R. and Singh, D. 1993. Integrated pest management in rapeseed-mustard crops in Haryana, Natn. Semin. Oilseeds Research and Development to India. Status and Strategies, Aug. 205, 1993. Hyderabad, pp. 117.

Singh, H.; Rohilla, H.R. and Kharub, S.S. 1987. Integration on chemical control of mustard aphid, *Lipaphis erysimi* (Kalt) and safety to pollinators. *Ann. Biol.,*3(1): 105-106.

Singh, S. and Sidhu, H.S. 1958. Control of mustard aphid by synthetic insecticides. *Indian Oilseeds J.,* 2: 24-31.

Singh, S. and Sidhu, H.S. 1959. A schedule for control of mustard aphid by some insecticides. *Indian Oilseeds. J.,* 3: 169-178.

Singh, W. and Mishra, D.S. 1986. Economics Injury level of mustard aphid, *Lipaphis erysimi* (Kalt) on mustard. In: Aphidology in India ed. B.K. Agrawala. A.R. Printers, Calcutta. pp105-112.

Singh, Y.P. 2005. Integrated insect pest management in rapeseed-mustard (including safe handling of pesticides, residues). In: Winter School Advances in Rapeseed-Mustard Research Tehchnology for sustainable production of oilseed. Deceber 15, 2004- Jan, 04, 2005. National Research Centre on Rapeseed-mustard, Bharatpur. 163-174.

Stray, P. and Ghosh, A.K. 1978., Further records of aphid parasitoid (*Hymenoptera: Aphididae*) from Meghalaya, India. *Oriental Insect.* 12(1): 77-80.

Sudhakar, T.R., Pandey, N.D. and Tewari, G.C. 1978. Antifeeding property of some indigenous plants against mustard sawfly, *Athalia proxima* Klug (Hymenoptera: Tenthredinidae). *Indian J. Agric. Sci.,* 48: 16-18.

Sundby, R.A. 1966. A comparative study of efficacy of three predatory insects, *Coccinella septempunctata, Chrysopa carnea* and *Syrpus ribesii* to different temperatures. *Entomophaga,* 11: 395-404.

Suri, S.M. 1986. Development of pest management programme in *Brassica* oilseeds in Kangra velley. Ph.D. Thesis. Y.S. Parmar University Hort. Forestry, Nauni (Solan). 153 p.

Suri, S.M., Dingh, D. and Brar, K.S. 1988. Estimation of losses in yield of brown sarson due to aphid in Kangra valley. I. Effect of crop growth stage and aphid feeding exposure. *J. Insect Sci.,* 1(2): 162-16.

Tandon, H.L.S. 1973. The crop nutrition-pest complex in India. *PANS.* 19: 372-380.

Tripathi, N.L.M. and Sachan, G.C. 1990. Effect of concentrations of insecticides on the growth and development of mustard aphid, *Lipaphis erysimi* (Kalt.). *Indian J. Ent.*, 52 (1): 63-68.

Tripathi, N.L.M. 1985. Effect of sublethal concentration of insecticides on mustard aphid,*Lipaphis erysimi* (Kaltenbach). Ph.D Thesis, GBPUA&T., Pantnagar, 200p.

Tripathi, N.L.M., Sachan, G.C., Verma, S.K. and Pathak, P.K. 1985. Relative efficacy of some synthetic pyrothroids and other insecticides against *Lipaphis erysimi* (Kalt.). *Pesticides.* 19 (12): 26-28.

Tripathi, N.L.M., Sachan, G.C., Verma, S.K. and Pathak, P.K. 1983. Studies on the compatibility of Dithano M-45 with insecticides for the control of rapeseed mustard pests. *Pestology,* 7(7): 5-7.

Tripathi, R.L. and Singh, S.V. 1964. Effect of sowing date variations on the incidence of mustard aphid, *Lipaphis erysimi* Kalt. on *Brassica campestris* var. brown and yellow sarson and *Brassica juncea,* rai *Indian J. Ent.,* 26(2): 251-252.

Tripathi, S.K. 1992. Population buildup of *Lipaphis erysimi* (Kaltenbach) and growth and development of *Bagrada cruciferarum* Kirk. on some crucifers, M. Sc. Thesis, G.B.P.U.A.&T., Pantnagar. P 57.

Tripathi, S.P. and Tripathi, C.P. 1991. Temperature plays a pivotal role in the survival of parasitoid, *Diaeretiella rapae* bioagent of *Lipaphis erysimi* (Kalt.). *J. Adv. 2001.,* 12(1): 68-69.

Udeaan, A.S. and Narang, D.D. 1986. A survey of mustard aphid, *Lipaphis erysimi* (Kalt.) population for resistance to insecticides in Punjab. Proc. 2nd Natn. Symp. Rec. Tri. Aphl. Std., 265.

Udeaan, A.S., Joia, S.S. and Chawla, R.P. 1991. Endosulfan residues in mustard seed at harvest. *J. Insect. Sci.,* 4(1): 99-100.

Veda, O.P. and Shaw, S.S. 1988. Population dynamics of mustard aphid, *Lipaphis erysimi* (Kalt.) and its parasite, *Diaeretus rapae* Mintosh. *Agric. Sci., Digest,* India, 8 (4): 189-190.

Verma, G.C. and Pandey, U.K. 1981. Studies on the effect of *Acorus calamus, Cinifuga foetida* and *Gynandropsis gynandra iextracts* against insect pests of cruciferous vegetables; painted bug, *Bagrada cruciferarum* Kirk. (Hemiptera: Pentatomidae). *Indian J. Agric. Sci.,* 52(4): 263-265.

Verma, R.K. 1987. Preference of *Athalia proxima* Klug to some *Brassica* cultivars and its mating and pupation behaviour, M.Sc. Thesis. G.B.P.U.A.&T., Pantnagar. P. 75.

Yadava, T.P. and Bhola, A.L. 1977. Achievements 1970-77. Oilseed section, Department of Plant Breeding, Haryana Agriculture University, Hisar, India, pp. 19-24.

Entomology : Novel Approaches, 2007
Eds.: *P.C. Jain and M.C. Bhargava,*(Page 425-451)
New India Publishing Agency, New Delhi
E-mail: newindiapublishingagency@gmail.com
web: www.bookfactoryindia.com

CHAPTER 21

ADVANCES IN MANAGEMENT OF STORED GRAIN PESTS

M.C. Bhargava, R.K. Choudhary and P.C. Jain
S.K.N. College of Agriculture, Rajasthan Agricultural University, Jobner

INTRODUCTION

In developing countries, cereals and legumes are the principal source of daily energy and protein intake. The major economic loss caused by grain infesting insects in not always the actual material they consume, but also the amount contaminated by them and their excreta which makes food unfit for human consumption. Today million of tones of the food grains are damaged by the ravages of the pests that not only cause quantitative but also qualitative changes in food grains. Stored product insects depress the food value of a product, introduce pathogens, parasites or toxins or make it less palatable. Insect infestation in high in the tropics and may lead 100 per cent damage in severe cases.

In order to cope with the requirement of the increasing population and food problem, at present as well as for the decades to come, the world will have to raise agricultural production very substantially, both by increasing the production and by reducing losses on account of post harvest spoilage especially of cereals and legumes which are the primary sources of food for men kind. This is a major problems during agricultural production, food transfer and processing. Million tonnes of grain worth several hundred corers of rupees every years are either damaged or loss for want of knowledge of scientific methods of saving the foodgrains during storage. In India, post-harvest losses caused by the unscientific storage, rodents, insects, micro-organisms, moisture etc. account for about 10 per cent of total foodgrains (Anonymous, 1971). A world survey by F.A.O. indicated about 5 per cent loss of cereals, in storage annually (Herford, 1952).

In India, out of the total production, about 70% is retained and stored by farmers for consumption, seed, feed and payment of wedges. Only about 30% is marketable surplus, which is handled by traders and government agencies. Comparatively higher quantity of grains stored by farmers and traders is not on sound basis and the extent of loss is on the high side. Whereas, the storage losses by government and semi government organizations like FCI, CWC, SWC etc., which are on scientific line, are minimum but the total storage capacity is very small (3-5% of total production). At rural level, grains are stored mostly from 3 months to 2 years, depending upon the needs.

Safe storage of Agricultural produce/grains is a problem in almost all the tropical countries because of high humidity and warm climate. Storage of Agricultural produce particularly cereals, pulses and oilseeds for a long duration is associated with the losses in quality, quantity, viability and monetary values. The main bio-deteriorating agents are rodents, insects, birds, mites and microorganisms. The microorganism can produce toxins, which are carcinogenous. Insect occupy a key position among these bio-agents. About 500 species of insects have been associated with stored grain products. These are mostly coleopterans (beetles and weevils) and a few species of lepidopterons (moths), psocopterans (psocids or booklice), thysanurens (bristle-tails), dictyopterans (cockroaches) and hymenopterans (bees, wasps and ants). Nearly 100 species of insects pests of stored products cause economic losses (Sinha, 1965; Sinha and Watters, 1985 and Prakash and Rao, 1995).

In order to develop economic and effective control measures for storage pests, detailed and accurate knowledge of their bio-ecology is essential under variable macro and micro ecological conditions, which would be helpful in the possible prediction of population levels and study of the various mortality factors regulating pest abundance. This will be helpful in developing and effective management strategy. These aspects need more intensive investigations, as abiotic factors such as temperature, relative humidity and moisture percentage of stored products play a vital role in the pest infestation.

Solar heat is frequently used in most parts of our country for drying the seeds without impairing their germination and sometimes to minimize the insect infestation in various types of stored products. The scope of harnessing the solar energy for disinfesting the stored products has ample potential in the country; some work on using bulky and costly contraptions for the purpose has been done. However, more work, especially on smaller scale, using simpler techniques for trapping solar energy for post-harvest protection, needs careful investigation (Anonymous, 1979).

One of the eco-friendly and economic approaches, to keep the stored food grains free from insect attack, would be using the plant products as grain protectants. There are encouraging reports on the use of certain indigenous plant products as grain protectant (Jotwani and Sircar, 1965; Saramma and Verma, 1971; Chander and Ahmed, 1983; Jacob and Sheila, 1990, Abdallah *et al.*, 2001; Hassan, 2001 and Bhargava and Meena, 2002).

IMPORTANT INSECT PESTS OF STORED PRODUCTS

The storage pests are categorized two types namely, the primary and secondary storage pests.

Primary storage pests : That can damage sound grains are called as primary storage pests e.g. rice weevil, lesser grain borer, khapra beetle, pulse beetle, Angoumois grain moth, rice moth, Indian meal moth, almond/fig moth etc.

Secondary storage pest : That damage broken or already damaged grains e.g. rust red flour beetle, flat grain beetle, saw-toothed grain beetle etc.

Table 1 Common insect pests of stored commodities

S.No.	Order/family	Scientific name	Common name
Coleoptera			
1.	Anobiidae	*Lasioderma serricorne* F. *Stegobium paniceum* L.	Cigarette beetle Durg store beetle
2.	Bruchidae	*Callosobruchus maculatus* F.	Pulse beetle
		C. analis	
		C. chinensis	
		Acanthoscelides obtectus Sag.	Bean beetle
		Bruchus pisorum L.	Pea beetle
		Caryedon serratus Oilv.	Tamarind beetle
3.	Bostrychidae	*Rhizopertha dominica* F.	Lesser grain borer
4.	Curculionidae	*Sitophilus granarius* L.	Granary weevil
		S. oryzae	Rice weevil
		S. zea maize Motsch	Maize weevil
5.	Cucujidae	*Cryptolestes acutangulus* Gyle.	Acute angled fungus beetle
		Laemophloeus minutus Oliv.	Flat grain beetle
		Oryzaephilus surinamensis L.	Saw toothed grain beetle
6.	Dermestidae	*Trogoderma granarium* Eversts	Khapra beetle
		Anthremus scrophulariae L.	Carpet beetle
		Attagemus piceus Oliv.	Black carpet beetle
7.	Tenebrionidae	*Tenebrio molitor* L.	Yellow meal worm
		Tribolium castaneum Herbst	Rust red flour beetle
		Tribolium confusum Jacq.	Confused flour beetle
		Tribolium destructor Uythen.	Large flour beetle
8.	Ptinidae	*Gibbium psylloides* Fab.	Hump spider beetle
Lepidoptera			
9.	Gelechiidae	*Gnorimoschama opreculella* Zell.	Potato tuber moth
		Pectinophora gossypiella Saownd	Pink ball worm
		Sitotroga cerealella Oliv.	Angoumois grain moth
10.	Galleriidae	*Aphomia gularis* Zell.	Stored nut moth
		Galleria mellonella L.	Wax moth
11.	Phycitidae	*Cadra cautella* Wlk.	Fig/almond moth

Contd. ...

Table 1 Contd. ...

		Ephestia elutella Hb.	Tobacco moth
		Plodia interpunctella Hb.	Indian meal moth
12.	Pyralidae	*Pyralis furinalis* L.	Grain snout moth
		Corcyra cephalonica Staint	Rice moth
13.	Tineidae	*Nemapogon granella* L.	European grain moth
		Tinea pellionalla L.	Case making cloth moth
		Tineola bisselliella Hum.	Webbing cloth moth
		Trichophaga tapetzella L.	Carpet moth

COMMON INSECT PESTS OF STORAGE

Khapra beetle, *Trogoderma granarium* Everts (Coleoptra : Dermestidae)

The native place of khapra beetle is India but now the pest is distributed to Europe, Australia, North America and other countries of Indian sub-continent. Adults (2-3 mm) oval shaped having no distinction of head, thorax and abdomen when observed dorsally. Antennae are clubed shaped. Larvae are straw coloured and hairy with dark brown hairy bands on each body segment and tuff forming tail at posterior end. It is a major pest of stored grains, reported from India (Cotton, 1952) and is very serious pest of wheat, but also found feeding on bajra, jowar, rice, gram, maize and pulses in storage. Only larval stage is destructive, adults being harmless. The larva start feeding on the germ portion of the grain and feed deep into them. This cause great economic loss both in quality and quantity. The germination of seed is also affected.

Female lay eggs 5-6 days after breeding. A single female lays about 100-120 eggs on grain surface or crevices. At 35°C, incubation period is 4-6 days but it can vary depending upon the temperature and humidity. The larvae is brownish white in colour, body covered with bundles of long reddish brown movable and erectile hairs on the posterior segments and forming a sort of tail in the posterior end. First stage larvae feeds on broken grains and debris resulting from the feeding of older larvae as it can not attack the whole grain. Larval period extends up to 20-25 days at 30°C whereas, pupal life is for 4-8 days. It is highly resistant to starvation. Under abnormal conditions, larva can survive without food for few years.

Rice weevil, *Sitophilus orgzae* Linn. (Coleoptra : Curculionidae)

Rice weevil is one of the most destructive primary pest of stored cereals such as rice, wheat, maize, oat, barley. Adults have characteristic snout or beak and elbow shaped antennae. Both adults as well as larvae cause damage. The female lay eggs on the grains by making a depression with the help of their mandibles. After an egg has been laid, the hole is sealed with a gelatinous secretion. Larvae of pest feed inside the seed and pupate there. On emergence; the adult weevil cut its way out of the grain. Initially seed is not killed but the vigour of the seedling is affected adversely. In case of heavy infestation seed is reduced to flour. Infestation of the pest takes place from field to store. The weevil destroy more than what they eat.

Female starts laying eggs after 4-5 days of emergence as adults and 300-400 eggs are laid by each individual. Generally only one egg is laid on one grain but two or more eggs can also be there. Eggs are translucent and white in colour. It measures 0.7mm long and 0.3mm broad. Female weevil makes a slit like opening with her mandibles and rostrum in soft part of the grain to lay egg in the hole and plugs it with secreted material from the ovipositor. Under optimum conditions eggs hatches in about 4-5 days but may takes 6-9 days also. The young tiny grub bores in to grain kernel. Grubs are white, curved, translucent, yellow or brown head and biting jaws. As soon as it emerges from egg, it starts feeding on the starchy materials of the seeds till it becomes fully-grown and leaves behind only an intact pericarp shell filled with frass. The grub stage lasts for 19-34 days and then pupates to a non feeding pupal stage after passing 2-3 days as prepupa. Pupal period lasts for 1 week but sometimes it is more depending on temperature and then adult emerges out after cutting grain. The adult weevil emerge out of grain is at ones ready for breeding. Grin with emergence hole becomes quite hollow and can float in water. At 70% RH and 28°±1°C, complete life cycle is completed within a month.

Lesser grain bores, *Rhizopertha dominica* (Fab.) (Coleoptra : Bostrychidae)

Like khapra beetle this pest is also native of Indian sub-continent but now it is reported from rest of the world. In adult the head downwards to the extent that it is almost hidden in a dorsal view. Antennae large, loose, three segmented and club shaped. Adults as well as larvae feed on the seed and reduce them to mere shells with many irregular messy waste flour spots. Among the various stored grain insect pests, it causes considerable damage to stored grain (Campbell and Singh, 1976). It has a long history of worldwide occurrence feeding on a variety of stored products like cereals, pulses, groundnut, kernels other edible substances and grocery stores because of this it is considered to be a one of the important pest of stored grains. It not only causes post-harvest losses in terms of quantity but also affect the quality through depletion of specific nutrients and contamination with uric acid and quinons.

Female lays 300-500 eggs in its total lifetime. Eggs are laid either on the seed surface near the embryo end which is soft and easy for young larvae to penetrate of interstices of the grain or in other parts of the stores like on cracks and crevices, begs, walls etc. Eggs are pear shaped and white but later on changes to pinkish opaque. The larvae hatches from eggs bore straight in to the grain immediately or feed on farinaceous material in the store for some period and then penetrate in to the grain in first or second instar. Incubation period is 4-7 days. Newly hatched larva is active and campodeiform. Larva undergo five moults. Full grown larvae is dirty white with light brown head and curved abdomen covered with tiny hairs. Larvae lasts for 35-40 days. With fall in temperature it becomes less destructive. A single kernel of wheat provides food for 5-6 larva. Profuse powdery substance is the characteristics of its damage. Pupal period is for 7-8 days. It completes life cycles normally in 6-8 weeks most destructive stages are larva and adult.

Pulse beetle, *Callosobruchus* spp. (Coleoptra: Bruchidae)

The pulse beetle is found all over the world but it is most common in Asia, North America and Europe. Adults are black, dark, 5-6mm in size and elytra do not covers the last abdominal segment and is called pygnidium. Antennae serrate type. All pulses and beans are attacked by this pest during July-October. The damage is caused by the grubs, which feed on the entire grain contents leaving only the shell behind. The attack of these beetles start in the field, from where the infestation is carried over to the storage. Severe losses due to this pest have been reported in several tropical and subtropical countries (Caswell, 1968) while in India, Gujar and Yadav (1978) recorded 30.2 to 55.7 per cent loss in seed weight and 17.0 to 53.5 per cent loss in protein content . In case of severe infestation cent per cent damage is caused by the pest (Pruthi and Singh, 1950).

Female lays about 80-100 eggs per individual attached to the substratum by means of transparent glue like substance. Freshly laid eggs are creamy but become white later on. Eggs period is for 3-7 days. Legg less larva is cylindrical, fleshy and with wrinkled body. It is white or creamy yellow in colour and remains inside the seed in curved position. Mouthparts are dark brown in colour. Newly hatched larva enters in the seed through the seed coat. Larval stages develop inside the whole pulse seed after hatching from egg. Seed germination is spoiled even during developing stages of beetle. Adults come out of the seed after pushing out a circular lid prepared by prepupal stage of larva. Generally one hole is there in one seed but it depends on the size of seed also.

Rust red flour beetle, *Tribolium castaneum* Herbst (Coleoptra : Tenebrionidae)

The rust red flour beetle is a worldwide pest of milled products, animal feed and processed foods. The pest attacks the already damaged seeds. Both the adults and larva cause damage. Adults are flat brown, 5-6mm in length and reddish brown in colour. Head, thorax and abdomen are distinct and antennae are club shaped. The rust red flour beetle is secondary seed feeder. Adults and larvae feed on seeds already damaged during harvesting, threshing, processing or by primary seed feeders. The larvae are negatively phototactic and are always found hidden in the feed. The embryo portion of seed is preferred for feeding. In severe infestation the flour turns grayish and moldy, and has a pungent disagreeable odour making it unfit for human consumption. The greatest damage is done during hot and humid monsoon season.

Individual female lays 400-450 eggs in its whole life. Eggs are sticky and are laid on the grain/debris of grains. They are small, cylindrical in shape, rounded at both ends and of wheatish colour. Incubation period is 5-10 days. Larvae are very active, cylindrical and pupate after 3-4 weeks. Pupation takes place on the grain surface, which lasts for 6-8 days. Adult has 4-5 months longevity and feeds throughout the life.

Cigarette beetle, *Lasioderma serricorne* F. (Coleoptera : Anobiidae)

It is a cosmopolitan insect but prefers warm environment. It feeds on large number of food varying from spices, chocolate, cocoa and tobacco leaves. Adult is round, small, oval, reddish brown beetle of about 2.5mm in length. Head is projected downwards with concealed antennae. Adult can fly freely. After a day of emergence, female starts laying eggs on grain or muslin cloths. Eggs are laid by female in folds and crevices of food material. Eggs are ovoid, elliptical, whitish becoming opaque and dull in colour just before hatching. A single female lays about 110 eggs, which hatches in 5-6 days and larval stage lasts for 25 days followed by pupal period of 5-7 days. Newly hatched larva are less than one mm in length and covered with fine hairs. Head of larva is yellowish, body semi transparent and whitish. Making small galleries in the host causes damage. After 25-30 days of larval life, it constructs a smooth lined cell in which it rests. Larvae lay quiescent in curved position within the cell, which gradually took pupal form. Newly formed pupa is glossy white but changes to reddish brown after a few days. Female adult is slightly bigger than males. Insect prefers high temperature.

Drug store beetle, *Stegobium paniceum* (Coleoptera : Anobiidae)

It is distributed in tropical, sub-tropical and temperate regions. It is primary pest of turmeric, ginger, pepper, coriander seed, cumin seed etc. Adults and grubs feed/destroy the seeds. Adults are short lived, light brown in colour, smaller than Cigarette beetle. Female can lay 50-80 eggs per individual in the powdered material, which hatched within 8-10 days. Larval life is for 4-5 weeks followed by 6-10 of pupal period. Minimum temperature for development is 17°C and development proceed fastest (7-5 times increase over 28 days) at 25-26°C (Howe, 1965). High humidity is favored.

Angoumois grain moth, *Sitotroga cerealella* (Oliver) (Lepidoptra : Gelechiidae)

The pest is of worldwide occurrence. Cereals like rice, wheat, maize, sorghum, barley, oats and millets are severely damaged. Adults are small pale brown with a wing span of 13-19mm and a length of 6-9mm; wings narrow with wide fringes; for wings buff-coloured, trapezoidal, and bear a few black spots; hind wings buff-coloured, and narrow sharply at apices; labial palps sickle shaped, curving upwards with last segment needle-like and commonly black in colour.

The pest ranks second after the rice weevil in economic importance. The losses caused by this pest to different cereals have been estimated ranging from 6 to 20 per cent (Anonymous, 1979) and in stored paddy from 3 to 12 per cent. Eggs are laid on seeds before harvest (when grains are in field at 'milky stage') and also in store. Larvae feed seeds from inside as it grows it extend the hole, which partly gets filled with the excreta, usually; about 30-50 per cent of the contents are consumed. Finally seeds are hollowed out with flap or trap door. With infestation the grains give out an unpleasant smell and presents an unhealthy appearance, each grain being covered with scales from moths. In bulk, infestation is confined to upper 30 cm depth only. Damaged endosperm results in reduced seedling vigor.

Each female is capable of laying nearly 120-350 eggs. Eggs are laid singly or in small clusters on the surface of the grain, which look white at early stage but changes to bright red later on. Small larvae hatch out after nearly one week and enter inside the grain through joints of the husk. Larvae feed inside the grain (endosperm) for about 3-4 weeks and then it pupates. Full grown larvae are nearly 5mm long, white in colour with yellowish head. Larva spines a silken cocoon inside the grain and changes into reddish brown pupa. Before pupation larvae cuts a small circular opening on the husk which is covered by silken cover to come out. Pupal stage is for 7 days. Development from egg to adult takes about 4 weeks but the developmental period is extended under poor environmental conditions. The growth of single and mixed population of *S. cerealella* and S. *zeamais* was studied on maize (Ayertey, 1979)

Almond moth, *Ephestia cautella* (Wlk.) (Lepidoptera : Phycitidae)

Almond moth is widely spread insect of the tropics and sub-tropics. It is a serious pest of figs, rough rice. It also attacks wheat, barley, sorghum, soybean and other oilseeds etc. Adult moth has a dark band on the forewing and in three fourth of the *Corcyra cephalonica.* It has grayish body, while at rest, its forepart is elevated giving a distinct slope to the wings. Female lay about 200-250eggs per individual, indiscriminately scattered in stores, in cracks, on grains or other surfaces. Eggs hatch within 3-4 days and measures <1mm in size, while larvae starts feeding on grains debris. Full grown larvae are 6-12mm and lasts for 4-5 weeks. Larvae spine silk profusely and at maturity they form small silken tubes among the food particles or in the grain wherein they remain lodged. Pupal life is for 7-10 days. Moths are generally more in rainy and humid season.

Rice moth, *Corcyra cephalonica* Stainton (Lepidoptera : Pyralidae)

It is widely distributed in all rice growing areas of the world and serious pest of rice and jowar. It flourishes well in humid climates and also attacks wheat, sorghum, barley, maize, millets, soybean and oilseeds. Larva is mainly responsible for damage. Young larva feeds on the broken grain or on the grains, which are already damaged by other insects. As the larvae grow old it becomes capable of feeding on entire grain. Besides polluting the grains with large quantities of frass and silken cocoons. Webbing together the food grains into large lumps occur. Presence of such webbing leads to serious public discrimination. Adults are 10-12 mm in length, readily distinguished by closed straight beak formed by palpi pointed forward, and antenna are not crossed over wings while insect is at rest. Female is larger than males and adult life is short 7-10 days. Head has tufts of hairs with grayish forewings and creamy hind-wings. Females lay about 200 eggs which are small, oval white and are laid anywhere on bags, walls or on grains. Young larva is creamy white with prominent light brownish yellow head. Larva webs silken shelter soon gets densly covered with broken grain and frass. Larval life is for 4-5 weeks after that it pupates inside the silken cocoon for 8-10 days.

Indian meal moth, *Plodia interpunctella* (Hubner) (Lepidoptera : Pyralidae)

The pest is reported all around the world. Moths have markings on its forewings, which are reddish brown with a copper luster on the distal portion and whitish gray on the proximal end. Pest cause damage to barley, wheat, maize, oats, rice, sorghum, groundnut, dried fruits, chocolate and beans. Only the larva of pest causes damage. Extensive damage may be caused by the active, crawling caterpillar, which may completely web over the surface of a heap of grains with silken thread. Larva feed inside seed and leaves, silken thread behind, thus making a silken web in the seeds, which is contaminated with frass.

Up to 300 eggs are laid by the female, but the precise number depends on the food taken by the female during larval life. Development from egg to adult at 30°C/70% RH, take about 25 days. The life cycle is completed by the fact that the development of a proportion of larvae is, with strains of the moth, and under certain conditions of temperature, prolonged by a pre-pupal resting stage or diapause. At this time metabolism, and therefore respiration, is very low and normal rate of application of control chemical, particularly fumigants, may be ineffective days (Allotey, 1982; Allotey, 1984 and Allotey and Goswami, 1990).

Grain mite, *Acarus siro* Linn. (Acarina : Acaridae)

It is important economic pest, eats only germ portion and imparts taint to the cereals and cereal products known as mistiness. It is very small just visible to eye particularly when moving with translucent shiny roundish body. The colour is pale straw to dark reddish brown. Female lays approximately 100 eggs. Eggs hatch into 6 legged larvae which moult into nymphs (hypopi). There are 1-3 instar and each active stage is followed by an inactive hypopi stage. Such inert hypopus is a 'cyst like stage' resistant to starvation and desiccation under optimum conditions egg to adult takes 9-12 days at 23°C and RH 70%. It is a very important pest in Himalayan region, in plains being mainly confined to post-monsoon period and quite common in pit storage and costal areas.

Losses caused by storage insects

The losses during post harvest operations are net losses and final, which can not be compensated, whereas growth loss of crops partially or more, compensated by increasing yield for the surviving plants. Further, the concepts and attitude of consumers towards the food grains and grain products is changed with the advancement of knowledge and so called availability of food material. Prior to the how know of ailments caused due to spoiled food in storage, consumer could swallow food containing insects and their fragments. But these days the changes has focused attention on the production of insect free foods and resulted in broad cooperative programmes to prevent insect infestation in raw commodity and final product.

Losses in storage are of multiple characters, including loss in weight, quality, nutritive and market values. Each of these types of losses may have different significance, which varies with people, time, place and with existing loss assessment methods and techniques. The losses are not limited to technical and physical factors only. The socio-economic realities do play an important role particularly in the developing countries. A precise, comprehensive and uniform method for expressing storage losses is yet not available. The process of determining weight changes has not been given due consideration for all the variants *viz*., moisture from atmosphere, moisture through insect metabolism, heating, moisture migration and presence of insect contaminants and socio-economic factors.

Losses due to storage insect infestation in food grains are as follows:

I. Quantitative or weight losses
II. Qualitative losses
III. Losses of seed viability
IV. Damage to storage containers

Quantitative/weight losses

Government of India committee for estimates of losses indicated that when insect damage was quantified, per cent losses in various food grains were 3 per cent in wheat, 2 per cent in rice, 2 per cent in jowar, 3 per cent in bajra, 5 per cent in millets, 5 per cent in gram and 5 per cent in pulses, and on as average it was 2.55 per cent. Similarly, Department of Agriculture, Government of Gujarat conducted survey in 1966 and reported 8.7 per cent losses in bajra and 9.84 per cent in maize. Srivastava *et al.* (1971) reported 0.2 to 3.6 per cent loss in weight due to insect damage in well managed cooperative seed stores. Further, Srivastava *et al.* (1973) found 9.7 per cent weight loss due to insect in a farm level storage practice in wheat and kernel damage up to 30 per cent. Girish and Jain (1974) recorded 0.06 to 9.7 per cent weight loss due to feeding of insects during 6 months of wheat storage. In Punjab, Doharey *et al.* (1975) conducted a preliminary survey in some selected grain markets and observed 0.03 to 4.02 per cent storage losses due to insects alone in 10 months of storage. Lal (1984) also studied the quantitative and qualitative losses in wheat due to storage insects at farm level in Madhya Pradesh and reported only 0.06 per cent insect damage in metal bin and 1.17 per cent damage in improved Kachchi Kothi after 12 months of storage in a Save Grain Campaign adopted village, while in non adopted villages the minimum weight loss was 2.7 per cent in Banda (out door) and maximum 11.2 per cent in Palai after 9 months of storage.

Qualitative losses

Qualitative losses caused by storage insects are categorized in four main heads as follows:

a) Loss of nutritive and caloric value
b) Chemical changes in grain content
c) Contamination of grain with pathogenic and toxicogenic microbes
d) Increase in moisture content

Loss of nutritive and caloric value

When storage insects feed and develop inside/on the grains also caused bio-chemical degradation. Bio-chemical changes in grains take places such as changes in carbohydrates, proteins, fats and vitamins.

Contamination of grain with pathogenic and toxicogenic microbes

Contamination of grains with pathogenic and toxicogenic microbes by the storage insects is found to cause loss of palatability, bad odour, allergy to human beings and food poisoning.

Increase in grain moisture content

Infestation of storage insects also enhanced the grain moisture during the course of their multiplication. Grain moisture of stored wheat increased from 14.6-14.8 to 17.6-23.0 per cent due to damage by *S. granarius* (Agrawal *et al.*, 1957). Insect infestation in stored rice is found to increase its moisture content and allow moulds to grow (Kauraw and Prakash, 1980). *S. oryzae* multiplication enhanced grain moisture from 14.0 to 23.6 per cent in milled rice (CR-1014) during 9 months of storage under controlled conditions. In these grains fungi also developed luxuriously and completely spoiled (Prakash *et al.*, 1987).

Losses of seed viability

Retention of seed viability during storage is essential to increase crop productivity. Insects attack to the seed, first damage the germ, which infact is much softer than rest of the kernels and as a result its germination is lost (Prakash and Rao, 1985). In paddy storage, seed viability deteriorates individually as well as collectively by the interactions of insects, fungi, grain moisture and temperature (Prakesh, 1983, 1994, Prakash ands Kauraw, 1982a b). But it is difficult to estimate the viability losses due to insect damage alone, where visible damage symptoms are not clear (Prakash and Rao, 1983). Grain boring insects like *S. cerealella* and *R. dominica* are also known to cause invisible loss of viability due to attack of their first instar larvae to the germ of the seed and sudden death without further development. This kind of seed damage can not be seen as there is no insect emergence hole in the seed and its weight is almost remain intact and thus difficult to separate out at the time of seed soaking in water before sowing (Prakash and Rao, 1995).

Germination of stored seed tends to decrease with the increase of insect infestation. Mookerjee (1969) reported 6.9 per cent loss of seed viability due to insects in stored paddy for 6 months. Similarly, 66 per cent germination of paddy seed was recorded when insect infestation was 16.61 per cent (Khare, 1972). Prakash and Kauraw (1982b) found losses of viability ranging from 3.67 to 41 per cent in paddy seed sample possessing 1.43 to 14.05 per cent infestation of storage insect during 3-9 months of storage under warm and humid climate in Orissa. In a study, Narsimhan *et al.* (1985) reported 20-28 per cent loss of paddy seed viability due to infestation of *S. cerealella* over 6-9 months of paddy storage in the top position of bamboo weaven containers and 8-10 per cent due to *R. dominica* in jute bags during 6 months of storage.

Un-accountable loss of paddy seed (Ratan) viability due to *S. cerealella* has also been recorded by Behera *et al.* (1990). According to them the seeds having no visible symptoms/emergence holes recorded 11.97 per cent loss in germination, whereas, total loss of viability was 26.54 per cent at visible grain damage of 14.57 per cent. Therefore, it was also suggested to consider this invisible loss of viability while certifying the seed.

Damage to storage containers

R. dominica is a notorious pest, which has the capacity to destroy wooden based storage containers and even polyethylene lined in the storage and also almost all local containers made of material other than cement and iron can be destroyed by the grain borer. This insect damage all these structures and materials by making the holes in them, which help other insects like *S. oryzae* and *Tribolium* spp to infest/penetrate through these holes inside the structures. Considerable damage to wooden framing in cargo ships was recorded by *R. dominica* (Purthi and Singh, 1950, Hall and McFarlane, 1961)

Sources of insect infestation

There are many sources that leads to infestation in storage *viz.,* left over seed from bins/stores or spilled seeds under the stacks, cleanings/sweepings of storage structure/ processing plants, old infested stores, old gunny bags, insect infested transport vehicles, storing insect infested seeds with fresh stock or *vice-versa,* entry of insects from neighboring stores and carry over of field infestation

Prevention of storage insect pests in storage

Prompt harvest, drying and through cleaning of the seed help to prevent entry of insects into the seed stores. Good sanitation in the processing plant and harvest machinery is crucial, since seed pests often breed in small remnants, deposits of old seeds in inaccessible places *viz.*, conveyer belts, screens, lift, hoppers etc. Hard to reach spots should be disinfected by fumigation. Destroy (burn or burry deep) the sweepings, clear trash, litter from out side the store/processing hall and remove the spilled seeds from under the stacks. Remove all the leftover seeds from store/bin; sweep down the walls, ceilings, rills, ledges and floor etc. Make necessary repairs of cracks and crevices. Use new gunny bags for fresh harvest. However, if old bags are to be used dip them in 0.1% malathion 50 EC. Drying, followed by cooling and ventilation of bulk stored seed, as well as temperature monitoring is important for long term storage. Use proper dried, disinfected dunnage. Bagged seeds must be kept at distance from walls to allow inspection, fumigation and for avoiding the seed to absorb moisture. To disinfect the stores, spray 0.5% malathion 50 EC (1:100) on floor, walls and ceiling or fumigate with aluminum phosphoide @ 21-30g or 10.0 liter EDCT/1000 feet^3. Spray inside and outside surfaces of the stores with one of the following insecticides to kill any insect that has remained in the store or insect that may crawl across these areas to damage the stored seed, 0.1% malathion 50 EC (1 part insecticide + 500 part water) or 0.1% fenvalerate 20 EC (1 part insecticide + 200 part water) or 0.1% cypermathrin 25 EC (1 part insecticide + 250 part water)

Management of storage insect pests
Use of physical factors

Temperature and Relative Humidity : Temperature and moisture are the two most important components of the environment and the rate of metabolism, growth, development, reproduction, general behaviour and distribution of insect pests are largely controlled by it. According to Mallis (1960) the rate of reproduction of insect pest was directly proportion to temperature and the moisture content of the grains in which they are propagating. Rose (1965) has found temperature, humidity, wind and moisture as some of the important factors for development of the insects. Fields (1992) proposed lethal, sub-optimal and optimal temperature ranges for many stored product insects; 25-33°C is optimal for growth and reproduction, while 13-25°C or 33-35°C are sub-optimal ranges at which insects are able to complete their development and produce offsprings and lastly at < 13 or > 35°C insects eventually die. Changes in metabolic rate caused by fluctuations of temperature have a direct bearing on two important aspects of insect life cycle; one is locomotion for performing crucial activities such as mating and feeding and the other is growth represented by development and metamorphosis. Thus temperature mediated physiological actions regulate population dynamic. Similarly, the effect of humidity on the development of storage insect pests is almost intimately associated as temperature and operates indirectly through the moisture content of grains. Temperature ranging from 20°C to 40°C accelerate the development of insects but above 42°C and below 15°C retards reproduction and development, while prolonged exposure to temperature above 45°C and below 10°C may kill the insects. It has been reported that all the immature stages of *Sitophilus* *oryzae, Rhizopertha dominica* and *Sitotroga cerealella* are killed when exposed to temperature of 80°C, 70°C and 60°C,respectively for about 4, 6 and 11 minutes, respectively. Although heating of grains to 50°C is lethal to many stored insect pests, it is not advisable as the grains lose their viability at such temperatures.

Solar energy : Utilization of solar energy has long been practiced as one of the effective method for minimizing pest infestation in stored grains (Dean, 1911 and Thompson, 1915). The use of heat energy was first practiced in Australia to disinfest wheat by exposing the grains for at least 3 times at a temperature range of 60–66°C (Winterbottom, 1922). Mookherjee *et al.* (1968) observed a solar heat temperature range of 45-55°C to eliminate the infestation of *C. maculatus* in moong, pea and gram. Dortbudak (1975) reported that exposing containers of lentil and black gram infested with *C. chinensis* to sunshine for 4 hrs (from 9.00 to 13.00) inhibited the rate of pest multiplication and reduced grain loss. Saxena *et al.* (1992) observed the effect of exposing 1,2 and 3 days old pupae of *T. granarium T. castaneum* and *C. chinensis* to temperatures of 30°, 40° and 45°C for 24, 48 and 72 h. Observation on adult emergence and their progeny revealed that, although all temperatures were effective in causing sterility, exposure at 45°C was particularly effective in suppressing adult emergence of *T. castaneum* and *C. chinensis*; adult emergence was totally prevented by 48 h and 72 h exposures at 45°C. Sharma *et al.* (1997) exposed the pupae of *R. dominica* to temperatures of 40° and 45°C for 24-72 hours and reported that the adult emergence was totally suppressed at 45°C for 72 hours exposure. Heating at 45°C completely prevented hatching in 1-day old eggs of *R. dominica* but longer heating was needed for older eggs. Singh (2000) evaluated the effect of solar energy *vis-à-vis* colour of the storage bag by studying the biology of *C. chinensis* on green gram. He reported that both black and blue colours were effective in harnessing the solar energy for *C. chinensis* control; 16 h exposure was enough to cause egg mortality while 24 h exposure for grub mortality. Yadav *et al.* (2002) conducted experiment to harness solar energy and to evaluate the effect of different colour of polythene storage bags (black, blue, gram and yellow) for the management of *R. dominica* in stored wheat. They reported that the mortality of test insect increased while seed germination decreased with the increase in exposure duration. Maximum infestation was eliminated by exposing the infected grains to solar radiation for 24 h in black polythene bag. Chander (2003) reported that exposure period of 24 hrs to solar energy in black and blue coloured bags was most effective to reduce infestation and adult emergence and to cause adult mortality of *R. dominica*.

Effect of different storage receptacles : Awaknavar *et al.* (1989) used storage structures *viz.,* metal bin, coaltar drum, plywood bin, earthen pot and gunny bag at farm level for storage of pulses against *C. chinensis* and stated that the performance of metal bin and coaltar drum was better compared to other structures with reference to insect infestation with a loss in weight of 3.2, 4.2, 4.8, 6.3 and 5.0 per cent in metal bin, coaltar drum, gunny bag, plywood bin and earthen pot, respectively. Senguttuvan *et al.* (1993) studied the effect of different storage containers on the incidence of *Corcyra cephelonica* in stored groundnut both under artificial and natural conditions. They reported significantly low dry mass loss and damaged kernels in gunny bag under artificial condition followed by polythene bag and urea bag. Higher damage was noticed on kernels stored in mud pots.

Under natural condition, low dry mass loss and damaged kernels were observed in gunny bag and urea bag. Chander (2003) noted significantly higher dry mass loss (8.53%) and damaged grains (32.43%) due to *R. dominica* when stored in mud pot in artificial condition while under natural condition, low dry man loss and damaged grains were recorded in urea bag followed by gunny bag, polythene bag, mud pot and cloth bag. Meena and Bhargava (2003a) studied the effect of different storage containers on the incidence of *C. cephalonica* in stored groundnut under both artificial and natural conditions and found significant differences in damage on kernels stored in different containers under artificial and natural conditions. Significantly higher dry mass loss (5.29%) and damage kernels (29.24%) was noticed in mud pot stored kernels. The number of adults varied from 2.33 in gunny bag to 4.00 in cloth bag. Under natural condition, low dry mass loss and damaged kernels were noticed in gunny bag followed by urea bag, cloth bag, mud pot and polythene bag.

Moisture: - Grains stored around 10 per cent moisture content escape from the attack of insects. However, khapra beetle can develop even below this moisture content. Over the moisture range of about 12-16 per cent, the rate of seed deterioration increases as the moisture content increases. The losses could be raped due to mould growth and/or due to heating.

Oxygen availability: The reduction in oxygen level in the grain containers is mainly due to the respiration of grains and also by the insects. The reduction in oxygen tension to a critical limit stops further multiplication of a particular species. The khapra beetle, which can thrive even at much lower moisture content, is very susceptible to the reduction in oxygen content. Reduction of O_2 availability can be achieved by addition of CO_2 or N_2.

Grain proofing: Some inert dusts are abrasive and scar the cuticle, causing desiccation and starvation, but the actual efficacy depends upon the particle size (maximum 10 mμ), particle hardness, particle shape (hard crystalline and angular substances are effective), humidity (efficacy of dusts slowed down at 70% RH or more). Some common dusts used for grain proofing are wood ash (1: 4 w/w), lime (burnt lime, hydrated lime, calcium carbonate @ 1:2 w/w), gypsum, calcium chloride (1:65) and other substances like talc, rock phosphate, magnetite, china clay, bentonite, activated pyrophyllite etc. are also used for grain proofing (Kushwa and Sharma, 1967).

Irradiation: With the discovery of X-ray by Roentgent in 1895 there have come a long way to understand the mechanism of radiation in living organism. Various type of ionizing radiation used is: gamma rays, X-rays, ultra violet rays, alpha ray and beta ray. The ã and X-ray have great penetrating power and are most commonly used for insects. Irradiation of food grains with gamma rays is effective against stored grain insect pests as it causes mortality as well as sterility in insects depending on the dose and time of exposure. The infrared rays and microwaves are also used in raising the temperature of store that disinfect the grains. The irradiation is measured either in *rad* (1 rad=100 erg/g or 10^{-2} Jule/kg of irradiate object) or in *Gray* {(Gy) (1Gy=100rad)}

In insect control the lethal and sterilizing effect are produced by ã-ray mainly from Co^{60}. It is possible to employ ã-rays in doses lethal to the population or in doses, which sterilize the irradiated insect. In general with increasing dosage of radiation the mortality of exposed insect increases and with further increase in dose the mortality takes place in a shorter time. A dose between 200-500 krads brings about complete mortality of stored grain. The egg and larval stages of insect are more susceptible to irradiation as compared to pupal and stage.

Effect of radiation on insects: When an insect is exposed to ionizing radiation, it being about many harmful effect *viz.,* effect at cellular level (increased membrane permeability, lysosomes ruptured, alteration in DNA and RNA and death of intercellular structures like mitochondria), effect on embryonic development (metaphase suffered and death of embryo), morphological abnormalities (deformalities of head, body segments, wings and other appendages and difficulties in moulting), effect on reproduction (brings sterility, reduction in fecundity and fertility and damage to oocytes or nurse cells), histopathological effects (damage to somatic tissues of mid gut, loss of peritrophic membrane and affect production of enzymes and hormones) and biochemical effects (inhibition of DNA synthesis in developing eggs, the alkylating azirinyl compounds react with DNA, RNA and protein and inhibition of malic acid, proteins and free amino acid in eggs) as described by Sharma and Single (2001).

Table 2 Radiation induced sterilization of insects

Insects	Stage	Radiation	Sterilizing dose (rad)	Reference
Acanthoscelides obsoletus	A	γ	20000	Pesson, 1963
Attagenus grandis	A	γ	15000	Davich and Lindquist, 1962
Callosobruchus chinensis	A	γ	42000	Quraishi and Metin, 1963
Ephestia cautella	P	γ	>16000	Pendlebury *et al.*. 1962
Ephestia kuhniella	P♂	γ	60000	Pelerents, 1963
Lasioderma serricorne	A	γ	16000	Pendlebury *et al.*. 1962
Plodia interpunctella	-	γ	>16000-50000	Pendlebury *et al.*. 1962
Rhizopertha dominica	A	γ	12000	Pendlebury *et al.*. 1962
Sitophilus granarius	A	X	7500-10000	Hoover *et al.*, 1963
Tribolium castaneum	A	X	6000	Erdman, 1962
		γ	16000	Crook, 1962
Trogoderma granarium	A, P♂	γ	15000	Harvey, 1963 and Kanso, 1962

A= Adult
P= Pupa

Sound waves: Very intense sound may carry sufficient energy to kill the insect. The death occurs due to heating caused by absorption and degradation of the energy. Sound of this intensity is difficult to be transmitted efficiently and as a means of raising the change insect behaviour, which differs remotes possibility in the controlling programme. Many insect cells are in the range of the ultrasonic. This is a sound of a frequency above about 20,000 cycles/second (below threshold to human ear). Infrasound (20 cycles) leads to setting up of resonance among the internal agents causing violent agitation and hemorrhage.

Centrifugal force for certain commodities can be used to away with the insect infestation. For example a mechanical device, called Entoleter at a speed of 2000-3000 rpm destroys insect infestation in milled products. However, this method is costly.

Chemical control measures

Among the present methods of control of stored grain pests chemical control is the most popular and the most effective one. Surface spray of bags as and when the insect pests are visible are done with insecticides such as malathion, primiphos-methyl, DDVP, fenitrothion etc. Malathion 50 EC is sprayed over the bags in the ratio of 1:100 @ 3 l/100m^2. Recently the deltamethrin 2.5 WP @ 3mg a.i./m^2 has been found very effective. Allay charging or spraying of allays between two grain stocks with chemicals like dichlorovos (DDVP) @ 0.05 per cent has been found very effective in preventing the cross infestation.

Dust formulations of the insecticides are sold rarely for use and contain 0.1 to 5.0 per cent a.i. The formulations contain additives, which increased the adhesion to the stored produce. Dust formulations are suitable for mixing with grain and for applying in layers in the stored produce (sandwich method) as well as for surface treatment of individual bag, bag stacks and stores.

Fumigation

Fumigation for pest control is a process of exposing pests or pest affected stocks to the fumes of chemicals at a lethal strength in a gas tight space. An ideal fumigant should be toxic to $<$300-400 g/m^3 to kill all the pests, its pesticidal properties should not be changed under storage period, it should not affect nutritive, cooking, biochemical and other properties of grains and grain products, should not leave odor or any harmful residue after fumigation, should not damage storage container, should not cause fire or explosion and should have high vapor pressure and should have high diffusion properties.

Characteristics of an ideal fumigant

An ideal fumigant should have following characters: (i) Volatile with deep penetration power. (ii) Extremely toxic to all stages of life. (iii) Able to distribute uniformly without settling down. (iv) No adverse reaction with commodities, no effect on nutritive, cooking, biochemical and other properties of grains and its products. (v) Less toxic to plants and mammals. (vi) Least absorption and high diffusion power. (vii) Should have distinct warning odour. (viii) Easy to handle and transportation. (ix) Should not least harmful residue. (x) Non-explosive. (xi) Should not be corrosive to equipments and house hold articles. (xii) No effect on viability of seeds. (xiii) No consumer hazards. (xiv) Easily and cheaply available. (xv) Known antidotes and treatments. (xvi) Least soluble in water. (xvii) Should be chemically stable during storage.

Ethylene dibromide (EDB): EDB is one of the most toxic chemicals to insect damaging stored food grains. The use of EDB, filled in ampoules has been found suitable for fumigation of food grains stored in household structures, holding small quantities, which can be made reasonably airtight. The fumigant is filled in glass ampoules and nozzles are sealed. Individual sealed ampoule is wrapped in cotton, blotting paper and then stitched in cloth bag. When ampoule with ECB liquid is broken and inserted into grain mass, the EDB changes into vapoures.

EDB$^+$: It is a mixture of EDB and carbon tetrachloride in the ratio of 1:8 w/w. It has a greater penetration power and hence can be used for bagged storage under gas proof covers also. It is equally effective in rural storage structures. EDB$^+$ is also available in ampoules in sizes of 11 ml. 22 ml. and 30 ml for small quantities of food grains stored in rural storage structures. Smaller sized ampoules (1-5 ml) are available for disinfestations of even a few kg of grains and pulses.

Durofume: It is a mixture of EDB and Methyl bromide in ratio of 1:1 w/w first prepared by CFTRI, Mysore. The mixture is highly effective for all stages of insects and even against resistant species of stored pests like *Tribolium* and *Trogoderma*. The dosage of durofume is 1-2 lbs/1000 ft^3 with exposure period of 48 hours. If temperature is above 26 ^{0}C the exposure period may be reduced to 36 hours. If temperature is below 18 ^{0}C the mixture may be altered to 1:3. For oilseeds the mixture is in ratio of 1:2 and dosage is 1.5-2.0 lbs/100m^3 with exposure period of 36-48 hours and for milled cereals 1:3 mixture is used @ 2-3 lbs/1000 ft^3 with exposure period of 48-72 hours.

Hydrogen Cyanide or Hydrocyanic Acid, (HCN): It is a volatile, colourless liquid with a bitter almond odour and a Sp. gr, 0.699 at 20^{0}C. The gas has Sp. gr. 0.9483 and is highly inflammable and explosive in mixtures above 5.6 per cent or above 4 lb/1000 cu. ft. In the market two forms of CaCN are available. Cyano gas contains 40 to 50 per cent CaCN and the gas is liberated slowly. The calcyanide dust contains 88% CaCN. For general fumigation the dosage is 1 lb sodium cyanide (98 per cent), 1½ pints H_2SO_4 and 3 pints water per 1000 to 1500 cu.ft.

Carbon bisulphide, (CS_2): It is a colourless liquid with a disagreeable odour due to traces of hydrogen sulphide. Sp. gr. 1.263 at 20^{0}C and vapour pressure 360 mm Hg at 25^{0}C. It is slightly soluble in water but soluble in organic solvents, highly inflammable and explosive in mixtures above 1 per cent in air. It is used for the control of stored product pest in go downs and pests of clothes, dress and draperies. The dosage is 10 lb of CS_2 per 1000 cu.ft. of space or 2 to 3 gallons per 1000 bushels of grain for 36 to 72 hr. The fumigant is highly phytotoxic.

Carbon tetra chloride, (CCl_4): A colourless liquid with Sp. gr. 1.595 at 20 ^{0}C and vapour pressure 114.5 mm Hg at 20^{0}C. It is almost insoluble in water but soluble in organic solvents. The gas has a Sp. Gr. of 5.31, is not inflammable or explosive and possesses pungent chloroform like odour. It is comparatively less toxid to insects and slow acting.

Ethylene dichloride or 1,2-Dichloroethane, (CH_2ClCH_2Cl): A colourless liquid with Sp. gr. 1.257 at 20^{0}C, vapour pressure 78 mm Hg at 25^{0}C. It is slow acting and death of insect takes place from one to three days after exposure. The trade names available are Brocide and Destruxol.

EDCT or Ethylene dichlorideCarbon tetrachloride mixture: Is a general mixture with 3 of ethylene dichloride and 1 of CCl_4 by volume. Such a mixture, either as gas or as liquid is free from fire hazard. The dosage is 10 to 18 lb/1000 cu.ft. in airtight stores for 24 hr.

Phosphine: It is hydrogen phosphide PH_3 and is widely used to fumigate grain, flour and cereals in godowns the commercial product is available as tablets or pellets of aluminum phosphide and ammonium carbamate. The CO_2 and ammonia released by ammonium carbamate prevents spontaneous ingnition of phosphine. A tablet weights 3g and measures 20 mm in diameter and 5 mm thick. Gives rise to 1 gm of phosphine.

ECO_2FUME: It is a product of M/S Cytee industries in USA. It is a non-flammable, premixed cylinderized mixture of phosphine and CO_2, which provide highly effective fumigation in both sealed and unsealed storage facilities. ECO_2FUME fumigant gas is a gaseous mixture of 2% phosphine by weight in 98% CO_2. It can be used in a wide variety of applications to attain insect free and dust free products, it doesn't affect real value of seeds. It can also be used to fumigate grains, dried fruits, cocoa and coffee beans, nuts, tobacco, cut flours and others items and it is well suited to fumigate structures and buildings.

PROFUME (Sulfuryl fluoride): It is a product of M/S Dow Agro-Sciences and is effective as methyl bromide on all postembryonic life stages of insects. It can be used for short or long (i.e. 4 -12 hr) fumigation to meet the time constraints. Unlike methyl bromide it does not deplete the ozone. Fumigated commodities less sorbs it. The gas is packed in cylinders and is non-flammable non-corrosive, odorless and has a high vapour pressure and quickly penetrate the fumigated area.

Acrylonitrile: The compound is highly inflammable, been never used alone and always mixed with another suitable materials such as CCl_4 to reduce the risk of fire/explosion. The mixing ratio of acrylonitril and CCl_4 is 34:66. It is highly toxic to human being. It does not affect the viability of seeds and not leave any kind of residues in the fumigated commodity.

Grain-O-Cide: It is a mixture of CS_2 and CCl_4 in 4:1 ration. The efficacy of mixture is directly proportional to its CS_2 content. The CCl_4 is mixed to reduce the fire hazard of CS_2. It has proved to satisfactory under Indian condition, and also acts as a useful soil fumigant. It is liposoluble and hence a nerve poison. It is used at a dosage of 30-40 kg/1000 quintal grains with exposure period of 7 days.

Precautions

(i) Foodstuffs are removed from the building to be fumigated.

(ii) Edibles to be fumigated should be properly safety to human lives, operation by trained personals with protective clothing.

(iii) Enclosure must be airtight as possible. Avoid windy or cold weather.

(iv) Use accurate dosages, proper exposure and ventilation for live plant fumigation.

(v) Optimum temperature for fumigation of live plant is 14-27 °C.

(vi) Optimum temperature for fumigation of stored grains/receptacles is 21-30°C.

(vii) Commodities containing fat should not be fumigated with EDB/MB.

(viii) Study literature of fumigant before the operation.

(ix) Keep first aid ready and wash hands after fumigation.

(x) Destroy empty containers.

(xi) Put "Poison Gas Danger" at the entry of building till proper ventilation.

(xii) Call a physician immediately in accidental poisoning.

(xiii) No smoke within 15 meter of fumigated area.

Poisoning symptoms and antidotes

If poisoning symptoms are noticed, carry out first aid and call a physician.

EDB: Liquid on skin may cause blisters. Inhalation cause pulmonary lesions and may lead to pulmonary oedema. Respiratory and circulatory depression may occur. Necrosis of the liver and kidney damage has been reported. The condition resembles chloroform poisoning and poisoning by other ethylene compounds, diseases of the central nervous system. Delayed pulmonary oedema liver and kidney damage will have to be distinguished from other causes for these conditions. Oxygen therapy is desired. If the poison has been swallowed, gastric lavage and sealing purgative may be used. Respiratory stimulants like nikethamine or caffeine are useful. After acute nervous symptoms subside, adrenergic blocking agents like dibenzyl betachloroethyl amine hydrochloride have been recommended to prevent kidney and liver damage.

Alluminium phosphide (Phosphine): The poisoning symptoms are feeling of fatigue, buzzing of ears, nausea, pressure in the chest, intestinal pain and diarrhea and vomiting. In medium to severe poisoning dry cough, choking attacks, intense thirst, reeling gait, severe pain in limbs, enlarged pupils and rapid on set of stupor occurs. Oxygen therapy is desired besides administration or a cardiac tonic and a drug to stimulate blood circulation. Under some circumstances blood transfusions or infusion of isotonic saline or glucose into the blood system are indicated. In the case of pulmonary oedema a hypertonic solution of glucose be injected.

EDCT: Signs and symptoms are produced mainly by depression in central nervous system which resembling the effect of chloroform. There may be respiratory depression and delayed pulmonary oedema. Kidney and liver damage may occur. If swallowed nausea followed by vomiting and abdominal pain, headache, dizziness, confusion drowsiness and occasionally convulsions may occur. Visual disturbance may precede central nervous system depression. Oxygen therapy is desirable. If the poison has been swallowed gastric lavage and saline purgative may be used. Dimercaprol (British anti-levicide) may be tried against ethylene dichloride. Respiratory stimulants like nikethamine or caffeine are useful. Steps should be taken to prevent lever damage. After acute nervous symptoms subside adrenergic blocking agents like dibenzyl batachloroethyl amine hydrochloride has been recommended to prevent kidney and liver damage.

Admixture of substances of plant origin

Traditionally many of plants are used against store grain pests. Although promising results have often been achieved in laboratory tests with plant material (botanicals), the efficacy under practical storage conditions varied as lot. Most methods have a limited effect but it provides satisfactory protection of the stored product when they are applied properly. There are many plant species which have toxic properties against insect pests of storage. The most commonly used plants are neem, karanj, undi, castor, lemongrass, mustard, coconut, sesamum, groundnut, mandhi, dharak, aak, datura etc. The plant part extract, oil and plant part powder are used to admixture with stored commodities to protected them from insects.

Jotwani and Sircar (1965) found that the powdered neem seed kernel when mixed with wheat seed at 1 to 2 parts per 100 parts of seed, protects wheat seed against, *S. oryzae, R. dominica* and *T. granarium* for at least about 269, 321 and 379 days, respectively. The germination of treated seed was not impaired. Chander and Ahmed (1983) studied the efficacy of some indigenous plant products against infestation of *S. oryzae, T. granuraim* and *C. chinensis* and reported that powder of *Acrous calamus* rhizome was the most promising against the three pest exhibiting a fairly high protection of grains even at the lowest concentration of 1 per cent. Jacob and Sheila (1990) reported that neem oil at the rate of 1.0 ml/100g seeds caused cent per cent mortality of adult beetle of *C. chinensis* and inhibited oviposition and hence prevented development of subsequent generations. Neem oil at 0.5 ml/100g seeds did not inhibit egg laying but prevented the development of progenies in the next generation. Khaire *et al.* (1993) studied the effect of 10 vegetable oils *viz.*, sunflower, castor, mustard, safflower, groundnut, palm, sesamum, neem, karanj and corn (maize) on ovipositional preference and egg hatching of *C. chinensis* on pigeonpea. They observed that all the treatments had an adverse effect on ovipositional preference compared with the control. Neem oil at a concentration of 1 per cent was superior to all the other treatments. Sharma and Bhargava (2001) tested different concentrations (0.25, 0.50, 1.0, 2.0, 3.0 and 5.0%) of plant extracts *viz.*, neem, karanj, undi and lemongrass against eggs of *C. cephalonica*. They found that the egg hatch inhibition increased with the increase in concentration of plant extracts. The LC_{50} values for egg mortality were 0.7079 ± 0.0581, 0.08954 ± 0.0566, 1.1041 ± 0.0597 and 1.4125 ± 0.0547 in lemongrass oil, neem extract, karanj extract and undi extract, respectively. Bhargava and Meena (2002) tested six vegetable oils *viz.,* castor (*Ricinus communis*), mustard (*Brassica juncea* L.), groundnut (*Arachis hypogaea*), sesamum (*Sesamum indicum*), coconut (*Cocos nucifera*) and sunflower (*Helianthus annus*) against *C. chinensis* in cowpea and found that castor oil @ 1.0ml/100g seeds was the most effective in inhibiting the oviposition (26.6 eggs/female) as against 79.4 eggs/female in untreated seeds. At 1.0ml/100g seeds, castor oil caused maximum reduction in egg viability (61.7%) followed by mustard oil (56.7%). The longevity of male and female adults decreased with the increase in doses of oil. The treatment of castor oil @ 1.0ml/100g seeds caused maximum reduction in adult emergence in F_1 generation (85.0%) followed by mustard oil (83.7%), groundnut oil (73.3%). No adverse effect of tested oils was observed on the germination of cowpea seeds upto 150 days of treatment. Meena and Bhargava (2003b) tested thirteen plant extracts/oils @ 0.1, 0.5 and 1.0 ml /100 g broken seeds to evaluate the effect on fecundity, egg viability and adult longevity of *C. cephalonica.* They reported that all the doses of test compounds were found to be effective reducing the fecundity and egg viability. The longevity male and female adults decreased when treated food was fed. The karanj extract was most effective and mehandi oil was least effective in reducing the fecundity, egg viability and longevity of adults.

Host resistance

With the advancement of science a lot of crop varieties are developed that shows varying degree of resistance against different insects pests. The resistance is due to the one or more biochemical, physiological, genetical, histological etc. characters of the plant. Presently there are many varieties of cereals and pulses are available that shows varying degree of resistance to insects.

Now these days there are some genetically modified (transgenic) plants are available that carry genes for insect resistance. The most common example is pea having foreign gene (α -AI-Pv) resistant to pulse beetle and tobacco having á-ai gene resistant to *Tenebrio molitor*.

Table 3 Varieties of some crop resistance to stored grain insect pests

Insect	Crop and varieties	Reference
S. oryzae	**Sorghum** : CSV-15, Lakadi, **Wheat :** K-65, K-136, K-143	Reddy *et al.* (2002) Tiwari and Sharma (2002)
R. dominica	**Wheat**: K-157, K-162, K-140 **Rice**: Cauvery, Bala, Jaya, IR-8 **Barley:** RS 6, RD 2627, K 719	Tiwari and Sharma (2002) Chander (2003)
C. chinensis	**Cowpea** : K-98, FTC-27, GC-2 **Pea** : RPG-3, BR-12, Rachna	Singh and Sharma (2003) Singh and Sharma (2003)
T. granarium	**Wheat:** Kalyansona, UP-301, WH-291	Jakhar (2005)

Biological control measures

Biological control of insect pests in storage godawns is not very encouraging. However, some of the hymenopterous parasites such as *Anisopteromalus calandrae* on *Sitophilus* spp., *Lariophagus distinguendus* on *Rhizopertha* and *Sitophilus* have been recorded parasitising various lepidopterous pests of grain stores. *Bacillus thuringiensis* has also been tried on various lepidopterous pests of grain stores. It is the first microorganism to be approved for use on stored grain pests. B.t. is compatible with most protectant, fumigants and seed fungicides. However, its wide use is yet to take up even in western countries.

Insect growth regulators: - Alarm, dispersal, aggregation and sex attractant pheromones are known in a number of storage pest but studies have been most valuable with sex pheromones in almond moth, Indian meal moth and with aggregation pheromone of lesser grain borer. There are three possible application of pheromones to control storage pest *viz.,* trapping with sex attractant, using aggregation pheromone with insect pathogen or chemosterilant and perfusing the warehouses atmosphere with pheromone to disrupt normal courtship behaviour of males. The compound 4,8-dimethyldecanal (or DMD) was found effective aggregation pheromone against *Tribolium* and compound sitophinone and grain volatiles (valeraldehyde, maltol, vanillin) were used in pheromonal control of *S. oryzae* (Phillips *et al.*, 1993). The adults of *T. castaneum* were also traped by synthetic lures *i.e.* PT 6 Allure and Trappit (Mullen, 1992) under laboratory conditions. The 1-pentadecene and 1,6-pentadecadiene are also found effective aggregation pheromone for *Tribolium* (Amaud *et al.*, 2002). Ismoratou and Coworkers (2003) reported that synthesis of (4R)-methynnonan-1-ol and (4R, 8RS)-dimethyldeconal from (5)-(+)-3,7-dimethyl 1-1,6-octadiene were effective when 4R used as sex pheromone against yellow meal worm and 4R, 8RS used as aggregation pheromone against rust red flour beetle.

Juvenile hormone analogue : Ambika and Abraham (1982) reported that topical application of methoprene (JHA) at a dose of 1.0 and 0.5μ/4th instar larva gave 100 per cent mortality by 42 and 90 days after treatment, respectively. Such larval treatment resulted in the formation of supernumerary larva which died within 7-9 days. Bhargava and Devaraj Urs (1993) reported the ovicidal properties of IGRs which hydroprene, methoprene and fenoxycarb against *C. cephalonica*. Sharma and Bhargava (2000) observed that larval duration, length and width of larva, pupal weight and pupal period increased due to fenoxycarb treatment. Reduction in adult emergence increased with the increase in concentration of fenoxycarb.

Table 4 List of predators/parasites attacked on stored pests

S. No.	Predator/Parasites	Attacking stage	Insect pest
Parasites			
1.	*Aximopsis javensis* Girault	Larva	*Araecerus fasiciculatus* (De Geer)
2.	*Anisopteromalus calandrae* (Howard)	Larva	*A. fasiciculatus, L. serricorne, R. dominica, Callosobruchus* spp. and *S. oryzae*
3.	*Antrocephalus aethiopicus* Masi	Larva	*C. cephalonica*
4.	*Antrocephalus mahensis* Masi	Larva	*C. cephalonica*
5.	*Aximopsis tephrosiae* Girault	Larva	*Araecerus fasiciculatus* (De Geer)
6.	*Cephalonomia gallicola* Ashmeed	Larva	*S. paniceum, L. serricorne*
7.	*Cephalonomia tarsalis* Ashmeed	Larva	*Oryzaephilus surinamensis*
8.	*Choetospila elegans* Westwood	Larva	*L. serricorne, R. dominica, Callosobruchus* spp., *S. granarius, S. oryzae* and *Trogoderma granarium*
9.	*Dimachus discolor* Walk	Larva	*S. paniceum*
10.	*Entedon* spp.	Larva	*S. paniceum*
11.	*Eurytoma tylodermatis* Ashmeed	Larva	*Araecerus fasiciculatus* (De Geer)
12.	*Habrobracon hebetor*	Larva	*Plodia interpunctella*
13.	*Habrobracon brevicornis* (Wesmael)	Larva	*C. cephalonic, Ephestia cautella*
14.	*Lariophagus distinguendus* (Forester)	Larva	*S. paniceum, L. serricorne, R. dominica, S. granarius* and *S. oryzae*
15.	*Pteromalus cerealellae* (Ashmeed)	Larva	*Sitotroga cerealella*

Contd. ...

Table 4 Contd. ...

16.	*Pteromalus pyrophilus* Kollar	Larva	*Sitotroga cerealella*
17.	*Ptinobius texanus* Crawford	Larva	*Araecerus fasiciculatus* (De Geer)
18.	*Pyxinia crystalligera*	Larva and Adult	*Dermestes* sp.
19.	*Trichogramma pretiosum* (Riley)	Egg	*Ephestia cautella, Plodia interpunctella*
20.	*Trichogramma cacoeciae* Marschal	Egg	*Sitotroga cerealella* and *Corcyra cephalonica*
21.	*Venturia canescens*	Larva	*Ephestia cautella, Plodia interpunctella*
22.	*Zatrophus incertus* (Ashmeed) L.	Larva	*S. oryzae*
Predatory mites			
23.	*Acarophenax* sp.	Egg	*Dermestes* sp.
24.	*Cheyletus* sp.	Egg and young larva	*L. serricorne, R. dominica*
25.	*Pyemotes ventricosus*	Egg	*Callosobruchus* spp., *T. granarium, Tribolium castaneum*
26.	*Pyemotes tritici*	Egg and young larva	*L. serricorne,*
27.	*Tenebroides mauritanicus*	All stages	*R. dominica*

References

Abdallah, S.A.; Badawy, H.M.A.; Barakat, A.A and El-Sabaay, T.N. 2001. Efficacy of certain vegetable oils as wheat grain protectants against the lesser grains borer, *Rhyzopertha dominica* (F.). *Bulletin of faculty of Agriculture*, Cairo University, 52 : 167-182.

Agrawal, N.S. and Christensen, C.M. and Hodson, A.C., 1957. Grain storage fungi associated with the granary weevil. *J. econ. Ent.*, 50: 312-315

Allotey, J. 1982. Competition between *Corcyra cephalonica* (Staint.) and *Ephestia cautella* (Wlk.) and study of radiosensitivity of the immature stages of *Corcyra cephalonic. M. Sc. Thesis, Univ. of Ghana, Legon*

Allotey, J. 1984. Activity pattern of *Corcyra cephalonica* (Staint.) and *Ephestia cautella* (Wlk.) and competition for pupation sites. *Utter Pradesh J. Zool.* 5: 150-155

Allotey, J. and Goswami, L. 1990. Comparative biology of the phycitid moths Plodia interpunctella (Hubn.) and *Ephestia cautella* (Wlk.) on some food media. *Insect Sci. and Its Applic.*, 11:209-215

Amaud, L.; Lognay, G.; Verscheure, M.; Leenaers, L.; Gaspar, C. and Haubrug, E. 2002. Is dimethyldeconal a common aggregation pheromone of *Tribolium* flour beetle. *J. Chem. ecol.* 28 (3) : 523-532.

Ambika, B. and Abraham, C.C. 1982. Effect of juvenile hormone analogue, methoprene (ZR-515) on development of eggs and larvae of *Corcyra cephalonica* Stainton (Lepidoptera : Pyralidae). *Kerala Agril. Res. J.*, 20 : 60-62.

Anonymous, 1971. Report of the committee on post harvest losses of food grain in India. *Min. of Fd. Agric. Govt. of India*. pp. 35.

Anonymous, 1979. Annual Progress Report of Harvest and Post harvest Technology, College of Technology and Agricultural Engineering, Udaipur (Raj.).

Atwal, A. S. and Dhaliwal, G.S. 1971. The rate of increase in number of *Trogoderma granarium* Everts on different varieties of wheat. *Indian J. Ent*. 33 (2) : 172-175.

Awaknavar, J.S.; Krishnamurthy, K.C.; Babu, C.K. and Ramakumar, M.V. 1989. Different storage structures for safe storage of pulses. *Mysore J. agri. Sci.*, 23 : 526- 532.

Ayertey, J.N. 1979. The growth of single and mixed laboratory populations of *Sitophilus zeamais* Motschulsky and *Sitotroga cerealella* (Oliver) on stored maize. *Res. Popul. Eco*., 21: 1-11

Behera, J.; Jagadiswari Rao and Prakash, A., 1991. Un-accountable loss of the paddy seed viability due to Angoumois grain moth, *Sitotroga cerealella* Oilv. *J. Appl. Zool. Res*., 2(1): 32-33

Bhargava, M.C. and Devaraj Urs, KC. 1993. Ovicidal effect of three insect growth regulators on *Corcyra cephalonica*. *Indian J. Pl. Prot*. 21 (2) : 195-197.

Bhargava, M.C. and Meena, B.L. 2002. Efficacy of some vegetable oils against pulse beetle, *Callosobruchus chinensis* (Linn.) on cowpea, *Vigna unguiculata* (L.). *Indian J. Plant Prot*., 30 : 46-50.

Campbell, A. and Sinha, R.N. 1976. Damage of wheat by feeding of some stored beetles. *J. Econ. Ent*., 69 : 11-13.

Caswell, G.H. 1968. The storage of cowpea in the northern state of Nigeria. *Proc. Agric. Soc. Nigeria*, 5 : 4-6.

Chander, H. and Ahmed, S.M. 1983. Potential of some new plant products as grain protectants against insect infestation. *Bull. Grain Tech*., 21 (3) : 179-188.

Chander, R. 2003. Host preference and bio-control studies of *Rhizopertha dominica* (Fab.) on barley and its management. Ph.D. (Thesis) submitted to Rajasthan Agricultural University, Bikaner, pp. 224.

Cotton, R.T. 1952. Insect–pest of stored grain and grain products. pp. 2.

Crook, L.J. 1962. The susceptibility of the red-rust flour beetle *Tribolium castaneum* (Herbst.) to gamma radiation. AERE-3889, *United Kingdom Atomic Energy Authority, Research Crop, Isotope Research Div*, Wantage, Berks, England, pp.-19.

Davich, T.B. and Lindquist, D.A. 1962. Exploratory studies on Gamma radiation for the sterilization of the boll weevil. *J. econ. Ent*. 55 (2) : 164-167.

Dean, G.A. 1911. Heat as a means of controlling mill insects. *J. Econ. Entomol*., 4 : 142-158.

Doharey, R.B.; Srivastava, P.K. and Giris, G.K. 1975. Studies on the assessment of losses in wheat in Punjab. *Bull. Grain Tech*., 13 (13): 159-161

Dortbudak, N. 1975. Studies on the biology, ecology and methods of control of *Bruchus eryii* F. on lentil around Mardin province. *Rew. Appl. Ent*., 65 : 738, 2740.

Erdman, H.E. 1962. X-rays effects on single and mixed species population of *Tribolium confusum* and *Tribolium castaneum* (Coleoptera : Tenebrionidae). HW-SA-2686, *General Electric Co. Hanford Atomic Products Operation*, Richland, Wash, pp. 8.

Fields, P.G. 1992. The control of stored product insects and mites with extreme temperatures. *J. stored Prod. Res*., 28 : 89-118.

Girish, G.K. and Jain, S.K. 1974. Studies on efficacy of neem kernel powder against stored grain pests. *Bull. Grain Tech*., 12: 226-228

Gujar, G.T. and Yadav, T. D. 1978. Feeding of *Callosobruchus maculatus* (Fab.) and *Callosobruchus chinensis* (Linn.) in green gram. *Indian J. Ent*., 40 : 108-112.

Hall, D.W. and McFarlane, J.A. 1961. Post harvest problems with paddy and rice in Bristish oversea Territories. 1961 pp. 28

Harvey, J.M. 1963. Irradiation of fruits and vegetables in a mobile cobalt 60 unit. *Division of Isotopes Development and Division of Biology and Medicine*, AEC, pp 55-57.

Hassan, T.A. 2001. Effect of three plant oils (Sesame, sunflower and castor) against stored grain insects (*Trogoderma granarium* and *Sitophilus granarius*). *University of Aden Journal of Natural and Applied Sciences*., 5 (1) : 103-110.

Herford, G.V.B. 1952. The infestation of stored food stuff by insect. *J. Sci. Food. Agri*., 3 : 1-11.

Hoover, D.L.; Floyd, E.H. and Richardson, H.D. 1963. Effects of 300 kv X-ray radiation on *Sitophilus oryzae*. *J. econ. Ent.* 56 (5) : 584-586.

Ismuratov, G.Yu; Yakovelva, M.P.; Galyautdinova, A.V. and Tolstikov, G.A. 2003. Synthesis of (4R)-methynnonan-1-ol and (4R, 8RS)-dimethyldeconal from (5)-(+)-3,7-dimethyl 1-1,6-octadiene. *Chemistry of Natural Compounds*, 39 (1) : 31-33.

Jacob, S. and Sheila, M.K. 1990. Treatment of green gram seeds with oils against the infestation of the pulse beetle, *Callosobruchus chinensis* L. *Plant Prot. Bull*., 42 : 9-10.

Jakhar, B.L. 2005. Bio-ecology and management of khapra beetle, *Trogoderma granarium* Everts on wheat. Ph. D. thesis, Raj. Agril. Univ., Bikaner, Campus- Jobner.

Jotwani, M.G. and Sircar, P. 1965. Neem seed as protectant against stored grain pests infesting wheat seed. *Indian J. Ent*., 27 : 160-164.

Kansu, A. 1962. Preliminary experiment of the sterilization of the pupae of the khapra beetle by irradiation with Gamma rays. *Z. angew Ent.* 42 (2) : 224-228.

Kauraw, L.P. and Prakash, A. 1980. Influence of storage fungi on the development of lesser grain borer in paddy cultivar CR-1084. *Seed Research* 8(2): 67-70

Khaire, V.M.; Kachare, B.V. and Mote, U.N. 1993. Effect of different vegetable oils on ovipositional preference and egg hatching of *Callosobruchus chinensis* Linn. on pigeonpea seed. *Seed Res*., 21: 128-130.

Khare, B.P., 1972. Insect pests of stored grains and their control in UP. *GBPUAT, Pantnagar, Res. Bull. No. 5* pp. 152.

Kushwaha, K.S. and Sharma, J.C. 1967. Safe storage of grains and other product from insect pests and rates. College of Agriculture, University of Udaipur, Udaipur, Rajasthan, India, pp. 14-23.

Lal, Sone, 1984. Assessment of insect damage in stored wheat at farm level and studies on the impact of introduced new storage technology. *Ph. D. Thesis submitted to Bhopal University, Bhopal,* pp.138

Mallis, A. 1960. Hand book of pest control, Mac-Nair-Dorland Company, New York 3rd Edition. pp. 620.

Meena, B.L. and Bhargava, M.C. 2003a. Effect of different storage containers on the incidence of *Corcyra cephalonica* in stored kernels of groundnut. *Indian J. Appl. Ent*., 17 : 41.43.

Meena, B.L. and Bhargava, M.C. 2003b. Effect of plant products on reproductive potential of *Corcyra cephalomica* Stainton (Lepidoptera : Pyralidae). *Ann. Pl. Protec. Sci*., 11 (2) : 196-200.

Mookerjee, P.B. 1969. Integrated control of storage insects. *Proc. Intern. Sem. On Integrated Pest Control, ESI-IARI*, New Delhi: 97-98

Mookherjee, P.B.; Jotwani, M.G.; Yadav, T.D. and Sircar, P. 1968. Disinfestation of stored seeds by heat treatment. *Indian J. Ent.,* 30 : 197-202.

Mullen, M.A. 1992. Development of a pheromone trap for monitoring *Tribolium castaneum*. *J. Stored Prod. Res.* 28 (4) : 245-249.

Narsimhan, K.S.; Gawal, M.H. and Doharey, R.B., 1985. Impact of containers on seed paddy. *Bull grain Tech*., 23(3): 269-277

Pelerents, C. 1963. Quelques resultants supplementaries Concernant L'influence Des Rayons Gamma Sor Les Chrysalides Et Les Ceufs De La Teigne De La Farine '*Ephestia kuehniella* Z.' *Reu. Agric.* 16 (2) : 305-320.

Pendlebury, J.B.; Jefferies, D.J.; Banham, E.J. and Bull, J.O. 1962. Some effects of Gamma radiation on the lesser grain borer (*Rhizopertha dominica* F.), tropical warehouse moth (*Cadra (Ephestia) cautella* Wlk.), Indian meal moth (*Plodia interpunctella* Hubn.) and the Cigarette beetle (*Lasioderma serricorne* F.). AERE-R-4003, United Kingdom Atomic Energy Authority. Research Groups. Isotope Research Div. Wantage, Berks, England, p. 23.

Pesson, P. 1963. Some experimental data on cobalt 60 radiation doses capable of arresting insect infestation of cereal and flour. *Food Irrad.* 3 (4) :A8-A21.

Phillips, T.W.; Jiang, X.L.; Burkholder, W.E.; Phillips, J.K. and Tran, H.Q. 1993. Behavioral responses to food volatiles by two species of stored product Coleoptera, *Sitophilus oryzae* (Curculionidae) and *Tribolium castaneum* (Tenebrionidae). *J. Chem. ecol.* 19 (1) : 723-734.

Pingale, S.V. 1953. Impregnation of jute bag container with insecticides for protecting stored food stuff. Efficacy of Piperonyl butoxide-pyrethrin. *Indian J. Ent.* 15 (4): 343-351

Pingale, S.V. 1954. Biological control of some stored grain pests by the use of bugs predator, *Amphibolus venator*. *Indian J. Ent.* 16 (4): 300-302

Prakash, A. 1983. Factors effecting losses due to insects pests and their management in rice storage ecosystem. *Proc. Natl. Symp. On "Insect Ecology and Resource management" on 2-4 October 1993 at S.D. (PG) College, Muzaffernagar,* UP, pp. 50-54

Prakash, A. and Jagadiswari Rao, 1983. Insect pest and their management in rice storage in India. *Presented in national symposium on maximizing and stabilization of yields on rainfed rice production system, held at CRRI, Cuttack, India,* February 23-25, pp. 17

Prakash, A. and Jagadiswari Rao, 1985. Losses due to insects in stored rice. *Bull. Grain Technol.*, 3(1): 77-82

Prakash, A. and Jagadiswari Rao, 1995. Insect pest management in stored rice ecosystems. Pp. 709-736. In Stored grain ecosystem. *Marcel Dekker Inc. New York, USA,* pp. 757

Prakash, A. and Kauraw, L.P., 1982a. Problems and quality control in paddy seed. *Seed and Farms,* 8(11): 13-16

Prakash, A. and Kauraw, L.P., 1982b. Studies on insects, mites and fungal infection in rural and farm stored rice in Orrisa. *J. Res. Assam Agric. Univ.* 3(2): 10-16

Prakash, A., 1994. Insect damage in stored rice and their management in India. *Presented in II National Conference of AZRA on Recent Trends in Plants, Animals and Human Pest Management : Impact on Environment at Madras Christian College, Madras on 27-29 December (Abstract) :* 20-25

Prakash, A.; Rao, J.; Pasalu, I.C. and Mathur, K.C., 1987. Rice storage and insect pest management. B.R. Publishing Corporation, New Delhi, pp. 337

Pruthi, H.S. and Singh, M. 1950. Pests of stored grain. *ICAR Misc. Bull.*, 57 : 1-57.

Purthi, H.S. and Singh, M. 1950. Pests of stored grains and their control. *Indian J. Agric. Sci. (Spl. Nomber)* 18(4): 1-52

Quraishi, M.S. and Metin, M. 1963. Radiosensitivity of various stages of *Callosobruchus chinensis* L. In "Radiation and radioisotopes applied to insects of Agricultural importance. Proceeding of symposium, Athens, 22-26 April, Vienna, International Atomic Energy Agency, pp. 479-484.

Rahman, K.A.; Sohl, G.S. and Sapra, A.N. 1945. Studies on stored grain pests in Punjab. VI Biology of Trogoderma granarium Everts. *Indian J. Agric. Sci.* 15: 85-92

Rahman, S, 1942. Storage of wheat in Punjab. *J. Agric. Sci.* 12: 567-587

Reddy, K.P.K.; Reddy, K.D. and Singh, B.U. 2002. Components of resistance in sorghum to the rice weevil, *Sitophilus oryzae*, Antixenosis. *I. J. Ent.*, 64 (3): 262-274.

Rose, D.J.W. 1965. Insect pests and climate, *PANS*., 11 : 169-176.

Saramma, P.U. and Verma, A.N. 1971. Efficacy of some plant products and magnesium carbonate as protectants of wheat seed against attacks of *Trogoderma granarium*. *Bull Grain Tech.* 9 (3) : 207-210.

Saxena, B.P.; Sharma, P.R.; Thappa, R.K. and Tikku, K. 1992. Temperature induced sterilization for control of three stored grain beetles. *J. Stored Prod. Res*., 28 : 67-70.

Senguttuvan, T.; Kareem, A.A. and Rajendran, R. 1993. Effect of different storage containers in storing groundnut against rice moth, *Corcyra cephalonica* Stainton (Galleriidae : Lepidoptera). *Bull. Grain Tech.*, 31 : 129-131.

Sharma, K.C. and Bhargava, M.C. 2000. Effect of juvenile hormone analogue (fenoxycarb) on rice moth, *Corcyra cephalonica* Staint. (Lepidoptera : Pyralidae). *Bull. Grain Tech.* 33 (1) : 59-65.

Sharma, K.C. and Bhargava, M.C. 2001. Ovicidal effect of some growth disrupting compounds on rice moth, *Corcyra cephalonica* Staint. (Lepidoptera : Pyralidae). *Indian J. Appl. Ent*., 15 : 24 –28.

Sharma, P.R.; Thappa, R.K.; Tikku, K.; Chand, D. and Saxena, B.P. 1997. Control of stored product moths and beetles by suboptimum temperatures. *Tropical Sci*., 37 : 28-34.

Sharma, S.S and Singal, S.K. 2001. Role of radiation in management of insect pests of stored products. *Advances in Storage and Seed Entomology,* pp. 190-198.

Singh, S. 2000. Qualitative and quantitative pest status and eco-friendly management of *Callosobruchus chinensis* Linn. on different pulses. Ph.D. (Ag.) Thesis Submitted to Rajasthan Agricultural University, Bikaner.

Singh, S. and Sharma, G. 2003. Preference of *Callosobruchus chinensis* on pea varieties. *I. J. Ent.*, 65 (2): 277-280.

Sinha, R.N. 1965. Insect associated with stored products in Canada. *Canada Insect Pest Rev. Suppl.*, 2:47

Sinha, R.N. and Watters, F.L. 1985. Insect pests of floure mills, grain elevators and feed mills and their control. *Agric. Branch, Agric. Canada Publ.* pp. 290

Srivastava, P.K.; Singh, K.; Tomer, R.P.S.; Girish, G.K. and Krishnamurthy, K. 1971. Storage losses in cooperative seed stores in UP. *Bull Grain Tech.*, 9(3): 177-185

Srivastava, P.K.; Tripathi, B.B.; Girish, G.K. and Krishnamurthy, K. 1973. Code of practice for storage of paddy in villages. *Bull Grain Tech*., 11(3): 206-210

Thompson, R.L. 1915. Sun heat v. weevils. *Rhodesia Agric. J. Salisbury*, 12 : 653-656.

Tiwari, R. and Sharma, V.K. 2002. Resistance to two major stored grain pests in wheat. *I. J. Ent.,* 64 (3): 247-253.

Winterbottom, D.C. 1922. Weevil in wheat and storage of grain in bags. *A record of Australian experience during the war period* (1915-1919). Govt. Printer, Adelaide.

Yadav, R. P.; Bhargava, M.C. and Yadav, J.P. 2002. Studies on the solar radiation as a means of eradicating *Rhizopertha dominica* (Fab.) in wheat. *Indian J. Plant Prot*., 30 : 26-28.

Entomology : Novel Approaches, 2007
Eds.: *P.C. Jain and M.C. Bhargava*,(Page 453-459)
New India Publishing Agency, New Delhi
E-mail: newindiapublishingagency@gmail.com
web: www.bookfactoryindia.com

CHAPTER 22

ROLE OF PREDATORY MITES IN IPM OF PHYTOPHAGOUS MITES

Ashok Sharma and V.K. Agarwal
Department of Entomology, SKN College of Agriculture, Jobner-303 329

INTRODUCTION

Indiscriminate use of acaricides not only causes the development of resistance in phytophagous mites but are also harmful to their natural enemies, therefore, poses a serious problem of pest resurgence. Hence, worldwide attention has been focused to utilize the natural biocontrol agents and minimize the use of chemical pesticides for the control of mite pests in agriculture with a view to develop a suitable IPM strategy. Among natural enemies of phytophagous mites, predatory mites have been found to be most effective and efficient tool in IPM programme because of the fact that many of these are not only abundantly available in nature but are also voracious feeders and have proven ability to suppress the pest mite population below economic injury level. Beside this, predatory mite can locate, run and catch hold of the pest mites even the pest mite is hiding in rolled leaves, under webs or even in the galls or erinium, etc. Some of the predatory mites require very low population of pest mites as prey, for their survival and multiplication, which is an added advantage for potential predation. Some predatory mites such as *Typhlodromus pyri* have good synchronization with the prey mite, *Panonychus ulmi* as, both hibernate and remain active in orchards.

It was Shimer (1868) who recorded for the first time *Hemisarcoptes malus* feeding on apple bark louse, *Lepidosaphes ulmi* (L.) which was later, transferred from eastern to western parts of Canada in 1917 for the control of the same pest and rated as a successful biological control agent. The biggest impact of biological control of mite pests in agriculture has been made by augmenting natural enemies, particularly phytoseiids for the use in orchard and green house system (Hussey and Scopes, 1977). In some agroecosystems, the impact of natural enemies, particularly the contribution of generalist mite predators, can be both dramatic and economical (Mc Murtry, 1983). Studies conducted to record the effect of nutrition on the biology of predatory mite (Zaher *et al*, 1969; Nangia and ChannaBasavanna, 1985; and Sharma *et al*, 1986) have also shown promising results.

There are large numbers of predatory mites potentially feeding on the harmful mites (Gupta, 1985, 1986, 2001; Haq and Ramani, 1992; Tagore and Putatunda, 1996, 1997 & 1998). A comprehensive account of the systematics of predatory mites of India has been given in detail by Gupta (2001). Typical representatives of predatory mites are found in phytoseiidae, stigmaeidae, cheyletidae, cunaxidae, Bdellidae, Anystidae, Erythraeidae,

Eupodidae, Raphignathidae and Tydeidae (Welbourn, 1983; Mc Murthy, 1984). The details of various predatory mites alongwith their prey species are given in Table 1.

Table 1 List of important predator mite species (*highly effective) along with their prey stage, host and locality in India

S. No.	Name of predator	Name of prey	Stage	Host	Locality
Phytoseiidae					
*1.	*Amblyseius ovalis*	*Oligonyehus coffeae*	All Stages	Tea	North-East India
*2.	*A. longispinosus*	*O. oryzae*	All Stages	Paddy	W. Bengal
		Tetranychus neocaledonicus	All Stages	Cucurbits	W. Bengal, Assam, Orissa, M.P.
*3.	*A. tetranychivorus*	*T. ludeni* *Raoiella Indica* *Coccids*	All Stages Eggs, Nymphs Eggs	Beans Arecanut Guava	Karnataka, Kerala, Kerala W. Bengal
*4.	*A. multidentatus*	*Aceria lycopersici* *T. neocaledonicus*	All Stages All Stages	Brinjal Brinjal	cosmopolitan cosmopolitan
*5.	*A. largoensis*	*Brevipalpus deleoni* *O. mangiferus*	All Stages Eggs	Tea Mango	Assam W.Bengal, Bihar
*6.	*A. finlandicus*	*Eotetronychus orientalis*	Nymphs, Eggs	Citrus	cosmopolitan
7.	*A. channabasavanni*	*R. indica*	Larva, Nymphs	Arecanut	Kerala
8.	*A. herbicolus*	*O.coffeae*	All Stages	Tea	South India
9.	*A. delhiensis*	*B. phoenicis*	Eggs	Guava	Punjab
10.	*A. fallacis*	*Schizotetranychus andropogoni*	All Stages	Paddy	W.Bengal, Orrisa
11.	*Indoseiulus ricini*	*T.cinnabarinus*	All Stages	Ornam-ental	Sikkim
12.	*A. alstoniae*	*Petrobia latens*	Nymphs, Adult	Wheat, Barley, Coriander, Mustard	Rajasthan, Punjab

Contd. ...

Table 1 Contd. ...

		E. orientalis	Nymphs, Egg	Citrus	W. Bengal, Punjab
13.	*A. aerealis*	*Paranychus citri*	Eggs	Citrus	Assam
14.	*A. coccineae*	*P. citri* *T. neocaledonicus*	Eggs Eggs, Nymphs	Mulberry Cucurbits	W. Bengal W. Bengal
15.	*Typhlodromus homalii*	*O.coffeae*	Eggs	Tea	Tamilnadu
16.	*T. rickeri*	*Tetranychus sp.*	Eggs	Ornamental Plants	W. Bengal
17.	*Phytoseius (Pennaseius) kapuri*	*Eotetranychus sp.*	Eggs, Nymphs	Ornamental Plant	W. Bengal
Stigmaeidae					
18.	*Agistemus fleschneri*	*A. mangiferae*	All Stages	Mango	Manipur
19.	*A. terminalis*	*A. theae*	All Stages	Tea	North-East India
20.	*A. heterophylla*	*Eotetranychus sp.*	Nymphs	Ornamental Plants	Arunachal Pradesh
21.	*E. coimbatorensis*	*Melanaspis glomerata*	Eggs	Sugarcane	Tamilnadu
22.	*Zetzellia languida*	*P.citri*	Eggs	Mulberry	W. Bengal
Bdellidae					
23.	*Bdella (H.) affinis*	*O. indicus*	Eggs	Arecanut	Andaman & Nicobar Islands
24.	*Bdella sp.*	*O. mangiferus*	Eggs	Mango	W. Bengal
Erythraeidae					
25.	*Leptus sp.*	*T.cinnabarinus*	All Stages	Ornamental Plants	W. Bengal
Tydeidae					
26.	*Pronematus fleschneri*	*E. orientalis*	Eggs	Citrus	Cosmopolitan
27.	*P. elongates*	*B. phoenicis*	Eggs	Guava	Assam
28.	*Tydeus* sp.	*A. litchii*	All Stages	Litchi	W. Bengal
Anystidae					
29.	*Anystis baccarum*	*Tetranychides*	-	-	Jammu & Kashmir

Contd. ...

Table 1Contd. ...

30.	*Tencateia sp. nov.*	*T. neocaledonicus*	All Stages	Betel Vine	W. Bengal
31.	*A. indicus*	*T. urticaria*	All Stages	Peach	Jammu & Kashmir
Cunaxidae					
32.	*Cunaxa setirostris*	*O. mangiferus*	All Stages	Mango	cosmopolitan
33.	*C. womersleyi*	*B. deleoni*	Nymphs	Papaya	W. Bengal
34.	*C. cynodonae*	*Schizotetranychus* sp.	Nymphs	Grass	Andaman & Nicobar Island
Cheyletidae					
35.	*Cheylatogenes ornatus*	*A. mangiferae*	All Stages	Mango	W. Bengal, Bihar, U.P.
		Larvacarus transitans	Nymphs, Adult	Ber	Rajasthan
36.	*C. fortis*	*O. punicae*	All Stages	Pomeg-ranate	Himachal Pradesh

*highly effective predator of mite species

Phytoseiidae: Among the predatory mites which have received maximum attention globally, phytoseiids are perhaps the most important one and various studies have been undertaken for exploration and utilization of these mites as biocontrol agents. These mites are whitish or creamish or reddish or light brownish, fast moving and abundantly available in nature.

The phytoseiids have several advantages over other predatory mites such as abundant availability, high fecundity, good searching ability, good dispersal rate, adoptability to different ecological niches, high degree of prey specificity, etc. Phytoseiids, prey all almost on tetranychids and some tarsonemids beside few tenuipalpids. The use of phytoseiids in biological control has been reviewed by Mc Murthy (1982). Moreover they can also be mass reared in the laboratory and released in the field when required (Mc Murthy and Scriven, 1965; and Koppert, 1980). A model of mass production technique of predatory mite, *Amblyseius, longispinosus* has been developed at Bangalore (Mallik *et al*, 1999). Among 173 species that are so far known in India, 18 species have been found in the field active and feeding upon pytophagous mites, of these, 8 species are major predators of high potential.

Stigmaeidae : The representative of this family are probably next to phytoseiids as far as the predatory efficiency is concerned. These are yellowish or reddish or light brownish, ovoid or elongate in shape, very active and occupy various habitats. *Eryngiopus coimbatorensis* had recently been observed to be an efficient predator of sugarcane scale insect. They are fairly abundant in nature and highly predaceous on eggs, nymphs and adults of phytophagous mites. Out of 25 species, only 6 have shown promise as predators.

Cheyletidae: There mites are free living predators, sluggish in nature and have yellowish or orange coloured body. A total of 15 species have been reported.

Cunaxidae: Cunaxids are very strong, mostly brighly coloured (yellowish/reddish/brownish) with thorny mouthparts. These mites posses strong and long legs and runs very fast. They are known to be active predators of phytophagous mites but in nature their number is limited. However, if they are artificially reared, mass cultured and released, possibly they can prove quite useful bioagents. Only one species has been identified in India having high predatory potentiality.

Bdellidae: Members of this group are fast mover, comparatively delicate and are not likely to be very effective in open cropping system with much fluctuating environment. Some species have been reported from India but no species of high potential has been noticed in field.

Anystidae: These mites are round, long legged, reddish, soft bodied and highly predacious. They start whirling as soon as they are disturbed. Although some species are important predators but practically nothing is known about their efficiency.

Some of the members of Erythraeidae and Tydeidae have also been found feeding on phytophagous mites.

CONSIDERATIONS FOR SUCCESSFUL EXPLOITATION OF PREDATORY MITES

Major aspects which are of significance in IPM of mite pests through predatory mites are:

1. Encouragement of already present predatory mites :
    - Discourage indiscriminate use of broad spectrum pesticides
    - Use only selected and bioeffective pesticides.
    - Laboratory mass culture and inoculative release in field of the beneficial mites from time to time.
    - Maintain an alternate food source during critical time when the pest species are scare.
2. If required, import and introduce predatory mites of the mite pest (augmentation of predatory mites).

   Suppression of phytophagous mites by naturally occuring predatory mites is well recognized but some time these bioagents fail to maintain pests below acceptable label of population levels. In this situation augmentation of host specific mites can provide a solution. Although over 40 species of predatory mites have been listed as important one for biological control, only few have been evaluated in augmentative release in India on crops.
3. IPM programmes based on pesticide resistant strains of predators :
   More recently, success has been achieved in developing pesticide resistant strain of phytoseiid predators for use in intensive managed agroecosystems (Hoy, 1991). This may stablize the mite fauna in certain ecosystem (Roush and Hoy, 1981 a 1981 b).

Case study: Phytoseiid mite, *Metaseiulus occidentalis* is the most important predator of spider mites of deciduous orchards and vineyards in western North America. It acquired resistant to some of insecticides through selection in orchard. To augment these naturally acquired resistance, *M. occidentalis* was artificially selected in the laboratory with permethrin and carbaryl and found that resultant strain retained their resistance to OP insecticides despite the fact that they were not subjected to selection with OP pesticides during the project. This work enabled to produce the predatory mites which could control spider mites even though insecticides had to be used. Further, permanent establishment of the resistant predator strains should allow to reduce the number of pesticidal applications. The successful genetic improvement of these phytoseiids could serve as a model for other taxonomic groups of natural enemies, including parasites and pathogens of pest arthropods, as well as predators.

REFERENCES

Gupta, S.K. 1985. *Hand book on plant mites of India.* Zoological survey of India, Calcutta, pp. 510

Gupta, S.K. 1986. Fauna of India. Acari: Mesostigmata : Family phytoseiidae. Zoological survey of India, Calcutta, pp. 346

Gupta, S.K. 2001. In: Yadav, P.R.; Chauhan, R.; Putatunda, B.N. and Chillar, B.S. (eds.), *Mites, Their Identification and Management.* Dept. of Entomology, HAU, Hissar. pp. 168-181

Haq, M.A. and Ramani, N. 1992. *Man, mites & Environment* (ed.) Anjengo Publication, Calicut, Kerala

Hoy, M.A. 1991. Genetic improvement of phytoseiids : Theory and practice. In : Dusbabek, F. and Bukva, V. (eds.) *Modern Acarology Vol. I.* pp. 175-184

Hussey, N.W., Scopes N.E.A. 1977. In : Ridgway R.L., Vinson S.B. (eds.), *Biological control by augmentation of natural enemies.* Plenum Press, New York, pp. 349-377

Koppert, P.C. 1980. *Proc. Int. Symp. IOBC/WPRS on Integrated Control in Agriculture and Forestry*, Vienna, 1979, pp. 484

Mallik, B.; Vaidya, R. and Kumar, M.H. 1999. Mass production of the predator *Amblyseins longispinosus* (Acari: Phytoseiidae)- A model. *J. Acarol.* 15 (1&2) : 15-17

Mc Murthy, J.M. 1982. In: Hoy M.A. (ed.), *Recent advances in knowledge of the phytoseiidae*. Division of Agricultural Sciences, University of California. Publication 3284, pp. 23.48

Mc Murthy, J.M. 1983. In: Hoy, M.A., Cunningham, G.L. and Khuston, L. (eds.), *Biological control of pest by mites.* Division of Agricultural Sciences, University of California, Berkeley, Special Publication 3304, pp. 21-26

Mc. Murthy, J.A. and Scriven, G.T. 1965. Insect production of phytoseiid mites. *J. econ. Ent.*, 58: 282-284

Nangia, N. and ChannaBasavanna, G.P. 1985. Feeding potential of *Amblyseius tetranychivorus* (Acari: Phytoseiidae), an indigenous predator on selected Tetranychid and Tenuipalpid mites. *Indian J. Acar.* 10(1&2) :75-81

Roush, R.T. and Hoy, M.A. 1981a. Genetic improvement of *Metaselulus occidentalis*: Selection with methomyl, dimethoate, and carbaryl and genetic analysis of carbaryl resistance. *J. econ. Ent.* 74:138-141

Roush, R.T. and Hoy, M.A. 1981b. Laboratory, glasshouse and field studies of artificially selected carbaryl resistance in *Metaselulus occidentalis. J. Econ.Ent.* 74:142-147

Sharma, N.K. and Sadana, G.L. 1986. Development of *Amblyseius finlandicus* (Acari: Phytoseiidae) on its prey, *Eutetranychus* (Acari: Tetranychidae) *J. Acarol.* 11 (1&2): 51-54

Shimer, H. 1868. *Trans Am. Entomol. Soc.* 1: 361-74

Tagore, A. and Putatunda, B.N. 1996. A review of phytoseiidae mites of India. *Res Bull (Sci.) Punjab Univ.,* 46:43-56

Tagore, A. and Putatunda, B.N. 1997. Distribution of phytoseiid mites (Acari: Mesostigmata: Phytoseiidae). *India. Agril. Rev.* 18(1): 13-26

Tagore, A. and Putatunda, B.N. 1998. A. review of predatory- prey relationship of phytoseiid mites controlling destructive mites in India. *Ad. BioSci.* 17(11):1-8

Welbourn, W.C. 1983. In : Hoy, M.A. Cunninughaun, G.L. and Knutson, L (eds.), *Biological control of pests by mites.* Division of Agricultural Sciences, University of California, Berkeley, Special Publication 3304, pp. 103-104

Zaher, M.A.; Wafa, A.K. and Shehata, K.K. 1969. Life history of the predatory mite, *Pytseiulus plumifer* and the effect of nutrition on its biology (Acarina: Phytoseiidae). *Entomol. Exp. Appl.* 12(4): 383-388 [illegible] Hort. [illegible] Trans. [illegible] Zool. [illegible]

[illegible], A. and [illegible], B. [illegible] 19[illegible]. [illegible] rates of [illegible] (Acari) [illegible] 10[illegible]–56.

[illegible], A. and [illegible], P. N. 19[illegible]. Distribution of phytoseiid mites [illegible] [illegible] *India* [illegible] vol. 26[illegible]

[illegible], A. and [illegible], P. N. 1988. A review of [illegible] prey [illegible] of phytoseiid mites [illegible] [illegible] 19[illegible]

[illegible], W. G. 19[illegible]. In: [illegible] and [illegible] (Eds.) [illegible] [illegible] pp. [illegible]

[illegible] and [illegible] [illegible] of the predatory mite [illegible] [illegible] [illegible] 38[illegible]

Entomology : Novel Approaches, 2007
Eds.: *P.C. Jain and M.C. Bhargava*,(Page 461-476)
New India Publishing Agency, New Delhi
E-mail: newindiapublishingagency@gmail.com
web: www.bookfactoryindia.com

CHAPTER 23

INTEGRATED PEST MANAGEMENT OF PHYTOPHAGOUS MITES

M.S. Dhooria and Manmeet B. Bhullar
Department of Entomology, Punjab Agricultural University, Ludhiana – 141 004

INTRODUCTION

Phytophagous mites have long been considered as potentially serious pests of a wide range of food and fibre crops, and ornamentals. They are covered under subclass Acari of class Arachnida and phylum Arthropoda. Mites unlike insects are 8-legged creatures and lack compound eyes, antennae and wings. Plant feeding mites measure from 200µ to 600µ. Majority of the plant feeding mites belong to suborder Prostigmata, and phytophagous mites belong to family Tetranychidae (spider mites); Eriophyidae (bud or gall mites) ; Tenuipalpidae (false spider mites); Tarsonemidae (broad mites) and Tuckerellidae, in the decreasing order of their importance. Before one can develop integrated management of phytophagous mites in different agri-horticultural ecosystems, knowledge about the following fundamentals of phytophagous mite management principles is a big must.

REASONS OF MITE OUTBREAKS

Prior to the commercial introduction of broad-spectrum insecticides like DDT and other organochlorine insecticides, spider mites did not commercially cause wide-spread damage under natural or semi-natural environments which were little influenced by man. This might be due to the natural processes of prey-predator relationship in undisturbed situations. Contrastingly in situations where broad-spectrum pesticides are commonly used, the natural enemies are more adversely affected than the spider mites, and an outbreak of the latter occurs. Besides becoming generally intensified with the introduction of modern insecticides, the spider mite problem in some cases has subsequently become even more serious due to development of resistance against these materials.

NATURE AND EXTENT OF DAMAGE

Spider mites damage protective leaf surfaces - the stomata, palisade and spongy parenchyma, and some may even inject toxic substances into the leaf and interferes with the vital processes of affected plants. Annual losses due to spider mite injury alone in USA during 1951-60 period have been reported to be as high as 10% in strawberries, 7% in mint and apples,

6% in hops, 3% in lime and snap beans, 2.5% in citrus, and 5% in floriculture crops in greenhouses (LeClerg, 1965). From India, 6-12% tea crop have been reported to be lost annually because of injury by spider mite, *Oligonychus coffeae* (Nietner) alone. Spider mite, *Oligonychus oryzae,* have been reported to reduce rice yield in India by about 25%. Although crop loss estimates are not available for many crops but the damage must be considerable because of strong host-plant relationships of mites.

Spider mites and worm mites which overwinter as adults are often readily distinguishable in colour or structure both from actively breeding summer forms. In worm mites, the winter forms are known as deuterogynes. In spider mite species that live on deciduous hosts, mites which are to undergo hibernation often possess a characteristic reddish or orange colouration, and are more deeply pigmented than normal summer females

REPRODUCTION

Reproduction in most spider mites involve arrhenotokous parthenogenesis with unmated females producing all male progeny and mated females producing both male and female offspring. Eggs are deposited on the foliage and upon hatching the developing mites pass through larva, protochrysalis, protonymph, deutochrysalis, deutonymph, teliochrysalis, and adult. The proto-, deuto- and teliochrysalis stages are inactive. Males emerge before females, guard female teliochrysalis and copulate with the adult female immediately upon her emergence. Both intrinsic (e.g., mite strain, colony density, age of population, females mating status), and extrinsic (e.g. temperature, relative humidity, light, level of predation, intra- and interspecific competition, host plant, pesticides etc.) influence the reproductive parameters affecting r_m. Factors affecting fecundity and survivorship will also affect r_m. The reproductive parameters are strongly influenced by temperature, relative humidity and host plant. Mori (1961) reported the temperature preference for two-spotted spider mite, *Tetranychus urticae,* to be 13-35°C. An understanding of the relationship between temperature and development rate is a prerequisite to the prediction of real population growth rates and to the modeling of predator-prey interaction.

OVERWINTERING

Spider mites overwinter either as diapausing adult females or as actively reproducing populations. The diapause condition occurs where winters are more harsh and overwintering host plants are less available. Diapausing females do not feed or oviposit ; they pass the winter in sheltered locations with high levels of mortality occurring among mites overwintering in soil. Non-diapausing populations often occur in glasshouses and in areas with moderate winter temperatures. Chapman *et al.,* (1952) reported an increase of 23.1% in numbers of apples on trees on which spider mites were controlled. Lienk *et al.,* (1956) reported that if mite feeding was uncontrolled for one season then it resulted in 75% bloom reduction in the following year in Cortland variety and 34% in Red Delicious variety of apple. The harvest yield reductions amounted to 64.8% and 35.8% respectively. A knowledge of the diapause status of overwintering populations in an area is important in understanding the dynamics of colonization of crops in the spring.

EFFECT OF ENVIRONMENT

The most severe mite problems are typically experienced under hot and dry conditions. Crop canopy temperatures in drought-stressed plants can be significantly higher than in non-stressed plants, leading to more rapid development strictly attributable to the effects of temperature. Moisture stress may also affect mite population growth independently of temperature through its effect on the physiology of the host plant.

Relative humidity can affect reproductive rate through its direct effects on mite development, survival and reproduction. *T. urticae* develops faster and produces more eggs under conditions of low (25-30%) than high (85-90%) relative humidity (Nickel, 1960). In addition, survival of immature stages, adult longevity and percentage egg hatch are all reduced under conditions of high relative humidity. The effects of low relative humidity on active stages may result from increased feeding rates to replace water loss (Crooker, 1985). Conversely, conditions of high relative humidity may interfere with mite feeding by preventing water loss through the cuticle (Boudreaux, 1958). Tulisalo (1974) found that simple misting of cucumber plants in greenhouses periodically resulted in drowning of mite, *T. cucurbitae,* in water droplets and helped in controlling the mite. Control of mites increased by adding a surfactant to the water.

NUTRITIONAL STATUS OF THE HOST PLANT

Fertilization affects the performance of *T. urticae*. The effects of nitrogen (N) fertilization and the ratio of nitrogen to phosphorous (P) and potassium (K) in relation to mite multiplication have been frequently reported. In general, increased levels of N fertilization improve host quality (Van de Vrie *et al.*, 1972). Rodriguez *et al.,* (1970) associated mite population increases on strawberry plants with increased levels of foliar nitrogen. Wermelinger *et al.,* (1985) reported the performance of *T. urticae* on bean and apple leaf-discs from plants grown at different levels of N fertilization, and found that nitrogen deficiency increased preimaginal development time and preoviposition period, and decreased adult female weight, fecundity and oviposition rates. On apple leaves, the above authors found that 50% reduction in leaf N was associated with 10-fold decline in fecundity. Plant water stress also alters the nutritional status of the host plant (Mellors and Propts, 1983) and tends to enhance susceptibility to spider mites (Hollingsworth and Berry, 1982). Feeding by spider mites results in increase in plant water stress and lead to an accumulation of soluble leaf carbohydrates which may play a part in enhancing mite reproduction (DeAngelis *et al.*, 1982)

PLANT PHENOLOGY

Plant phenology can also profoundly affect host-plant suitability for spider mites, with outbreak populations often occurring coincidentally with flowering and fruiting stage of the host plant (Huffaker *et al.*, 1969). Such association between densities of *T. urticae* and host physiology has been commonly reported on strawberry (Poe,1979), corn (Margiles and Kennedy, 1984), soybean (Cadapan, 1976) and peanut (Boykin, 1983). Margolis and Kennedy (1984) found that change from vegetative to reproductive growth involves a redistribution of water, sugars, nitrogenous compounds and inorganic ions among plant parts. Such changes may account for the enhanced suitability of the host during particular phenological stages. A thorough understanding of the relationship between plant phenology and suitability for spider mite population growth is important for predicting the potential for mite outbreaks.

CROPPING PATTERNS, CULTURAL AND PRODUCTION PRACTICES

The pest situation of spider mites is unquestionably related to various agricultural practices. Virtually all aspects of crop production can influence spider mite population dynamics. In general, the improved case of crops has contributed to the frequency of spider mite outbreaks. Quality of host plants, sources of alternate hosts, soil mineral water relations, biochemical composition of leaves and pesticide effects, are all agricultural parameters that influence spider mite outbreaks. Agriculture is a dynamic system and will continue to change as the need for more efficient food production increases. Planting an area with a single crop provides extensive food supplies for a mite pest and limits the reservoirs of mite enemies and competitors. Monoculture predisposes rapid development of pest populations over extensive areas before the regulating agencies can effectively operate. Large acreage plantings increase the difficulties of employing pest management by predators and at the same time provide more convenient situations for application of pesticides. New varieties and increased plant vigour are an integral part of modern agricultural production and mainly serve as important factors contributing to spider mite outbreaks. Host quality, in general, contributes to mite population increases (Wrensch and Young, 1978). Good leaf quality during oviposition and development resulted in relatively faster developmental rates and higher survival of offspring.

IRRIGATION

Irrigation has been reported as an alternative control measure in some situations. Kattes (1976) observed a significant reduction in spider mite, *Oligonychus pratensis,* in irrigated maize in Texas (USA). Chandler *et al.,* (1979) concluded that significant rainfall and/or irrigation can limit the development of spider mite populations (*Oligonychus pratensis* and *Tetranychus cinnabarinus*) on corn. Wilson *et al.,* (1984) reported that sprinkler irrigation also helps in controlling spider mites. In almonds and citrus, irrigation have been reported to reduce the impact of mite feeding on trees (Hoy, 1985). Overhead irrigation may significantly increase the relative humidity to levels conducive to epizootics of fungi pathogenic on spider mites. Irrigation may also provide some dust control in the orchard (Muma, 1970). Irrigation may create an environment less suitable for spider mites, it may also create appropriate conditions for mite predators.

WEEDS

Mite outbreaks in economic crops usually result from spider mite infestations in other non-crop sources. Infestations in cotton and other crops, for example, frequently originate from nearby field margins, dooryard ornamentals, home gardens, adjacent perennials, susceptible plants left over in the offseason or the crops kept as ratoon, or from early season crops (Leigh, 1985). Manipulation of these sources of infestation can be utilized as an additional management tool (Croft and McGroarty, 1973 and Brandenberg and Kennedy, 1982). Strip cropping and adding habitat diversity have also been reported to reduce spider mite build-up and encourage phytoseiid and stigmaeid predatory mite populations on cotton (Elbadry, 1979).

CULTURAL PRACTICES

Ploughing kills overwintering populations and result in delayed population build-up on peppermint (Hollingsworth and Berry, 1982). In strawberry crop, pruning and removal of non-functioning leaves reduces the two-spotted spider mite population by over 50% (Oatman *et al.*, 1967). Such practices although are used for agronomic reasons, in fact they provide an additional benefit by reducing the need for acaricide applications.

Most successful work involving the use of cultural practices has occurred in greenhouses. The environment of the greenhouse can be closely controlled and utilized in creating an environment that is less than optimal for spider mite outbreaks. Sanitation in the greenhouse is the first step (Hussey and Scopes, 1985). In addition, because high humidity and moisture reduce spider mite outbreaks, programmes utilizing a mister water spray on greenhouse crops have shown potential (Tulisalo, 1974). In greenhouse, diapause also can be manipulated by controlling day-length.

INERT DUSTS

Crops grown near dusty roads are the first to become heavily infested with spider mites. These inert dusts may be lethal to predaceous insects and mites either directly or indirectly. They may cause death through dessication. Dusts also seem to help the development of mites on smooth surfaces of plants by providing bases for attaching webs (Jeppson *et al.*, 1975)

CROP RESISTANCE

Resistance to spider mites is present to varying degrees in many crops (De Ponti, 1977). Breeding for resistance against mites have been reported in few crops including cassava, cotton, strawberry, cucumber and peanuts. High level of resistance have been found in strawberry and cucumber. Resistant germplasm have been identified in peanuts (Johnson *et al.*, 1982 and Johnson and Campbell, 1982). In case of castor, no bloom varieties have been reported resistant to *T. urticae* attack as compared to double and triple bloom varieties (Palanisamy, 2001).

Khanna and Ramanathan (1947) found that sugarcane varieties with spinous outgrowths on the leaves were susceptible to *Oligonychus indicus* and those without hairs were immune. Resistance to *Tetranychus cinnabarinus* in tomato varieties (Stoner *et al.*, 1968) and in *Solanum pennelli* and *Lycopersicon hirsutum* (Gentile *et al.*, 1969) was attributed to the glandular hairs. Resistant varieties had more number of glandular hairs on the leaves than susceptible ones. Peters and Berry (1980) observed a positive correlation between pubescence and oviposition of *T. urticae* on hops. In case of cucumber, varieties resistant to *T. urticae* were found to contain more cucurbitacins. Leaf structure have been shown to be related to mite damage, for example pear and plum varieties that had a thin cuticle exhibited more damage by mite, *Panonychus ulmi*. Leaf surface texture and contours (ridges, depressions etc.) are important in the reproductive potential of a mite : strong webbers like *T. urticae* will develop more uniformly heavy populations and more quickly on smooth plant surfaces if the surfaces are covered with a lint than if left untreated. On untreated surfaces, initial development is slow and more restricted to depressions and rough places, where suitable anchorage for webbing is available. Weak webbers like *P. ulmi* and *P. citri,* under harsh situations tend to colonize along and remain close to the midribs and veins (Van De Vrie *et al.*, 1972).

PESTICIDE RESISTANCE

Utilization of integrated pest management approaches for pesticide use is essential in reducing the incidence of pesticide resistance (Georghiou and Taylor, 1986). Pesticides provide cheap insurance against crop failure and their costs have decreased relative to other population inputs (Miranowski and Carlson, 1986). Since spider mites particularly two-spotted spider mite, *Tetranychus urticae,* has a high reproductive rate, rendering it an excellent candidate for rapid development of resistance. Various attempts to delay the development of resistance have met with some success. Fitness, or the ability of the resistant genes to persist under reduced selection pressure, is quite strong (Cranham and Helle, 1985). However, if chemical selection is stopped at an early stage, natural selection will generally tend to favour non-resistant individuals. Limited selection can be achieved by less frequent pesticide applications using pesticides with short residual activity and by leaving integrated refuges for susceptible migrants.

PATHOGENS

Fungi are the most frequently observed and reported on plant feeding mites. *Entomophthora* infections have been reported on *Panonychus, Tetranychus* and *Tyrophagus* mites. *Hirsutella thompsoni* which was commercially produced under name 'Mycar' has been recommended against spider mite infestations in glasshouses (Poinar and Poinar, 1998). Success of fungal preparations is based on their dependency on humidity for conidial germination and problems of survival of conidia in the field. Infection by non-occluded viruses have been reported on *Panonychus ulmi* and *P. citri*, infested mites turn dark red and often die with their legs extended downwards and their midguts contain bifringent crystals. No mite-specific bacteria have been isolated to date.

INTEGRATED MANAGEMENT OF INDIAN PHYTOPHAGOUS MITES

From India, there are many reports of mites as pests in different agri-horticultural systems, and their different control strategies mainly involving chemical control, natural enemies, plant resistance and manipulation of various agronomic practices. Information regarding pest status of mites on different crops, vegetables, plantation crops and ornamental plants alongwith the technologies developed for adoption is given in Table 1.

Table 1 Major mite problems of different crops reported from India and technologies developed for their management

Crop	Mite problem	Technology developed for adoption
Vegetables		
Brinjal, cucurbits and okra	Spider mites (*Tetranychus cinnabarinus, T. neocaledonicus, T. ludeni, T. macfarlanei, and T. urticae*) infests severely these crops particularly during flowering and fruiting period, and attack is more severe during summer months when the weather is hot and dry(Dhooria and Mann, 1989; Dhooria *et al.*, 1987 and Ghai, 1976).	1. Dicofol (0.04%), wettable sulphur (0.5%), and dimethoate (0.05%) effectively control these mites (Anonymous, 2000). 2. Synthetic pyrethroids like cypermethrin, fenvalerate, fenpropathrin, and carbaryl results in resurgence of mites on these crops (Dhooria and Mann, 1989). 3. Okra varieties i.e. Pusa-7 and Pusa Sawani are relatively less susceptible to mite attack (Anonymous, 2000). 4. Early sown okra crops exhibited less mite incidence while late sown crops are more susceptible to mites 5. Phytoseiid mite, *Amblyseius multidentatus* found as a good predator of pest mite, *T. ludeni* (Anonymous, 2000). 6. Bitter-gourd leaves found unsuitable for feeding and breeding of pest mites (Anonymous, 2000). 7. By increasing frequency of irrigation during hot and dry months, mite incidence is significantly reduced (Dhooria, 1994).
Chilli	Tarsonemid mite, *Polyphagotarsonemus latus,* causes severe damage to chilli (both *Capsicum annum* and *C. fruitescens*). Attack is more severe on young terminal growth, downward curling of leaf margins and even necrosis of growing points occur. Severely affects flowers and fruit formation, yield is considerably reduced.	1. Application of wettable sulphur, ethion, diafenthiuran and dimethoate all @ 0.05% proved very effective against *P. latus* (Anonymous, 2000). 2. Upto 32% yield loss reported in chilli variety Suryamukhi (Mukherjee *et al.*, 1987). 3. Aqueous extracts of plants, *Aloe* sp., and *Lippia nodiflora* at 5% proved quite effective against the mite (Channa Basvanna and Viraktamath, 1989). 4. Mite damage was lesser on chilli crop planted in red soils than in clay soils (Mukherjee *et al.*, 1987).
Cowpea	Spider mites, *T. cinnabarinus* and *T. ludeni,* severely infests cowpea and considerably reduced yield in terms of green pods and dry grains.	1. Application of ethion (0.1%) effectively controls these mites (Ghai, 1976).

Contd. ...

Table 1 Contd. ...

		2. Early sown crop is more damaged by mites than late sown (Mukherjee *et al.*, 1987).
Fruit trees		
Citrus	Spider mite, *Eutetranychus orientalis* ; false spider mite, *Brevipalpus phoenicis* ; and tarsonemid mite, *P. latus,* are among major mite pests of different citrus species. *E. orientalis* infests predominantly dorsal leaf lamina and produces yellowish-white symptoms. Infestation of *B. phoenicis* results in circular rind disorder on "Kinnow" fruits resulting in lower market value (Dhooria *et al.*, 1997). *P. latus* infests nursery plants and hinders their growth, their infestation on lemon fruits results in rough corky surface (Dhooria, 1996).	1. Soil application of aldicarb @ 0.5g.a.i./ plant reported most effective against *E. orientalis.* Dimethoate (0.05%) proved effective against *P. latus* on nursery plants (Dhooria, 1983). 2. Cannabis and castor plants are alternate hosts of *E. orientalis,* and should be removed from surroundings of citrus orchards. 3. Monocrotophos, ethion and dicofol (all at 0.075%) found most promising against mite pest (Dhooria, 1983). 4. Phytoseiid mites found to control pest mites under untreated conditions (Anonymous, 2000).
Pear	*T. urticae* and *E. orientalis* seriously infests leaves during summer months and causes leaf scorching and leaf fall. Both species have very wide host range (Dhooria, 1994).	1. Frequency of irrigation during summer months should be increased (Dhooria, 1994). 2. Dicofol (0.037%), ethion (0.05%) and dimethoate (0.03%) are very effective against mites (Dhooria, 1994). 3. Cannabis and castor – alternate hosts, should be destroyed (Dhooria, 1994).
Fig	Spider mite, *Eotetranychus hirsti,* is a major pest of fig trees and its serious infestation results in too much leaf-fall and reduced yields (Anonymous, 2000).	1. Phosphamidon (0.025%) results in significant reduction of mite (Anonymous, 2000).
Coconut	Eriophyid mite, *Aceria guerronis,* infests fruits and peak mite population found during Feb. – Mar. Mite feeding scars the surface of nut and reduces market value of tender coconuts. Size of kernel tends to reduce and this hampers	1. Wettable sulphur and dicofol sprays reported effective against mites (Anonymous, 2000). 2. Coconut varieties "Kenthali" and "British Solomon Island" suffered comparatively lower damage while variety "Tiptur tall" was most susceptible (Mukherjee *et al.*, 1987).

Contd. ...

Table 1 Contd. ...

	proper development of the nut. Mites commonly infested nuts of size 3cm long into 5cm diameter and above. Depleted growth of kernel causes copra loss upto 25%.	
Litchi	Eriophyid mite, *Aceria litchii,* produces velvety brown erineum on undersurface of leaf. Mites probably carry the spread of parasitic algae, *Cephaleuros virescens* and provide them entry in leaf tissue through feeding punctures (Mukherjee *et al.*, 1987).	1. Application of dimethoate or wettable sulphur controls this mite (Anonymous, 2000). 2. Varieties "Casba" and "Elachi" in Bengal and "Bedana" in Bihar are highly susceptible while "Piazi" in West Bengal and "China" in Bihar are most resistant.
Sugarcane, maize and sorghum		
	Spider mite, *Oligonychus indicus,* is a serious pest of these crops during summer months. Another spider mite, *Schizotetranychus andropogani,* produces small circular webs on under surface of leaves during April – October (Anonymous, 2000).	1. "Baru", "Kahi" and some other grasses which act as alternate hosts should be destroyed (Anonymous, 2000).
Pulses		
Moong bean and mash bean	Tarsonemid mite, *P. latus,* is a most serious pest of moong and mash bean during comparatively hot and humid months (July-August). *T. cinnabarinus* infests these crops in May-June when conditions are hot and dry (Dhooria, 1996).	1. Dicofol (0.03%) is highly effective against *P. latus* . Dimethoate and oxydemeton methyl both @ 0.05% are very effective against *T. cinnabarinus* (Dhooria and Mann, 1989).
Pigeon pea	Spider mites, *Schizotetranychus cajani* and *Eotetranychus broodryki* sometimes seriously infests and reduces yield. Eriophyid mite, *Aceria cajani,* is an important vector of pigeon pea sterility mosaic virus	1. Dicofol and ethion (both @ 0.05%) provided good control of pigeon pea mites (Anonymous, 2000). 2. Plants growing under shady conditions are more damaged by spider mites, so do not grow pigeon pea near wind breaks (Mukherjee *et al.*, 1987).

Contd. ...

Table 1 Contd. ...

	disease and results in considerable yield reduction. Disease affected plants developed light green or chlorotic leaves (Anonymous, 2000 and Choudhary and Dhillon, 1992).	3. Pigeon pea sterility mosaic incidence was more when pigeon pea was intercropped with sorghum than in pure crop (Mukherjee *et al*., 1987). 4. "Bahar" variety is comparatively resistant to disease while "BDN-1" was most susceptible (Anonymous, 2000, (Mukherjee *et al*., 1987)
Plantation crops		
Tea	Spider mite, *Oligonychus coffeae* ; false spider mite, *B. phoenicis* and eriophyid mites, *Calcarus carinatus* and *Acaphylla thea* are the major mite pests. Attack of *O. coffeae* resulted in reddish discolouration which is visible from distance (Anonymous, 2000).	1. Both dicifol and ethion @ 0.05% are very effective against *O. coffeae* (Anonymous, 2000, Choudhary and Dhillon, 1992). 2. Proper pruning of twigs should be done (Mukherjee *et al*., 1987).
Ornamentals		
Roses and marigold	Spider mite, *T. urticae,* is a pest of roses under polyhouse conditions. *T. urticae* and *E. orientalis* are serious pests on marigold (Gupta, 1991).	1. Vertimec and dicofol successfully controlled mites on rose (Gupta, 1991). 2. Dicofol (0.037%) sprays proved effecbtive against mites on marigold (Dhooria and Mann, 1989).

Incorporating resistant predators into an IPM system can lead to reduced number of sprays (Croft, 1982). In the absence of resistant predators, control with short residual sprays would require more frequent applications and thus would provide selection pressure equal to that from longer-persisting treatments. Rotation of pesticide compounds apparently delays the development of resistance only in terms of calendar time and not in terms of number of applications (Cranham and Helle, 1985). Cultural practices can be used to reduce selection pressure and to enhance the presence of susceptible mites in the cropping system.

STRATEGIES FOR FUTURE LINE OF RESEARCH

Work on following strategies should be strengthened so as to effectively manage the mite pests in different agri-horticultural systems :

- Studies on harmful effects of broad-spectrum pesticides on natural enemies of mites need thorough investigations so as to properly understand resurgence of mites on different crops.
- Role of natural enemies of mites (both predaceous mites and insects) in regulating pest mites needs to be thoroughly established in different cropping systems.

- Attempts should be made to develop pesticide resistant strains of predaceous mites so that they can be utilized effectively on high value crops under polyhouse conditions along with the use of different pesticides.
- Work on selective use of miticides, proper timings and application rates of pesticides needs strengthened so that problems of acaricide resistance may not haunt immediately.
- Facilities should be generated for mass multiplication of potential natural enemies so that they can be properly utilized for safe and cost effective control of different mite problems particularly on high value crops in the greenhouses.
- Use of pyrethroid insecticides for general pest control should be restricted as these pesticides do not kill pest mites but readily kill natural enemies, resulted in mite flare-ups under many situations.
- Work on development of economic threshold values for different mite pests on economic crops should be initiated on priority basis so as to minimize the use of pesticides.
- Surveillance for different mite pest problems under different agrosystems should be done thoroughly.
- Work on finding more potential natural enemies and their utilization in controlling mites needs strengthened.

References

Anonymous. 2000. *Proceedings of seventh group meeting of All India coordinated Research Project on Agricultural Acarology held at Kalyani (W. Bengal)*, April 5-6, 2000.

Boudreaux, H.B. 1958. The effect of relative humidity on egg laying, hatching and survival in various spider mites. *J. Ins. Physiol.*, 2 : 65-72.

Boykin, L.S. 1983. *Ecology of the two-spotted spider mite, Tetranychus urticae Koch, on Peanut in North Carolina.* PhD dissertation. North Carolina State University, Raleigh, 97pp.

Brandenburg, R.L. and Kennedy, G.G. 1982. Intercrop relationships and spider mite dispersal in a corn/peanut agroecosystem. *Entomol. Experiment. Appl.*, 32 : 269-276.

Cadapan, E.P. 1976. *The Effect of the Two-spotted Spider Mite and several Insects on the yield of soybeans.* Ph.D dissertation. University of California, Berkeley. 97pp.

Chandler, L.D., Archer, T.L., Ward, C.R. and Lyle, W.M. 1979. Influences of irrigation practices on spider mite densities on field corn. *Environ. Entomol.*, 8 : 196-201.

ChannaBasavanna, G.P. and Viraktamath, C.A.(eds). 1989. *Progress in Acarology* Vol.2. Oxford and IBH Co, New Delhi.

Chapman, P.J. ; Lienk, S.E. and Curtis, O.F. 1952. Responses of apple trees to mite infestations. I. *J. Econ. Ent.*, 45 : 815-821.

Choudhary,S. and Dhillon, R.1992. *Geography : World and India* Dhillon publications,New Delhi. pp.308

Cranham, J.E. and Helle, W. 1985. Pesticide resistance in Tetranychidae. In : *Spider Mites their Biology, Natural Enemies and Control* (W. Helle and M.W. Sabelis, Eds.), Elsevier, New York. 1B, pp.405-421

Croft, B.A. 1982. Arthropod resistance to insecticides : a key to pest control failures and successes in North American apple orchards. *Entomol. Experiment. Appl.*, 31 : 88-110.

Croft, B.A. and McGroarty, D.L. 1973. A model study of acaricide resistance, spider mite outbreaks and biological control patterns in Michigan apple orchards. *Environ. Entomol.*, 2 : 633-638.

Crooker, A. 1985. Embryonic and juvenile development. In : *Spider Mites, Their Biology, Natural Enemies and Control* (W. Helle and M.W. Sabelis, Eds.). Elsevier, New York. 1A, pp 149-163

DeAngelis, J.D., Larson, K.C., Berry, R.E. and Krantz, G.W. 1982. Effects of spider mite injury on transpiration and leaf water status in peppermint. *Environ. Entomol.,* 11 : 975-978

DePonti, O.M.B. 1977. Resistance in *Cucumis sativus* L. to *Tetranychus urticae* Koch. 1. The role of plant breeding in integrated control. *Euphytica* 26 : 633-640

Dhooria,M.S. 1983. Persistence of toxicity of pesticides applied to soil against *Euteranychus orientalis* on malta seedlings. *Indian J .Acarol.,* 8 : 13-21.

Dhooria, M.S. 1994. An outbreak of two spotted spider mite, *Tetranychus urticae* Koch on sand pear, *Pyrus communis,* and its control. *Pest Mng. Econ. Zool.,* 2 : 127-130.

Dhooria, M.S. 1996. Observations on host-range of leaf-curl mite, *Polyphagotarsonemus latus* (Banks) (Acari : Tarsonemidae) in Punjab. *J. Insect Sci.,* 9(1) : 9-11.

Dhooria,M.S.and Mann,G.S.1989. Role of insecticides in inducing mite-outbreaks. In : *'Soil Pollution and Soil Organism'*. Ashish Publishing Co, New Delhi. pp.45-62.

Dhooria, M.S. ; Mann, G.S. and Bhalla, J.S. 1997. An unusual rind-disorder of 'kinnow' mandarin found associated with false spider mite, *Brevipalpus phoenicis* (Geijskes) (Acarina : Tenuipalpidae) in Punjab. *Proc. Natn. Symp. Citriculture* pp. 371-372.

Dhooria, M.S. ; Singh, D. ; Sukhija, B.S. and Bhathal, G.S. 1987. Differential response of watermelon cultivars for resistance to red spider mite, *Tetranychus cinnabarinus* (Boisd.). *J. Res. Punjab Agric. Univ.,* 24 : 590-594.

ElBadry, E.A. 1979. Management of mite pests of cotton in Egypt. In : *Recent Advances in Acarology* (J.G. Rodriguez, Ed.), Academic Press, New York. pp.49-57

Gentile, A.G. ; Webb, R.E. and Stoner, A.K. 1969. *Lycopersicon* and *Solanum* spp resistant to the carmine and the two-spotted spider mite. *J. Econ. Ent.,* 62 : 834-36.

Georghiou, G.P. and Taylor, C.E. 1986. Factors influencing the evolution of resistance. In : *Pesticide Resistance, Strategies and tactics for management,* National Academy Press, Washington DC. pp. 159-169.

Ghai S.1976. Phytophagous mites and their control in India. *Proc. Nat. Acad. Sci. India,* 46(B) : 49-599.

Gupta, S.K.1991. *Mites of agricultural importance in India and their management. Technology Bulletin No 1,* ICAR, pp.18. All India Coordinated Research Project on Agricultural Acarology UAS,Bangalore.

Hollingsworth, C.S. and Berry, R.E. 1982. Two-spotted spider mite (Acari : Tetranychidae) in peppermint : population dynamics and influence and influence of cultural practices. *Environ. Entomol.,* 11 : 1280-1284.

Hoy, M.A. 1985. Almonds (California). In : *Spider Mites, their biology, natural enemies and control* (W. Helle and M.W. Sabelis, Eds.), Elsevier, New York. 1B, pp.299-310.

Huffaker, C.B., van De Vrie, M. and McMurtry, J.A. 1969. The Ecology of tetranychid mites and their natural control. *Ann. Rev. Ent..,* 14 : 125-174.

Hussey, N.W. and Scopes, N.E.A. 1985. Greenhouse vegetables (Britain). In : *Spider Mites, their biology, natural enemies and control* (W. Helle and M.W. Sabelis, Eds.), Elsevier, New York. 1B, pp.285-297.

Jeppson, L.R.; Keifer, H.H. and Baker, E.W. (1975) *Mites injurious to Economic plants.* University of California Press, Berkeley, 614pp.

Johnson, D.R. and Campbell, W.V. 1982. Variation in the foliage nutrients of several peanut lines and their association with damage received by the two-spotted spider mite, *Tetranychus urticae. J. Georgia Entomol. Soc.,* 17 : 69-72.

Johnson, D.R., Campbell, W.V. and Wynne, J.C. 1982. resistance of peanuts to the two-spotted spider mite (Acari : Tetranychidae). *J. Econ. Entomol.,* 75 : 1045-1047.

Kattes, D.H. 1976. *Selected factors influencing the abundance of spider mites (Acari : Tetranychidae) in grain sorghum on the Texas High Plains.* M.S. thesis. Texas A&M University, College Station. pp 66.

Khanna, K.L. and Ramanathan, K.R. 1947. Studies on the association of plant characters and pest incidence. 1. Structure of leaf surface and mite attack in sugarcane. *Proc. Nat. Inst. Sci. India,* 13 : 232-234.

LeClerg, E.L. 1965. *Losses in agriculture.* U.S. Dept. Agr., A.R.S. handbook no. 291, pp 120.

Leigh, T.F. 1985. Cotton. In : *Spider Mites, their biology, natural enemies and control* (W. Helle and M.W. Sabelis, Eds.), Elsevier, New York. 1B, pp.349-358.

Lienk, S.E. ; Chapman, P.J. and Curtis, O.F. 1956. Responses of apple trees to mite infestations. II. *J. Econ. Ent.,* 49 : 350-353.

Margolis, D.C. and Kennedy, G.G. 1984. Population of the two-spotted spider mite, *Tetranychus urticae*, to host phenology in corn and peanut. *Entomol. Experiment. Appl.,* 36 : 193-196.

Mellors, W.K. and Propts, S.E. 1983. Effects of fertilizer level, fertility balance and soil moisture on the interaction of two-spotted spider mites (Acarina : Tetranychidae) with radish plants. *Environ. Entomol.,* 12 : 1239-1244.

Miranowski, J.A. and Carlson, G.A. 1986. Economic issues in public and private approaches to preserving pest susceptibility. In : *pesticide resistance, strategies and tactics for management.* National Academy Press, Washington, DC. pp.436-448.

Mori, H. 1961. Comparative studies on the thermal reaction in four species of spider mites. *J. Faculty of Agriculture Hokkaido University,* 51 : 574-591.

Mukherjee, A,B., Somchoudhry, A.K. and Sarkar, P.K.(eds). 1987. *Contributions to Acarological Researches in India.* Dept. of Agricultural Entomology, BCKV, Kalyani, W.Bengal. pp. 445pp.

Muma, M.H. 1970. Preliminary studies on environmental manipulation to control injurious insects and mites in Florida citrus groves. *Proceedings Tall Timbers Conference on Ecological Animal Control by habitat management,* Tall Timbers Research Station, Tallahassee. pp. 23-40

Nickel, J.L. 1960. Temperature and humidity relationships of *Tetranychus desertorum* Banks with special reference to distribution. *Hilgardia,* 30 : 41-100.

Oatman, E.R., McMurtry, J.A., Shorey, H.H. and Voth, V. 1967. Studies on integrating *Phytoseiulus persimilis* releases, chemical applications, cultural manipulations and natural predation for control of two-spotted spider mite on strawberry in southern California. *J. Econ. Entomol.*, 60 : 1344-1351.

Palanisamy, S. 2001. Role of host plant resistance in the management of phytophagous mites. In : *Biodiversity and management of phytophagous mites* (Eds. Rabindra *et al.* 2001). TNAU, Coimbatore. pp. 139-147.

Peters, K.M. and Berry, R.E. 1980. Resistance of hop varieties to two-spotted spider mite. *J. Econ. Entomol.* 73 : 232-234.

Poe, S.L. 1979. *Tetranychus urticae* population development on vegetative and senescent host plants. *Proc. Fourth Int. Congr. Acarol.,* 4 : 215-220. Akadémiai Kiadó, Budapest.

Poinar, G. and Poinar, R. (1998) Parasites and pathogens of mites. *Ann. Rev. Ent.* 43 : 449-69.

Rodriguez, J.G., Chaplin, C.E., Stolitz, L.P. and Lasheen, A.M. 1970. Studies of resistance of strawbwrries to mites. I. Effects of plant nitrogen. *J. Econ. Entomol.,* 63 : 1856-1858.

Stoner, A.K. ; Frank, J.A. and Gentile, G.A. 1968. The relationship of glandular hairs on tomatoes to spider mite resistance. *Proc. Amer. Soc. Hort. Sci.,* 93 : 532-38.

Tulisalo,U. 1974. Control of two-spotted spider mite, *Tetranychus urticae* Koch by high air humidity or direct contact with water. *Annales Entomologici Fenniei,* 40 : 158-162.

Van De Vrie, M., McMurtry, J.A. and Huffaker, C.B. 1972. Ecology of tetranychid mites and their natural enemies : A review. III. Biology, Ecology and Pest status, and Host-Plant relations of Tetranychids. *Hilgardia,* 41(13) : 343-432.

Wermelinger, B., Oertli, J.J. and Delucchi, V. 1985. Effect of host plant nitrogen fertilization on the biology of two-spotted spider mite, *Tetranychus urticae. Entomol. Experiment. Appl.,* 38 : 23-28.

Wilson, L.T., Gonzalez, D., Leigh, T.F., Maggi, U., Foristiere, C. and Goodell, P. 1983. Within-plant distribution of spider mites (Acari : Tetranychidae) on cotton : a developing implementable monitoring program. *Environ. Entomol.,* 12 : 128-134.

Wrensch, D.L. and Young, S.S.Y. 1978. Effects of density and host quality on rate of development, survivorship and sex-ratio in the carmine spider mite. *Environ. Entomol.*, 7 : 499-501.

□□□

PLATE 1: MAJOR MITE PEST OF PUNJAB

Spider mites, *Teranychus urticae*

Citrus mite, *Eutetranychus orientalis*

Wheat mite, *Petrobia latens*

Eriophyid mite, *Cisaberoptus kenyae*

Leaf-curlmite, *Polyphagotarsonemus latus*

PLATE 2 : MITE DAMAGE ON VARIOUS CROPS

Brinjal (Tetranychidae)

Okara (Tetranychidae)

Kinnow (Tenuipalpidae)

Chilli (Tarsonemidae)

Rose (Tetranychidae)

Pear (Tetranychidae)

Entomology : Novel Approaches, 2007
Eds.: *P.C. Jain and M.C. Bhargava*,(Page 477-497)
New India Publishing Agency, New Delhi
E-mail: newindiapublishingagency@gmail.com
web: www.bookfactoryindia.com

CHAPTER 24

BIODIVERSITY OF LEAFHOPPERS (HEMIPTERA: CICADELLIDAE) OF THE INDIAN SUB-CONTINENT

C. A. Viraktamath
Department of Entomology, University of Agricultural Sciences
GKVK, Bangalore 560 065

INTRODUCTION

Leafhoppers belong to the family Cicadellidae and constitute one of the largest families in the exopterygotan Hexapoda. Recent estimates of species for the world range between 35,000-45,000 species though little over 22,600 species are now described (Dietrich, 2004a). The leafhoppers belong to the order Hemiptera and suborder Auchenorrhyncha.

Leafhoppers can readily be recognized from other members of the Auchenorrhyncha by the presence of two or more rows of spines on hind tibiae and by the pronotum not extending back over abdomen. Their close relative, the Membracidae, have pronotum extending back over the abdomen. From planthoppers (Delphacidae) with which they are often confused, they can be recognized by the position and structure of antennae. The leafhopper antennae are placed between eyes compared to beneath the eyes in planthoppers; the second segment of the antenna (pedicel) is small without externally visible sensillae in leafhoppers compared to enlarged pedicel with prominent sensillae in the planthoppers. The movable large spur at the apex of hind tibia in plant hoppers is absent in the leafhoppers.

Fabricius described the first Indian leafhopper, a member of the subfamily Ledrinae under the name *Membracis planata* (present name *Epiclinata planata)* in 1794. This was followed by description of two other Ledrinae namely, *Ledra serrulata* Fabricius and *L. mutica* Fabricius in 1803. It was in 1889 that the three notorious pests of mango, *Idioscopus clypealis* (Lethierry), *Idioscopus niveosparsus* (Lethierry) [now *nitidulus* (Walker)] and *Amritodus atkinsoni* (Lethierry) were described by Lethierry under the genus *Idiocerus*. Two prominent works namely that of Melichar (1903) on Ceylonese fauna and of Distant (1908, 1916 and 1918) on the leafhopper fauna of the Indian subcontinent under the fauna of British India series, paved way for the understanding of the leafhoppers of the region. While Melichar dealt with 130 species of leafhoppers from Sri Lanka, Distant dealt with

561 species for the subcontinent. Pruthi (1930, 1934a, 1934b, 1936 and 1940) was the first Indian leafhopper worker and described 76 species and also revised the taxa described by Distant.

CATALOGUES AND CHECKLISTS

Information on the world leafhoppers is well documented by two important works. Metcalf's "Catalogue of the Homoptera Fascicle VI, Cicadelloidea" published in 17 parts deals with information on leafhoppers from 1758-1955 (Metcalf, 1962-68, 1964). Oman, Knight and Nelsons' "Leafhoppers (Cicadellidae) a bibliography, generic checklist and index to the world literature 1956-1985" updates the knowledge till December 1985 (Oman *et al.*, 1990). The world leafhoppers are grouped in nearly 40 subfamilies and 98 tribes. However, there is no agreement in the number of subfamilies recognized among the leafhopper workers. Studies are underway to rationalize the higher classification (subfamilies and tribes) of leafhoppers with the help of morphological, behavioral and even molecular techniques, which can be used in phylogenetic analysis. In this paper, arrangement suggested by Oman *et al.* (1990) is followed for convenience.

Very few regional checklists of leafhoppers are published from the subcontinent; more significant ones are the following: 39 species for West Bengal (Ghosh and Ghosh, 1994), 123 species along with host plants for Karnataka (Viraktamath and Viraktamath, 1995), 154 species for Kerala (Mathew, 2004). Jacob *et al.* (2000) and Reddy and Rama Subba Rao (2001) reported on the leafhopper fauna of oilseed and vegetable crops in Andhra Pradesh. Ahmed *et al.* (1980) and Ahmed and Naheed (1982) dealt with the leafhoppers of vegetables and fruit crops in Pakistan. Sohi and Dworakowska (1984) reviewed the subfamily Typhlocybinae and provided a checklist for the Indian species. Later, Sohi (1998) updated it up to 1997 with 568 species under 146 genera.

SUB-FAMILIES

Of the 40 subfamilies recognized in the family Cicadellidae by Oman *et al.* (1990), 25 occur in the Indian subcontinent. There are more than 1350 species of leafhoppers known to-date from the Indian subcontinent (Viraktamath, 2005) (Table 1). Rao (1990) provided a key to some of these subfamilies following older concept of the subfamilies. Dietrich (2004b) provided a key to the world subfamilies and tribes and also the diagnosis of subfamilies in his web page on leafhoppers. Viraktamath (2005) gave key for the subfamilies and tribes occurring on the Indian subcontinent.

The characters used to recognize the subfamilies and tribes include those of head, thorax, front and hind wing venation, hind leg spinulation and characters of the male and female genitalia.

Table 1 Diversity of leafhoppers of the world and the Indian subcontinent

Subfamily	Genera World	Genera Indian subcontinent	Species World	Species Indian subcontinent
1. Acostemminae	7	1	20	1
2. Adelungiinae	13	1	60	2
3. Agalliinae	37	6	600	28
4. Aphrodinae	8	3	20	3
5. Arrugadinae	1	0	3	0
6. Austroagalloidinae	1	0	10	0
7. Bythoniinae	1	0	5	0
8. Cicadellinae	368	25	3290	92
9. Coelidiinae	126	5	1300	55
10. Deltocephalinae	750	73	6500	296
11. Drakensbergeninae	1	0	2	0
12. Errhomeninae	10	0	70	0
13. Euacanthellinae	1	0	2	0
14. Eupelicinae	10	2	100	8
15. Eurymelinae	33	0	120	0
16. Evnsiolinae	1	0	3	0
17. Hylicinae	14	8	40	10
18. Iassinae	50	4	700	19
19. Idiocerinae	90	9	720	41
20. Koebeliinae	1	0	5	0
21. Ledrinae	70	9	450	49
22. Macropsinae	26	5	550	36
23. Makilinginae	1	0	25	0
24. Megophthalminae	7	0	40	0
25. Mukariinae	2	2	10	3
26. Neocoelidiinae	18	0	130	0
27. Nioniinae	3	1	12	1
28. Nirvaninae	40	8	150	30
29. Penthimiinae	46	5	245	29
30. Phereurhininae	3	0	5	0
31. Phlogisinae	2	1	2	1
32. Gyponinae	90	0	1300	0
33. Selenocephalinae	60	15	330	48
34. Signoretiinae	2	2	25	5
35. Stegelytrinae*	6	3	26	3
36. Tartessinae	38	1	200	1
37. Tinterominae	1	0	3	0
38. Typhlocybinae	470	146	5200	569
39. Ulopinae	33	6	180	14
40. Xestocephalinae	7	2	200	9
Total	**2448**	**343**	**22653**	**1353**

*Note: Viraktamath (2005) did not list any genera and species for this subfamily.

The position of ocelli and facial sutures are also used. These characters can be readily seen with the help of a stereo-binocular microscope that gives a magnification upward of 40X.

For male and female genitalia study, the abdomen need to be removed from the specimen and macerated in 10% caustic potash, washed in distilled water and preserved in glycerol in a microvial. The genitalia can be examined either under stereo-binocular microscope or a research compound microscope.

Twenty-five subfamilies that occur on the Indian subcontinent are briefly described along with the included tribes in some large subfamilies.

Acostemminae are a small subfamily occurring in south India and Sri Lanka. Only one species *Acostemma walkeri* Kirkaldy occurs on the subcontinent. It breeds on *Dodonea viscosa.* The adults are also collected on *Santalum album, Webera carymbosa, Canthium didymium, Ziziphus aenoplia* and *Pterolobium indicum* (Pruthi, 1934a). The subfamily includes medium to large sized (10-15 mm long) leafhoppers usually green in color with clypeus delimited from the frons by a keeled epistomal sulcus. Male valve is fused laterally with the pygophore and subgenital plates are also fused with each other. Linnavuori and Quartau (1975) redefined the subfamily and treated the Afrotropical genera and species.

Adelungiinae are largely Palaearctic group of leafhoppers breeding on xeric plants of the genera *Salsola, Halophila* etc. in the desert region. Two species of the genus *Platyproctus* Lindberg occur in Pakistan and India (Punjab and Rajasthan). These are medium sized (5-8 mm) leafhoppers with short and broad heads. Ocelli are on face; the lateral frontal sutures end in antennal pits. Anterior margin of pronotum is exposed behind eyes. Forewings have numerous cross veins. This group was earlier treated as a tribe in Agalliinae. Emeljanov (1975) and Al-Ne'amy and Linnavuori (1982) revised the group for the Palaearctic and Afrotropical regions, respectively.

Agalliinae breed on herbs, shrubs and grasses. Some species of *Austroagallia* use economically important crops like potato, cowpea, horse gram, groundnut, Lucerne, *Gossypium arboreum, Cajanus cajan, Zea mays, Phaseolus vulgare, Oryza sativa*, *Sesamum indicum, Piper betel* and weeds such as *Leucas aspera, Achyranthes aspera, Parthenium hysterophorus* etc. The subfamily occurs in all zoogeographical regions except Australian where it has been recently colonized by a single species, *Austroagallia torrida* Evans. These are small to medium sized (2-8 mm long) cylindrical leafhoppers with short and broad heads. Ocelli are on face and frontal sutures do not extend beyond antennal pits. Valve is fused laterally with the pygophore. Forewings usually do not have accessory cross-veins. The species of the genera *Austroagallia* Evans and *Agallia* Curtis are common and are found throughout the subcontinent. The species of *Durgades* Distant are restricted to the Sub-Himalayan region in India, Pakistan, Nepal and Bhutan. This genus shows sexual dimorphism - the females being paler and larger compared to darker and smaller males. Viraktamath (2004a) revised the genus *Durgades* for the subcontinent whereas the species of the genus *Austroagallia* have been dealt with by Viraktamath and Sohi (1980). Some species of *Austroagallia* and *Nehela* White show great variation in size and color among the populations of the same species collected across distribution range of the species.

Aphrodinae are grass feeders and are Holarctic in their distribution. These are small to medium sized (3-6 mm long) leafhoppers with ocelli on crown near margin but distant from eyes. Forewing venation is complete. Hind wing marginal vein is often incomplete. Male subgenital plates are ligulate. *Gurawa* Distant, *Baileyus* Pruthi and *Stroggylocephalus*

Flor are the genera that occur on the subcontinent. The species of the first genus are found throughout the subcontinent; the species of latter two genera are restricted to Sub-Himalayan region (Pruthi, 1930 and Rao, 1989).

Cicadellinae breed on herbs, shrubs, trees and grasses in moist habitats and under shrubs. The group is worldwide in distribution with poor representation from the Palaearctic and Australian regions. These include medium to large (4-14 mm long), brightly colored leafhoppers with swollen faces to accommodate powerful pharyngeal muscles required to pump large quantities of xylem fluids from the plants. Ocelli are on crown closer to hind margin of head. Frontal sutures reach ocelli. Wing venation is complete, inner apical cell is usually narrow and elongate. The subfamily has three tribes Cicadellini, Mileewini and Evacanthini. The latter two tribes are some times considered as independent subfamilies (Young, 1968). The tribe Cicadellini is a large group with anterior margin of the head not carinate and the frontoclypeus without median carina. These are very brightly colored leafhoppers feeding on under story vegetation in moist forests. Some species of the genus *Cofana* Melichar are associated with rice and sugarcane. The species of *Atkinsoniella* Distant, *Anatkina* Young, *Bothrogonia* Melichar are commonly found in coffee and cardamom plantations in south India whereas the species of *Kolla* Distant are found breeding on new leaves of shrubs in open forest or edges of forests. Young (1968, 1977, 1979 and 1986) in a series of papers revised the world Cicadellinae. The Evacanthini are characterized by having carinae on the anterior margin of head, frontoclypeus is also with a median carina. The species of the genus *Dussana* Distant are common on the under story vegetation in south India and those of *Pythamus* Melichar in Sri Lanka. Species of *Evacanthus* Le Pelletier and Serville are Holarctic with extension of a few species in the Sub-Himalayan region. The genera *Onukia* Matsumura, *Cunedda* Distant, *Vangama* Distant are Sub-Himalayan. The last mentioned genus has greatly produced head in front of eyes and sexually dimorphic. Both nymphs and adults are found on low moist vegetation and the nymphs have a brush of black hair at the tip of abdomen that they move sideways when resting on vegetation. Mileewini (Mileewanini of some authors) is a small tribe with dark colored typhlocybine-like (see below) leafhoppers with pale areas on forewing. Head is rather smooth, with ocelli on crown and the frontal suture reaching ocelli. They have two anteapical cells and the clavus is small occupying at most half the length of forewing. Species of the genus *Ujna* Distant are common in the Peninsular India and Sri Lanka whereas the species of *Mileewa* Distant are common in the Sub-Himalayan region.

Coelidiinae breed on trees, shrubs and grasses. Some of the Indian species have been collected from *Dodonea viscosa, Santalum album, Ziziphus aenoplea* and also by sweeping grasses. The subfamily is found in all the zoogeographical regions except Europe. These are medium to large (6-12 mm long) stoutly built, cylindrical leafhoppers. Head has a narrow crown and the frontoclypeus is parallel sided. Ocelli are on crown or near anterior margin of crown. Antennal ledges are absent or weak. Subgenital plates are elongate and well sclerotized. Aedeagus is long and tubular. Forewing with appendix that goes round the apex. In some cases, early instar nymphs may feed on grasses and then switch over to herbs or shrubs as they grow (De Long, 1971). *Calodia kirkaldyi* Nielson (earlier misidentified as *Coelidia indica* Distant) is a known vector of sandal spike in south India

(Rangaswamy and Griffith, 1941). Nielson (1977 and 1982) in a series of papers revised the world fauna of Coelidiinae.

Deltocephalinae are small to medium sized (2-8 mm long) leafhoppers with ocelli on front margin of head often close to eyes than to each other. Lateral frontal sutures reach ocelli. Antennal ledges are reduced or absent. Forewing venation is complete and the claval region is with at least one cross-vein connecting outer claval vein with claval suture. Male valve is triangular and not fused with lateral margins of pygophore. This subfamily occurs in all zoogeographical areas and is the largest as far as the numbers of species are concerned. It also includes species of great economic importance. Of the 151 species of leafhopper vectors reported, 95 species belong to this subfamily (Nielson, 1985). In India, species of *Nephotettix* Matsumura, *Deltocephalus* (*Recilia*), *Orosius* Distant and *Hishimonus* Ishihara are of importance as they serve as vectors of plant pathogens of agriculturally important crops (Table 2).

The Deltocephalinae are usually divided into 23 tribes (Oman *et al.*, 1990) and there is an urgent need to rationalize the classification of tribes on a worldwide basis. Of these, 16 tribes occur in India. Athysanini are the most common leafhoppers breeding on shrubs, trees and grasses. The male connective is usually Y-shaped. Chiasmusini (including Doraturini) are grass feeders. They are either brachypterous or macropterous and have a specialized aedeagus that is movably articulated at base. The species of *Nephotettix* are important pests of rice as they not only suck the sap form rice but also transmit plant disease organisms and Ghauri (1971) revised the genus. *Exitianus* Ball is a common grass inhabiting genus found throughout the world and has been revised by Ross (1968). Cicadulini is very similar to Athysanini but has hair-like setae confined to dorsal margin of male subgenital plates. They feed on trees and shrubs. Deltocephalini are the most common on grasses and have linear connective fused with aedeagus. It is a complex group and needs to be thoroughly revised on a worldwide basis. Dash and Viraktamath (1998) recently dealt with the species of *Deltocephalus* from India and Nepal. The genus *Paramesodes* Ishihara also a grass feeder has been revised by Wilson (1983). Fieberiellini are largely Holarctic and only *Phlogotettix indicus* Rao occurs in Nepal and northeast India (Rao, 1989). The anterior arms of the connective are strongly divergent forming an obtuse angle and the male style has a reduced apophysis. Goniagnathini include robust dark brown leafhoppers with short and broad heads having the male subgenital plates and valve fused to form a median plate. Only the genus *Goniagnathus* Fieber with 10 species occurs on the subcontinent (Dash and Viraktamath, 2001). The Grypotini are pine-feeding group of largely Holarctic leafhoppers. The head is short and broad. The antennae are more anterior and the ocelli are more than thrice own diameter away from adjacent eye. Clypellus is narrow and more than thrice longer than greatest width, tapered and acute distally. Ghauri and Viraktamath (1987) described two new genera and six new species of this group and placed them in Paraboloponinae, and Zhang and Webb (1996) transferred these to Grypotini. At present three genera (*Pinopona* Viraktamath and Sohi, *Shivapona* Ghauri and Viraktamath and *Sohipona* Ghauri and Viraktamath) and seven species are known from the Sub-Himalayan region of the subcontinent (Viraktamath and Sohi, 1998). The Hecalini are depressed grass feeding leafhoppers usually green with bright orange or red stripes. Head is strongly produced and the female second valvulae are strongly humped sub-basally. Morrison (1973) revised the Oriental Hecalini. Rama Subba Rao and Ramakrishnan (1990) and Dash and Viraktamath (1997) dealt with the Indian Hecalini. There are two genera (*Glossocratus* Fieber and *Hecalus* Stål) and 15 species are reported from India. Macrostelini (including Balcluthini) are an economically important group of leafhoppers transmitting viruses and phytoplasmas of cereals and millets. They are characterized by plumose setae on male and female pygophore and their typhlocynine-like facies. Knight and Webb (1993) dealt with the phylogeny of this group. Webb (1987) revised the genus *Cicadulina* China for the world and Webb and Vilbaste (1994) dealt with the genus *Balclutha* Kirkaldy for the Indian subcontinent. This group includes species that are grass feeders and show very wide distribution across zoogeographical region and some of them such as *Balclutha incisa* (Matsumura) are cosmopolitan. Occinirvanini are a group of leafhoppers that were placed in the subfamily Nirvaninae earlier and now rightly placed in Deltocephalinae. Head is strongly produced and antennal bases are closer to apex of crown than to eyes. Only one endemic genus and species, *Omaranus typicus* Distant occurs in Sub-Himalayan region of the subcontinent. This species is highly sexually dimorphic so much so that Distant (1918) placed the two sexes in two different monotypic genera. Opsiini are another economically important group breeding on herbs, shrubs and trees and are involved in the transmission of phytoplasmas in the subcontinent. The group is characterized by the presence of paired aedeagal shafts with paired gonopores. The genus *Hishimonus* along with two related genera were dealt with by Knight (1970a, 1970b and 1973). Ghauri (1966) and Viraktamath *et al.* (1987) revised the genera *Orosius* Distant and *Hishimonoides* Ishihara, respectively. The genera *Circulifer* Baker and *Neoaliturus* Distant are important vectors of plant viruses and spiroplasmas, and both the genera occur in the subcontinent. This group is known by 8 genera and 29 species from the subcontinent. Paralimnini are exclusively grass feeding deltocephalines with racket shaped male connective and forewings with cross-vein r-m_1 at or basad of R fork. They are slender to very robust leafhoppers. The apophysis of style in some genera is heavily sclerotized and pigmented. The genera *Paralevicephalus* Ishihara and *Hengchunia* Vilbaste are some of the common genera. Webb and Heller (1990) and Asche and Webb (1994) dealt with this group and provided a checklist for the Afrotropical and Oriental regions. Platymetopiini are mainly Holarctic group and very few species occur on the subcontinent. The group has depressed and slightly produced crown. Forewings have supernumerary reflexed veins. The group is similar to Athysanini. The Scaphoideini are also related to Athysanini. The male connective has well developed paired paraphyses and the front wings have vein R strongly curved. Some of the species are brightly colored and have tufts of backwardly directed setae in both the sexes. Viraktamath and Mohan (2003) revised the genus *Scaphoideus* for the subcontinent and it contains 33 described species. The Scaphytopiini include leafhoppers with expanded genae that are visible dorsally behind eyes. This group shows considerable variation in male genitalia. The aedeagus may have single or paired shaft and gonopores as in the case of *Japananus* Osborn and *Masiripius* Dlabola. The male valve and subgenital plates also show varying degrees of fusion (Viraktamath, 2004b). Viraktamath and Anantha Murthy (1999) revised the group for the subcontinent and there are 9 genera and 13 species. The species breed both on trees (*Japananus* Ball, *Univagris* Viraktamath and Ananatha Murthy, or shrubs (*Masiripius*) and on monocots (*Varta* Distant, *Curvimonus* Viraktamath and Anantha Murthy). The

Stenometopiini are exclusively grass feeding leafhoppers. Many species of this group are iridescent green or blue in life and lose that color once they are killed. They have a very uniform male and female genitalia but the shape of the head is very variable from short to greatly produced as in the subgenus *Cymbopogonella* Viraktamath of the genus *Doratulina* Melichar. A careful study of the group by Vilbaste (1965) resulted in the synonymy of 10 genera described by Distant (1908 and 1918) and 2 genera by Pruthi (1930 and 1936) with *Doratulina*. This group is also in need of revision.

Eupelicinae are medium sized to large (4-10 mm long) depressed seed-like elongate leafhoppers with strongly produced sptatulate heads. Ocelli are on the crown near margin. Forewing veins are raised. Male valve is fused with pygophore. Male styles are weakly sclerotized. Two genera *Bumizana* Distant with 8 species that includes seed-like grass feeding species and *Chloropelix* Lindberg with one species breeding probably on annual herbs are known from the subcontinent (Viraktamath and Viraktamath, 1989). The group is worldwide in distribution except the Neotropical region.

Hylicinae are the large robust dark brown leafhoppers with variably produced heads and ocelli are on crown. The lateral frontal sutures extend to ocelli. Both head and pronotum often possess bizarre projections. Scutellum is large. Abdomen is flattened with sharp lateral margins exposed. Wings are usually held flat over abdomen. Forewing appendix is large, corrugated and extends around the apex as far as costal margin. The body and wings are sparsely covered with scales and hairs. Viraktamath and Webb (1991) provided a key to the Indian genera and also revised the genus *Traiguma* Distant. They breed on shrubs and trees in moist forests.

Iassinae are medium to large (4-12 mm long) usually green or brown leafhoppers with short heads. Ocelli are on upper part of face below anterior margin of crown. Lateral frontal sutures do not reach ocelli. Both head and thorax are usually with transverse striae. Forewings often are setose, venation is complete and the inner apical cell elongate and membranous. Hind wing veins R_{4+5} and M_{1+2} usually confluent. Male sternum 8 is enlarged and largely concealing ligulate subgenital plates. Female second valvula is with 2-3 widely spaced dorsal teeth. The tribes Krisnini and Iassini occur on the subcontinent. The Krisnini are larger leafhoppers with submarginal vein on the hind wing extending over to vannal lobe that is not the case in the tribe Iassini. The genera *Krisna* Kirkaldy and *Batracomorphus* Lewis are found through out the subcontinent. The species of *Krisnini* breed on trees and shrubs such as *Piper betel* whereas Iassini breed on trees, shrubs and herbs. A few species of *Batracomorphus* breed on plants like *Dodonea viscosa, Ziziphus mauritania, Ficus, Parthenium, Morus alba* etc. *Batracomorphus angustatus* (Osborn) a very widespread species across Oriental and Australian regions is also common on the subcontinent. *Batracomorphus brunomaculatus* (Evans) extends its range from Africa to India. The genera *Gessius* Distant (Krisnini) and *Trocnadella* Pruthi (Iassini) occur in the Sub-Himalayan region. Linnavuori and Qaurtau (1975) dealt with the Afrotropical Iassinae. Kameswara Rao and Ramakrishnan (1980) and Kumar (1983) dealt with the species of the Indian subcontinent.

Idiocerinae occur on all zoogeographical regions. These are small to medium sized leafhoppers with short and broad heads. Ocelli are on face and distant from eyes. Lateral frontal sutures reach ocelli. Forewing venation is complete with large appendix. Male valve is fused with the pygophore. Both antennae and apical segment of labium are expanded in males of some species of *Amritodus* Anufriev and *Idioscopus* Baker. There is considerable host specificity for egg laying and development of early instars. Most species of *Idioscopus* and *Amritodus* use Anacardiaceae plants for breeding; the species of *Balocha* Distant use *Syzigium* (Myrtaceae) where as those of *Idiocerus* Lewis use *Cedrus* on the subcontinent. *Busoniomimus majunathi* Viraktamath and Viraktamath breeds on *Garcinia* and is attended by ants. Ghauri (1975), Kameswara Rao and Ramakrishnan (1979 and 1983), Maldonado-Capriles (1961, 1964 and 1977), Viraktamath (1973, 1976, 1979a and 1979b, 1980a, 1998a), Viraktamath and Murphy (1980), Viraktamath and Viraktamath (1985), Viraktamath and Sohi (1997), Viraktamath and Hongsaprug (1989-90) dealt with the Idiocerinae of the subcontinent. The genera *Amritodus*, *Balocha* and *Idioscopus* are found throughout the subcontinent whereas those of *Busonia* Distant, *Idiocerus* and *Tasnimocerus* Ghauri are found in the Sub-Himalayan region.

Ledrinae are medium to large sized (6-15 mm) rather cosmopolitan group containing brown or green leafhoppers with head usually produced and often spatulate. The ocelli are on crown and distant from margin. Lateral frontal sutures are partially or completely obsolete. Forewing venation is complete often with numerous supernumerary veins. Hind femora have three stout short macro-setae apically. Male subgenital plates are ligulate and partially retracted into enlarged eighth sternum. The nymphs are paper-thin and are oppressed to leaves or tree trunks or branches. Most species exhibit high degree of sexual dimorphism. The species of *Ledra* Fabricius, *Ledrospsis* White, *Confucius* Distant and *Petalocephala* Stål are most common. They breed on trees and shrubs. The species of *Petalocephala* use mango, guava and *Polyalthea* as breeding hosts.

Macropsinae are found in all zoogeographical regions and include small to medium sized (3-8 mm long), cylindrical leafhoppers with short broad beads. Head, pronotum and scutellum are rugose punctate. Ocelli are on face, distant from eyes. Lateral frontal sutures are reduced or absent. Pronotum usually strongly produced in front. Forewing venation is complete. Hind wing usually with three apical closed cells (R_{2+3} absent). Male valve is transverse and fused with pygophore. Male subgenital plates are ligulate. Species of *Macropsis* in south India breed on species of *Acacia*, *Ziziphus* and *Leucas*. Many of these species are gregarious and are attended by ants (Hymenoptera: Formicidae). Hamilton (1980) made a world revision of the genera of the Macropsinae. Kameswara Rao and Ramakrishnan (1980) and Viraktamath (1980b, 1981 and 1996) dealt with the fauna of the subcontinent.

Mukariinae are small, depressed shining black leafhoppers with head either produced or not. The ocelli are on crown near margin and the anterior margin of crown is transversely striate. Forewing venation is reduced. The male valve is triangular. Aedeagus is with paired shafts and gonopores. The two genera *Buloria* Distant and *Mukaria* Distant are very similar and occur on the subcontinent. The nymphs are yellow and breed on bamboo leaves.

Nirvaninae are medium sized depressed leafhoppers with black or bright red or orange markings. Head is strongly produced and often sptatulate. The ocelli are on crown usually near margin. Lateral frontal sutures are not extending to ocelli. Forewing venation is reduced with fewer cross veins. Front with a basal short ridge or carina. The male valve is small, laterally fused with pygophore. Both the tribes Balbillini and Nirvanini occur on the subcontinent. The Balbillini have fore and middle tibiae flattened, the head is notched in front of eyes so that the scape is visible in dorsal aspect. The Nirvanini have fore and middle tibiae rounded and head is entire in front eyes. *Nirvana pallida* Melichar and *Nirvana suturalis* Melichar are the most common species found on the subcontinent. They breed on trees, shrubs herbs and grasses. Viraktamath and Wesley (1988) dealt with the Nirvaninae of the subcontinent.

Nioninae are small to medium sized punctate shining black, cylindrical leafhoppers with very short and broad head. The ocelli are on face distant from eyes. Gena are broad. Pronotum is produced anteriorly. Claval veins are confluent in the middle. The male subgenital plates are triangular. The nymphs are yellow or reddish brown, smooth breeding on grasses. Only one species of the tribe Magnentiini, *Magnentius clavatus* (Pruthi) occurs on the Indian subcontinent.

Penthimiinae are medium sized depressed rather ovoid leafhoppers. Head is produced with transverse carinae on the anterior margin. The ocelli are either on anterior margin (*Neodartus* Melichar) or on crown near anterior margin of head. Lateral frontal sutures either extending to ocelli or not. Face is usually concave in profile. Forewing venation is complete with large appendix that goes round the apex. Claval cross veins are present. The male valve and subgenital plates are triangular. *Neodartus acocephaloides* Melichar and *Penthimia maculosa* Distant are the common and polyphagous leafhoppers of this subfamily and show considerable variation in coloration. Linnavuori (1977) dealt with the Afrotropical Penthimiinae and a similar treatment is urgently required for the Indian subcontinent.

Phlogosinae are a small group of medium sized cylindrical, chocolate brown leafhoppers. The head, thorax and basal half of forewing have coarse circular pits. Ocelli are on crown. The frontoclypeus is swollen. Pronotum is gibbous. Forewing venation is complete. This subfamily is similar to Signoretiinae but lacks carinae on head. It is distributed in the tropical Africa and China. One undescribed species and genus occurs in the Western Ghats of India (Viraktamath and Webb, unpublished record).

Selenocephalinae are medium to large (6-12 mm long), rather depressed leafhoppers. Heads are either short or produced and have anterior marginal transverse carinae. The ocelli are near margin of crown. Usually the pedicel of antennae is chocolate brown, antennal ledges are either poorly developed (Paraboloponini and Selenocephalini) or well developed (Drabescini). Three tribes occur on the subcontinent. Drabescini are large often dark brown leafhoppers with large appendix in the forewing. The Paraboloponini are green or yellow with sptatulate heads and antennae are nearly half body length. They breed on trees and shrubs. The genus *Dryadomorpha* Melichar breeds on *Ziziphus* whereas species of *Megabyzus* Distant have been collected on low vegetation on forest floor. Selenocephalinae have shorter antennae that arise near middle or lower corner of eyes. Species of the genus *Tambocerus* Zhang and Webb, *Carvaka* Distant and *Megabyzus* Distant are common. Webb (1981) revised the Paraboloponini of the Asian region. Zhang and Webb (1996) revised the group for the Oriental, Palaearctic and Australian regions. Viraktamath (1998b) dealt with the Paraboloponini of the Indian subcontinent.

Signoretiinae are slender, elongate, medium sized cylindrical leafhoppers. Head, pronotum, scutellum and forewings are covered with coarse circular pits. Crown of head has marginal transverse carinae. The frontoclypeus and clypellus are with a median longitudinal keel. Ocelli are on margin near eyes. Pronotum extends back covering most of the scutum. Forewing venation is complete with raised veins and the claval veins are connected by cross veins. Lateral frontal sutures extend to ocelli. These breed on herbs and shrubs in moist habitat. Genera *Signoretia* Stål and *Preta* Distant occur on the subcontinent. Anufriev (1971) dealt with the Oriental Signoretiinae.

Stegelytrinae are Palaearctic with Sub-Himalayan distribution of species that occur on the subcontinent. These are medium to large sized (5-11 mm long) robust light brown leafhoppers with heads narrower than pronotum. Ocelli are on crown near anterior margin. Lateral frontal sutures reaching ocelli. Forewing venation is reticulate with large appendix. Hind tibiae are with additional rows of setae near base. Patches of hair-like setae occur on head and thorax. This group was earlier placed in the subfamily Coelidiinae but now separated (Nielson, 1977). Genera *Kunasia* Distant (Burma), *Placidus* Distant (northwest India) and *Cyrta* Melichar (Nepal) are known to occur on the subcontinent with one species each. Zhang and Wei (2002a and 2002b) and Zhang *et al*. (2002) revised these genera for the world.

Tartessinae are large (9-11 mm long) leafhoppers with short and broad beads. The pronotum is produced well beyond outer margin of eyes. Ocelli are just below the crown. Front wing venation is complete. Hind wing has submarginal vein that extends over to vannal lobe as in Krisnini. Hind femoral setal formula is 2+2. This is an Australian group of leafhopper with only one species, *Tartessus ferrugineus* Walker recorded from Burma. They breed on trees and shrubs. Identity of the species from Burma needs to be confirmed.

Typhlocybinae are small, delicate, often brightly colored leafhoppers. Head often produced. The ocelli, when present, are on anterior margin of head. Forewings are without cross veins basad of apical cells. Hind basitarsomere is acutely pointed apically. These show worldwide distribution. They breed on agricultural crops, trees, shrubs, herbs and grasses. Many are mesophyll feeders and some feed on phloem. Many species especially those of *Empoasca* Walsh, *Amrasca* Ghauri, *Thaia* Ghauri, *Empoascanara* Distant, *Zygina* Fieber etc. include species of great economic importance. The subfamily is divided into six tribes and all of them occur in the subcontinent. The Alebrini include those typholocybines that have an appendix. In Dikraneurini the hind wings have the submarginal vein extending laterad and basad along costa beyond apex of R+M. In Erythroneurini, the anal veins in the hind wings are fused. In Empoascini, the submarginal vein in the hind wing reaches wing apex and fuses with R or R+M or M_{1+2}. In Zyginellini, the submarginal vein in the hind wing is confluent with Cu1 near fusion of M_{3+4} and Cu1; vein M_{3+4} and Cu1 fuses up to wing apex and R+M is present forming one open apical cell. In Typhlocybini, the submarginal vein is confluent with Cu1; M_{3+4} and Cu1 are separate apically, R+M may or may not be present, thus two or thee apical cells present. The Typhlocybinae of the subcontinent have received considerable attention by Dworakowska (1977, 1979, 1980,1981a and 1981b), Pruthi (1940), Sohi (1977, 1978 and 1998), Ramakrishnan and Menon (1971, 1972a, 1972b 1973, 1974), Mathew and Ramakrishnan (2002), Sharma (1979 and 1984) and Thapa (1983, 1985). Sohi and Dworakowska (1984) prepared a catalogue of the Typhlocybinae of India, which has been updated up to 1997 by Sohi (1998). Mahmood (1975), Mohmood and Ahmed (1969) and Ahmed (1969a, 1969b and 1970) in a series of papers dealt with Typhlocybinae of Pakistan.

Ulopinae are small to medium sized (2-8 mm long) leafhoppers with elytra-like forewings. Head is weakly to strongly produced. The ocelli are on crown in usually well-developed post-frontal sutures. Head, thorax and forewings are with coarse circular pits. Pro-episternum is entirely exposed. The spinulation on hind tibia is usually reduced. The Mesargini breed on trees and shrubs whereas the Ulopini are found often breeding on plant parts that are at soil level or on roots of plants. These are found in all the zoogeographical regions. The species of *Mesargus* Melichar and *Ulopa* Fallén are common in the subcontinent.

Xestocephalinae include small to medium sized leafhoppers distributed in all the zoogeographical regions. This subfamily is usually considered as a subfamily of Deltocephalinae. The coloration is usually brown or mottled with brown. Head is produced and rounded. The ocelli are near anterior margin distant from eyes. Lateral frontal sutures are extended just ventro-mesad of ocelli. The antennal ledges are well developed. Forewing venation is complete. Subgenital plates are usually ligulate. The female ovipositor is strongly arcuate. Both the tribes Xestocephalini and Portanini occur on the subcontinent. They occur on shrubs and trees. The nymphs of some *Xestocephalus* Van Duzee feed on grass roots. The adults are common at light.

ECONOMIC IMPORTANCE

Leafhoppers are one of the most important groups of insects that utilize plants for their growth and development. They suck plant sap and thus reduce the vitality of the plant. Some species such as, cotton leafhopper, *Amrasca biguttula biguttula* (Ishida), mango leafhopper *Amrasca splendens* Ghauri inject toxins thus affecting the growth of the plant and also flower and fruit production. The female leafhoppers insert their eggs in plant tissue thus causing oviposition injury. These injuries also form avenue for the entry of plant-pathogens. More important losses caused by leafhoppers are indirect. Several species act as vectors of plant pathogens such as viruses, spiroplasmas, phytoplasmas and bacteria (Table 2).

The relationship between the leafhopper vectors and the pathogens that they transmit is very intimate. Except for rice tungro wherein the virus is transmitted in a semi-persistent manner, other pathogens either circulate or multiply in their vectors. The pathogens also are entirely dependent on leafhopper vectors for their transmission. In some instances as in the case of rice dwarf virus, the pathogen is transmitted trans-ovarially. There are little over 151 species of leafhoppers in the world that transmit pathogens (Nielson, 1979 and 1985) and they are second only to aphids in their role as vectors of plant-pathogens.

LIFE CYCLE

The leafhoppers undergo three stages – egg, nymph and adult. The eggs are small, banana shaped and are inserted in plant tissue with the help of saw-like ovipositor. They may be inserted in the leaf lamina, leaf veins (as in species of *Empoasca*, *Amrasca*, *Empoascanara*), or shoots (as in *Amritodus*, *Macrospsis*) or stem (as in *Ledra* and *Petalocephala*). Eggs may be placed either singly or in groups. In some groups of leafhoppers as in the case Proconini of Cicadellinae the egg groups are covered by waxy material by the female during oviposition (Rakitov, 2004). The incubation period depends on species and prevailing temperature. After hatching the nymph feeds on the host plant. Some of the species of leafhoppers are very specific as for as egg laying hosts are concerned and their nymphs may feed on wider range of plants. The nymphs undergo five moults to reach adult stage.

Table 2 Leafhopper vectors of plant-pathogens in India

Leafhopper vector*	Pathogen transmitted
Hishimonus phycitis(Distant)	Little leaf of brinjal (Phytoplasma)
Orosius albicinctus Distant	Phyllody of sesame (Phytoplasma)
	Phyllody of gram (Phytoplasma)
	Potato purple top roll (Phytoplasma)
Nephotettix virescens (Distant)	Rice tungro virus
	Rice yellow dwarf (Phytoplasma)
Nephotettix nigropictus (Stål)	Rice tungro virus
	Rice yellow dwarf (Phytoplasma)
Deltocephalus dorsalis	Rice stunt (Phytoplasma)
(de Motschulsky)	Orange leaf virus
	Ragi streak virus
Cicadulina bipunctata (Melichar)	Ragi mottle streak virus
Cicadulina chinai Ghauri	
	Sandal spike (Phytoplasma)
Calodia kirkaldyi Nielson	
(= *Coelidia indica* auct)	
	Sandal spike (Phytoplasma) suspected
Mesargus albimaculata (Distant)	Purple top roll of potato
Alebroides nigroscutellatus(Distant)	Witches broom of potato

*Based on Nielson (1979 and 1985)

The adults feed on a variety of plants thus making it difficult to recognize their true egg laying or nymphal hosts. Some species of leafhoppers are highly host specific, for example, the mango leafhopper, *Idioscopus clypealis* breeds only on the inflorescence stalk of mango whereas other species namely, *Idioscopus nitidulus, Amritodus atkinsoni* and *Amrsca splendens* though specific to mango can breed both on new flush of leaves and inflorescence stalk (Viraktamath, 1989). *Amrasca biguttula biguttula* can however, feed and breed on a number of hosts of several families such as Solanaceae, Cucurbitaceae, Malvaceae, Asteraceae etc.

In temperate region the leafhoppers undergo hibernation either in egg or late larval or adult stage. Female has to mate in order to lay eggs. There is a single case of parthenogenetic reproduction in leafhoppers (Oman, 1949).

BEHAVIOR OF LEAFHOPPERS

Both adults and nymphs of leafhoppers show very peculiar behavior of spreading secretions of the Malpighian tubules that contain lipids. Two types of behavior are seen in nymphs namely, anointing and bathing.

In anointing the nymphs immediately after moulting produce a viscous fluid at the tip of abdomen, which is collected by the hind legs and smeared on the ventral surface of the abdomen. Then the tip of the abdomen along with viscous fluid is bent over to the thoracic region so that the viscous liquid spreads over the dorsum of the abdomen and legs are used to spread this liquid over dorsal regions of thorax, head and face till it dries up.

In bathing, the nymph produces large amount of viscous fluid and attaches it to the substratum on which it is feeding. Then it plunges in this fluid and twists and turns its body till entire body is smeared with the fluid.

In adults the viscous fluid is picked up by the hind legs and deposited on the forewings. In some species there are special areas in the costal region of forewing specialized to hold the fluid (wax field of earlier worker). From here the material is rubbed to all parts of the body including ventral surface of the front wings by vigorous movement of hind legs. The rows of setae on the hind tibiae acting as comb and brush and help in spreading the material now called brochosomes. It has now been shown by Rakitov (2004) that the function of series of setae on hind tibiae is spreading the brochosomes on different body parts of the leafhoppers. It is suggested that the brochosomes play an important role of waterproofing the surface of leafhoppers in addition they may also provide protection against invasion of pathogens.

ANT ASSOCIATION

It is a common site to see ants attending aphids, coccids and membracids. The ants attend very few leafhoppers. The leafhoppers attended by ants are often gregarious providing a continuous source of food (honeydew) to attending ants and are not disturbed by the movements of the ants. Some of the Indian species of leafhoppers attended by the ants are *Hishimonus viraktamathi* Knight breeding on *Terminalia* sp. (Knight, 1973), species of *Macropsis* breeding on *Acacia* (Viraktamath, 1980b), *Busoniomimus manjunathi* breeding on *Garcinia.* It is possible that attending ants provide protection to leafhoppers from their natural enemies as they do in the case of aphids.

COMMUNICATION

The two sexes of leafhoppers communicate through substrate-transmitted sound (Claridge, 1985). The sound is produced by both the sexes, though males have more elaborate sound producing organs (tymbal). In order to communicate, both the sexes should either be on the same leaf or leaves on which they are sitting should be in physical contact with each other. The sound produced by each species of leafhoppers is very distinctive and hence can be used in species identification.

CONCLUSIONS

Taxonomy of insects is complicated because of species complexes that are externally often very similar and one often fails to recognize them unless he is trained and uses the right type of techniques. In some leafhopper taxa such as species of *Circulifer* and *Neoaliturus,* molecular techniques need to be employed to unravel species complexes. It is estimated that, as in other groups of insects, we know only 40-50% of our leafhopper fauna. The number of trained taxonomists is also dwindling day by day as fewer young students are choosing taxonomy as their professional career. Need of the hour is to create interest among the younger generation so that they take up taxonomy as a profession and give them the best training available. It is also necessary to develop checklists, distribution maps and illustrated keys for identification of known species. A number of leafhopper taxa need to be revised for the subcontinent so that they can be identified. This is especially important for the vector species wherein accurate identification of the species, subspecies and even the biotype is very essential as in some cases ability to transmit pathogen may be influenced by the prevalence of biotypes. Development of insect collection based database goes a long way in promoting export and import of agricultural produce. Similarly, phylogenetic studies of different taxa need to be undertaken using morphological, biochemical, and behavioral, ecological and molecular characters. Leafhoppers form an ideal subject for ecological and behavioral studies, assessing ecological health of a habitat and also understanding the conservation needs of the region.

References

Ahmed, M. 1969a. Studies of tribe Dikraneurini (Typhlocybinae: Cicadellidae: Homoptera) from West Pakistan. *Pakistan Journal of Zoology* 1: 55-63.

Ahmed, M. 1969b. Some recent studies of *Empoasca* complex in Pakistan, their bearing on tribal classification of subfamily Typhlocbinae (Homoptera: Cicadellidae). *Agriculture Pakistan* 20(4): 451-455.

Ahmed, M. 1970. Studies on the genus *Erythroneura* Fitch (Erythroneurini: Cicadellidae) in West Pakistan. *Pakistan Journal of Zoology* 2: 29-42.

Ahmed, M. and Naheed, R. 1982. Leafhopper infestations of vegetable and fruit plants in Sind-Pakistan during 1979-80. *Pakistan Journal of Scientific and Industrial Research* 25(4): 124-125.

Ahmed, M., Baluch, Ahmed, M. and Ahmed, M. 1980. Abundance and diversity of typhlocybine leafhoppers on vegetable plants in Pakistan. *Pakistan Journal of Scientific and Industrial Research* 23 (1&2): 41-47.

Al-Ne'amy, K. T. and Linnavuori, R. E. 1982. On the taxonomy of the subfamily Adelungiinae (Homoptera: Cicadellidae). *Suomen Hyonteistieteellinen Aikakauskirja* 48(4): 109-115.

Anufriev, G. A.1971. Study of leafhoppers from the tribe Signoretiini (Homoptera: Auchenorrhyncha) with descriptions of two new species of *Signoretia* Stal from Western Africa. *Bulletin de l'Academie Polonaise des Sciences. Serie des Sciences Biologiques* 19(11): 721-726.

Asche, M. and Webb, M. D. 1994. Review of the southern Palaearctic and Palaeotropical leafhopper genus *Hengchunia* Vilbaste (Homoptera, Cicadellidae). *Tijdschrift voor Entomologie* 137(2): 143-154

Claridge, M.F. 1985. Acoustic behaviour of leafhoppers and planthoppers: Species problems and speciation. In: Nault, L.R. and Rodriguez, J.G. eds. *The Leafhoppers and Planthoppers*. pp. 103-125. John Wiley & Sons, New York, 500pp.

Dash, P.C. and Viraktamath, C. A. 1997. New species of grass feeding hecaline leafhopper genera *Glossocratus* and *Hecalus* (Hemiptera: Cicadellidae) from India. *Journal of the Bombay Natural History Society* 94: 127-138.

Dash, P.C. and Viraktamath, C. A. 1998. A review of the Indian and Nepalese grass feeding leafhopper genus *Deltocephalus* (Homoptera: Cicadellidae) with description of new species. *Hexapoda* 10: 1-59.

Dash, P.C. and Viraktamath, C. A. 2001. Deltocephaline leafhopper genus *Goniagnathus* (Hemiptera: Cicadellidae) in the Indian subcontinent with descriptions of four new species. . *Journal of the Bombay Natural History Society* 98: 62-79.

DeLong, D.M. 1971. The bionomics of leafhoppers. *Annual Review of Entomology* 16: 179-210.

Dietrich, C. H. 2004a. Phylogeny of the leafhopper subfamily Evacanthinae with a review of Neotropical species and notes on related groups (Hemiptera: Membracoidea: Cicadellidae). *Systematic Entomology* 29: 455-487.

Dietrich, C.H. 2004b.http://www.inhs.uiuc.edu/dietrich/leafhom.html.

Distant, W.L. 1908. Rhynchota- Homoptera. In: Bingham, C.T. ed. *The Fauna of British India including Ceylon and Burma* 4, 501 p.

Distant, W.L. 1916. Rhynchota, Homoptera. Appendix. In: Shipley and Marshall, G.A.K eds. *The Fauna of British India including Ceylon and Burma* 6, vii+248 p.

Distant, W.L. 1918. Rhynchota- Homoptera: Appendix. Heteroptera: Addenda. *In*: Shipley and Marshall, G.A.K eds. *The Fauna of British India including Ceylon and Burma* 7, viii, 2pp 210 p.

Dworakowska, I. 1977. On some north Indian Typhlocybinae. *Reichenbachia Staatliches Museum für Tierkunde, Dresden* 16(29): 283-306.

Dworakowska, I. 1979. On some Typhlocybinae from India and adjoining areas (Homoptera, Auchenorrhyncha: Cicadellidae). *Reichenbachia Staatliches Museum für Tierkunde, Dresden* 17(18): 143-162.

Dworakowska, I. 1980. On some Typhlocybinae from India (Homoptera, Auchenorrhyncha, Cicadellidae). *Entomologische Abhandlungen und Berichte aus dem Staatlichen Museum für Tierkunde in Dresden* 43(8): 151-201.

Dworakowska, I. 1981a. On some Typhlocybinae from India, Sri Lanka and Nepal (Homoptera, Auchenorrhyncha, Cicadellidae). *Entomologische Abhandlungen und Berichte aus dem Staatlichen Museum für Tierkunde in Dresden* 44(8): 153-202.

Dworakowska, I. 1981b. On the Typhlocybini from India and Nepal (Homoptera, Auchenorhyncha, Cicadellidae, Typhlocybinae). *Bulletin de l'Academie Plonaise des Sciences. Serie des Sciences Biologiques* 28(10&11): 593-602.

Emeljanove, A.F. 1975. Revision of the tribe Adelungiini. *Entomologicheskoe Obozrenie* (Moscow) 54: 383-390.

Fabricius, J.C. 1794. Ryngota. Entomologia Systematica emendata et Aucta. Secundum classes, ordines, genera, species adiectis synonimis, locis, observtionibus descriptionibus 4: 1-472. Haftniae, Proft.

Fabricius, J.C. 1803. Rhyngota. Systema Rhyngotorum, secundum ordines, genera, species, adiectis synonymis, ocis, observationibus, descriptionibus. 314 pp.

Ghauri, M. S. K. 1975. A new genus and species of idiocerine leafhopper (Cicadelloidea, Auchenorrhynhca, Homoptera) on pear tree from Simla. *Novelle Revue d'Entmologie* 5(3): 287-290.

Ghauri, M.S.K. 1966. Revision of the genus *Orosius* Distant (Homoptera: Cicadelloidea). *The Bulletin of the British Museum Natural History Entomology Series* 18: 231-252.

Ghauri, M.S.K. 1971. Revision of the genus *Nephotettix* Matsumura (Homoptera: Cicadelloidea: Euscelidae) based on the type material. *Bulletin of Entomological Research* 60: 481-512.

Ghauri, M.S.K. and Viraktamath, C. A. 1987. New Paraboloponinae from the Sub Himalayan region (Insecta, Homoptera, Cicadelloidea, Iassidae). *Reichenbachia Staatliches Museum für Tierkunde, Dresden* 25: (12) 47-58.

Ghosh, L.K. and Ghosh, M. 1994. Insects: Hemiptera: Cicadellidae. Pp. 29-68. In: *Fauna of West Bengal, Part-5 (Insecta: Hemiptera)*. Zoological Survey of India, Calcutta.

Hamilton, K.G.A. 1980. Contributions to the study of the World Macropsini (Rhynchota: Homoptera: Cicadellidae). *Canadian Entomologist* 112: 875-932.

Jacob, P.S., Ramasubba Rao, V. and Punnaiah, K.C. 2000. Leafhopper fauna associated with oilseed crops in Andhra Pradesh, India. *Pest Management and Economic Zoology* 8: 11-27.

Kameswara Rao, P. and Ramakrishnan, U. 1979. Two new species of *Balocha* (Idiocridae, Homoptera). *Journal of the Bombay Natural History Society* 76: 342-345.

Kameswara Rao, P. and Ramakrishnan, U. 1979. Two new species of *Macropsis* (Homoptera: Cicadellidae: Macropsinae) from northern India. *Oriental Insects* 13: 307-309.

Kameswara Rao, P. and Ramakrishnan, U. 1980. Two new species of the genus *Batracomorphus* (Homoptera: Cicadelloidea: Iassidae). *Entomon* 5: 43-46.

Kameswara Rao, P. and Ramakrishnan, U. 1983. Description of *Balocha tricolor* Distant (Idiocrinae: Homoptera). *Bulletin of Entomology* 24(1): 21-23.

Knight, W.J. 1970a. A revision of the genus *Hishimonus* Ishihara (Hom., Cicadellidae). *Suomen Hyonteisteellienen Aikakauskirja* 36: 125-139.

Knight, W.J. 1970b. Two new genera of leafhoppes related to *Hishimonus* Ishihara (Hom., Cicadellidae). *Suomen Hyonteisteellienen Aikakauskirja* 36: 174-182.

Knight, W.J. 1973. A new species of *Hishimonus* Ishihara (Hom., Cicadellidae) attacking *Terminalia* spp. In India, with comments on the relationship of the genus to *Cestius* Distant. *Suomen Hyonteisteellienen Aikakauskirja* 39: 153-156.

Knight, W.J. and Webb, M.D. 1993. The phylogenetic relationships between virus vector and other genera of macrosteline leafhoppers, including descriptions of new taxa (Homoptera: Cicadellidae: Deltocephalinae). *Systematic Entomology* 18: 11-55.

Kumar, A.R.V. 1983. A revision of Indian *Batracomorphus* Lewis (Homoptera: Cicadellidae: Iassinae). *Thesis Abstract, Haryana Agricultrual Unviersity* 9(4): 331.

Lethierry. L.F. 1889. Description of three new Homoptera. *Journal of Asiatic Society of Bangal* 58: 252-253.

Linnavuori, R. 1977. Revision of the Ethiopian Cicadellidae (Hemiptera-Homoptera): Penthimiinae. *Etudes du Continent Africain*, Fascicle 4: 1-76

Linnavuori, R. and Quartau, J. 1975. Revision of the Ethiopian Cicadellidae. (Hemiptera-Homoptera. Iassinae and Acroponinae. *Etudes du Continent Africain, Fascicle* 3: 1-167.

Mahmood, S. H. and Ahmed, M. 1969. Studies of tribe Alebrini (Typhlocybinae: Cicadellidae) in East Pakistan. *Sind University Research Journal Science Series* 4(1&2): 85-91.

Mahmood, S.H. 1975. A revision of the leafhoppers (Cicadellidae: Homoptera) of Pakistan and adjoining countries of the Oriental region. *First Annual Report, Department of Zoology, University of Karachi, Pakistan.* Mimeo. 24pp.

Maldonado-Capriles, J. 1961. Studies on Idiocerinae leafhoppers: I. *Idiocrinus* Baker, 1915, synonym of *Balocha* Distant, 1908, and notes on the species of *Balocha*. *Proceedings of the Entomological Society of Washington* 63(4): 244-246.

Maldonado-Capriles, J. 1964. Studies on Idiocerinae leafhoppers: II. The Indian and Philippine species of *Idiocerus* and the genus *Idioscopus* (Homoptera: Cicadellidae). *Proceedings of the Entomological Society of Washington* 66: 89-100.

Maldonado-Capriles, J. 1977. Studies on Idiocerinae leafhoppers: XVI. *Busonia* Distant and an allied new genus from the Oriental region (Cicadellidae: Idiocerinae). *Pacific Insects* 17*: 491-501.*

Mathew, G. 2004. Biodiversity documentation for Kerala, Part 7: Insects. *Kerala Forest Research Institute Handbook* 17, 1-281 pp. Peechi, Kerala.

Mathew, M. P. and Ramakrishnan, U. 2002. Five new species of *Zyginellini* (Cicadellidae: Typhlocybinae) from India. *Shashpa* 9(2): 109-120.

Melichar, L. 1903. *Homopteren- Fauna von Ceylon*. 248 p.

Metcalf, Z.P. 1962-68. *General Catalogue of the Homoptera Fascicle VI Cicadelloidea parts 1-17*. Agricultural Research Service, U.S. Department of Agriculture.

Metcalf, Z.P. 1964. *General Catalogue of the Homoptera Fascicle VI Bibliography of the Cicadelloidea (Homoptera: Auchenorrhyncha)*. Agricultural Research Service, U.S. Department of Agriculture.

Morrison, W. P. 1973. A revision of the Hecalinae (Homoptera: Cicadellidae) of the Oriental region. *Pacific Insects* 15: 379-438.

Nielson, M. W. 1977. A revision of the subfamily Coelidiinae (Homoptera: Cicadellidae). II. Tribe Thagriini. *Pacific Insects Monograph* 34. 218 p.

Nielson, M. W. 1985. Leafhopper systematics. In: Nault, L.R. and Rodriguez, J.G. eds. *The Leafhoppes and Planthoppers.* Pp. 11-39. John Wiley & Sons, New York. 500p.

Nielson, M.W. 1979. Taxonomic relationships of leafhopper vectors of plant pathogens. *In:* Maramarosch, K. and Harris, K. eds. *Leafhopper Vectors and Plant Disease Agents*. pp 3-27. Academic Press, New York. 654 p.

Nielson, M.W. 1982. A revision of the subfamily Coelidiinae (Homoptera: Cicadellidae). VI. Tribe Coelidiini. *Pacific Insects Monograph* 38. 318 p.

Oman, P. W. 1949. The Nearctic leafhoppers (Homoptera: Cicadellidae). A generic classification and check list. *Memoirs of the Entomological Scociety of Washington*. 3: 1-253.

Oman, P.W., Knight, W.J. and Nielson, M.W. 1990. *Leafhoppers (Cicadellidae) A bibliography, generic checklist and index to the world literature 1956-1985*. C.A.B. International Institute of Entomology, 368p.

Pruthi, H.S. 1930. Studies on Indian Jassidae (Homoptera). Part I. Introduction and description of some new genera and species. *Memoirs of the Indian Museum* 11: 1-68.

Pruthi, H.S. 1934a. Entomological investigations on the spike disease of sandal (14). Jassidae (Homopt.). *Indian Forest Records, Entomology Series* 19(4): 1-30.

Pruthi, H.S. 1934b. Studies on Indian Jassidae (Homoptera). Part II. Description of the genotypes of some of the genera founded by W.L. Distant, with a revision of the genus *Moonia* Distant. *Memoirs of the Indian Museum* 11: 69-99.

Pruthi, H.S. 1936. Studies on Indian Jassidae (Homoptera). Part III. Descriptions of some new genera and species, with first records of some known species from India. *Memoirs of the Indian Museum* 11: 101-131.

Pruthi, H.S. 1940. Descriptions of some new species of *Empoasca* Walsh (Eupterygidae, Jassoidea) from North India. *Indian Journal of Entomology* 2(1): 1-10.

Rakitov, R. A. 2004. .Powdering of egg nests with brochosomes and related sexual dimorphism in leafhoppers (Hemiptera: Cicadellidae). *Zoological Journal of the Linnean Society* 140(3): 353-381.

Ramakrishnan, U. and Menon, M.G.R. 1971. Studies on Indian Typhlocybinae (Homoptera: Cicadellidae. 1. Five new genera and a new record of Dikraneurini. *Oriental Insects* 5: 455-469.

Ramakrishnan, U. and Menon, M.G.R. 1972a. Studies on Indian Typhlocybinae (Homoptera: Cicadellidae. 2. Ten new species of Typhlocybini. *Oriental Insects* 6: 111-129.

Ramakrishnan, U. and Menon, M.G.R. 1972b. Studies on Indian Typhlocybinae (Homoptera: Cicadellidae. 3. Three new genera with four new species of Typhlocybini. *Oriental Insects* 6: 183-192.

Ramakrishnan, U. and Menon, M.G.R. 1973. Studies on Indian Typhlocybinae (Homoptera: Cicadellidae. 4. Seven new genera and fourteen new species of Erythroneurini. *Oriental Insects* 7: 15-48.

Ramakrishnan, U. and Menon, M.G.R. 1974. Studies on Indian Typhlocybinae (Homoptera: Cicadellidae. 5. The tribe Erythroneurini. *Oriental Insects* 8: 433-452.

Ramasubba Rao, V. and Ramakrishnan, U. 1990. The Indian species of *Hecalus* with descriptions of three new species (Homoptera: Cicadellidae). *Oriental Insects* 24: 385-397.

Rangaswamy, R.S. and Griffith, A.L. 1941. Demonstration of *Jassus indicus* (Walk.) as a vector of the spike disease of sandal (*Santalum album* Linn.). *Indian Forester* 67: 387-394.

Rao, K. R. 1989. Descriptions of some new leafhoppers (Homoptera: Cicadellidae) with notes on some synonymies and imperfectly known species from India. *Hexapoda* 1:59-83.

Rao, K. R. 1990. Studies on a small collection of leafhoppers (Homoptera: Cicadellidae) from Khasi Hills, Meghalaya. Occasional paper No. 127, Zoological Survey of India, Calcutta, 123p.

Reddy, P.P. and Ramasubba Rao, V. 2001. Leafhopper fauna associated with vegetable crops of Andhra Pradesh in India. *Entomon* 26: 121-130.

Ross, H.H. 1968. The evolution and dispersal of the grassland leafhopper genus *Exitianus*, with keys to the Old World species (Cicadellidae: Hemiptera). *The Bulletin of the British Museum (Natural History) Entomology* 22: 1-30.

Sharma, B. 1979. *Indodikra* gen. nov. along with four new Typhlocyinae from Nagaland, India (Auchenorrrhyncha: Homoptera: Cicadellidae). *Proceedings of the Indian Science Congress* 66: 5.

Sharma, B. 1984. On some Typhlocybinae from Kishtwar Jammu and Kashmir India with description of five new species. *Reichenbachia Staatliches Museum für Tierkunde, Dresden* 22(3): 31-40.

Sohi, A. S. 1998. Taxonomic status of Indian Typhlocybinae (Homoptera: Auchenorrhyncha: Cicadellidae). *Journal of Insect Science* 11(2): 85-109.

Sohi, A.S. 1977. New genera and species of Typhlocybinae (Homoptera: Cicadellidae) from north-western India. *Oriental insects* 11: 347-362.

Sohi, A.S. 1978. Keys to the tribes and Indian species of Dikraneurini of subfamily Typhlocybinae (Homoptera: Cicadellidae). *Journal of Research Punjab Agricultural Universiry* 15(2): 182-186.

Sohi, A.S. and Dworakowska, I. 1984. A review of the Indian Typhlocybinae (Homoptera: Cicadellidae) from India. *Oriental Insects* 17: 159-213.

Thapa, V.K. 1983. Descriptions of two new genera and a few new records of leafhoppers (Typhlocybinae, Cicadellidae, Homoptera) from Nepal. In: Knight, W.J., Pant, N.C., Robertson, T.S. and Wilson, M. R. eds. *Proceedings of the 1st International Workshop on Biotaxonomy Classification and Biology of Leafhoppers and Planthoppers (Auchenorrhyncha) of Economic Importance.* pp. 173-177. London, 4-7 October 1982. Commonwealth Institute of Entomology, 56 Queen's Gate, London. 500pp.

Thapa, V.K. 1985. Some empoascan leafhoppers (Homoptera, Cicadellidae, Typhlocybinae) from the Kathmandu Valley, Nepal. *Journal of Entomological Research* (New Delhi) 9(1): 65-74.

Vilbaste, J. 1965. On the genus *Aconura* Leth. (Homoptera, Jassidae). *Notulae Entomologicae* 45: 3-12.

Viraktamath, C. A. 1973. A new species of Idiocerinae (Cicadellidae: Homoptera) on *Semecarpus anacardium* L. *Oriental Insects* 7: 133-135.

Viraktamath, C. A. 1976. New species of idiocerine leafhoppers from India with a note on male *Balocha astuta* (Melichar) (Homoptera: Cicadellidae: Idiocerinae). *Mysore Journal of Agricultural Sciences* 10: 234-244.

Viraktamath, C. A. 1979a. *Jogocerus* gen. nov. and new species of idiocerine leafhoppers from southern India (Homoptera: Cicadellidae). *Entomon* 4: 17-26.

Viraktamath, C. A. 1979b. Four new species of *Idioscopus* (Homoptera: Cicadellidae) from southern India. *Entomon* 4: 173-181.

Viraktamath, C. A. 1980a. Notes on *Idioscopus* species (Homoptera: Cicadellidae) described by Dr H.S. Pruthi, with description of a new species from Meghalaya, India. *Entomon* 5: 227-231.

Viraktamath, C. A. 1980b. Indian Macropsinae (Homoptera: Cicadellidae) I. New species of *Macropsis* from south India. *Journal of Natural History* 14: 319-329.

Viraktamath, C. A. 1981. Indian Macropsinae (Homoptera: Cicadellidae). II. Species described by W.L. Distant and description of new species from the Indian subcontinent. *Entomologica Scandinavica* 12: 295-310.

Viraktamath, C. A. 1996. New Oriental Macropsinae with a key to species of the Indian subcontinent (Insecta: Auchenorrhyncha: Cicadellidae). *Entomologische Abhandlungen, Stätliches Museum für Tierkunde, Dresden* 57(7): 183-200.

Viraktamath, C. A. 1998a. A revision of the idiocerine leafhopper genus *Amritodus* (Hemiptera: Cicadellidae) breeding on mango. *Entomon* 22 (1997): 111-117.

Viraktamath, C. A. 1998b. A revision of the leafhopper tribe Paraboloponini (Hemiptera: Cicadellidae: Selenocephalinae) of the Indian subcontinent. *The Bulletin of the British Museum of Natural History (Ent. Series)* 67 (2): 153-208.

Viraktamath, C. A. 2004a. Revision of the Agalliinae leafhopper genus *Durgades* (Hemiptera: Cicadellidae) along with description of six new species. Pp. 363-380. *In*: Rajamohana, K., Sudheer, K., Girish Kumar, P. and Santhosh, S. eds. 2004. *Perspectives on biosystematics and biodiversity, Prof. T.C. Narendran Commemoration volume,* Sersa Publisher, Calicut, Kerala, India, xviii+666pp.

Viraktamath, C. A. and Anantha Murthy, H.V. 1999. A revision of the leafhopper tribe Scaphytopiini from India and Nepal (Insecta, Hemiptera, Cicadellidae, Deltocephalinae). *Senkenbergiana biologica (Germany)* 79: 39-55.

Viraktamath, C. A. and Hongsaprug, W. 1989-90. Two new species of *Idioscopus* (Homoptera: Cicadellidae) from Nepal and Thailand. *Zoologica Orientalis* 6 and 7: 1-5.

Viraktamath, C. A. and Mohan, G.S. 2003. A revision of the deltocephaline leafhopper genus *Scaphoideus* (Hemiptera: Cicadellidae) from the Indian subcontinent. *Zootaxa* 578: 1-48.

Viraktamath, C. A. and Murphy, D. H. 1980. Description of two new species with notes on some Oriental Idiocerinae (Homoptera: Cicadellidae). *Journal of Entomological Research* 4: 83-90.

Viraktamath, C. A. and Sohi, A.S. 1994. Three new species of *Idiocerus* (Hemiptera: Cicadellidae) from north India. *Entomon* 19: 23-28.

Viraktamath, C. A. and Sohi, A.S. 1998. A new grypotine leafhopper genus and species from the Indian subcontinent (Hemiptera: Cicadellidae). *Journal of Insect Science* 11 (2): 114-116.

Viraktamath, C. A. and Viraktamath, S. 1989. New species of *Bumizana* Dist. and *Chloropelix* Lindb. From Indian subcontinent (Insecta: Homoptera: Auchenorrhyncha, Cicadellidae). *Reichenbachia Staatliches Museum für Tierkunde, Dresden* 27 (4): 19-26.

Viraktamath, C. A. and Webb, M.D. 1991. A review of the Indian Hylicine leafhoppers of the genus *Traiguma* Dist., with description of a new species (Insecta, Homoptera, Auchenorrhyncha, Cicadellidae). *Reichenbachia Staatliches Museum für Tierkunde, Dresden* 28 (3): 123-133.

Viraktamath, C. A. and Wesley, C.S. 1988. Revision of the Nirvaninae (Homoptera: Cicadellidae) of the Indian subcontinent. In: Research in the Auchenorrhyncha Homoptera: A tribute to Paul W. Oman. *Great Basin Naturalist Memoirs No.* 12, 182-223.

Viraktamath, C. A., Anantha Murthy, H.V. and Viraktamath, S. 1987. The deltocephaline leafhopper genus *Hishimonoides* in the Indian subcontinent (Homoptera). *Journal of Natural History, London* 21: 1225:1236.

Viraktamath, C.A. 1989. Auchenorrhyncha (Homopterq) associated with mango, *Mangifera indica* L. *Tropical Pest Management* 35: 431-434.

Viraktamath, C.A. 2004b. A revision of the *Varta-Stymphalus* generic complex of the leafhopper tribe Scaphytopiini (Hemiptera: Cicadellidae) from the Old World. *Zootaxa* 713: 1-47.

Viraktamath, C.A. 2005. Key to the subfamilies and tribes of leafhoppers (Hemiptera: Cicadellidae) of the Indian subcontinent. *Bionotes* 7(1): 20-24; (2): 44-49.

Viraktamath, C.A. and Sohi, A.S. 1980. Notes on the Indian species of *Austroagallia* (Homoptera: Cicadellidae). *Oriental Insects* 14: 283-289.

Viraktamath, S. and Viraktamath, C. A. 1985. New species of *Busoniomimus* and *Idioscopus* (Homoptera: Cicadellidae: Idiocerinae) breeding on mango in south India. *Entomon* 10: 305-311.

Viraktamath, S. and Viraktamath, C.A. 1995. The leafhoppers (Homoptera: Cicadellidae) and their host plants in Karnataka. *Karnataka Journal of Agricultural Sciences* 8: 249-255.

Webb, M. D. 1981. The Asian, Australasian and Pacific Paraboloponinae (Homoptera: Cicadellidae). *The Bulletin of the British Museum (Natural History) Entomology series* 43: 39-76.

Webb, M. D. and Heller, F.R. 1990. The leafhopper genus *Pseupalus* in the Old World Tropics, with a check-list of the Afrotropical and Oriental Paralimnini (Homoptera: Cicadellidae: Deltocephalinae). *Stuttgarter Beitrage zur Naturkunde Serie A (Biologie)* 452: 1-10.

Webb, M. D. and Vilbaste, J. 1994. Review of the leafhopper genus *Balclutha* Kirkaldy in the Oriental region (Insecta: Homoptera: Auchenorrhyncha: Cicadellidae). *Entomologische Abhandlungen (Dresden)* 56(1): 55-87.

Webb, M.D. 1987. Species recognition in *Cicadulina* leafhoppers (Hemiptera: Cicadellidae), vectors of pathogens of Gramineae. *Bulletin of Entomological Research* 77(4): 683-712.

Wilson, M. R. 1983. A revision of the genus *Paramesodes* Ishihara (Homoptera, Auchenorrhyncha: Cicaellidae) with descriptions of eight new species. *Entomologica Scandinavica* 14: 17-32.

Young, D.A. 1968. Taxonomic study of the Cicadellinae (Homoptera: Cicadellidae) Part I. Proconiini. *U.S.National Museum Technical Bulletin* 261. 287pp.

Young, D.A. 1977. Taxonomic study of the Cicadellinae (Homoptera: Cicadellidae) Part 2. New World Cicadellini and the genus *Cicadella. North Carolina Agricultural Experiment Station Technical Bulletin* 239. 1135pp.

Young, D.A. 1979. A review of the leafhopper genus *Cofana* (Homoptera: Cicadellidae). *Proceedings of the Entomological Society of Washington* 81: 1-21.

Young, D.A. 1986. Taxonomic study of the Cicadellidae (Homoptera: Cicadellidae). Part 3. Old World Cicadellini. *North Carolina Agricultural Experiment Station Technical Bulletin* 281. 639pp.

Zhang, Y. and Webb, M. D. 1996. A revised classification of the Asian and Pacific selenocephaline leafhoppers (Homoptera: Cicadellidae). *The Bulletin of the Natural History Museum Entomology Series* 65(1): 1-103

Zhang, Y. and Wei, C. 2002a. A systematic study on the genus *Placidus* Distant (Homoptera: Cicadellidae). *Entomologia Sinica* 9: 63-72.

Zhang, Y. and Wei, C. 2002b. Study on the Oriental leafhopper genus *Kunasia* Distant (Homoptera: Cicadellidae). *Entomotaxonomia* 24: 83-88.

Zhang, Y., Wei, C. and Sun, G.H. 2002. A systematic study on the genus *Cyrta* Melichar (Homoptera: Cicadellidae). *Entomotaxonomia* 24: 27-44.

Entomology : Novel Approaches, 2007
Eds.: *P.C. Jain and M.C. Bhargava,*(Page 499-508)
New India Publishing Agency, New Delhi
E-mail: newindiapublishingagency@gmail.com
web: www.bookfactoryindia.com

CHAPTER 25

ROLE OF HONEY BEES IN CROP POLLINATION

H.D. Kaushik, R.S. Jana and R.S. Jaglan
Department of Entomology, CCS Haryana Agricultural University
Hisar-125 004, Haryana

INTRODUCTION

India is the largest developing economy of the world and about 70 per cent of its population is dependent directly or indirectly on this profession. To meet the food requirement of the ever-multiplying population of this country, agriculture production has to be maximized by using genetically improved varieties, fertilizers improved implements, seed, and other agronomical technologies. Beside these improvements the agriculture production has remained static for the last decade. For further improvements in agricultural production we have to adopt comprehensive strategies. One of such strategies is the pollination management that is a very vital process of agriculture crop production. Pollination is a process involving reproductive parts of a flower and an external agent in the form of a vector. According to Prof. E.C. Martin, the value of bee pollination of fruits, vegetables and seeds in the United States of America totaled very close to $8 billion in the 1960s; the figure 20 years later is probably close to $20 billion.

Pollination is an important step in the reproduction of seed plants for the transfer of pollen grains (male gametes) to the plant carpel, the structure that contains the ovule (female gamete). The receptive part of the carpel is called a *stigma* in the flower of angiosperms and a *micropyle* in gymnosperms The study of pollination brings together many disciplines, such as botany horticulture entomology and ecology. Pollination is important in horticulture because most plant fruits will not develop if the ovules are not fertilised

TYPES OF POLLINATION

The process of pollination requires pollinators as agents that carry or move the pollen grains from the anther to the receptive part of the carpel. Methods of pollination, categorized by pollinator type, are:

- ***Entomophily***: pollination by insects e.g
  - — Bee pollination on Sunflower
  - — Bee pollination on Eucalyptus

- ***Zoophily***: pollination by animals such as birds or bats
  - Hummingbird
- ***Anemophily***: pollination by wind
  - very common in grasses
  - Sweet chestnut
  - Conifers
- ***Hydrophily***: pollination by water
  - Mesophytes like Ribbonweed

Some flowers are pollinated using buzz pollination.

Pollination process in plants

Pollination is a process involving flowers (the pollen source and stigma receiver) and a pollination vector (an agent carrying pollen to the stigma). The study of pollination is an interdisciplinary field involving botany (the plant side) and some other field (the vector side). it is an ecological discipline

The gymnosperms used wind as a pollination vector. The pollen grains (endosporic microgametophytes) were carried by wind to the pollination droplets in the micropyles of the naked ovules. Here in angiosperms, the ovules are held inside the folded megasporophyll (the carpel) and so the pollen tube has a greater role. It must grow from the stigma to the micropyle and then penetrate the nucleolus (megasporangium) to deliver the sperm cells to the egg and central cell.

Pollination cues

The plant must first attract the vector. This is usually the function of the perianth. The attraction cues might be visual or olfactory.

Visual: Showy petals or sepals with obvious shape, size, and color for the vector's vision are important.

Bees have vision that is shifted toward the blue end of our visible spectrum. They do not see red colors but do see into the near ultraviolet. Typically the red itself is attractive to other potential pollinators, but there is likely some other color pattern that works for the bee. It may be a pattern in the UV that we cannot see which actually attracts the bee.

Olfactory: Some vectors have limited vision but extensive ability to find a flower by its fragrance. Flowers produce volatile chemicals that diffuse and are carried by air movements through an environment. A vector that can recognize this odor and fly up the concentration gradient of this fragrance, can easily find the next flower of a particular species. Flowers have evolved a wide array of odors to assist in attracting vectors. Humans have limited appreciation of and vocabulary for fragrances, but other animals are quite discriminating. Flowers that specialize in attracting flies, for example, are famous for their fetid (manure) aroma.

Flower Shape Consideration

The flower has to be designed to accommodate the vector and prevent pollen and nectar robbers from stealing the rewards. The shape can also make a flower more attractive. Flowers pollinated by hovering vectors generally hang down and have long styles and filaments.

Role of Rewards in pollination

The vector is usually intelligent enough to avoid the energy waste of behaviors that do not result in some kind of reward. The plant must reward the vector's visit. This will result in training the vector to concentrate its efforts on visiting other flowers of the same species and achieving the pollination goal. The vector's reward is usually one of three things:

Nectar: Somewhere in the flower is some secretary tissue called the nectary. Its location varies from species to species. This tissue is connected to phloem and secretes the sweet liquid known as nectar. This is a carbohydrate rich droplet that is used as an energy source for vectors. Hummingbirds must consume vast quantities of nectar to continue their high-energy mode of flight. It is the sole energy source for most butterflies. Bees collect the nectar and evaporate it down to make honey for winter supplies. Nectar is rich in carbohydrate but is a weak source of most other nutrients-trace vitamins and minerals. Thus the person who tells you their kids get not sugar but you watch them pour honey all over their cereal is just plain wrong-honey is essentially sugar with just a little water added.

Pollen: Pollen itself is decent vegetable food. It contains protein, starch, oil, and other nutrients. It is far richer than nectar in vitamins and minerals too. For beetles and bees, the consumption and collection of pollen is critical. It is their basic protein supply. Fortunately, the pollinators are not very careful in cleaning off sticky pollen that collects on their bodies in each visit. Thus they arrive at a second flower with lots of pollen from the first flower they visit. The position of stigma and vector allows this pollen to be removed and stuck to the stigma.

Bee Behavior: Behavior can also be a reward that gets a repeat visit by a vector. The *Paphiopedilum* orchids provide a barroom place for an insect to "get high" and that is the only reward. The flies must like the experience and come back for more achieving pollination. Another example includes the *Ophrys* orchid sex pheromone used by the female wasps whose mate's virgin attempts at copulation are with the orchid flower that makes the pheromone scent. Thus the flowers attract both sexes, but only one pollinates the other collects!

International Pollination System (IPS)

IPS specializes in what we call the "managed bees"; that is, bees that exist and pollinate in a controlled fashion in man-made environments. The most commonly used pollinators include the Alfalfa Leaf cutting Bee, the Alkali Bee, the Bumble Bee, the Blue Orchard Bee, and of course, the Honeybee. Contemporary research is focused on a number of additional bees, which might prove to be viable managed pollinators in coming years. A key criterion in the concept of "managed bees" is that they must be readily available in large numbers either through purchase or lease.

There are over 20,000 species of bees in the world and several have been developed as crop pollinators. While the Honeybee is still the "queen" of the pollinators, non-Apis bees are becoming increasingly important in agriculture. IPS specializes in working with these non-Apis bees. With the threats that Africanized Bees and Varroa and tracheal mites represent to the Honeybee industry, it is vital that we continue to develop the non-Apis bees as crop pollinators.

Pollination in agriculture

Pollination management is a branch of horticulture that seeks to protect and enhance present pollinators and often involves the culture and addition of pollinators in monoculture situations, such as commercial fruit orchards. The largest managed pollination event in the world is in Californian almond orchards, where nearly half (about one million hives of the US honeybees are trucked to the almond orchards each spring. New Yorks apple crop requires about 30,000 hives; Maine's blueberry crop uses about 50,000 hives each year.

Bees are also brought to commercial plantings of cucumbers squash melons, strawberries and many other crops. Honeybees are not the only managed pollinators: other species of bees are also raised as pollinators. The alfalfa leafcutter bee is an important pollinator for alfalfa seed in western United States and Canada Bumblebees are increasingly raised and used extensively for greenhouse tomatoes and other crops.

The ecological and financial importance of natural pollination by insects to agricultural crops improving their quality and quantity, becomes more and more appreciated and has given rise to new financial opportunities. The vicinity of a forest or wild grasslands near agricultural crops, such as apples can improve their yield by about 20%. This may result in forest owners demanding payment for their part in the improved results.

Pollination also requires consideration of pollenizers. (The terms "pollinator" and "pollenizer" are often confused: a ***pollinator*** is the agent that moves the pollen, whether it be wind, bees, bats, moths, or birds; a ***pollenizer*** is the plant that provides the pollen.) Some plants are self-fertile or self-compatible and can pollinate themselves. Other plants have chemical or physical barriers to self-pollination and need to be cross-pollinated: with these self-infertile plants, not only pollinators must be considered but pollenizers as well. In pollination management, a good pollenizer is a plant that provides compatible, viable and plentiful pollen and blooms at the same time as the plant that is to be pollinated.

Pollination can be cross-pollination with a pollinator and an external pollenizer, self-pollenization with a pollinator, or self-pollination without any pollinator:

- Cross-pollination (***syngamy***): pollen is delivered to a flower of a different plant. Plants adapted to outcross or cross-pollinise have taller stamens than carpels to better spread pollen to other flowers.
- Self-pollenization (***autogamy***): pollen moves to the female part of the same flower, or to another flower on the same individual plant. This is sometimes referred to as self-pollination, but this is not synonymous with autogamy. Clarity requires that the term "self-pollination" be restricted to those plants that accomplish pollination without an external pollinator (example: the stamens actually grow into contact with the pistil to transfer the pollen). Most peach varieties are autogamous, but not truly self-

pollinated, as it is generally an insect pollinator that moves the pollen from anther to stigma. Plants adapted to self-fertilize have similar stamen and carpel length.

- ***Cleistogamy***: pollination that occurs before the flower opens is always self-pollination. Some *cleistogamous* flowers never open, in contrast to ***chasmogamous*** flowers that open and are then pollinated. Cleistogamous flowers must of necessity be self-compatible or self-fertile plants. Other plants are self-incompatible. These are end points on a continuum, not absolute points.

Hybridization is effective pollination between flowers of different species of the same genus, or even between flowers of different genera (as in the case of several orchids.

Peaches are considered self-fertile because a commercial crop can be produced without cross-pollination, though cross-pollination usually gives a better crop. Apples are considered self-incompatible, because a commercial crop must be cross-pollinated. Remember that most fruits are grafted clones, genetically identical. An orchard block of apples of one variety is in effect all one plant. Growers now consider this a mistake. One means of correcting this mistake is to graft a limb of an appropriate pollenizer (generally a variety of crabapple) every six trees or so.

To attract pollinators, some flowers, when viewed under ultraviolet light, have patterns to help pollinators find the nectaries. These are called nectar guides.

Pollination of food crops has become an environmental issue, due to two cross trends. The trend to monoculture means that greater concentrations of pollinators are needed at bloom time than ever before, yet the area is forage poor or even deadly to bees for the rest of the season. The other trend is the decline of pollinator populations, due to pesticide misuse and overuse, new diseases and parasites of bees, clearcut logging, decline of beekeeping, suburban development, removal of hedges and other habitatfrom farms, and public paranoia about bees. Widespread aerial sprayingfor mosquitoes due to West Nile fears is causing an acceleration of the loss of pollinators.

USA solution to the pollinator shortage, so far, has been for commercial beekeepers to become pollination contractors and to migrate. Just as the combine harvesters follow the wheat harvest from Texas to Manitoba, beekeepers follow the bloom from south to north, to provide pollination for many different crops.

In Africa, many crops depend on the wind for their pollination, including cereals: millet, guinea corn, maize and rice. Plantain, coco yam and yam also do not need insects to pollinate them. But most leguminous vegetables; cash crops such as coffee, cola nut, cocoa, coconut, palm, cashew and shea butter; fruits such as the mango and citrus; and many other plants cannot be pollinated without insects.

The insect forages for nectar and pollen from several thousands of plant species and in the process pollinates a wide variety of crops important for the survival of life. While bees may visit many species of plants in a day, a forager may also constantly visit one plant, sometimes for several days, until there is no more nectar or pollen to collect.

Pollen is a vital food for the brood of the honeybee. The bee needs it in the hive, but the flower also needs the bee to fertilize it. The bright colour and sweet odour of the flower combine to attract insects. The honeybee, which has a powerful sense of smell and a keen sense of sight, is easily attracted to the flower. In the process of gathering both nectar and pollen, it incidentally transfers some pollen from the flower's male organs, the anthers, to its female organ, the stigma. The pollen germinates, and penetrates to the plant's ovaries, where the seeds are formed.

The honeybee starts its foraging activity between 5:00 and 5:15 A.M. Flight usually depends on the weather and temperature. The honeybee will not leave the hive if the temperature is below 14°C or if the wind speed is above 30 kilometers per hour. To be effective, therefore, pollination should take place in warm, clear weather, but too much heat has the same adverse effect as too much cold. If the weather is hot, dry and windy, the flower's stigma may dry out, so that pollen deposited does not germinate. Pollen may also not be available in conditions of continuous rainfall, since flowers are usually scarce during the rainy season. Thus, areas of frequent rainfall offer few flowers for bee activity.

The honeybee is the only insect that can successfully be moved from farm to farm. It is estimated that five average colonies (about 50 000 bees each) can work a two-hectare plantation. Moving colonies and siting them close to the farm achieve this.

How to move bees?

The beekeeper must study carefully the temperament of his bees, and keep a record of the seasons when they are friendly or aggressive in his locality, because during certain periods of the year, bees can be transported easily without precautionary measures against aggression and stinging.

A motorized lorry is an indispensable tool for the professional beekeeper. Bees cannot be moved by head-load. The top-bar hive cannot be transported over long distances on poor roads or tracks, and therefore hives that are to be transported frequently must be fitted with movable frames. The hives should have a ventilation hole about two centimeters in diameter, covered with mosquito mesh. Their bottom board should be coated with termite-repellent paint.

The hives should be light enough to be loaded and unloaded easily. Combs with capped honey should be removed to reduce weight, while uncapped combs should be retained in the hive. Before loading, hives should be inspected and all unwanted openings sealed with wood and wax, since they can create problems if the bees come through them and cluster outside the hive. However, bees tend to cluster quietly in moving vehicles and rarely attempt to fly away or cause any harm until movement stops.

In general, bees are less troublesome riding at night. Hives should be loaded in the evening, when most of the bees have returned from foraging. The bees clustered at the entrance are smoked to drive them into the hive. The entrance is then sealed, and the ventilation hole is opened to allow air to circulate through the hive. Hives can be stacked one above another. They should be packed closely and neatly, and secured with a trucer's rope. The journey and the reinstallation of the hives should take place at night: hives should be unloaded and installed before sunrise, because the bees will be ready to leave shortly after that time. To assist in the night work, a flashlight with a red bulb, or covered with red cellophane, is a useful tool, because bees cannot detect red light and are therefore not disturbed by it.

Why we need Pollination programmes?

With the advent of newer techniques in agriculture crop production the human knowingly or unknowingly disturbed the natural floral system. Furthermore, intensive agriculture and clean culture has destroyed natural food sources and nesting sites of wild pollinators. Hence there is a dire need for management programmed for bees to handle this issue. Fruit-growers should be encouraged to keep their own bees to ensure that their crops are properly pollinated. Growers who for any reason do not wish to keep bees should be advised to use the services of a beekeeper that has colonies to move about for the purpose.

In order for a pollination programme to work successfully, it is important that the farmer and the beekeeper work in close collaboration, because both have an essential role to play. Yell before he needs the bees (where bees are scarce, several months' notice may be necessary), the farmer should contact the beekeeper to make detailed arrangements. At this time they should agree on a number of points, and it is often advisable for the agreement to be written down in a simple contract. Among the things they should agree on, these are possibly the most important:

1. The number and the strength of the colonies to be provided. Rented colonies should be queen-right, with at least four brood combs
2. The dates of delivery and removal of the colonies;
3. The distribution of the hives on the farm
4. The fee and schedule of payments. This is negotiable and depends on a number of factors, including proximity and transport costs, the quantity and quality of honey produced from the crop, and the risks, if any, involved in the operation. Due to the fact that he can expect two different benefits from the operation, the beekeeper should be prepared to charge moderate prices:
5. Ownership of the honey produced. Normally, the honey belongs to the beekeeper. If not, his fee is likely to be higher;
6. The right of the grower to verify that the colonies are up to strength. However, the grower should never attempt to open hives without the consent of the beekeeper. Colonies can be simply inspected by counting the number of returning foragers: on a warm, bright day during the flowering period, 100 bees returning to the hive within one minute is usually regarded as very good;
7. Restrictions on the use of insecticides or pesticides toxic to bees, not only during the rental period but also for a specified period before it begin. This is important because the farmer owes it not only to the beekeeper but also to himself to be sure that the colonies are strong, and therefore that he does not kill bees by the wrong use of pesticides;
8. Precautions to be taken against bush fires;
9. Liability for random stinging by bees, for vandalism, for livestock damage, and for theft of hives.

Honeybee species useful in pollination

In India, four honeybee species are found *viz. Apis dorsata* (giant honeybee), *A. florea* (small honeybee), *A. cerana indica* (Indian honeybee) and *A. mellifera* (Italian honeybee). The former two are wild species and categorized as wild insect pollinators. Moreover, these cannot be domesticated and not considered useful in pollination programmes. The latter two species can easily be domesticated in wooden bee hives. They are considered industrious and amenable to human handling. Moreover, they can be moved to various locations in required numbers and considered ideal agent for pollination programmes. However, pollination efficiency of particular honeybee species depends on number and strength of colony, time of moving, their distribution in the field, prevalence and attractiveness of field crops and weather conditions. Therefore, honeybee can only compensate scarcity of wild pollinators in the field as these are readily available whenever and wherever required in sufficient numbers. Furthermore, honeybees do not restrict themselves for pollination services to a particular floret species or genes but extend this service to large number of agricultural crops. Such peculiar features lead the honeybees to become most dependable and ideal pollinator in agro and forest ecosystems.

Pollination programmes *vis-a-vis* crop productivity

Honeybees play vital role in augmenting the production of agricultural and Horticultural Crops through pollination (20 times more value of the produce) besides the honey that is hive product having the secondary importance. However, the effect of pollination services rendered by honeybees increase the yield but also improve the early setting of seeds and quality of their produce as reported, published and documented by various workers throughout the world. Based on above information, it has been confirmed that many crops of economic importance are very much benefited by honeybee pollination (Table 1&2). Similarly, the number of bee colonies varies that are required for pollination. Generally, 2-3 bee colonies per hectare are needed for pollination which varies with crop to crop depending on many factors such as abundance and attractive of flowers, competition among various pollinating insects, crop and its locality etc. On the basis of area under particular crop, the number of colonies required for its pollination and monitory gains thereof can be estimated (Table 3).

Table 1 Crops known to require or benefit from insect pollination

Fruit crops		Seed crops		
		For foods, oils & fibers	For propagation	
Almond	Mango	Beans(Horse)	Alfalfa	Onion
Apple	Muskmelon	Beans (Limas)	Asparagus	Okra
Apricot	Nectarine	Buckwheat	Broccoli	Pepper
Avocado	Okra	Celery	Brussels sprout	Pumpkin
Blackberry	Papaya	Cotton	Buckwheat	Radish
Cherry	Peach	Flax	Cabbage	Rape
Cranberry	Pear	Mustard	Cauliflower	Rutabaga
Cucumbers	Pepper	Rape	Celery	Squash
Currant	Persimon	Raya	Clovers	Sunflower

Contd. ...

Table 1 Contd. ...

Dewberry	Plum & Prune	Sunflower	Collard	Sweet clover
Egg plant	Pumpkins	Toria	Cotton	Trefoil
Gooseberry	Quince		Coriander	Turnip
Grape	Squash		Cucumber	Vetches
Guava	Strawberry		Egg plant	Water melon
Loquat	Watermelon			

Table 2 Increases in yield by bee pollination over self-pollination (% increase by bee pollination over self pollination)

Orchard crops	**% increase in yield**	**Vegetables**	**% increase in yield**
Apple varieties	180 to 6950	Asparagus	12405
Pears	240 to 6014	Cabbage	100 to 300
Plums	6.7 to 2739	Carrot	9 to 135
Cherry	56 to 1000	Onion	354 to 9878
Raspberry	291 to 463	Radish	22 to 100
Persimmon	20	Turnip	100 to 125
Litchi	4538 to 10246	**Miscellaneous crops**	
Citrus varieties	21 to 411	American cotton	5 to 20
Orange	471,900	Buckwheat	63
Strawberry	38, 68	Coffee	17 to 83
Guava	70, 140	Egyptian cotton	16 to 24
Legume seeds		Fennel	100
Alfalfa	23 to 19733	Niger	24 to 173
Berseem & others	23 to 33, 150	**Oilseeds**	
Clovers		Brown mustard	13 to 222
Broad beans	7 to 90	Linseed	2 to 40
Dwarf beans	3 to 21	Rai	18
Kidney beans	500 to 600	Rape	12 to 139
Runner beans	21 to 1100	Sunflower	72 to 82
		Toria	66 to 220
		White mustard	128 to 152

Table 3 Requirement of honeybees colonies for pollination

Crop	Colonies/ha	Crop	Colonies/ha
Alfalfa	10 to 12	Medicago (Lucerne or alfalfa)	6 to 18
Almond	2 to 4	Melon	2.5
Apple	2 to 3	Onion	8 to 10
Berseem	3 to 4	Pears	4 to 6
Buckwheat	3	Pumpkins	0.2
Carrot	6	Rapeseed	4
Cauliflower	2	Red clover:	
Cherry	2.5	Late flowering crop	3
Citrus	2.5	Early flowering crop	6 to 8
Cotton	1.0 to 2.5	Sunflower	2.5
Cucumbers	6 to 20	Watermelon	2.5
Hairy vetch	3.5 to 7.5	White clover	0.2 to 1.0
Kiwi fruit	8		
Ladino clover	2.5 to 3.5		

Entomology : Novel Approaches, 2007
Eds.: *P.C. Jain and M.C. Bhargava*,(Page 509-533)
New India Publishing Agency, New Delhi
E-mail: newindiapublishingagency@gmail.com
web: www.bookfactoryindia.com

CHAPTER 26

INTEGRATED MANAGEMENT OF SOIL INSECTS

N.G. Kumar
Department of Entomology, University of Agricultural Sciences
GKVK, Bangalore-560 065, Karnataka

INTRODUCTION

Many insets inhabit the soil and participate in different activities which will result in changes in soil physico-chemical properties. Thus soil is a very good habitat for some insects. Several insects spend either complete or few stages of their life cycle in soil. Soil provides protection from predators for many soil dwelling insects especially in immature stages. Few insects also undergo diapauses in soil. Soil insects include phytophagous, scavengers, predators and parasites. Among these, soil macro fauna especially termites and white grubs are very important pests of many crops and their activity also brings many changes in soil ecosystem.

TERMITES

Termites (Order: Isoptera) are eusocial, polymorphic and live in small to large colonies. The colony consists of different castes *viz*., functional reproductive (queen and king), sterile workers and soldiers, which are morphologically distinct. The tasks of the latter two castes include foraging, nest building and care of eggs and young ones and defense. All species maintain symbiotic relationships with microorganisms, which aid in digestion. Few species build the often-spectacular mounds or arboreal nests and others build subterranean nests. They constitute important group of soil fauna and play a major role in the arid and semiarid tropics. Worker castes are dominant and are responsible for major functions in the colony. During the process of food collection, the workers generally attack the plants that are weakened due to some internal growth factors or environmental causes such as drought or poor soil conditions (Roonwal, 1978).

Distribution of Termites

Termites are mostly found in tropical and subtropical regions and a few species occur in temperate regions. Their geographical range of 45-48 °N to 45° S includes roughly two thirds of the worlds land surface. Vegetation, temperature and moisture determine the distribution of these insects. Seven families and fourteen sub-families are recognized and they comprise more than 2500 species belonging to 268 genera (Krishna, 1970). Of these families of the order Isoptera, four families *viz.*, Kalotermitidae, Hodotermitidae, Rhinotermitidae and Termitidae are predominant in India.

The termite fauna of India has been studied from Assam (Roonwal and Chhotani, 1962), West Bengal (Maiti, 1975), Rajasthan (Roonwal, 1982a: Roonwal and Bose, 1964), Himalayas (Chatterjee and Thakur, 1967), Andaman and Nicobar Islands (Roonwal and Bose, 1970) and South India (Bose, 1984). In India occurrence of 224 species of termites have been reported (Kumar, 2005). Termites are found in all the agro climatic regions of India. Family Termitidae comprising of nearly three fourth of the known species in the world has a comparatively poor representation in India. The termites of the subfamily Macrotermitinae have the largest representation in India. Of the several genera found in India, the genus *Odontotermes* is most widely distributed (Thakur, 1981).

How termites damage plants?

Termites are herbivores and detritivores variously involved in the comminuting and decomposing of vegetable matter. The workers, both for their own consumption and for the dependent castes, collect food. Gathering of food in-group by termites is known as foraging. Foraging activity may be seen on grasses, trees, leaf litter, bark, dung or any cellulose materials under the earthen sheeting. The majority of the mound building and subterranean species construct underground galleries, which extend several meters from the nest. Workers emerge from these galleries and forage on the surface by constructing earthen runways, which are extended depending on the availability and requirement of food. Foraging activity may be seen during cooler hours of the day or some times in the nights during summer months. This activity is affected by various abiotic, biotic factors and agricultural practices.

Termites have been viewed as harmful insects because the loss caused to agriculture is to the extent of Rs.280 million (Fletcher, 1912: Mehta and Verma, 1968). Agarwala (1955) and Gangwar *et al.* (2003) estimated a loss of 2.5 and 13.6 % in sugarcane tonnage, respectively and 4.5% in sugar output due to damage by *Microtermes obesi* alone. In Uttar pradesh, 30-50 % eye buds have been reported to be destroyed by termites resulting in yield reduction. Termites have been reported to destroy several field and horticultural crops *viz.* wheat, maize, Jowar, barley, paddy, groundnut, castor, soybean, pearlmillet, tobacco, sugarcane, cotton, coconut, Arecanut, Bengal gram, chilli, peas, brinjal, potato, turnip, jute, tapioca, tea, betelvine, mango, citrus, grapevine, cocoa, cabbage etc. (Singh and Soneja, 1966;Avasthy, 1967;Singh and Sandhu, 1968;Reddy, 1968;Shinde, 1971;Agarwal, 1972;Tripathi, 1972;Saxena, 1972;Nath, 1972;Vevai, 1974;Saxena and Rizvi,1974;Nair,1975;Singh,1977;Gupta and Sharma,1977;Srivastava *et al.,* 1978 ;Nair,1978 and Subash Chandra *et al.,*2003).

Kumar (1991) has assessed the damage to finger millet (4.12%), maize (45.27), Wheat (1.5%), sorghum (1%), cowpea (2.65%), redgram (0.5%), green gram (1.62%), rice bean (3.7%), sunflower (10.5%), castor (20.85%), soybean (35.6%), Niger (38.5%) and mulberry (5.25%) due to feeding by *Odontotermes horni*. Further, its incidence was also noticed on mango (15.50%), cashew (45.6%), coconut (10.5%), Arecanut (2%), citrus (4.5%), sapota (6.5%), pomegranate (2%), drumstick (8%), rose, croton, bougainvilla, tecoma, gulmohar and *Cassia* sp.

Termites have been gaining ground as important pests of several crops in the country. Earlier, cyclodienes were used to combat these termites. At the same time, several termites' species continue to bother different crops as a result the pesticide industry is marketing several alternate pesticides targeting these termites. But, none of the recent non-persistent chemicals are capable of protecting the crops for longer duration. After banning of cyclodienes, very little attention has been paid to develop chemicals that can control the termites.

Management of termites

Cultural methods

Termites may sometimes damage healthy plants, but unhealthy, stressed plants are generally more susceptible (Sen-Sarma, 1986). Cultural control therefore includes those procedures which help to maintain or enhance plant vigour and which generally are good agricultural practice.

Water stress

Levels of rainfall and irrigation may be related to termite attack on plants because levels of termite foraging activity are frequently related to soil moisture levels, deficiency or excess of water stress plants and encourage attack.

In India, termite attack on tea is greater during the raining season (Srivastava and Butani, 1987) and attack on maize and groundnut was greater in experimental fields with higher soil moisture levels where termite foraging activity was greater (Reddy and Sammaiah, 1988). Various measures may reduce attack in circumstances where plants are subject to water stress. Watering of nursery stock just prior to transplanting out provides immediate supply of water in the event of an unexpected dry period and ensures cohesion of the soil root ball. Planting at the correct time so that crops/trees are not subject to drought also helps to ensure plant vigour (Harris, 1971).In Nigeria, Johnson *et al.* (1981) suggested the use of short season varieties of groundnut and early planting to avoid lack of rain towards harvest.

In irrigated areas regular irrigations will limit periods of water stress. Frequent irrigation may therefore reduce attack (Kumarsinghe and Ranasinghe, 1988). More frequent irrigation reduced attack by *Microtermes obesi* in many field crops *viz.* maize, wheat, sugarcane, groundnut helps in reducing the termite attack (Verma, 1980; Verma *et al.*, 1981; Ashok Kumar and Veeresh, 1990). Jayanthi *et al.* (1993) observed more activity of termites on the drip-irrigated plots compared to surface irrigated plots of groundnut.

Crop rotation

Intensive cropping, especially monocropping for long periods reduces soil fertility and structure. Under such conditions, crops will be less vigorous and therefore, more susceptible to attack. Crop rotation, especially including fallow periods, to reduce detrimental effects on soil fertility and structure is in principle, to be recommended.

Damage to plants

Both dry-wood and subterranean termites frequently gain access to living plant tissue in plantation crops and forestry trees through wounds or fire damage (Harris, 1971).

In plantation crops, many authors have recommended that dead wood be removed and cut back to the living tissue and pruning be carried out carefully with clean cuts and in a manner to avoid die back (Harris, 1971; Sands, 1977; Sivapalan *et al.*, 1977). Pruning cuts and accidental wounds should be treated with paint or tar (Kashyap *et al.*, 1984). Such procedures are labour-intensive; they may be cost-effective in plantation crops such as tea but not in forestry (Sen-Sarma, 1986).

Inorganic fertilizer

Fertilizer enhances plant vigour and should therefore reduce susceptibility to termite attack (Singh and Brar, 1988). Fuller (1912) reported that fertilizer gave good protection from termite attack on trees in Australia. Sivapalan *et al.* (1977) showed that excessive use of nitrogenous fertilizer in tea encouraged fast growing soft tissue particularly susceptible to attack by the dry-wood termites *Glyptotermes dilatatus.*

Manipulating termite number and behaviour, Crop residues and other organic matter

The addition or removal of organic matter from the field has both been suggested as methods of reducing levels of attack. Essentially, the conflicting principles are (i) that removal of residues and other debris from the field will reduce potential termite food supplies and hence lead to a reduction in termite numbers and subsequent attack and (ii) that leaving residues in the field or adding further organic matter will provide alternative food to which the termites will be attracted, thereby reducing levels of attack (Brown, 1962: Rajagopal *et al.*, 1996).

In established plantations, the removal and destruction of dead wood and termite-infested trees is recommended as a control measure (Srivastava and Butani, 1987) and should be done before the next rainy season to prevent release of large numbers of alates. Grass from soil within a 30 cm radius of the collar of the bushes, removal of termite earthen sheeting and hallowed branches, covering of cut surface with bituminous protective paint (Banerjee, 1967).

Breaking up termite galleries

Termite damage is not generally a problem in crops planted on deep cracking soils because the frequent cracking prevents adequate building and maintenance of mounds, runways and galleries (Lee and wood, 1971). Artificial breaking up of the soil may have a similar effect, reducing termite foraging activity.

Kumar (1998) observed less termite activity after deep Ploughing (disc and cultivator) operations, which altered the soil structure up to 20 to 25 cm depth and the termite galleries formed during the post harvest period. In addition to this, inter-cultivation also slightly interfered with the termite activity. Termite activity increased as the settling of soil particle and compactness of the soil increased due to rainfall.

Other practices

High density sowing: Many farmers sow seed at higher rate than recommended in order offsetting anticipated losses due to termites; surviving plants may these be thinned to the appropriate number (Wood and Cowie, 1988).

Time of harvesting: Crops are often more seriously damaged towards harvest than earlier in the season (Sands, 1977) although this depends on the species of termite involved. Prompt harvest may therefore reduce yield loss. Delay in harvesting can also lead to yield loss through lodging (e.g. in maize and sunflower) or pod/peg damage in groundnut.

Biological control

A wide range of vertebrate eats termites and invertebrate predators who are natural influence on termite numbers and population dynamics is partially understood in small number of cases.

Predators

The predation hazards confronting termites are enormous. They are prayed upon by a wide range of predators, specialist or opportunistic. Attack is mainly upon individuals outside the nest, foraging neuter castes or alate reproductive leaving the nest to found new colonies.

Predation of alate reproductive colony-founding swarms

This is entirely opportunistic since the flights are seasonal and short-lived and cannot provide a regular food supply. Arthropod predators include scorpions, spiders, centipede, dragonflies, cockroaches, mantids, crickets, beetles, flies, ants and wasps (Wood and Sands, 1978). It is likely that ant's account for a significant proportion of the probable near 100% mortality of swarming alates. Even when pairs have successfully dug into the soil they are liable to predation by several species of predators. Vertebrate predator of alate swarms includes reptiles, amphibians, birds and mammals including man.

Opportunistic predation of foraging sterile castes

These predators include all those groups, which attack alate although the relative importance differs, birds being much less important than ants. In Zambia, the main opportunist ant predators of surface foraging species were *Myrmicaria eumenoides* and *Pheidole megacephala* (Sheppe, 1970).

Among vertebrate opportunist birds and mammals are best documented. Clearly, surface foraging termites, such as harvester termites which feed in the open, or those Macrotermitinae which feed under thin surface sheets of soil are occasionally in the open, suffer from most from opportunistic predation, whether this is by ants, birds or mammals. Beeson (1941) claimed that transplanting *Solenopsis* sp. to nurseries of free seedlings in India controlled termites. Specialized invertebrate predators of foraging termites seem to be confined to a few species of Ponerinae and Myrmicinae ants.

Termites are more vulnerable to their natural enemies during foraging, expansion of the nests and swarming. Ants *viz., Leptogenys processionalis, Camponotus sericeus, Anoplolepis longipes* and *Oecophylla smaragdina* have been observed to predating termite at different activity. Other predators include lizards, frogs, birds, dragonflies, robberflies, assassin bugs, carabid beetles, sloth bear etc. (Rajagopal, 1979; Kumar, 1991).

Pathogens

The distinction between naturally occurring microbial symbionts, commensals, parasites and pathogens is difficult without knowledge of the ecology of the termites. Many of the microorganisms that have been found in association with termites are not significant pathogens and hence not potentially useful as biological control agents. Termites have been considered as good insect for pathogenic control because they live in a conducive environment, humid, crowded and intimate association. Successful control may be unlikely in general because of the well-known termite behaviour of isolating infected colony members (Logan and Abood, 1990; Wood and Sands, 1978), thus preventing spread of the pathogen through the colony.

Bacteria

The mutualistic association between termites and their dense and diverse bacterial gut flora is well known best only rarely have pathogenic bacteria been isolated from termites. Khan *et al.*(1977) isolated a naturally occurring termite strains of *Bacillus thuringiensis* Berliner from *Bifiditermes beesoni* (Gardner) and demonstrated its pathogenicity to other termite species. It can be passed between individual termites by trophallaxis and live termites can pick up infection from treated cadavers. There are differences in susceptibility among termite species. However, Kakaliyev and Saparliyev (1975) have assessed its efficacy in the field, concluding that it held little promise as a control agent.

Fungi

Termites are natural hosts for many species of obligate ectoparasitic fungi, while many other fungal species are associated more generally with them. Hanel and Watson (1983) have attempted to control *Nasutitermes exitiosus* (Hill), which damage forest trees in Australia, by dusting or spraying individual termites with *Metarhizium anisopliae.* Although the fungus spread through the colony and survived for up to 15 weeks in some cases, the results for colony mortality were unconvincing. Quarles (1999) observed 92% of control of termites (*Reticulitermes flavipes)* by using commercial formulation of the fungus *M. anisopliae* after one week.

Nematodes

Pionar (1975) listed the termite species in which natural nematode infections, frequently due to *Rhabiditis spp*. had been recorded. Nematodes enter their insect hosts both via natural opening (mouth, anus, spiracles) and by active penetration of the cuticle and intersegmental membranes. They have been tested against a number of soil pests in the field but the results have not been consistent (Poinar, 1979 and Georgis and Poinar, 1983). In contrast, application of a suspension of *Heterorhabiditis* sp. to individual branches of tea bushes, infested with the dry wood termite *Glyptotermes dilatatus* in Srilanka, suggested that infection could spread through the colony and give effective and economic control (Danthanarayana and Vitarana, 1987). Although host termites could detect the nematode and seek ways to avoid them by extending their earthen tunnels to the outside of the nest, as considerable number of the termites had been parasitised before escaping.

Queen removal

Removal of queens and /or destruction of the nest have been suggested frequently for the control of mound building termites, especially the fungus growing Macrotermitinae (Kakde, 1985; Rajagopal, 1982). The nest is readily identified and the royal chamber easy to locate. Dawkins (1949) advised lighting a fire in the mound after queen removal.

Plant insecticides

Locally available parts of plants, plant extracts and other substances have frequently been claimed to be effective in termite control, although they have received little rigorous assessment in the field. A wide range of plants is toxic or repellent to insects or has antifeedant properties. Plant insecticides should come from locally available plants or those, which are easy to grow, preferably on poor quality land so that they do not compete with other crops. In addition, they should have low toxicity to non-target organism's especially beneficial insects and humans.

Beeson (1941) gave recipes for mixture of plant extracts, which prevented termite foraging on trees and attack on wound or fire damage in India. Gambir mixture made from the leaves of *Uncaria gambir* or the dried aqueous extract of *Acacia catechu* mixed with oil from *Canarium strictum*, *Hopea* sp. or *Shorea* sp. painted on wounds or fire damage to trees prevents invasion by termites. Gondal fluid (Castor oil cake with plant extracts from *Gardenia gummifera* (or *G. lucidia*), *Ferula jaeschukearna* and aloes (*Agave vera, A. cantela* or *A. angustifolia*) and soaked for two weeks) painted round the base of a tree gives protection against termite foraging for about eight months. Giridhar *et al.* (1988) showed that *Calotropis latex* protected wooden pegs for four months. Sharma *et al.* (1990) observed castor oil + castor cake (1:1) was more effective against termites compared to castor oil, castor cake, neem oil and neem leaves. Parihar (1994) noticed effective control of *Microtermes* and *Odontotermes* in the field with *Calotropis procera* leaves and seed kernels of *Azadiracta indica*. Further, Manager Singh *et al.*, (2002) concluded that the cane sett dipping in 20 or 15% solution of *Calotropis procera* extract or soil treatment with 2% solution were effective against termites. On the contrary, Varma *et al.*, (1995) reported that neem cake mixed with soil did not protect the roots of Eucalyptus seedlings from attack by subterranean termites in field plots on sandy loam soil. Further, they noticed significantly greater termite damage to seedlings when these were treated with neem cake and deoiled neem cake (50 and 100 g/kg soil) than in the control plots.

Wood ash and other materials

Wood ash heaped around the base of the trunk has been recorded as preventing termite infestation of coffee bushes (Khashyap *et al.*, 1984). It is said to be effective in protecting tree seedlings if mixed in to forestry nursery beds or applied as a layer below polythene planting tubes (Beeson, 1941). In India, water containing decomposed fish; tobacco and salt or the washings from a bearskin were reputed to keep termites from mango trees (Anon., 1898). Kerosene, diesel and crude oil have been recommended to prevent attack on timber and on tree bark (Beeson, 1941; Giridhar *et al.*, 1988).The addition of crude fuel or fish oil emulsion to irrigation water is reputed to reduce damage to tree seedlings and sugar cane (Beeson, 1941; Rao, 1951; Butani, 1967; Roonwal, 1979). Coal tar, traditionally used to protect the cut ends of sugar cane setts in India is ineffective as termite attack on unprotected buds (Rao, 1951, Butani, 1967) but painted round trunks of fruit trees, it can prevent foraging on bark and subsequent ring barking (Giridhar *et al.,* 1988).

Physical barriers

A layer of sand under forestry or tree crop nursery beds might provide some protection. Germinating coconuts are protected by this method or by covering them with sand rather than soil (Kashyap *et al.,* 1984). Beeson (1941) suggested digging a deep trench around tree nurseries. This may provide some protection from species with large colonies and long surface or near-surface foraging galleries (eg. *Macrotermes* spp) but will not attack by species foraging from deep in the soil below the nursery beds (*Microtermes* spp.).

Chemical control

Effective control of termites can be achieved in various situations by the following methods

Wheat

- Soil application of carbofuran (Yadav *et al.,* 1992) or chlorpyriphos 20 EC @ 1.25 kg a.i./ha (Madan *et al.*, 1998; Sharma *et al.*, 2002) or 25 kg /ha of methyl parathion 2% dust or Endosulphan 4%dust or quinalphos 1.5% dust (Kumawat, 2001) before sowing or
- Seed treatmènt with chlorpyriphos 20EC @ 4.5 ml /kg or endosulphan 35 EC @7.3 ml /kg or lindane 20 EC @ 7.5 ml /kg (Kumawat, 2001) are effective.The earlier two chemicals were also effective at 0.9 and 2.4 g a.i. /kg seed, respectively (Rana *et al.*, 2001).

Sugarcane

- Soil application of carbofuran(1 kg a.i./ha) (Gangwar *et al* .,2003) or chlorpyriphos 20EC or endosulphan 35 EC at 1.25 kg a.i./ha (Maan Singh *et al.*, 2001) or phorate 10G @ 2.5 kg a.i./ha or chlorpyriphos 20EC @ 1 kg a.i./ha or chlorpyriphos 15 G @ 2.5 kg a.i. /ha (Singh and Singh ,2002; Manager Singh *et al* .,2003) or
- 2) Sett dipping in Imidachloprid 70WS @ 0.1 -0.2% solution(Singh and Singh ,2002; Santharam *et al.*,2002; Manager Singh *et al.*, 2003 and Saroj Jaipal and Dharam Singh,2003) or
- 3) Spraying of Imidachloprid 200SL at 250-375 ml/ha on sugarcane setts (Santharam *et al.*,2002)

Maize

- Seed treatment with endosulphan 35 EC @ 3g/kg (Sharma *et al.*, 2003) is effective.
- Fipronil granules or chlorpyriphos granules application as a side dressing four weeks after planting significantly reduces termite's infestation (Rickert *et al.*, 2003).

Forest plantations

Soil application of HCH or carbaryl (0.25, 0.5 and 10%) (Gul and Chaudhary, 1989: Thakur *et al.,* 1989) or carbosulfan controlled release granules (10g/plant) (Canty and Harrison, 1990) are effective against termites. Other promising insecticides *viz.* cypermethrin, fenvalerate, permethrin, imidacloprid, fipronil against termites in the forest plantations (Wagner, 2003).

Mounds

Application of Deltamethrin (1 and 3 g a.i./mound) or fenthion (1 g a.i./mound) (Mariconi *et al.*, 1990) or chlorpyriphos 24 EC (10ml/l) or Abamectin 1.8% EC (3.3 to 6.6 ml/l) were effective against mounds of *Cornitermes cumulans* (Valerio *et al.,* 1994).

Insect growth regulators

Field trials using JHAs against termites failed to yield an increase in soldiers or a colony decline.Derivatives of benzoylphenyl ureas are known to inhibit the chitin synthesis of insects. Testing with diflubenzuron (Dimilin) on field colonies of *Microcerotermes* sp., provided inconclusive results (Faragalla *et al.*, 1985). Laboratory studies have indicated that CSIs such as diflubenzuron or lufenuron (SU and Scheffrahn, 1996) inhibited the ecdysis of *R. flavipes* but caused virtually no effect on *S. formosanus*. Majority of field studies using baits containing hexaflumuron confirmed its efficacy against field colonies of *Coptotermes* and *Reticulitermes* sp. (Su *et al.*, 1997 and Forschler and Ryder, 1996).

WHITE GRUBS

White grubs form an important component both in number of species and diversity of habit in our country. White grub is the common name used in many countries as the larvae of beetles belonging to the family Scarabeidae and in particular to those of economic importance to agricultural crops (Ritcher, 1958). The white grubs are considered as one of the major crop pests in India. Improper application of organic manure, delay in planting and harvesting, leaving crop residue in farm, soil degradation have promoted the white grub problem (Villa Lobos, 1999: Umeh *et al.*, 2001). A negative correlation is reported between the soil organic matter content and the abundance of white grubs (Gomez *et al.*, 1999).

Adults are commonly known as chaffer or May-June beetles, which belong to the super family Scarabaeoidea, order Coleoptera. They are characterized by their lamellate antennae. Adults are sluggish, vary in size from few millimeters to 1.5 cm. Body colour vary from brown or black to metallic green, blue or yellow luster. They are nocturnal and attracted to light. Few are diurnal and may be found on flowers or fruits. The adult emergence occurs once a year after the summer rains. The shorter adult activity (above ground) meant for the purpose of mating and feeding. The rest of the life is spent inside the soil which may be one to several years, as in the case of rain beetles in Eastern America where the immature stage takes nine or more years to become adults (Ritcher, 1958). Majority of the root grubs belongs to the subfamilies *viz.* Melolonthinae, Rutelinae and Dynastinae of the family Scarabaeidae.

The white grubs are 'C' shaped, have hypognathous head with mandibles projecting downwards and a pair of three or four segmented antennae, three thoracic and ten abdominal segments with eleven kidney shaped spiracles on the three thoracic and first eight abdominal segments. All white grubs have three larval stages.

Key to family Scarabaeidae

1. Antennae not elbowed, eight to ten segmented, the joints of club very thin and closely adopted ....Family Scarabaeidae.

   The larval maxillary stridulatory teeth usually present, if absent, the distal segment of the antennae not reduced; anal opening transverse, angulate or Y-shaped, stridulatory organs present or absent on legs.

Economic importance

White grubs as pests of crop were first reported by Stebbing (1902, 1914) from Punjab. Further, Sharp (1903), Lefroy (1909), Fletcher (1919), Beeson (1919), Anstead (1919), Isaac (1920) have reported from several parts of our country. Several reports of damage to cultivated crops have been recorded in our country (Agarwal *et al.,* 1965; Avasthy, 1965, 1967; Avasthy and Kishan singh, 1973; Abraham and Kurian, 1970; Banerjee and Mukherjee, 1960; Batra *et al.,* 1973; Misra and Tyagi, 1989; Pillai *et al*., 1993; Satyavir and Parihar, 1993; Bhat *et al*., 1994; Misra, 2003 and Singh *et al*., 2005)

White grubs are grouped into three based on feeding habits *viz*

- Grubs feeding on dead organic matter.
- Grubs normally feeding on organic matter but also live roots in absence of organic matter.
- Grubs feeding on preferably on live roots of plants (De Fluiter, 1941).

The white grubs of the subfamilies Melolonthinae and Rutelinae are destructive to many cultivated crops. There are nearly 200 species of white grubs, which are injurious to cultivated crops in India (Pillai *et al.,* 1993; Satyavir and Parihar, 1993; Bhat *et al.,* 1994 and Bhagat and Kashyap, 1997). Thirty-three species of white grubs belonging to 8 sub families are known to feed on 51 host plants in four hill districts of Garhwal region (650 to 3870 m altitude). White grub fauna are absent in temperate (>2900MSL) and alpine zone (>3500 MSL) (Singh *et al.,* 2003). The grub of Holotrichia has more than 120 economically important species (Frey, 1971b). *H. serrata* and *H. consanguinea* are most widely distributed in India. The earlier species is found throughout India, whereas the latter one restricted to the Northern India (Singh *et al*., 2005). Other white grubs are *Leucopholis, Schizonycha, Anomala, Adoretus etc.* Some Dynastid adult's *viz.* *Oryctes* spp., *Heteronychus* spp. are pests of crops.

Abundance and distribution

The abundance and distribution of the beetles depends upon the species involved, the adult food and its location in relation to the emergence place (Forbes, 1916; Veeresh, 1978). Adults of *Holotrichia* spp. can fly from 100 m to 2.5 km in search of food (Avasthy, 1965; Rai *et al.,* 1969). White grubs are more abundant around adult host plants. Some species are confined to limited agro climatic regions of heavy rainfall areas (Veeresh, 1977a and Singh *et al.*, 2005). *Leucopholis* species are found in the Western Ghats, Sub-Himalayan range and Eastern India. *H. nilgiria* is dominant in coffee plantations of Western Ghats where the adult host *Ficus racemose* occurs. *H. reynoudi* dominant around *Moringa oleifera.* Pheromone activity and light are also influencing factors for the distribution of white grubs (Pal, 1977; Veeresh, 1974 and 1977a). Apart from these, cultivated soils are most suitable for the survival of white grubs. Red loamy soil and sandy loam soils are found to be best suited for the multiplication of white grubs. However, gravelly hard soil can provide shelter to smaller white grubs *viz. Serica, Antoserica, Apogonia* etc. Well-aerated soil with sufficient quantity of organic matter attracts adult for egg laying. In moisture stress conditions the grubs have been found going deep up to 150 cm. The grubs can withstand drought conditions and also shortage of food up to 120 days. Soil moisture, adult maturity and rain coupled with atmospheric humidity are important factors for the adult emergence and survival (Veeresh, 1977a).

Bionomics

In general, the white grubs species of India have one year life cycle but vary in the timings of different life stages depending upon the climatic conditions Adults emerge from the soil immediately after dusk on the receipt of first rains during April to July depending upon the area and quantity of rain (Avasthy, 1964; Majumdar and Teofia, 1965; David and Kalra, 1964; Kalra and Kulsheshtha, 1961; Srivastava and Khan, 1963; Veeresh, 1977a and b; Yadava, 1981). Mating takes place immediately after emergence on any plant. Mating lasts from few to 15 minutes. Female beetles deposit their eggs singly in an earthen cell inside the soil at depth varying from few to 10 cm. the fecundity varies from 30 to 120 depending upon the species (Gupta and Avasthy, 1960; Srivasava and Khan, 1963; Veeresh, 1977a). The incubation period varies from 7 to 14 days. Generally, the white grubs pass through three larval instars (Ritcher, 1958). Immediately after hatching, the grubs feed on organic matter. The second and third instars are root feeders, 'C' shaped. The larval period takes place up to 167 days.

The grubs can move up to 125 cm depth, depending upon the root zone, type of soil, temperature, host plant and the grub's species (Nigam and Avasthy, 1978; Yadava, 1981). The pupal period varies from 10 days to 3-4 months depending upon temperature (Ritcher, 1958). Pupation generally occurs during November-December. Adults will not come out of the cell until the next summer rain in April to July.

Nature of damage

White grubs can feed on any root or underground stem. The white grubs can cause a reduction of 15 t 20 tonnes of sugarcane per acre (Mortorel and Gaud, 1965; Gruner, 1975). In the absence of roots around the stem, the grubs may feed on the bark of the tree all round up to 15 cm deep from soil surface (Veeresh, 1977a). Grubs with heavy losses also damage potato, tobacco, yam and colocasia sometimes.

Management of white grubs

For managing the population of white grub, efforts can be directed against beetles or grubs. The *kharif* crops are cultivated in rainfed as well as irrigated conditions. In rainfed condition, the crops are sown with the onset of monsoon in June-July and this coincides with the emergence of beetles. Where irrigation is available, the farmers usually go for early (pre-monsoon) sowing of the crops. Some of the practices, which are applicable to rainfed conditions, may not be suitable for advance planted crops, as the sowing time does not coincide with the beetle emergence. Looking to the variations in the conditions and suiting to these requirements, the management strategies have been devised separately for monsoon sown and early sown (pre-monsoon) situations. Nevertheless, the beetle control strategy is suitable for both the situations. The farmers may choose any of the practices mentioned below suiting to their condition and requirement. The various strategies have been described below:

Natural control

In nature, natural enemies *viz.* parasites, predators and microbes (fungi, bacteria and viruses) are responsible in bringing down the population of white grubs (Davis, 1919; Niklas, 1960, 1969). Continuous flooding for more than a week and prolonged drought at the time of adult emergence are also causing natural decline of white grub's population (Veeresh, 1977a).

Autocidial control

The pheromone, isolated from the abdominal glands of calling female beetles of *H. consanguinea* (Leal *et al.,* 1996) have a unique property of acting as aggregation pheromone and attracting both the sexes from a distance of 15 meters. Use of this pheromone has made beetle control operation easier, economical and less hazardous as only one tree within a radius of 15 m is required to be sprayed with insecticide and loaded with pheromone. For this purpose a single tree from a group of host trees within a radius of 15 m could be selected and sprayed with 0.05 per cent monocrotophos or carbaryl 0.2 per cent during day time and pheromone dispensers be placed (2-3 dispensers/tree) on this host tree in the evening continuously for three evenings after beetle emergence.

The beetles, which are attracted by the pheromone, settle and feed on the sprayed host trees get killed soon by contact as well as stomach action of the insecticide and are not able to return to the soil next morning for egg-laying. By pheromone loading, number of trees to be sprayed is drastically reduced, which makes the beetle control operation more effective, eco-friendly, economical and less time consuming. Generally it is argued that the beetle control will be effective only when the entire village/region are sprayed with insecticide through community efforts. But this is not true, as the beetles do not fly long distances, individual effort by a farmer shall also provide protection to his crops.

Cultural methods

Repeated ploughing or hoeing after the first summer rains kills the grubs gradually (Andrews, 1919, Veeresh, 1977a). Rotation of crops with sugarcane and avoiding ratooning give satisfactory control of grubs in sugarcane fields (Jepson, 1956; Avasthy, 1967; David and Kalra, 1966). Flooding associated with puddling and keeping the field in this condition for a week will kill all the grubs (Veeresh, 1977a). Sowing quick growing crops *viz.* groundnut,

French bean, maize, sorghum as trap crops followed by first summer rains can help to concentrate grubs in limited areas.

Mechanical method

Hand collection of adults on adult host plant as a measure of control is very effective. It has to be done collectively by the farmers from the first day of the first summer rains (Prasad and Thakur, 1959; Gupta, 1958; 1973; Sri Pathi Rao, 1966; Veeresh, 1974 and 1977a). This could be accomplished by jerking the host trees at night time (around 9 PM), collecting the fallen beetle and killing them in five per cent kerosenised water. Normally the beetles of white grubs are attracted towards light in very small numbers. The beetles which drop from the trees, because of one reason or the other, are drawn towards light. As such, light trap may not be use to control them but it may be used as a tool for population monitoring. However, in the hills of Uttar Pradesh, light trap can be employed as an effective management tool for *Anomala dimidiata*, which is highly phototactic.

Biological control

The beetles are preyed upon by birds like black dragon, blue jay, crow and cuckoo which stay late in the evening up to the time of beetle's emergence. Insectivorous bats prey upon the beetles during dusk when they are on flight just before settling on the host trees. Bran owl also picks up some beetles. Toad, hedge hog and some lizards may feed on beetles, particularly those of small size. Dogs also fondly pick the fallen beetles. The grubs are picked up by crows, mynas and egrets which follow the plough. Such birds need encouragement and protection. Two cross plaguing in presence of the birds may reduce 70 per cent of the grubs in the field. Carabid, *Scarites indus*, burrows in soil and feeds on grubs. A mite of genus *Sancassania* has been observed as efficient predator of eggs.

Large majority of natural enemies of white grubs have been reported from many countries (Table-1). However, few parasites *viz. Campsomeris collaris* (Chandy kurian and Abdulla koya, 1974), *Scolia oryctophaga* , *Tiphia vernalis, T. pullivora* (Covery and Langford, 1955), *Dexilla vacua, D. rustica* (Gyrisco *et al.*, 1954) have shown promise against white grubs. Among the predators *Bufo marinus* (Walcott, 1950), pigs, birds play an important role in checking the white grub population.

Bateria, fungi, nematodes and rickettsiae are the major entomogenous pathogens of white grubs causing epizootics in natural ecosystem. Among these *Bacillus popilliae, B. thuringiensis, Metarrhizium anisopliae, Beauveria bassiana, B. brongniarti, Steinernema spp.* and *Heterorhabditis spp.* are promising in managing white grubs.

Soil application of 75 or 150X10^{12} spores/ha of *Beauveria bringniartii* + 1.2 kg isofenphos methyl 40% resulted in 80% control of white grubs in groundnut fields (Veeresh, 1977a and Deng *et al.*, 1995). The nematode *Heterorhabditis bacteriophora* are more virulent at (14090 IJ's/100 g soil grub) against third instar grubs of *Maladera insanabilis* (Ashoka Bhatnagar *et al.*, 2004).

Chemical control

Use of kerosene (Alwood, 1888) was the earliest attempt against white grubs. Further, crude naphthalene was also used as soil disinfectant against white grubs (Anstead, 1919).

Table 1 Natural enemies of white grubs.

Host (1)	Location (2)	Parasite Species (3)	Family (4)	Reference (5)
Insect parasites				
White grubs	India	*Scolia pustulata* *S. aureipennis* *S. wagneri* *Camposmeris collaris* *Tiphia hirsuta*	Scoliidae	Channabasavanna, 1954
White grubs	India	*Adaprilia flaviseta* *Tiphia maiurea* *T. pullivora*	Pyrgotidae	Gardener and Parker, 1940
Popillia sp		*T.vernalis* *T matura* *Tpopillivora*		Covery and Langford, 1955
Anomala sp. *Seria sp.*	Fareast	*Tiphia spp.*		Covery and Langford, 1955
Holotrichia serrata	Bihar, Coimbatore India	*Campromeris collaris* *Scolia aureipennis*		Gupta and Avasthy, 1960. Avasthy, 1964
Sugarcane white grubs	Mauritius Barbados	*T. parallela*		Walcott, 1950
Sarabaeid larvae	North Africa Wallis island	*Microphthalma europaea*		Cocherean, 1970
Leucophollis sp.	Kerala, India	*Campromerilla collaris*		Candy Kurian and Abdulla Koya, 1974.
Oryctes tarandus	Mauritius	*Scolia oryctophaga*		Gyrisco *et al.*, 1954.
Amphimallon majalin	USA	*Dexilla vaua* *D. rustica*		
Anomala orientalis	USA	*Scolia manilae*		Swezey, 1928.
(1)	(2)	(3)	(4)	(5)
Insect predators				
White grubs	USA, Australia	Asilidae, Tabanidae, Carabidae, Histeridae, Formicidae and Elatreridae		Davis, 1919, Gyrisco *et al.*, 1954, Ritcher, 1958.
Rhinoceros grubs	Assam, India	*Pheropsophus sobrinos var. desbodesi*		Rao and Manjunath, 1964.
White grubs	USA	*Pterostiothus haloites*		Lee Seaton, 1939.
June beetle adult	Rajasthan, India	*Anthia sexguttata*		Rai *et al.*, 1969.
European chafer		*Craspedonotus libialis* *Harpalus pennylvanicus* *H. erratious*		Gyrisco *et al.*, 1954.
Vertebrate predators				
Sugarcane white grubs	Puesto Rico	*Bufo marinus*		Walcott, 1950.
Grubs		Pigs, rows		Forber, 1916.
Eggs And Grubs	Puesto Rico	Lizard *Ameiva exsul*		Walcot, 1950

Contd. ...

Table 1 Contd. ...

Grubs	India	Indian Hoopoe *Upupa epopa indica* Crow *Corvus splendens* Mynah *Acridotheres tristis*		Avasthy, 1965; Kalra and Kulshrestha, 1961; Yadava *et al.*, 1973; Veeresh, 1977a; Kushwaha and Allah noor, 1974, Shah and Garg, 1988.
Microbial diseases				
Japanese beetle		*Bacillus popillae* type A. *Bacillus lentimorbus* type B.		Dutky, 1940; Hurpin, 1967; Choo Hoyul *et al.*, 2000.
Austrilian scarabaeids	Australia	*Bacillus lentimorbus var. Australia* *Bacillus euloomarachae*		Dumbleton, 1945.
(1)	**(2)**	**(3)**	**(4)**	**(5)**
Holotrichia serrata	South India	*Bacillus spp.*		David and Alexander, 1975; Veena kumari and Veeresh, 1982; Battu, 1978; David *et al.*, 1973 Patel *et al*, 1967, Shinde and Sharma, 1971; Veeresh, 1980.
Holotrichia grubs	India, Europe	*Bacillus thuringiensis* *Bacillus spp.*		Sekhar and Venkataramaiah, 1964; Franz, 1961.
White grubs	india	*Metarrhizium anisopliae* *Beauveria tenella* *B. bassiana*		Gupta and Avasthy, 1960; Rao and Vijayalakshmi, 1959; Nirula, 1957; Nirula *et al.*, 1955; Choo Hoyul *et al.*, 2000.
White grubs	Karnataka, India	*Bacillus brongniartii* *Rikettsiella melolonthae* *Entomopoxvirus malolonthae*		Ranganathaiah *et al.*, 1973; Veeresh, 1977a; Deng *et al.*, 1995. Hurpin and Robert, 1977
Holotrichia serrata	Australia	*Neoplectana carpocapsae* *Steinernema glaseri* *Steinernema longicaudum*		
White grubs	North Germany	*Heterorhabditis bateriophora*		Choo Hoyul *et al.*, 2000; Ehlers *et al.*, 2003; Ashoka Bhatnagar *et al.*, 2004.

The actual chemical control of white grubs started after the advent of chlorinated hydrocarbon insecticides. Several scientists have found organochlorine insecticides (chlordane, dieldrin, BHC, aldrin) were effective against grubs (Prasad and Thakur, 1959; Bharadwaj *et al.*, 1976; Bhat and Chacko, 1979; Bhatnagar *et al.*, 1974; Kushwaha, 1976; Kalra and Kulshrestha, 1961; Srivastava and Khan, 1963). However, these studies indicated development of resistant to chlorinated hydrocarbons (Tashiro *et al.*, 1971). And the time of application are important for which biology of the species must be known (Tashiro and Fiori, 1969). Incorporation of 100 kg BHC 10% dust mixed in equal quantity of FYM, drilled in rows before sowing of kharif crop was the common practice followed for food crops except groundnut (Veeresh, 1980).

Phorate 10G or Quinalphos 5G or carbofuran 3G at 2.5 kg a.i. per hectare and chlorpyriphos 20 EC at 1.25 kg a.i. /ha have given good control of white grubs provided the time of application is synchronized with the emergence of adults and egg laying (Veeresh, 1977b). Adequate irrigation must be given after treatment in case of management of grubs on standing crop. These insecticides have little effect on grownup grubs at the recommended dosage under dry land condition.

Foliar application of 0.2% carbaryl or 0.4% DDT or 0.05% monocrotophos or 0.03% quinalphos on adult host plants immediately after the first rains was effective against adults (Yadava *et al.,* 1977).

Soil application of phorate or carbofuran (3.00 kg a.i./ha) or chlorpyriphos (0.4 kg a.i./ha) at the time of earthing up of potato/ tuber crop was effective against white grubs (Bhagat *et al.,* 2001; Pillai *et al.,* 1993; Misra and Tyagi, 1989; Misra, 2003). Soil application of 200g of diazinon or phorate per nursery bed of teak was effective against white grubs (Meshram *et al.,* 1990).

Integrated management of white grubs

- Collection of beetles after first summer shower i.e. 7.30 pm onwards on adult food plants.
- Use of heavier equipment such as the mould board plough (for deep ploughing) to expose the white grubs to natural enemies (Oliveira *et al.,* 2000).
- Soil application of well decomposed farm yard manure (Gomez *et al.,* 1999).
- Soil application of 25 kg of phorate 10G or carbofuran3G or carbaryl 4% dust or 35 kg of γ-BHC before sowing or planting or application of 500 l of chlorpyriphos 20EC (10ml/l) solution on soil surface after first shower.
- If the white grub damage was noticed in the previous year, while sowing treat the groundnut seed with chlorpyriphos 20EC (15ml/kg) before sowing.

References

Abraham, V.A. and Kurian, C. 1970. Save your coconut palm from the ravages of white grubs. *Coconut Bull.,* 1: 3-4.

Agarwal, R. A.; Chaudary, P.P. and Bhatia, H. L. 1965. White grubs attack sugarcane in Punjab. *Indian Sugar J.*, 9: 254-255.

Agarwal, R.A. 1972. Problems of termites of sugarcane in India. In: Termite problems in India. Roonwal, M.L. (Editor), Council of Scientific and Industrial Research, New Delhi. pp. 46-55.

Agarwala, S.B.D. 1955. Control of sugarcane termites (1943-1953). *J. Eco. Ent.*, 48: 533-537.

Alwood, W. D. 1888. Kerosene emulsion as a remedy for white grubs. *Insect life, U. S. Dept. of Agri.*, 1: 48-49.

Andrews, E. A. 1919. Insect control. *Rept. Proc. 3rd Ent. Mtg.*, Pusa, 1: 28-30.

Anon. 1898. White ants as a pest of agriculture – a note by the settlement officer of Balaghat, Central Provinces, prefixed by certain passages reprinted from the Indian Museum Notes, on termites in relation to crops etc. Entomological Series No. 7, the Agricultural Ledger 1897 No. 18, 434, Calcutta, Office of the Superintendent, Government Printing.

Anstead, R. D. 1919. Insect pests of South India and their control. *Rept. Proc. 3rd Ent. Mtg.*, Pusa. 1: 329-330.

Ashok Kumar,C.T. and Veeresh,G.K.1990. Foraging activity of the subterranean termite,Microtermes obesi (Isoptera : Termitidae).In: Soil insect and environment (eds.Veeresh *et al.*,). Proc.XIth Int. Conf. IUSSI, Bangalore, pp 575.

Ashoka Bhatnagar; Vijay-Shinde and Bareth, S. S. 2004. Evaluation of Entomopathogenic nematodes against white grub, *Maladera insanabilis* Brenaka. *Int. J. Pest management,* 50(4) : 285-289.

Avasthy, P. N. 1964. White grubs in sugarcane. *Proc. Pl. Prot. Seminar,* New Delhi.

Avasthy, P. N. 1965. The problem of white grubs on sugarcane in India. *Proc XII Cong. Int. Soc. Tech., Peurto Rico.,* 1321-1333.

Avasthy, P. N. 1967. Sugarcane pests in India, their control (a review), *PANS,* 13: 111-117.

Avasthy, P. N., and Krishan Singh. 1973. Integrated control of sugarcane pests and diseases. *Indian Sug.,* 23: 529-531.

Banerjee, B. 1967. Control of *Microcerotermes*, a termite attacking live wood of tea. Serials, *Tocklai Experimental Station,* 77: 4.

Banerjee, S. N. and Mukerjee, A. L. 1960. Notes on *Alissonotum impresicollis* Arrow, a new pest of sugarcane in West Bengal. *Proc. IV Conf. Sug. Res. And Dev. Workers, India,* Waltair, 513-520.

Batra, R. C.; Bindra, O. S. and Sodhi, B. S. 1973. New record of some chafer beetles as pests of grape vine, *Indian. J. Ent.,* 35: 177.

Battu, G. S. 1978. A note on the use of milky disease bacterium against white grubs *Lachnosterna (Holotrichia) consanguinea* Blanchard. *Indian J. Ent.,* 40: 348-355.

Beeson, C. F. C. 1919. Report of the Imperial Entomologist. *Scient. Rept. Agric. Res. Inst.* Pusa, 1918. 84-116.

Beeson, C.F.C. 1941. A guide to the control of termites for forest officers. Indian Forest Records (New series), *Entomology,* 4: 44-90.

Bhagat, R. M. and Kashyap, N. P. 1997. A preliminary study on white grub of Dhauldhar Foot hills (Himachal pradesh) and adjoining areas. *Indian J. Ent.,* 59(3): 337-340.

Bhagat, R. M.; Kashyap, N. P. and Viji, C. P. 2001. Management of white grubs through insecticides. *Pesticide Research,* 13(2): 247-249.

Bharadwaj, S.C.; Yadava, C. P. S. and Yadava, S. R .S. 1976. Susceptbility of white grubs *Holotrichia consanguinea* adults to various insecticides. *Indian J. Ent,.* 38: 225-229.

Bhat, O. K.; Kaul, V. and Bhagat, K. C. 1994. Incidence of pests associated with the rhizosphere of tomato in Jammu. *Annals of Pl. Prot. Sci.,* 2(2); 23-26.

Bhat, P. K. and Chacko, M. J. 1979. Further trials with granular insecticides against cockchafer attacking coffee. *Indian Coffee,* 40: 120-125.

Bhatnagar, O. K.; Gupta, N. N. and Viswakarma, S. L. 1974. Efficacy of various pesticides against white grubs (*Holotrichia serrata*) of groundnut. *Pesticides,* 9: 22-24.

Bose, G. 1984. Records of the Zoological survey of India. Miscellaneous Publication Occasional Paper No. 49. *Zoological Survey of India*. pp 270.

Brown, K.W. 1962. Termite control research in Uganda with particular reference to control of attacks in Eucalyptus plantations, 9 pp. Eight British Commonwealth Forestry Conference. Entebbe, Government Publication, Uganda Protectorate.

Butani, D.K. 1967. Sugarcane termites – a resume. *Indian Sugar,* 17: 543-549.

Canty, C. and Harrison, P.G. 1990. Transfer of controlled release insecticide technology for protection of forest seedlings from termite attack. *Agricultural technology transfer*, No.54: 183-189.

Chandy Kurian and Abdulla Koya.1974. White grubs in relation to coconut. *White grubs Newsletter,* 1: 24-26.

Channabasavanna, G. P.1954. Scoliid parasites on chafer grubs. *Ann. Rep. Dept. of Agri., Mysore.*, 1953-1954.

Chatterjee, P.N. and Thakur, M.L. 1967. Contributions to the knowledge of systematics of North-Western Himalayan termites fauna (Isoptera : Insecta) III-Systematic account of the survey. Indian Forest records, *Entomology* (New Series), 11(1) : VI.

Choo-Ho Yul ;Lee-Dongwoon ; Lee, S. M.; Lee, T. W.; ChoriwooGeun; Chung Young Ki and Sung Young Tak.2000. Turfgrass insect pests and natural enemies in gulf courses, *Korean J. Appl. Ento.*, 39(3): 171-179.

Cocherean, P.1970. Breeding in New Caledonia of *Microphthalma europaea* Egg. (Diptera: Tachinidae) on an alternative host *Protaetia fusca* Herbst (Coleoptera: Scarabaeidae). *Entomophaga,* 15: 281-285.

Covery, E. N. and Langford, G. S.1955. The Japanese beetles retardation programme in Maryland. *Extension Bull. Uni. Maryland.* pp. 156.

Danthanarayana, W. & Vitarana, S.L. 1987 Control of the live wood tea termite. *Glyptotermes dilatatus* using *Heterorhabditis sp.* (Nemat.). *Agriculture, Ecosystems and Environment,* 19:333-342.

David, H. and Alexander, K. C. 1975. Natural occurrence of milky disease bacterium, *Bacillus popilliae* Dusky on white grubs in India, *Curr. Sci.,* 44: 819-820.

David, H.; Alexander, K. C. and Ananthanarayana, K. 1973. Milky disease bacterium against the control of white grub, *Holotrichia serrata* Fabr. infesting Sugarcane white grub, *Holotrichia serrata. Curr. Sci.,* 43: 695-696.

David, M. and Kalra, A. N. 1964.White grubs damage to sugarcane crop in Hospet area of Mysore state. *Indian J. Sug. Res. Dev.,* 8: 428.

David, M. and Kalra, A. N. 1966. Observation on *Holotrichia (Lachnosterna) serata* F., a beetle pest of sugarcane crop in Hospet area of Mysore state. *Indian sugar,* 16: 195-198.

Davis, J. J. 1919. Contribution to knowledge of the natural enemies of *Phyllophaga. Illinois State Nat. Hist. Survey Bull.,* 13: 53-138.

Dawkins, H.C. 1949. Timber planting in the Terminalia wood land of Northern Uganda. *Empire Forestry Review,* 28, 226-247.

De Fluiter, H. J. 1941. *Arch. Rubber Nod India,* 25: 167-227.

Deng, C. S.; Zhang, A. W.; Nong, X. Q.; Gao, S.; Li, J. G.; Jiang, H. L.; Hu, J. W.; Tian, J. X. and Li, J. Y. 1995. Field experiment on *Beauveria brongiartii* for controlling white grubs in groundnut fields. *Chinese J. Biological control.,* 11(2): 56-59.

Dumbleton, L. J.1945. Bacterial and nematode parasites of soil insects. *New Zealand J. Sci. Technol,.* 27: 76-81.

Dutky, S. R. 1940. Two new spore forming bacteria causing milky disease of Japanese beetle larvae. *J. agric. Res,.* 61: 57-58.

Ehlera, R. U.; Susurluk, A.; Ficher, R.; Koppenhofer, A.; Barth, M. and Strauch, O. 2003. New nematodes strain classes gaps in the biological control of white grubs. *DgaaE-Nachrichten,* 17(1): 13.

Farghal,A.I.; Darwish,Y.A. and Abdel Galil,F.A.1987.Relative susceptibility of four insect species to some bacterial insecticides.*Assiut J.Agri.Sci,* 18:21-30.

Fletcher, T. B. 1919. Annotated list of Indian crop pests *Rept. Proc. 3rd Ent, mtg.,* Pusa, pp. 33-417.

Fletcher, T.B. 1912. Termites or white ants. *Agriculture Journal of India,* 7 : 219-239.

Forbes, A. S. 1916. A general survey of the May beetles (*Phyllophaga*) of Illinois. III *Agril. Expt. State Bull.,* 186: 227.

Forschler, B.T. and Ryder. (Jr.). 1996.Subterranean termite, *Reticulitermes spp*(Isoptera:Rhinotermitidae)colony response to baiting with hexaflumuron using a prototype commercial termite baiting system.*J.Entomol.Sci.,* 31(2):143-151.

Franj, J. M. 1961. Biological control of pest insects in Europe. *Ann. Rev. Ent.,*. 6: 183-200.

Frey, G. 1971b, Bestimmunngstabellc der indischen and Ceyloneshen Artan der Gattung *Holotrichia* Hope. *Ent arb. Mus. Fary.* 22: 206-225.

Fuller, C. 1912. White ants in Natal. 13 pp. Pretoria, Union of South Africa Department of Agriculture.

Gangwar, S. K.;Tewari, R. K.; Lakshman Lal and Lal, L. 2003. Monitoring insect pest incidence, yield losses and their management in late planted sugarcane. *Proc. 65th Ann. Convention of the Sugar technologists. Assoc. of India, Orissa.* 186-195.

Gardener, T. R. and Parker, L. B. 1940. Investigations of the parasites of *Popillia japonica* and related Scarabaeidae in the far East from 1929 to 1933 inclusive. *U. S. Dept. of Agric. Tech. Bull.,* 738: 1-36.

Georgis, R. & Poinar, G.O. 1983. Effect of soil texture on the distribution and infectivity of *Neoaplectana glaseri* (Nematoda: Steinernematidae). *Journal of Nematology*, 15: 329-332.

Giridhar, G; Vasudevan, S. & Vasudevan, P. 1988. Antitermites properties of *Calatropis latex. Pesticides,* 22(1): 31-33.

Gomez, B.; Villalobos, F. J.; Ruiz, I.; Cartro, A. and Valle, J. 1999. The white grub complex (Coleoptera: Melolonthidae) in maize in Los Altor de chipar, Mexico: its relation with the time of agricultural use and soil organic matter. *Folia. Entomological Mexicana,* 107: 1-20.

Grace, J.K. 2003. Approaches to biological control of termites. *Sociobiolog,*. 41 (1): 115-121.

Gruner, L. 1975. Sequential sampling of population of white grubs and forecasting damage in fields of sugarcane. *Annals de Zoologic Ecologic Animale.,* 7: 505-524.

Gul, H. and Chaudhry, M.I. 1989. Efficacy of insecticides against termites infesting bark of living trees.*Proc.Pakistan congress of Zoology*, No. 9 :61-63.

Gupta, B. D. 1958. A new fluorescent light trap. *IISR Newsletter,* 4: p. 17.

Gupta, B. D. and Avasthy, P. N. 1960. *Insect pests of sugarcane and their control.* Indian central sugarcane committee, New Delhi: 9-34.

Gupta, K. M. 1973. Neem leaves attract white grub beetles. *Indian J. Ent*., 35: 276.

Gupta,V.S. and Sharma, S. 1977. *Package of pratices-Kharif1977*, Hissar, India, Haryana Agri.uni.126 pp.

Gyrisco, G. G.; Whitecomb, W. H.; Burrage, R. H.; Logothelis, C. and Schwardi, H. H. 1954. Biology of the European chafer, *Amphimalon majalis* Razonmowsky (Scarabaeidae). *Cornel. Agri. Expt. Sta. Memoir.,* 328: 35.

Hanel, H. & Watson, J.A.L.1983. Preliminary field tests on the use of *Metarhizium anisopliae* for the control of *Nasutitermes exitiosus* (Hill) (Isoptera: Termitidae). *Bulletin of Entomological Research,* 73:305-313.

Harris, W.V. 1971.Termites: Their recognition and control. Second Edition, 186 pp. London, Longmans.

Hurpin, B. 1967. Investigations on the epizootiology of the milky disease of *Melolontha melolontha* caused by *Bacillus popillae. Anhals Epiphy,* 18: 127-73.

Hurpin, G.and Robert, P.M. 1977. Effects on a natural population of cockchafers *Melolontha melolontha* introduction of *Rickettisiella melolonthae* and *Entomopoxvirus melolonthae*. *Entomophaga,* 22: 85-91.

Isaac, P.V. 1920. Some recently noted South Indian Melolonthidae of economic importance. *Rept. Proc. Third Entomol. Mtg.* Pusa, 1028-1029.

Jayanthi, MJ., Singh, K.M. and Singh, R.N. 1993. Differential influence of drip and surface irrigation on pest complex of groundnut. *Indian J. Ent.,* 55: 124-131.

Jepson, W.F. 1956. The biology and control of the sugarcane chafer beetle in Tanganyka. *Bull. Ent. Res.*, 47:377-397.

Johnson, R.A.;Lamb, R.W. and Wood, T.G. 1981. Termite damage and crop loss studies in Nigeria – a survey of damage to groundnuts. *Tropical Pest Management,* 27, 325-342.

Kakaliyev, K. and Saparliyev, K. 1975. Research in pathogenicity of entobacterine for controlling termites in nature. *Izvestiya Akademii Nauk Turkmenskoi SSR (Seriya Biologicheskikh Nauk)*, 6: 39-41. (in Russian).

Kakde, J.R. 1985. Sugarcane production. New Delhi, Metropolitan. pp.384

Kalra, A.N. and Kulashreshtha, J.P. 1961. Studies on the biology and control of *Lachnosterna consanguinea* Blanch. a pest of sugarcane in Bihar. *Bull. Ent. Res.*, 52: 577-587.

Kashyap, R.K.; Verma, A.N. and Bhanot, J.P. 1984. Termites of plantation crops, their damage and control. *Journal of Plantation Crops,* 12: 1-10.

Khan, K.I.;Fazal, Q. and Jafri, R.H. 1977. Pathogenicity of locally discovered Bacillus thuringiensis strain to the termites. *Heterotermes indicola* (Wassman) and *Microcerotermes championi* (Snyder). *Pakistan Journal of Scientific Research,* 29: 12-13.

Krishna, K. 1970. Taxonomy, physiology and distribution of termites.pp.127-152.In: Krishna, K. and Weesner, F.M. (eds.), *Biology of termites*, Vol.2.New York and London,Academic Press.

Kumar, N.G. 1991. Studies on the ecology of the subterranean termite, *Odontotermes horni* (Wasmann) (Isoptera: Termitidae) and its foraging effects on the soil nutrient status. Ph.D. Thesis, University of Agricultural Sciences, Bangalore 254 pages.

Kumar, N.G. 2005. Status of termite research in India. In.(Ramakrishnan *et al.* eds.,) *Soil Biodiversity, ecological processes and landscape management.* Oxford and IBH pub.Co.pvt.Ltd.New Delhi.pp.25-40.

Kumar, N.G. 1998. Effect of agricultural practices on the termite activity in soybean agro-ecosystem. National Symposium on "Diversity of Social Insects and Other Arthropods and the Functioning of Ecosystems". 7-9, March, College of Horticulture, Mudigere, UAS, Bangalore.

Kumarasinghe, N.C. and Ranasinghe, M.A.S.K.1988.Incidence of termite damage in sugar cane grown in Sri Lanka. *Beitrage zur Tropischen Landwirtshaft and Veterinarmedizin* ,6:-307.

Kumawat, K. C. 2001.Evaluation of some insecticides against field termites, *Odontotermes obesus* Holm. in wheat, *Triticum aestivum. Annals of Pl. Prot. Sci,* 9 (1): 51-53.

Kushawaha, K.S. 1976.White grubs control-a state wide campaign in Rajasthan. *Univ. UdaipurRes. J,* 14: 57-58.

Kushawaha, K.S. and Allah Noor.1974. Bird predators of white grubs in Rajasthan. *Univ. Udapipu Res. J.*,14: 57-58.

Leal, W.S.; Yadava, C.P.S. and Vijayvergia, J.N. 1996. Aggregation of the scarab beetle *Holotrichia consanguinea* in response to female released pheromone suggested secondary function hypotheses for semiochemical. *J. Chem. Ecol.*, 22 : 1557-1566.

Lee Seaton. 1939.A feeding record of *Pterostichus (Poecilus) chalcites* (Say) upon June beetle eggs and grubs. *J. Econ. Ent.*, 32: 151-152.

Lee, K.E. and Wood, T.G. 1971. Termites and soils. London, Academic Press. pp.251

Lefroy, H. M.1909.Indian *Insect life*. Thacker Spink, Calcutta, p. 786.

Logan, J.W.M. andAbood, F.1990.Laboratory trials on the toxicity of hydramethylon (Amdro: AC 217,300) to *Reticulitermes santonensis* Feytaud (Isoptera: Rhinotermitidae) and *Microtermes lepidus* Sjostedt (Isoptera: Termitidae). *Bull. Ent. Res.*, 80:19-26.

Maan Singh;Dharam Singh and Madan, Y. P. 2001.Evaluation of different soil insecticides for the control of termites in sugarcane. *Indian Sugar,* 51 (6): 365-368.

Madan, Y.P.;Maan Singh and Singh, M. 1998.Evaluation of some soil insecticides for termites and shoot borer control in sugarcane. *Indian Sugar,*49 (7): 515-518.

Maiti, P.K.1975.Two new species of termites (Isoptera) from West Bengal, India. *Oriental Insects,*9: 139-149.

Majumdar, N. and Teotia, T.P.S.1965.Biology of *Holotrichia serrata* Fab. a serious pest of palas (*Butea monosperma). Indian J. Ent,* 27: 135-159.

Manager Singh and Singh, N. B. 2003.Effect of insecticides on the infestation of termites on emerging shoots and millable canes. *Indian J. Ento,* 65 (1): 28-33.

Manager Singh;Singh, N. B. and Singh, M. 2002.Bud damage due to termites in sugarcane. *Cooperative Sugar,*33 (8): 655-658.

Mariconi,F.A.M.;Hamamura,R.andRangel,M.C.1990.Deltamethrin and fenthion in an assay against the mound termite *Cornitermes cumulans* (Kollar)(Isoptera:Termitidae).Revista-de-Agricultura-piraciaba, 65(3): 273-278.

Mathur, R.N.1962Enemies of termites (White ants).In;Termites in the humid tropics,New Delhl symposium 1960, pp.135-139,UNESCO, Paris.

Mehta, P.R. and Varma, B.K.1968.Plant Protection, Ministry of Food, Agriculture Community Development and Co-operation, New Delhi. 587 pp.

Meshram, P. B.;Pathak, S. C. and Jamaluddin 1990.Effect of some soil insecticides in controlling the major insect pests in teak nursery. *Indian Forester.,*116(3): 206-213.

Misra, S. S. 2003.Assessment of available losses in potato yield by managing white grubs, *Holotrichia spp.* at higher hills of Himachal pradesh, India. *Indian J. appl. Zoological Res.,* 14(1): 13-17.

Misra, S. S. and Tyagi, G. C. 1989.Efficacy of soil insecticides against white grubs, *Holotrichia spp.* infesting potatoes in Shimla hills. *Indian J. Pl. Prot.,*17(2): 283-286.

Mortorell, F. and Gaud S. M. 1965.Status of important insect pest of sugarcane in Puerto Rico and their control. *Proc. XII. Cong. Int. Soc. Sug. Tech. Puerto Rico.* 1268-1278.

Nair, M.R.G.K. 1975.Insects and mites of crops in India. New Delhi, Indian Council of Agri.Res.405pp.

Nair, M.R.G.K. 1978.Crop pests of Kerala and their control, Kerala, India, Kerala Agri.Uni.162pp.

Nath, D.K.1972. Occurrence of some little known insects on citrus plants in West Bengal. *Indian J.Agril.Res,*6: 249-254.

Nigam, P.M. and Awasthy B.K. 1978.Survey of white grub *Holotrichia consanguinea* Blanch. at different depths of soil. *Indian J. Ent,* 40: 90-91.

Niklas, O.F. 1960.Locality effect and natural enemies as factors limiting the larval population of *Melolontha* of the forest region of Lorsch Hesse. *Mitt. Biol. BurdeswfBeri,* 101:60.

Niklas, O.F. 1969.Transovarial transmission of *Rickettsiella melolonthae* Phialill, through two generations of *Amphimallon scolistitiale. Entomophaga.* 14: 225-228.

Nirula, K.K. 1957.Observation on the green muscardine fungus in population of *Oryctes rhinoceros* L. *J. Econ. Ent.,* 50: 767-770.

Nirula, K.K.;Radha, K. and Menon, K.P.V. 1955.A green muscardine disease of *Oryctes rhinoceros* L. *Indian coconut J.,* 9: 3-10.

Oliveira, L.J.;Hoffmann, Cango, C. B. and Garcia, M. A. 2000.Effect of soil management on the white grub population and damage in soybean. *Perquira- Agropecuaria Brassileira,* 35(5): 887-894.

Pal, S.K. 1977. Relative abundance of Scarabaeid beetles on light traps. *Indian J. Ent.*, 39: 197-200.

Parihar, D. R., 1994. Termites management in arid zone of Rajasthan, India. *Pest management and economic Zoology,* 2(1): 81-84.

Patel, R.M.;Patel, G.G. and Vyas, H.N. 1967.Further observation on the biology and control of white grubs *Holotrichia* sp. in soil affecting groundnut in Gujarat. *Indian J. Ent,* 29: 170-176.

Pillai, K. S.;Palaniswami, M. S.;Rajanna, P.;Mohandas, C. and Jayaprakas, C. A. 1993.Pest management in tuber crops. *Indian Horticulture,* 38(3): 20-23.

Poinar, G.O. 1975. Entomogenous nematodes. 317 pp. Leiden, Brill.

Poinar, G.O. 1979.Nematodes for biological control of insects. 277 pp. Boca Raton, Florida, CRC Press.

Prasad, B.K. and Thakur, C. 1959.White grub *Lachnosterna consanguinea* Blanch a new menace to sugarcane. *Indian J. Ent.,* 21: 184-189.

Quarles, W. 1999. Baiting subterranean termites. *IPM-Practitioner,* 19 (10): 1-9.

Rai, B.K.;Joshi, H.C.;Rathore, Y.K.;Dutta, S.M. and Shinde, V.K.R. 1969.Studies on the bionomics and control of white grub *Holotrichia consanguinea* Blanch. In Lalsot, Dist. Jaipur, Rajasthan. *Indian J. Ent.*, 31: 132-142.

Rajagopal, D. 1979.Ecological studies of the mound building termite, *Odontotermes wallonensis* (Wasmann) (Isoptera: Termitidae) Ph.D. Thesis, University of Agricultural Science, Bangalore 205 pp.

Rajagopal, D. 1982. Relative incidence of termites on exotic species of Eucalyptus in Karnataka. *Myforest,* 18: 9-13.

Rajagopal, D.;Kumar, N.G. and Veeresh, G.K. 1996. Effect of farm yard manure on the incidence of termites in maize crop. National Seminar on organic farming and sustainable agriculture, 9-11 October, University of Agricultural Science, Bangalore.

Rana, J. S.;Ombir and Dahiya, K. K. 2001. Management of termite, *Microtermes obesi* (Holm.) in wheat, *Triticum aestivum* through seed treatment. *Annals of Biology,* 17 (2): 207-209.

Ranganathaiah. K.G.;Veeresh, G.K. and Govindu, H.C. 1973. A new entomogenous fungus on the root grub *Hololtrichia serrata* from Mysore. *Curr. Sci,* 42: 432-433.

Rao, G.N. 1951.Control of termites in sugarcane. *Current Science,* 20, 330-331.

Rao, G.N. and Vijayalakshmi, U. 1959. A note on the occurrence of certain parasitic fungi on insect pests of sugarcane. *Curr. Sci.,*42: 432-433.

Rao, V.P. and T.M. Manjunatah. 1964. A new record of the carabid beetle *Pheropsophus sobrinus* (Dej) var. *desbordesi* (Maindr.) as a predator of Rhinoceros beetle in Inda. *Tech. Bull. Commonw. Inst. Boil. Control,* 4: 40-42.

Reddy, D.B. 1968. *Plant protection in India*. Bombay and London, Allied publishers.454pp.

Reddy, M.V. and Sammaiah, C. 1988. *Odontotermes brunneus* (Hagen.) (Termitidae: Isoptera) as a new pest of maize and groundnut. *Entomon,* 13 : 47-50.

Ritcher, P.O. 1958. Biology of Scarabaeidae. *Ann. Rev. Ent,* 2: 311-334.

Roonwal, M.L .and Bose, G. 1970. Taxonomy and zoogeography of the termite fauna of Andaman and Nicobar Islands, Indian Ocean. *Records of the Zoological Survey of India,* 62 : 109-170.

Roonwal, M.L .and Chhotani,O.B. 1962. Indian wood-destroying termites.*J.Bombay Natural History Society,* 63 : 354-364.

Roonwal, M.L. 1978. Bioecological and economical observations on termites of peninsular India. *Zeitshrift fur Angewandte Entomoloigie,* 85 : 15-30.

Roonwal, M.L. 1979.Termite life and termite control in tropical south Asia. 177 pp. Jodhpur, Scientific Publishers.

Roonwal, M.L. 1982. Fauna of the Great Indian Desert. In: Desert Resources and Technology. Vol. I, Alam, S. (Editor) Scientific Publishers and Geotech. Academia, Jodhpur. pp. 1-86.

Roonwal, M.L. and Bose, G. 1964. Fauna of Rajasthan, India. Part 4. A Check list of Rajasthan termites. *Records of Zoological Survey of India,* 61 : 437-450.

Sands, W.A. 1977. The role of termites in tropical agriculture. *Outllook on Agriculture,* 9, 136-143.

Santharam. G.;Kumar, K.;Kuttalam, S. and Chandrasekaran, S. 2002. Bioefficacy of imidacloprid against termites in sugarcane. *Sugar Tech.,*4 (3-4): 161-163.

Saroj-Jaipal and Dharam Singh.2003. Bioefficacy of imidacloprid and Amrutgaurd against termites and shoot borer in sugarcane crop. *Indian Sugar,* 53(9): 709-716.

Satya vir and Parihar, D. R. 1993. Insect pests of *Acacia senegal* (L.) Willd. and *Acacia tortilis* (Forssk. Hyne) in the Thar Desert. *MyForest.*,29(2): 105-111.

Saxena A.P. and Rizvi, S.M.A. 1974. Insect pest problems of potatoes in India.*J.Indian potato Assoiation,* 1:45-50.

Saxena D.K. 1972. Arthropod pests of soybean in Madhya Pradesh and their control.*Pesticides,*6 (4):17-19.

Sekhar, P.S. and Venkataramaiah, G.H. 1964. Susceptibility of the cockchafer *Holotrichia* to *Bacillus thuringiensis* Berl., *Curr. Sci,* 23: 405-406.

Sen-Sarma, P.K. 1986. Economically important termites and their management in the Oriental region. pp. 69-102 In: Vinson, S.B. (Ed.). Economic impact and control of social insects. New York, Praeger.

Shah, N. K. and Garg, D. K. 1988. Exposures of root grubs during ploughing and their predators by birds. *Indian J. Ent,* 50(1): 131-132.

Sharma, D. C.;Katoch, K. K. and Kashyap, N. P. 2002. Relative efficacy of different insecticides to *Odontotermes spp.* and *Agrotis* spp. in wheat. *Insect Environment,* 8 (1): 10-11.

Sharma, R. K.;Kirti Sharma and Sekhar, J. C. 2003.Evaluation of plant protectants on damage and yield of rainfed maize by termites. *Pesticide Res. J,* 15 (1): 36-39.

Sharma, S.;Vasudevan, P. and Madan, M. 1990.Insecticidal value of castor (*Ricinus communis*) against termites. *International Biodeterioration,*27 (3): 249-254.

Sharp, D. 1903.Lamellicorn Coleoptera from the Nilgiri hills. *Ann. Mag. Nat. Hist,* 11:15.

Sheppe, W. 1970.Invertebrate predation on termites of the African savanna. *Insectes Sociaux,*17, 205-218.

Shinde, C.B. 1971.Termites –a dangerous insect of agriculture. *Farmer and Parliament,* 6 (10):23.

Shindhe, V.K.R. and Sharma, S.K.1971.*Bacillus popillae* pathogenic to *Lachnosterna consanguinea* Blanch. *J. Econ. Ent,* 64: 1301-1302.

Singh, A. P; Bhandari, R. S. and Verma, T. D. 2005.Important insect pests of poplam in agro-forestry and strategies for their management in North-Western India. *Agroforestry systems,* 63(1): 15-26.

Singh, B.P. 1977.Package *of practices for Kharif crops of Punjab*.Ludhiana, India, PAU.16pp.

Singh, M. P.;Bisht, R. S. and Mishra, P. N. 2003.Distribution of white grubs fauna in Garhwal hills of Western Himalayas. *Indian J. Ento,* 65(2): 217-221.

Singh, S. K. and Singh, G. 2002.Comparative evaluation of chemical and botanical insecticides against termites. *Entomon,* 27 (2): 153-160.

Singh,D. and Brar, D. S. 1988.Growth and yield of rainfed wheat as affected by seed treatment with aldrin and fertilizer use. *J. Res. Punjab Agril. Uni,* 25 (2): 188-192.

Singh,O.P. and Sandhu,J.S.1968.Ecconomic effectiveness of certain soil pestiides application to the sugarcane crop .*Labdev J. Sci. and Tech.Bull,* 6:104-109.

Singh,O.P. and Soneja,P.C.1966. Telodrin for control of sugarcane pests and for increasing cane yield*J.Res.Punjab Agri.Uni,* 3:41-49.

Sivapalan, P.;Senaratne, K.A.D.W. and Karunaratne, A.A.C. 1977.Observations on the occurrence and behaviour of live-wood termites (*Glyptotermes dilatatus*) in low country tea fields. *Pest Articles and News Summaries,* 23:5-8.

Sripathi Rao, B. 1966.Cockchafer attack in Malayan rubber plantation and integrated control. *Eleventh Pacific Science, Congress,* Tokyo, pp. 54-62.

Srivastava, B.K. and Khan, K.M. 1963.Bionomics and control of *Holotrichia insularis* (Melolonthinae: Coleoptera). *Indian J. Ent,* 25:347-354.

Srivastava, K.P. and Butani, D.K. 1987, Insect pests of tea in India and their control. *Pesticides,*21(2) : 16-21.

Srivastava, M.P.;Srivastava,V.K. and Katyal,S.K.1978,Package of Practices,Rabi1978-79.hissar,Haryana Agri.Uni.148pp

Stebbing, E.P. 1902.*Departmental notes on insects that affect forestry,*1: 1-149.

Stebbing, E.P. 1914.*Indian forest insects of economic importance, Coleoptera.* 16: 648.

Su Nan Yao and Scheffrahn, R. H. 1996. Comparative effects of two chitin synthesis inhibition, hexaflumuron and lufenuron. In a bact matrix against subterranean termites (Isoptera: Rhinotermitidae). *J. Econ. Ent,* 89 (5): 1156-1160.

Su Nan Yao;Ban, P. M. and Scheffrahn, R. H. 1997. Remedial baiting with hexaflumuron in above-ground stations to control structure infesting population of the Formosan subterranean termites (Isoptera; Rhinotermitidae). *J. eco. Ent,* 60 (3): 809-817.

Subhash-Chandra;Aggarwal, P. K.;Naveen Kalra and Swaruparani, D. N. 2003.Changes in pest profiles in rice-wheat cropping system in Indo-Gangetic plains. *Annals of Pl. Prot. Sci,* 11(2): 258-263.

Swezey, O.H. 1928. Present status of certain insect pests under biological control in Hawaii. *J. Econ. Ent,* 21: 669-676.

Tashiro, H. and Fiori, B. J. 1969. Suggesting reduction application rates to European chafer. *J. econ. Ent,* 62: 1179-83.

Tashiro, H.; Personius, K. K.; Zinter, D. and Zinter, M. 1971. Ressistance of the European chafer to cyclodine insecticides. *J. econ. Ent,* 21: 669-676

Thakur, M.L. 1981. Revision of the termite genus *Odontotermes Holmgren* (Isoptera : Termitidae : Macrotermitinae) from India. *Indian Forest Records,* 14 : 183.

Thakur, M.L.; Sunil Kumar; Anil Negi and Rawat, D. S. 1989. Chemical control of termites in Eucalyptus hybrid. *Indian Forester,* 115 (10): 733-743.

Tripathi, R.L. 1972. Termite problems in jute and allied fibre crops.pp.44-45.In.Roonwal, M.L. (ed.).Termite problems in India. New Delhi,CSIR.

Umeh, V. C.; Youm, O. and Waliyar, F. 2001. Soil pests of groundnut in Sub-Saharan Africa- a review. *Insects science and its application,* 21(1): 23-32.

Valerio,J.R.;Buainain Alves,C.M. and olivera,M..M.1994.Evaluation of insecticides for the control of mound building termite *Cornitermes cumulans*(Kollar)(Isoptera:Termitidae) in pasture.*Anais-da-Sociedade-Entomologica-do-Brasil,* 23(1):19-24.

Varma, R. V; Balagopalan, M. and Nair, K. S. S, 1995. Field evaluation of neem cake as a possible protectant against subterranean termites attacking *Eucalyptus* seedlings. *Annals of Forestry,* 3 (2): 178-180.

Veenakumari, K. and Veeresh, G.K. 1982. Evaluation of different methods on inoculation for the production of milky white disease caused by the bacterium, *Bacillus popilliae* Dutky on the *Holotrichia serrata* Fabricius. *J. Soil. Biol. Ecol,* 2: 1-8.

Veeresh, G.K, 1977b. Bionomics and control of white grubs in India. *Sugar news,* 9:44-56.

Veeresh, G.K. 1977a. Studies on the root grubs of Karnataka with special reference to bionomics and control of *Holotrichia serrata* F. (Coleoptera: Melolonthidae). *UAS, Monograph Series No.* 2, pp. 87.

Veeresh, G.K. 1978. The rise and fall of white grubs in India. *Bull. Ent,* 19: 222-225.

Veeresh, G.K. 1980. Possibilities of biological control of white grubs in sugarcane. In: *Biological control of sugarcane pests in India.* Tamil Nadu State Federation of Co-operative Sugar Factories Ltd., Madras, (eds) Sithanatham and A.R. Solyappan, pp. 79-84.

Veeresh, G.K.1974. War against white grubs. *Curr. Res,* 3:37-38.

Verma, A.N. 1980. Effect of number of irrigations on termite damage in wheat crop. Haryana Agricultural University. *Journal of Research,* 10: 564-565.

Verma, A.N.; Khurana, A.D. and Malik, B.S. 1981. Effect of different levels of irrigations and fertilizers on termite damage and yield of wheat crop. In: Progress in Soil Biology and Ecology in India. Veeresh, G.K. (Editor). *University of Agricultural Science Tech. Ser.* 37: 285-291.

Vevai, E.J. 1974. Use of fertilizer-cum-pesticide mixture.*Pesticides*, 8 (9):60-62.

Villalobos, F. J. 1999. The sustainable management of white grubs (Coleoptera: Melolonthidae) pest of corn in 'ELCFelo' biosphere Reserve, Tamaulipas, Maxico. *J. Sustainable Agri,* 14(1): 5-29.

Wagner, T.L. 2003. U.S. Forest service termiticides tests. *Sociobiology*,41 (1):131-141.

Walcott, G.N. 1950. The white grubs. *Auit Naturalist,* 84: 183-193.

Wood, T.G andCowie, R.H. 1988. Assessment of on-farm losses in cereals in Africa due to soil insects. *Insect Science and its Application, 9* : 709-716.

Wood, T.G andSands, W.A. 1978. The role of termites in ecosystems pp. 245-292 in Brian, M.V. (Ed.) Production ecology of ants and termites. Cambridge, Cambridge University Press.

Yadav, J. L.; Madan, Y. P. and Chaudhary, J. P. 1992. Effect of insecticidal treatments on sugarcane production. *Haryana Agril. Uni. J. Res,* 22 (3): 145-151.

Yadava, C. P. S.; Saxena, R. .; Mishra, R. K. and Dadheech, L. N. 1977. Population management of white grubs *Holotrichia consanguinea* Blanch in an agro-eosystem. *Indian J. Ent.*,39: 205-210.

Yadava, C.P.S., 1981. Integrated control of white grubs. *Final Research Report of ICAR Scheme*, Dept. of Entomology, University of Udaipur, Rajasthan. 1981.

Yadava, C.P.S.; Pandey, S.N.; Bhardwaj, S.C. and Yadav, S.K.S. 1973. Predation of white grubs (*Holotrichia* sp.) by birds. *Indian J. Ent,* 35: 169.

Author Index